90 0636425 X

D1756330

WITHDRAWN
FROM
UNIVERSITY OF PLYMOUTH
LIBRARY SERVICES

Plastics Engineered Product Design

Dominick Rosato and
Donald Rosato

ELSEVIER

UK	Elsevier Ltd, The Boulevard, Langford Lane, Kidlington, Oxford OX5 1GB, UK
USA	Elsevier Inc, 360 Park Avenue South, New York, NY 10010-1710, USA
JAPAN	Elsevier Japan, Tsunashima Building Annex, 3-20-12 Yushima, Bunkyo-ku, Tokyo 113, Japan

Copyright © 2003 Elsevier Ltd.

All rights reserved. No part of this publication may be reproduced, stored in a retrieval system or transmitted in any form or by any means: electronic, electrostatic, magnetic tape, mechanical, photocopying, recording or otherwise, without permission in writing from the publishers.

British Library Cataloguing in Publication Data

Rosato, Dominick V.
    Plastics engineered product design
    1.Plastics 2.Engineering design 3.New products
    I.Title II.Rosato, Donald V. (Donald Vincent), 1947-
    620.1'923

    ISBN 1856174166

No responsibility is assumed by the Publisher for any injury and/or damage to persons or property as a matter of products liability, negligence or otherwise, or from any use or operation of any methods, products, instructions or ideas contained in the material herein.

Published by
Elsevier Advanced Technology,
The Boulevard, Langford Lane, Kidlington, Oxford OX5 1GB, UK
Tel: +44(0) 1865 843000
Fax: +44(0) 1865 843971

Typeset by Land & Unwin, Bugbrooke
Printed and bound in Great Britain by Biddles Ltd, Guildford and King's Lynn

UNIVERSITY OF PLYMOUTH

900636425X

Contents

Preface, acknowledgement

The proliferation of plastic products in all aspects of modern society continues unabated. New products are more demanding in their applications and require a higher level of design that addresses both mechanical design aspects for product performance as well as the plastic engineering aspects of design for manufacturing. A cross-fertilization of these two disciplines is required to address both aspects. This book will address the analytical approach for traditional mechanical design within the mechanical engineering field, and at the same time, point out behavior and constraints that arise because of specific plastic material, plastic processing and plastic product design limitations that would reduce part quality or process efficiency.

Using the first principles of physics and mechanics, as well as plastic material behaviors and properties that are time and temperature dependent, design problems will be illustrated showing the loading analysis for static and dynamic conditions. Engineering practices that extend material behavior from the simple application of Hooke's Law to short and long term loading as a function of time, temperature, and environmental conditions such as humidity. Application of superposition will be illustrated to accomplish this task. Problems will also consider applications such as for static and dynamic loads in different situations. In each case the underlying assumptions of the problem analysis are given. Basic principles point out the underlying fundamentals, while more advanced mathematical, analytical, or computer based techniques of interpretation highlight the value of refined analysis, if warranted by economic benefits.

True insight into the field of plastics product design will be gained from the dual approach that has been outlined and the use of appropriate laws of physics, mechanics, and material science. For the mechanical engineer the book will be a valuable asset because it treats plastic material selection for end use applications where factors such as thermal, chemical, electrical, optical, and environmental properties are important. The mechanical engineer will also gain an understanding of the manufacturing constraints imposed by mold and die designs as well as the processes used to manufacture plastic products. The plastic engineer will gain a better understanding of the principles of stress analysis, failure modes in structures, and the use of computer based finite element methods for in depth stress and deformation calculations. This book will provide the means that both can expand their expertise from the synergistic effect of combining both disciplines.

This book will provide many fundamentals with their required details so that the reader can become familiar and put to use the different design approaches. Reviews relate to fabricating over 35,000 plastics available worldwide to produce the many millions of different products that are used worldwide.

Information is concise and comprehensive. Engineering and non-engineering principles reviewed have been in use worldwide and are published in many different forms that are included in the bibliography. The book also lists commercial software sources as well as material databases. The reader, with or without design or engineering experience, can understand these principles. It will be invaluable to the most experienced designers or engineers, as well as providing a firm basis for the novice. It meets the designer's goal that is essentially an exercise in predicting product performances.

Its unique approach will expand and enhance your knowledge of plastic technology. Plastic ranges of behavior are presented to enhance one's capability in fabricating products to meet different performances, low cost requirements, and profits. Important basic concepts are presented such as understanding the advantages of different materials and product shapes. This full presentation provides the background needed to understand performance analysis and the design methods useful to the designer. It provides an important tool for approaching the target "get-to-market-right-the-first-time."

Patents or trademarks may cover information presented. No authorization to utilize these patents or trademarks is given or implied; they are discussed for information purposes only. The use of general descriptive names, proprietary names, trade names, commercial designations, or the like does not in any way imply that they may be used freely.

A practical approach was used to obtain the information contained in this book. While information presented represents useful information that can be studied or analyzed and is believed to be true and accurate, neither the authors nor the publisher can accept any legal responsibility for any errors, omissions, inaccuracies, or other factors. The authors and contributors have taken their best effort to represent the contents of this book correctly.

In preparing this book to ensure its completeness and the correctness of the subjects reviewed, use was made of the authors' worldwide personal, industrial, and teaching experiences totaling about a century. Use was also made of worldwide information from industry (personal contacts, material and equipment suppliers, conferences, books, articles, etc.) and major trade associations. The authors have taken their best effort to represent the contents of this book correctly.

The Rosatos
2003

ACKNOWLEDGEMENT

Special and useful contributions in preparing most of the figures and tables in this book were provided by David P. DiMattia. David is an experienced graphics art director specializing in marketing, product promotion, advertising, and public relations. He handles the design and production services for a number of consumer and business-to-business accounts.

About the authors

Dominick V. Rosato

Since 1939 has been involved worldwide principally with plastics from designing-through-fabricating-through-marketing products from toys-through-commercial electronic devices-to-aerospace & space products worldwide. Experience includes Air Force Materials Laboratory (Head Plastics R&D), Raymark (Chief Engineer), Ingersoll-Rand (International Marketing Manager), and worldwide lecturing. Past director of seminars & in-plant programs and adjunct professor at University Massachusetts Lowell, Rhode Island School of Design, and the Open University (UK). Has received various prestigious awards from USA and international associations, societies (SPE Fellows, etc.), publications, companies, and National Academy of Science (materials advisory board). He is a member of the Plastics Hall of Fame. Received American Society of Mechanical Engineers recognition for advanced engineering design with plastics. Senior member of the Institute of Electrical and Electronics Engineers. Licensed professional engineer of Massachusetts. Involved in the first all plastics airplane (1944/RP sandwich structure). Worked with thousands of plastics plants worldwide, prepared over 2,000 technical and marketing papers, articles, and presentations and has published 25 books with major contributions in over 45 other books. Received BS in Mechanical Engineering from Drexel University with continuing education at Yale, Ohio State, and University of Pennsylvania.

Donald V. Rosato

Has extensive technical and marketing plastic industry business experience from laboratory, testing, through production to marketing, having worked for Northrop Grumman, Owens-Illinois, DuPont/ Conoco, Hoechst Celanese, and Borg Warner/G.E. Plastics. He has written extensively, developed numerous patents within the polymer related industries, is a participating member of many trade and industry groups, and currently is involved in these areas with PlastiSource, Inc., and Plastics FALLO. Received BS in Chemistry from Boston College, MBA at Northeastern University, M.S. Plastics Engineering from University of Massachusetts Lowell (Lowell Technological Institute), and Ph.D. Business Administration at University of California, Berkeley.

1 OVERVIEW

Introduction

This book provides information on the behavior of plastics that influence the application of practical and complex engineering equations and analysis in the design of products. For over a century plastics with its versatility and vast array of inherent plastic properties as well as high-speed/low-energy processing techniques have resulted in designing and producing many millions of cost-effective products used worldwide. The profound worldwide benefits of plastics in economics and modern living standards have been brought about by the intelligent application of logic with modern chemistry and engineering principle.

Today's plastics industry is comprised of both mature practical and theoretical technology. Improved understanding and control of materials and manufacturing processes have significantly increased product performances and reduced their variability. Performance requirements for these products can be characterized in many different ways. Examples meeting different commercial and industrial market requirements worldwide include:

1. light weight,
2. flexible to high strength,
3. provide packaging aesthetics and performances,
4. excellent appearance and surface characteristics without using secondary operations,
5. degradation resistance in different environments,
6. performance in all kinds of environments,
7. adapt well to mass production methods,
8. wide range of color and appearance,
9. high impact to tear resistance,
10. decorative to industrial load bearing structures,

11. short to very long service life, degradable to non-degradable,
12. process virgin with recycled plastics or recycled alone,
13. simple to complex shapes including many that are difficult or impossible to form with other materials,
14. breathable film for use in horticulture,
15. heat and ablative resistance,
16. and so on.

There is a plastic for practically any product requirements, particularly when not including cost for a few products. One can say that if plastics were not to be used it would be catastrophic worldwide for people, products, communications, and so on with a major economic crisis because much more expensive materials and processes would be used.

Materials can be blended or compounded to achieve practically any desired property or combination of properties. The final product performance is affected by interrelating the plastic with its design and processing method. The designer's knowledge of all these variables is required otherwise it can profoundly affect the ultimate success or failure of a consumer or industrial product. When required the designer makes use of others to ensure product success.

Plastic plays a crucial and important role in the development of our society worldwide. With properties ranges that can be widely adjusted and ease of processing, plastics can be designed to produce simple to highly integrated conventional and customized products. While it is mature, the plastics industry is far from having exhausted its product design potential. The worldwide plastics industry offers continuous innovations in plastic materials, process engineering, and mechanical engineering design approaches that will make it possible to respond to ever more demanding product applications (Fig. 1.1).

Innovation trends emerging in plastics engineering designs are essentially combinations and improvements of different processes, combinations and improvements of different materials, integration of a wide range of functions within a single product, reduced material consumption, and recyclability of the materials employed. At the same time, rising requirements are being placed on design efficiency, product quality, production quality, and part precision, while costs are expected to be reduced wherever possible. This combination of objectives is achievable by factors such as process-engineering innovations that reduce the number of process steps.

The basic and essential design exercise in product innovation lies in predicting performances. This includes the process of devising a product that fulfills the total requirements of the end user and satisfies

Figure 1.1 Flow-chart from raw materials to products (Courtesy of Plastics FALLO)

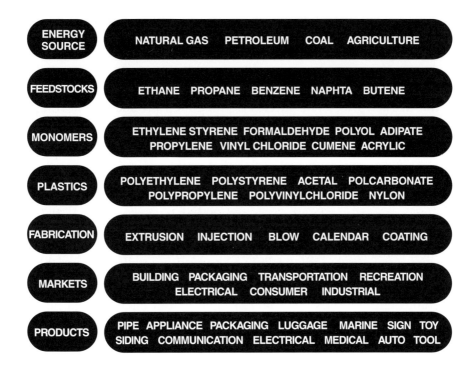

the needs of the producer in terms of a good return on investment (ROI). The product designer must be knowledgeable about all aspects of plastics such as behavioral responses, processing, and mechanical and environmental load stresses. Product loads range from short-time static, such as tensile, flexural, torsion, etc., to long time dynamic, such as creep, fatigue, high speed loading, motion control, and so on. In this book, plastics design concepts are presented that can be applied to designing products for a range of behaviors.

An inspired idea alone will not result in a successful design. Designing is, to a high degree, intuitive and creative, but at the same time empirical and technically influenced. Experience plays an important part that requires keeping up to date on the endless new developments in materials and processes. An understanding of one's materials and a ready acquaintance with the relevant processing technologies are essential for converting an idea to an actual product. In addition, certain basic tools are needed, such as those for computation and measurement and for testing of prototypes and/or fabricated products to ensure that product performance requirement are met. A single individual designer may not have all of these capabilities so inputs from many reliable people and/or sources are required.

Inputs from many disciplines, both engineering and non-engineering, may be required when designing a product such as a toy, flexible package, rigid container, medical device, car, boat, underwater device, spring, pipe, building, aircraft, missile, or spacecraft. The conception of such products usually requires coordinated inputs from different specialists. Input may involve concepts of man-machine interfaces (ergonomics), shape, texture, and color (aesthetics). Unless these are in balance, the product may fail in the market place. The successful integrated product is the result of properly collecting all of the required design inputs.

While plastic product design can be challenging, many products seen in everyday life may require only a practical, rather than rigorous approach. They are not required to undergo sophisticated design analysis because they are not required to withstand high static and dynamic loads (Chapter 2). Their design may require only the materials information in conventional data sheets from plastic material suppliers. Examples include containers, cups, toys, boxes, housings for computers, radios, televisions, electric irons, recreational products, and nonstructural or secondary structural products of various kinds like the interiors in buildings, automobiles, and aircraft. The design engineer will need to know when to use the practical approach, the rigorous approach, or a combination approach.

Plastics do not only have advantages but also have disadvantages or limitations. Other materials (steel, wood, etc.) also suffer with disadvantages or limitations. Unfortunately there is no one material (plastic, steel, etc.) that can meet all requirements thus these limitations or faults are sometimes referred to incorrectly as disadvantages. Note that the faults of materials known and utilized for hundreds of years are often overlooked; the faults of the new materials are often over-emphasized.

Iron and steel are attacked by the elements of weather and fire [815°C (1500°F)] but the common practice includes applying protective coatings (plastic, cement, etc.) and then forgetting their susceptibility to attack is all too prevalent. Wood is a useful material yet who has not seen a rotted board, wood on fire, etc. There is cracked concrete and so on. Regardless of these and many other disadvantages, lack of perfection does not mean that any steel, wood, or concrete should not be used. The same reasoning should apply to plastics. In many respects, the gains made with plastics in a short span of time far outdistance the advances made in these other materials.

Recognize that modern design engineering has links with virtually every technical area; material, mechanical, electrical, thermal, processing, and packaging to name a few. Any attempt to explain engineering by referring to the special rules for each area would mean that the engineer would need to have a thorough knowledge of each special field. This is not possible in the current state of technology. It is the case, however, that there are certain common concepts behind these specialized areas. Similar features exist among many consumer and industrial products. These features can be described by using a standard procedure, and the fundamental laws of engineering apply to all products, irrespective of the different forms of materials and equipment involved. Proper applications are required.

Materials of construction

Plastics comprise many different materials based on their polymer structure, additives, and so on. Practically all plastics at some stage in their fabrication can be formed into simple to extremely complex shapes that can range from being extremely flexible to extremely strong. Polymers, the basic ingredients in plastics, are high molecular weight organic chemical compounds, synthetic or natural substances consisting of molecules. Practically all of these polymers use virtually an endless array of additives, fillers, and reinforcements to perform properly during product fabrication and/or in service. There are many compounded base polymer combinations so that new materials are always on the horizon to meet new industry requirements that now total over 35,000 plastics worldwide. Table 1.1 provides examples of their manufacturing stages from raw materials to products. Examples of the diversified use of different plastics are shown in Figs. 1.2 and 1.3.

Plastic, polymer, resin, elastomer, and reinforced plastic (RP) are some-what synonymous terms. The most popular term worldwide is plastics. Polymer denotes the basic material. Whereas plastic pertains to polymers or resins (as well as elastomers, RPs, etc.) containing additives, fillers, and/or reinforcements. An elastomer is a rubber-like material (natural or synthetic). Reinforced plastics (also called plastic composites) are plastics with reinforcing additives such as fibers and whiskers, added principally to increase the product's mechanical properties but also provides other benefits such as increased heat resistance and improved tolerance control.

There are thermoplastics (TPs) that melt (also called curing) during processing. Cure occurs only with thermoset plastics (TSs) or when a TP is converted to a TS plastic and in turn processed. The term curing TPs occurred since at the beginning of the 20th century the term

Table 1.1 Examples of stages in plastic manufacturing

Basic Chemicals

Petroleum is converted to petrochemicals such as ethylene, benzene, propylene and Acetylene.

Monomers

Petrochemicals plus other chemicals are converted into monomers such as styrene, ethylene, propylene, vinyl chloride, and acrylontrile.

Polymerization

One or more monomers are polymerized to form polymers or copolymers such as polyethylene, polystyrene, polyvinyl chloride, and polypropylene.

Compounding

Additives, fillers, and/or reinforcements are mixed with polymers (referred to as plastics) providing different properties and/or different fabricating methods for plastics. Hundreds of different materials are used such as heat stabilizers, color pigments, antioxidants, inhibitors, and fire retardants.

Processing

Plastics are formed into different shapes such as sheets, films, pipes, buckets, primary and secondary structures (boats, cars, airplanes, bridges, etc.), toys, housings, and many thousand more products. Basically heat and pressure are used to shape these products that usually are in finished form. Processes used include extrusion, injection molding, blow molding, thermoforming, compression molding, spraying, rotational moldings, reaction injection molding, and filament winding.

Finishing

In certain applications a finishing step is required on the fabricated part such as printing, bonding, machining, etc.

curing was accurately used for TSs. At that time TSs represented practically all the plastic used worldwide. Thus TPs took on the incorrect term curing even though there is no chemical reaction or curing action.

Appreciate the polymer chemist's ability to literally rearrange the molecular structure of the polymer to provide an almost infinite variety of compositions that differ in form, appearance, properties, cost, and other characteristics. One must also approach the subject with a completely open mind that will accept all the contradictions that could make it difficult to pin common labels on the different families of plastics or even on the many various types within a single family that are reviewed in this book. Each plastic (of the 35,000 available) has specific performance and processing capabilities.

Figure 1.2   Use of plastics in recreational products range from unsophisticated types to high performance types such boats (Courtesy of Plastics FALLO)

Figure 1.3 Boeing 777 uses different types of plastics that include high performance reinforced plastics

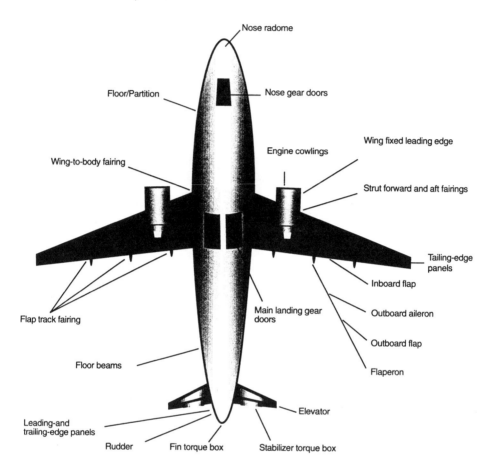

There are many different routes that the starting materials for plastics can take on the way to the user. In this book, we are concerned with those plastics that are supplied to the processor in the form of granules, powder, pellets, flake, or liquids and in turn they are transformed into plastic products. However, the same starting materials used to make these plastics can take other routes and end up in the textile industry (nylon fibers share common roots with a molded nylon gear; acrylic fibers share common roots with acrylic sheet for glazing; etc.), paint industry, adhesives industry, and other industries meeting their special requirements.

Worldwide total plastic consumption is over 154 million ton (340 billion lb) with about 90wt% thermoplastics (TPs) and 10% thermoset

(TS) plastics. USA and Europe consumption is about one-third each of the world total.

These two major classifications of thermoplastics (TPs) and thermosets (TSs) in turn have different classifications such as virgin or recycled plastics. Virgin plastics have not been subjected to any fabricating process. NEAT plastics identify plastics with Nothing Else Added To. They are true virgin polymers since they do not contain additives, fillers, etc. However they are rarely used since they do not provide the best performances. Thus the technically correct term to identify the materials is plastics. Of the 35,000 types available worldwide there are about 200 basic types or families that are commercially recognized with less than 20 that are popularly used. Examples of these plastics are shown in Table 1.2.

Table 1.2 Summation of the plastic families with their abbreviations

Acetal (POM)
Acrylics
  Polyacrylonitrile (PAN)
  Polymethylmethacrylate
   (PMMA)
Acrylonitrile butadiene styrene
 (ABS)
Allyl diglycol carbonate
Alkyd
  Diallyl isophthalate (DAIP)
  Diallyl phthalate (DAP)
Aminos
  Melamine formaldehyde (MF)
  Urea formaldehyde (UF)
Cellulosics
  Cellulose acetate (CA)
  Cellulose acetate butyrate (CAB)
  Cellulose acetate propionate
   (CAP)
  Cellulose nitrate
  Ethyl cellulose (EC)
Chlorinated polyether
Epoxy (EP)
Ethylene vinyl acetate (EVA)
Ethylene vinyl alcohol (EVOH)
Fluorocarbons
  Fluorinated ethylene propylene
   (FEP)

Polytetrafluoroethylene (FTFE)
Polyvinyl fluoride (PVF)
Polyvinylidene fluoride
 (PVDF)
Ionomer
Liquid crystal polymer (LCP)
  Aromatic copolyester (TP
   polyester)
Melamine formaldehyde (MF)
Nylon (or Polyamide) (PA)
Parylene Phenolic
Phenolic
  Phenol formaldehyde (PF)
Polyamide-imide (PAI)
Polyarylether
  Polyaryletherketone (PAEK)
  Polyaryl sulfone (PAS)
Polyarylate (PAR)
Polycarbonate
Polyester
  Saturated polyester (TS
   polyester)
  Thermoplastic polyesters (TP
   polyester)
   Polybutylene terephthalate
    (PBT)
   Polyethylene terephthalate
    (PET)

Table 1.2 continued

Polyetherketone (PEK)	Styrene
Polyetheretherketone (PEEK)	Acrylic styrene acrylonitrile (ASA)
Polyetherimide (PEI)	Acrylonitrile butadiene styrene
Polyimide (PI)	(ABS)
Thermoplastic PI	General-purpose PS (GPPS)
Thermoset PI	High-impact PS (HIPS)
Polymethylmethacrylate (or	Polystyrene (PS)
acrylic) (PMMA)	Styrene acrylonitrile (SAN)
Polyolefin (PO)	Styrene butadiene (SB)
Chlorinated PE (CPE)	Sulfone
Cross-linked PE (XLPE)	Polyether sulfone (PES)
High-density PE (HDPE)	Polyphenyl sulfone (PPS)
Linear LDPE (LLDPE)	Polysulfone (PSU)
Low-density PE (LDPE)	Urea formaldehyde (UF)
Polyallomer	Vinyl
Polybutylene (PE)	Chlorinated PVC (CPVC)
Polyethylene (PE)	Polyvinyl acetate (PVAc)
Polypropylene (PP)	Polyvinyl alcohol (PVA)
Ultra-high molecular weight PE	Polyvinyl butyrate (PVB)
(UHMWPE)	Polyvinyl chloride (PVC)
Polyurethane (PUR)	Polyvinylidene chloride (PVDC)
Silicone (SI)	Polyvinylidene fluoride (PVF)

Within these 20 popular plastics there are five major TP types that consume about two-thirds of all TPs. Approximately 20wt% are low density polyethylenes (LDPEs), 15% polyvinyl chlorides (PVCs), 10% high density polyethylenes (HDPEs), 15% polypropylenes (PPs), 8% polystyrenes (PSs). Each has literally many thousands of different formulated compounds and different processing and performance behaviors. These basic types, with their many modifications of different additives/fillers/reinforcements, catalyst systems, grafting, and/or alloying provide different processing capabilities and/or product performances. As examples there are the relatively new generation of high performance metallocene and elastomeric plastics providing different modifications.

Thermoplastics

TPs are plastics that soften when heated and upon cooling harden into products. TPs can be repeatedly softened by reheating. Their morphology, molecular structure, is crystalline or amorphous. Softening temperatures vary. The usual analogy is a block of ice that can be softened (turned back to a liquid), poured into any shape mold or die, then cooled to become a solid again. This cycle repeats. During the heating cycle care must be taken to avoid degrading or decomposition of the plastic. TPs generally offer easier processing and better adaptability to complex designs than do TS plastics.

There are practical limits to the number of heating and cooling cycles before appearance and/or mechanical properties are drastically affected. Certain TPs have no immediate changes while others have immediate changes after the first heating/cooling cycle.

Crystalline & Amorphous Polymers
The overall molecular physical structure of a polymer identifies its morphology. Crystalline molecular structures tend to have their molecules arranged in a relatively regular repeating structure such as acetal (POM), polyethylene (PE), polypropylene (PP), nylon (PA), and polytetrafluoroethylene plastics. The structures tend to form like cooked spaghetti. These crystallized plastics have excellent chemical resistance. They are usually translucent or opaque but they can be made transparent with chemical modification. They generally have higher strength and softening points and require closer temperature/time processing control than the amorphous TPs.

Polymer molecules that can be packed closely together can more easily form crystalline structures in which the molecules align themselves in some orderly pattern. Commercially crystalline polymers have up to 80% crystalline structure and the rest is amorphous. They are identified technically as semicrystalline TPs. Polymers with 100% crystalline structures are not commercially produced.

Amorphous TPs have no crystalline structure. Their molecules form no patterns. These TPs have no sharp melting points. They are usually glassy and transparent, such as acrylontrile-butadiene-styrene (ABS), acrylic (PMMA), polycarbonate (PC), polystyrene (PS), and polyvinyl chloride (PVC). Amorphous plastics soften gradually as they are heated during processing. If they are rigid, they may become brittle unless modified with certain additives.

During processing, all plastics are normally in the amorphous state with no definite order of molecular chains. TPs that normally crystallize need to be properly quenched; that is, the hot melt is cooled to solidify the plastic. If not properly quenched, they become amorphous or partially amorphous solids, usually resulting in inferior properties. Compared to crystalline types, amorphous polymers undergo only small volumetric changes when melting or solidifying during processing. This action influences the dimensional tolerances that can be met after accounting for the heating/cooling process and the design of molds or dies.

Crystalline plastics require tighter process control during fabrication. They tend to shrink and warp more than amorphous types due to their higher melting temperatures, with their relatively sharp melting point, they do not soften gradually with increasing temperature but remain hard until a given quantity of heat has been absorbed, then change rapidly into a low-viscosity liquid. If the correct amount of heat is not applied properly during processing, product performance can be drastically reduced and/or an increase in processing cost occurs. With proper process control this is not a problem.

During the melting process as the symmetrical molecules approach each other within a critical distance, crystals begin to form. They form first in the areas where they are the most densely packed. This crystallized area becomes stiff and strong. The noncrystallized, amorphous, area is tougher and more flexible. With increased crystallinity, other effects occur such as with polyethylene (crystalline plastic) there is increased resistance to creep.

Liquid Crystalline Polymers
A special classification of TPs are liquid crystalline polymers. They are self-reinforcing because of densely packed fibrous polymer chains. Their molecules are stiff, rod-like structures organized in large parallel arrays in both the melt and solid states. They resist most chemicals, weathers oxidation, and can provide flame resistance, making them excellent replacements for metals, ceramics, and other plastics in many product designs. They are exceptionally inert and resist stress cracking in the presence of most chemicals at elevated temperatures, including the aromatic and halogenated hydrocarbons as well as strong acids, bases, ketones, and other aggressive industrial products. Regarding flammability, LCPs have an oxygen index ranging from 35 to 50% (ASTM). When exposed to an open flame, they form an intumescent char that prevents dripping.

When injection molded or extruded the molecules align into long, rigid chains that in turn align in the direction of flow. Thus the molecules act like reinforcing fibers giving LCPs both very high strength and stiffness. LPCs with their high strength-to-weight ratios are particularly useful for weight-sensitive products (Table 1.3). They have outstanding strength at extreme temperatures, excellent mechanical property retention after exposure to weathering and radiation, good dielectric strength as well as arc resistance and dimensional stability, low coefficient of thermal expansion, excellent flame resistance, and easy processability.

Table 1.3 Liquid crystal polymer properties compared to other thermoplastics

Property	Crystalline	Amorphous	Liquid crystalline
Specific gravity	Higher	Lower	Higher
Tensile strength	Higher	Lower	Highest
Tensile modulus	Higher	Lower	Highest
Ductility, elongation	Lower	Higher	Lowest
Resistance to creep	Higher	Lower	High
Maximum usage temperature	Higher	Lower	High
Shrinkage and warpage	Higher	Lower	Lowest
Chemical resistance	Higher	Lower	Highest

Their UL (Underwriters Laboratory) continuous-use rating for electrical properties is as high as 240°C (464°F), and for mechanical properties it is 220°C (428°F) permiting products to be exposed to intermittent temperatures as high as 315°C (600°F) without affecting performance properties. Their resistance to high-temperature flexural creep is excellent, as are their fracture-toughness characteristics.

Because of their structure they provide special properties such as greater resistance to most solvents and heat. They have the lowest warpage and shrinkage of all the TPs. Unlike many high-temperature plastics, LCPs have a low melt viscosity and are thus more easily processed resulting in faster cycle times than those with a high melt viscosity thus reducing processing costs.

Thermosets

Outstanding properties of TS plastic products are their substantially infusible and insoluble characteristic along with resistance to high temperatures, greater dimensional stability, and strength. TSs undergo a crosslinking chemical reaction by techniques such as the action of heat (exothermic reaction), oxidation, radiation, and/or other means often in the presence of curing agents and catalysts. However, if excessive heat is applied, degradation rather than melting will occur.

TSs are not recyclable because they do not melt when reheated, although they can be granulated and used as filler in other TSs as well as TPs. An analogy of TSs is that of a hard-boiled egg that has turned from a liquid to a solid and cannot be converted back to a liquid. As shown in Fig. 1.4, TSs are identified by A-B-C-stages during the curing process. A-stage is uncured, B-stage is partially cured, and C-stage is fully cured. Typical B-stage is TS molding compounds and prepregs,

Figure 1.4 Thermoset A-B-C stages from melt to solidification

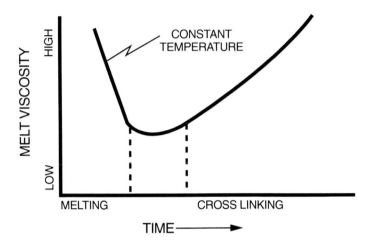

which in turn are processed to produce C-stage fully, cured plastic material products.

TSs generally cannot be used alone in primary or secondary structural applications; they must be filled with additives and/or reinforcements such as glass or wood fibers, etc. These compounds provide dimensional product precision and certain other desirable properties for use in certain products. There are TSs particularly suitable as substitutes for metals in products that have to meet severe demands such as high temperature with the added advantage of offering a very good cost reduction. Applications include kitchen appliances, heat-shield for an electric iron, collectors and a wide variety of circuit breaker housings in electrical devices, and automotive parts including headlamp reflectors, brake servo units, brake pistons, pump housings, valve caps, pulleys, and so on. Compression and transfer molding (CM and TM) are the two main methods used to produce molded products from TSs.

Within the TS family there are natural and synthetic rubbers, elastomers, such as styrene-butadiene, nitrile rubber, millable polyurethanes, silicone, butyl, and neoprene. They attain their properties through the process of vulcanization. Vulcanization is the process by which a natural rubber or certain plastic elastomer undergoes a change in its chemical structure brought about by the irreversible process of reacting the materials with sulfur and/or other suitable agents. The crosslinking action results in property changes such as decreased plastic flow, reduced surface tackiness, increased elasticity, greater tensile strength, and considerably less solubility.

Crosslinked Thermoplastics

TPs can be converted to TSs to improve or change properties. TPs can be crosslinked by different processes such as chemical and irradiation. Polyethylene (PE) is a popular plastic that can be crosslinked; it is identified as XLPE or PEX. Crosslinking is an irreversible change that occurs through a chemical reaction, such as condensation, ring closure, addition, and so on. Cure is usually accomplished by the addition of curing (crosslinking) agents with or without heat and pressure.

For TP systems such as PE, chemical or irradiation techniques have been used as the crosslinking technology; this is the recognized standard for manufacturing industrial materials such as cable coverings, cellular materials (foams), rotationally molded articles, and piping.

Enhancement of properties is the underlying incentive for the commercial development of crosslinked TPs. Crosslinking improves resistance to thermal degradation, cracking by liquids and other harsh environments, and creep and cold flow, among other improvements. The primary commercial interest has been in aliphatic polymers, which includes the main olefins polyethylene and polypropylene, also popular are polyvinyl chloride (PVC) and acrylates. Crosslinked films with low shrinkage and high adhesion properties have been used in such applications as pressure-sensitive adhesives, glass coatings, and dental enamels.

Reinforced Plastics

The term reinforced plastic (RP) refers to composite combinations of plastic, matrix, and reinforcing materials, which predominantly come in chopped and continuous fiber forms as in woven and nonwoven fabrics. Other terms used to identify an RP include: glass fiber reinforced plastic (GFRP), aramid fiber reinforced plastic (AFRP), boron fiber reinforced plastic (BFRP), carbon fiber reinforced plastic (CFRP), graphite fiber reinforced plastic (GFRP), etc.

In addition to fabrics, reinforcements include other forms such as powders, beads, and flakes. Both TP and TS plastics are used in reinforced plastics. At least 90wt% use glass fiber materials. At least 55wt% use TPs. RPs using primarily TS polyester plastics provides significant property and/or cost improvements compared to other composites. Primary benefits of all RPs include high strength, directional strength, lightweight, high strength-to-weight ratio, creep and fatigue endurance, high dielectric strength, corrosion resistance, and long term durability.

Both reinforced TSs (RTSs) and reinforced TPs (RTPs) can be characterized as engineering plastics, competing with engineering unreinforced TPs. When comparing processability of RTSs and RTPs, the RTPs are usually easier to process and permit faster molding cycles with efficient processing such as during injection molding. Higher performing fibers that are used include high performance glass (other than the usual E-glass), aramid, carbon, and graphite. Also available are whisker reinforcements with exceptional high performances (Fig. 1.5).

Figure 1.5 High performance whisker reinforcements compared to other materials

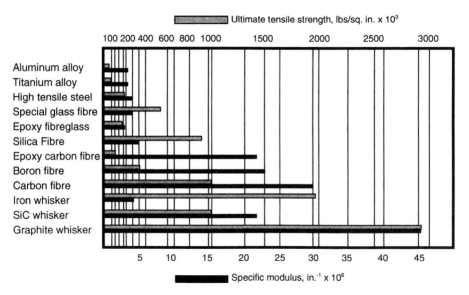

Fiber orientations have improved to the extent that 2-D and 3-D RPs can be used to produce very high strength and stiff products with long service lives. RTPs even with their relatively lower properties compared to RTSs are used in about 55wt% of all RP products. Practically all RTPs with short glass fibers are injection molded at very fast cycles, producing high performance products in highly automated environments.

RPs can be characterized by their ability to be molded into either extremely small to extremely large structurally loaded shapes well beyond the basic capabilities of other materials or processes at little or no pressure. In addition to shape and size, RPs possess other characteristics that make them very desirable in design engineering. The other characteristics include cost reduction, ease of fabrication, simplified installation, weight reduction, aesthetic appeal, and the potential to be combined with many other useful qualities.

Their products have gone worldwide into the deep ocean waters, on land, and into the air including landing on the moon and in spacecraft. In USA annual consumption of all forms of RPs is over 3.9 billion lb (1.8 billion kg). Consumption by market in million lb is aerospace at 24, appliances/business machines at 210, building/construction at 775, consumer at 253, corrosion at 442, electrical/electronic at 390, marine at 422, transport at 1268, and others at 116.

The form the RP takes, as with non-reinforced plastics, is determined by the product requirements. It has no inherent form of its own; it must be shaped. This provides an opportunity to select the most efficient forms for the application. Shape can help to overcome limitations that may exist in using a lower-cost material with low stiffness. As an example underground fuel tanks can include ribs to provide added strength and stiffness to the RP orientation in order to meet required stresses at the lowest production cost.

The formability of these products usually leads to one-piece consolidation of construction products to eliminate joints, fasteners, seals, and other potential joining problems. As an example, formed building fascia panels eliminate many fastenings and seals. Examples of design characteristics gained by using RP materials are presented as follows:

Thermal Expansions
Nonreinforced plastics generally have much higher coefficients of linear thermal expansion (CLTE) than conventional metal, wood, concrete, and other materials. CLTEs also vary significantly with temperature changes. There are RPs that do not have these characteristics. With certain types and forms of fillers, such as graphite, RPs can eliminate CLTE or actually shrink when the temperature increases.

Ductilities
Substantial yielding can occur in response to loading beyond the limit of approximate proportionality of stress to strain. This action is referred to as ductility. Most RPs do not exhibit such behavior. However, the absence of ductility does not necessarily result in brittleness or lack of flexibility. For example, glass fiber-TS polyester RPs do not exhibit ductility in their stress-strain behavior, yet they are not brittle, have good flexibility, and do not shatter upon impact. TS plastic matrix is brittle when unreinforced. However, with the addition of glass or other fibers in any orientation except parallel, unidirectional, the fibers arrest

crack propagation. This RP construction results in toughness and the ability to absorb a high amount of energy. Because of the generally high ratio of strength to stiffness of RPs, energy absorption is accomplished by high elastic deflection prior to failure. Thus ductility has been a major factor promoting the use of RPs in many different applications.

Toughness

The generally low-specific gravity and high strength of reinforcement fibers such as glass, aramid, carbon, and graphite can provide additional benefits of toughness. For example, the toughness of these fibers allows them to be molded into very thin constructions. Each fiber has special characteristics. For instance, compared to other fiber reinforcements, aramid fibers can increase wear resistance with exceptionally high strength or modulus to weight.

Tolerances/Shrinkages

TSs combined with all types of reinforcements and/or fillers are generally more suitable for meeting tight dimensional tolerances than are TPs. For injection molded products they can be held to extremely close tolerances of less than a thousandth of an inch (0.0025 cm) effectively down to zero (0.0%). Achievable tolerances range from 5% for 0.020 in. (0.05 cm), to 1% for 0.500 in. (1.27 cm), to 1/2% for 1.000 in. (2.54 cm), to 1/4% for 5.000 in. (12.70 cm), and so on.

Some unreinforced molded plastics change dimensions, shrink, immediately after molding or in a day or a month due to material relaxation and changes in temperature, humidity, and/or load application. RPs can significantly reduce or even eliminate this dimensional change after molding.

When comparing tolerances and shrinkage behaviors of RTSs and RTPs there is a significant difference. Working with crystalline RTPs can be yet more complicated if the fabricator does not understand their behavior. Crystalline plastics generally have different rates of shrinkage in the longitudinal, melt flow direction, and transverse directions. In turn, these directional shrinkages can vary significantly due to changes in processes such as during injection molding (IM). Tolerance and shrinkage behaviors are influenced by factors such as injection pressure, melt heat, mold heat, and part thickness and shape. The amorphous type materials can be easier to balance.

Compounds

Commercial RP compounds are available in several forms: pellets for injection molding or extrusion, unidirectional tape for filament winding and similar applications, sheets for stamping and compression molding, bulk compounds for compression molding, and so on. There are RTP elastomeric materials that provide special engineered products such as conveyor belts, mechanical belts, high temperature or chemical resistant suits, wire and cable insulation, and architectural designed shapes. Common categories of RP compounds are reviewed.

Prepregs

Preimpregnated materials usually are a compound of a reinforcement and a hot melt or solvent system. Prepreg also includes wet systems without solvent using TS polyester. They are stored for use at a latter time either in-house or to ship to a fabricator. The plastic is partially cured, B-stage, ready-to-mold material in web form that may have a substrate of glass fiber mat, fabric, roving, paper, cotton cloth, and so forth. With proper temperature storage conditions, their shelf life can be controlled to last at least 6 months.

Sheet Molding Compounds

A ready-to-mold material, SMC represent a special form of a prepreg. It is usually a glass fiber-reinforced TS polyester resin compound in sheet form. The sheet can be rolled into coils during its continuous fabricating process. A plastic film covering, usually polyethylene, separates the layers to enable coiling and to prevent contamination, sticking, and monomer evaporation. This film is removed before the SMC is charged into a mold, such as a matched-die or compression mold.

Depending on product performance requirements, the SMC consists of additional ingredients such as low-profile additives, cure initiators, thickeners, and mold-release agents. They are used to enhance the performance or processing of the material. Glass fibers are usually chopped into lengths of 12 mm (0.5 in.) to at least 50 mm (2 in.). The amount can vary from 25 to 50wt%. The usual ratio is based on performance requirements, processability, and cost considerations.

Bulk Molding Compounds

Also called dough molding compounds (DMCs), bulk molding compounds (BMCs) are mixtures usually of short 3 mm to 3 cm ($\frac{1}{8}$ to 1$\frac{1}{4}$in.) glass fibers, plastic, and additives similar to the SMC compound. This mixture, with the consistency of modeling clay, can be produced in bulk form or extruded in rope-like form for easy handling. The extrudate type is called a "log" that is cut to specific lengths such as 0.3 cm (1 ft).

BMC is commercially available in different combinations of resins, predominantly TS polyesters, additives, and reinforcements. They meet a wide variety of end-use requirements in high-volume applications where fine finish, good dimensional stability, part complexity, and good overall mechanical properties are important. The most popular method of molding BMCs is compression. They can also be injection molded in much the same way as other RTS compounds using ram, ram-screw, and, for certain BMC mixes, conventional reciprocating screw.

Commodity & Engineering Plastics

About 90wt% of plastics can be classified as commodity plastics (CPs), the others being engineering plastics (EPs). The EPs such as polycarbonate (PC) representing at least 50wt% of all EPs, nylon, acetal, etc. are characterized by improved performance in higher mechanical properties, better heat resistance, and so forth (Table 1.4).

Table 1.4 Thermoplastic engineering behaviors

Crystalline	Amorphous
Acetal	*Polycarbonate*
Best property balance	Good impact resistance
Stiffest unreinforced thermoplastic	Transparent
Low friction	Good electrical properties
Nylon	*Modified PPO*
High melting point	Hydrolytic stability
High elongation	Good impact resistance
Toughest thermoplastic	Good electrical properties
Absorbs moisture	
Polyester (glass-reinforced)	
High stiffness	
Lowest creep	
Excellent electrical properties	

The EPs demand a higher price. About a half century ago the price per pound was at 20¢; at the turn of the century it went to $1.00, and now higher. When CPs with certain reinforcements and/or alloys with other plastics are prepared they become EPs. Many TSs and RPs are EPs.

Elastomers/Rubbers

In the past rubber meant a natural thermoset elastomeric (TSE) material obtained from a rubber tree, hevea braziliensis. The term elastomer developed with the advent of rubber-like synthetic materials. Elastomers identify natural or synthetic TS elastomers (TSEs) and thermoplastic elastomers (TPEs). At room temperature all elastomers basically stretch under low stress to at least twice in length and snaps back to approximately the original length on release of the stress, pull, within a specified time period.

The term elastomer is often used interchangeably with the term plastic or rubber; however, certain industries use only one or the other terminology. Different properties identify them such as strength and stiffness, abrasion resistance, oil resistance, chemical resistance, shock and vibration control, electrical and thermal insulation, waterproofing, tear resistance, cost-to-performance, etc.

Natural rubber with over a century's use in many different products and markets will always be required to attain certain desired properties not equaled (to date) by synthetic elastomers. Examples include transportation tires, with their relative heat build-up resistance, and certain types vibrators. However, both synthetic TSE and TPE have made major inroads in product markets previously held only by natural rubber. Worldwide, more synthetic types are used than natural. The basic processing types are conventional, vulcanizable, elastomer, reactive type, and thermoplastic elastomer.

Plastic behaviors

A knowledge of the chemistry of plastics can be used to help with the understanding of the performance of designed products. Chemistry is the science that deals with the composition, structure, properties and transformations of substances. It provides the theory of organic chemistry, in particular our understanding of the mechanisms of reactions of carbon (C) compounds.

The chemical composition of plastics is basically organic polymers. They have very large molecules composed of connecting chains of carbon (C), generally connected to hydrogen atoms (H) and often also oxygen (O), nitrogen (N), chlorine (Cl), fluorine (F), and sulfur (S). Thus, while polymers form the structural backbone of plastics, they are rarely used in pure form. In almost all plastics other useful and important materials are added to modify and optimize properties for each desired process and/or product performance application.

The chemical and physical characteristics of plastics are derived from the four factors of chemical structure, form, arrangement, and size of the polymer. As an example, the chemical structure influences density. Chemical structure refers to the types of atoms and the way they are joined to one another. The form of the molecules, their size and disposition within the material, influences mechanical behavior. It is possible to deliberately vary the crystal state in order to vary hardness or softness, toughness or brittleness, resistance to temperature, and so on. The chemical structure and nature of plastics have a significant relationship both to properties and the ways they can be processed, designed, or otherwise translated into a finished product.

Morphology/ Molecular Structure/Mechanical Property

Morphology is the study of the physical form or chemical structure of a material; that is, the physical molecular structure. As a result of morphology differences among polymers, great differences exist in mechanical and other properties as well as processing plastics.

Knowledge of molecular size and flexibility explains how individual molecules behave when completely isolated. However, such isolated molecules are encountered only in theoretical studies of dilute solutions. In practice, molecules always occur in a mass, and the behavior of each individual molecule is very greatly affected by its intermolecular relationships to adjacent molecules in the mass. Three basic molecular properties affect processing performances, such as flow conditions, that in turn affect product performances, such as strength or dimensional stability. They are (1) mass or density, (2) molecular weight (MW), and (3) molecular weight distribution (MWD).

Densities
Absolute density (d) is the mass of any substance per unit volume of a material. It is usually expressed in grams per cubic centimeter (g/cm^3) or pounds per cubic inch (lb/in^3) (Table 1.5). Specific gravity (s.g.) is the ratio of the mass in air of a given volume compared to the mass of the same volume of water. Both d and s.g. are measured at room temperature [23°C (73.4°F)]. Since s.g. is a dimensionless quantity, it is convenient for comparing different materials. Like density, specific gravity is used extensively in determining product cost vs. average product thickness, product weight, quality control, and so on. It is frequently used as a means of setting plastic specifications and monitoring product consistency.

In crystalline plastics, density has a direct effect on properties such as stiffness and permeability to gases and liquids. Changes in density may also affect other mechanical properties.

The term *apparent density* of a material is sometimes used. It is the weight in air of a unit volume of material including voids usually inherent in the material. Also used is the term *bulk density* that is commonly used for compounds or materials such as molding powders, pellets, or flakes. Bulk density is the ratio of the weight of the compound to its volume of a solid material including voids.

Table 1.5 Comparing densities of different polyethylene thermoplastics

Type	Density, g/cm³ (lb/ft³)
LDPE	0.910–0.925 (56.8–57.7)
MDPE	0.926–0.940 (57.8–58.7)
HDPE	0.941–0.959 (58.7–59.9)
HMWPE	0.960 & above (59.9 & above)

Molecular Weights
MW is the sum of the atomic weights of all the atoms in a molecule. Atomic weight is the relative mass of an atom of any element based on a scale in which a specific carbon atom (carbon-12) is assigned a mass value of 12. For polymers, it represents a measure of the molecular chain length. MW of plastics influences their properties. With increasing MW, polymer properties increase for abrasion resistance, brittleness, chemical resistance, elongation, hardness, melt viscosity, tensile strength, modulus, toughness, and yield strength. Decreases occur for adhesion, melt index, and solubility.

Adequate MW is a fundamental requirement to achieve desired properties of plastics. If the MW of incoming material varies, the fabricating and fabricated product performance can be altered. The greater the differences, the more dramatic the changes that occur during processing.

Molecular Weight Distributions
MWD is basically the amounts of component polymers that make up a polymer (Fig. 1.6). Component polymers, in contrast, are a convenient term that recognizes the fact that all polymeric materials comprise a mixture of different polymers of differing molecular weights. The ratio of the weight average molecular weight to the number average molecular weight gives an indication of the MWD.

One method of comparing the processability with product performances of plastics is to use their MWD. A narrow MWD enhances the performance of plastic products. Wide MWD permits easier processing. Melt flow rates are dependent on the MWD. With MWD differences of incoming material the fabricated performances can be altered requiring resetting process controls. The more the difference, the more dramatic changes that can occur in the products.

Viscosities and Melt Flows

Viscosity is a measure of resistance to plastic melt flow. It is the internal friction in a melt resulting when one layer of fluid is caused to move in

Figure 1.6 Examples of narrow and wide molecular weight distributions

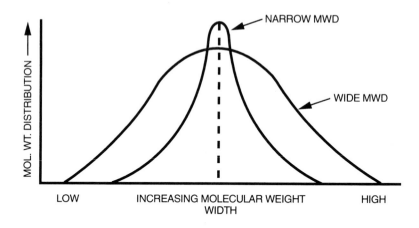

relationship to another layer. Thus viscosity is the property of the resistance of flow exhibited within a body of material. It is the constant ratio of shearing stress to the rate of shear. Shearing is the motion of a fluid, layer by layer, like playing cards in a deck. When plastics flow through straight tubes or channels they are sheared: the viscosity expresses their resistance.

The melt index (MI) or melt flow index (MFI) is an inverse measure of viscosity. High MI implies low viscosity and low MI means high viscosity. Plastics are shear thinning, which means that their resistance to flow decreases as the shear rate increases. This is due to molecular alignments in the direction of flow and disentanglements.

Newtonian/non-Newtonian
Viscosity is usually understood to mean Newtonian viscosity in which case the ratio of shearing stress to the shearing strain is constant. In non-Newtonian behavior, typical of plastics, the ratio varies with the shearing stress. Such ratios are often called the *apparent viscosities* at the corresponding shearing stresses. Viscosity is measured in terms of flow in Pa·s (P) with water as the base standard (value of 1.0). The higher the number, the less flow.

Melt Index
The melt indexer (MI; extrusion plastometer) is the most widely used rheological device for examining and studying plastics (principally TPs) in many different fabricating processes. It is not a true viscometer in the sense that a reliable value of viscosity cannot be calculated from the

measured flow index. However, the device does measure isothermal resistance to flow, using standard apparatus and test methods that are standard throughout the world. The standards used include ASTM D 1238 (U.S.A.), BS 2782-105°C (U.K.), DIN 53735 (Germany), JIS K72 IO (Japan), ISO RI 133/R292 (international), and others.

The standard apparatus is a ram type plasticator which at specified temperatures and pressure extrudes a plastic melt through the die exit opening. The standard procedure involves the determination of the amount of plastic extruded in 10 minutes. The flow rate, expressed in g/10 min., is reported. As the flow rate increases, viscosity decreases. Depending on the flow behavior, changes are made to standard conditions (die opening size, temperature, etc.) to obtain certain repeatable and meaningful data applicable to a specific processing operation. Table 1.6 lists typical MI ranges for the certain processes.

Table 1.6 Examples of melt index for different processes.

Process	MI range
Injection Molding	5–100
Rotational Molding	5–20
Coating Extrusion	0.1–1
Film Extrusion	0.5–6
Profile extrusion	0.1–1
Blow molding	0.1–1

Rheology & Mechanical Analysis

Rheology and mechanical analysis are usually familiar techniques, yet the exact tools and the far-reaching capabilities may not be so familiar. Rheology is the study of how materials flow and deform, or when testing solids it is called dynamic mechanical thermal analysis (DMTA).

During rheometer and dynamic mechanical analyses instruments impose a deformation on a material and measure the material's response that gives a wealth of very important information about structure and performance of the basic polymer. As an example stress rheometers are used for testing melts in various temperature ranges. Strain controlled rheology is the ultimate in materials characterization with the ability to handle anything from light fluids to solid bars, films, and fibers.

With dynamic testing, the processed plastic's elastic modulus (relating to energy storage) and loss modulus (relative measure of a damping ability) are determined. Steady testing provides information about creep and recovery, viscosity, rate dependence, etc,

Viscoelasticities

Understanding and properly applying the following information to product design equations is very important. A material having this property is considered to combine the features of a so-called perfect elastic solid and a perfect fluid. It represents the combination of elastic and viscous behavior of plastics that is a phenomenon of time-dependent, in addition to elastic deformation (or recovery) in response to load.

This property possessed by all fabricated plastics to some degree, indicates that while plastics have solid-like characteristics such as elasticity, strength, and form or shape stability, they also have liquid-like characteristics such as flow depending on time, temperature, rate, and amount of loading. The mechanical behavior of these viscoelastic plastics is dominated by such phenomena as tensile strength, elongation at break, stiffness, rupture energy, creep, and fatigue which are often the controlling factors in a design.

Processing-to-Performance Interface

Different plastic characteristics influence processing and properties of plastic products. Important are glass transition temperature and melt temperature.

Glass Transition Temperatures

The T_g relates to temperature characteristics of plastics (Table 1.7). It is the reversible change in phase of a plastic from a viscous or rubbery state to a brittle glassy state (Fig. 1.7). T_g is the point below which plastic behaves like glass and is very strong and rigid. Above this temperature it is not as strong or rigid as glass, but neither is it brittle as glass. At and above T_g the plastic's volume or length increases more rapidly and rigidity and strength decrease. As shown in Fig. 1.8 the amorphous TPs have a more definite T_g when compared to crystalline TPs. Even with variation it is usually reported as a single value.

The thermal properties of plastics, particularly its T_g, influence the plastic's processability performance and cost in different ways. The operating temperature of a TP is usually limited to below its T_g. A more expensive plastic could cost less to process because of its T_g location that results in a shorter processing time, requiring less energy for a particular weight, etc. (Fig. 1.9).

The T_g generally occurs over a relatively narrow temperature span. Not only do hardness and brittleness undergo rapid changes in this temperature region, but other properties such as the coefficient of thermal expansion and specific heat also change rapidly. This pheno-menon has been called second-order transition, rubber transition, or

Table 1.7 Range of T_g for different thermoplastics

Plastic	°C	°F
Polyethylene	−120	−184
Polypropylene	−22	−6
Polybutylene	−25	−13
Polystyrene	95	203
Polycarbonate	150	302
Polyvinyl Chloride	85	185
Polyvinyl Fluoride	−20	−4
Polyvinylidene Chloride	−20	−4
Polyacetal	−80	−112
Nylon 6	50	122
Polyester	110	230
Polytetrafluoroethylene	−115	−175
Silicone	−120	−184

Figure 1.7 Thermoplastic volume or length changes at the glass transition temperature

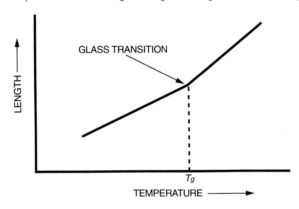

Figure 1.8 Change of amorphous and crystalline thermoplastic's volume at T_g and T_m

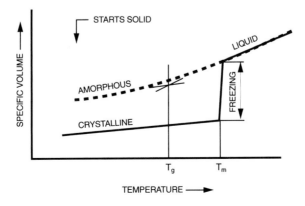

Figure 1.9 Modules behavior with increase in temperature (DTUL = deflection temperature under load). (Courtesy of Bayer)

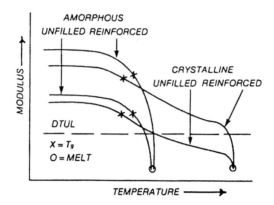

rubbery transition. The word transformation has also been used instead of transition. When more than one amorphous transition occurs in a plastic, the one associated with segmental motions of the plastic backbone chain, or accompanied by the largest change in properties, is usually considered to be the T_g.

Important for designers to know that above T_g, many mechanical properties are reduced. Most noticeable is a reduction that can occur by a factor of 1,000 in stiffness.

Melt Temperatures
Crystalline plastics have specific melt temperatures (T_m) or melting points. Amorphous plastics do not. They have softening ranges that are small in volume when solidification of the melt occurs or when the solid softens and becomes a fluid type melt. They start softening as soon as the heat cycle begins. A melting temperature is reported usually representing the average in the softening range.

The T_m of crystalline plastics occurs at a relatively sharp point going from solid to melt. It is the temperature at which melts softens and begins to have flow tendency (Table 1.8). They have a true T_m with a latent heat of fusion associated with the melting and freezing process, and a relatively large volume change during fabrication. Crystalline plastics have considerable order of the molecules in the solid state indicating that many of the atoms are regularly spaced. The melt strength of the plastic occurs while in the molten state. It is an engineering measure of the extensional viscosity and is defined as the maximum tension that can be applied to the melt without breaking.

Table 1.8 Crystalline thermoplastic melt temperatures

Plastic	°C	°F
Low Density Polyethylene	116	240
High Density Polyethylene	130	266
Polypropylene	175	347
Nylon 6	215	419
Nylon 66	260	500
Polyester	260	500
Polyarylamide	400	755
Polytetrafluoroethylene	330	626

The T_m is dependent on the processing pressure and the time under heat, particularly during a slow temperature change for relatively thick melts during processing. Also, if the melt temperature is too low, the melt's viscosity will be high and more costly power required processing it. If the viscosity is too high, degradation will occur. There is the correct processing window used for the different melting plastics.

Processing and Moisture

Recognize that properties of designed products can vary, in fact can be destructive, with improper processing control such as melt temperature profile, pressure profile, and time in the melted stage. An important condition that influence properties is moisture contamination in the plastic to be processed. There are the hygroscopic plastics (PET, etc.) that are capable of retaining absorbed and adsorbed atmospheric moisture within the plastics. The non-hygroscopic plastics (PS, etc.) absorb moisture only on the surface. In the past when troubleshooting plastic's reduced performance was 90% of the time due to the damaging effect of moisture because it was improperly dried prior to processing. At the present time it could be at 50%.

All plastics, to some degree, are influenced by the amount of moisture or water they contain before processing. With minimal amounts in many plastics, mechanical, physical, electrical, aesthetic, and other properties may be affected, or may be of no consequence. However, there are certain plastics that, when compounded with certain additives such as color, could have devastating results. Day-to-night temperature changes is an example of how moisture contamination can be a source of problems if not adequately eliminated when plastic materials are exposed to the air. Moisture contamination can have an accumulative effect. The critical moisture content that is the average material moisture content at the end of the constant-rate drying period, is a function of material properties, the constant-rate of drying, and particle size.

Although it is sometimes possible to select a suitable drying method simply by evaluating variables such as humidities and temperatures when removing unbound moisture, many plastic drying processes do not involve removal of bound moisture retained in capillaries among fine particles or moisture actually dissolved in the plastic. Measuring drying-rate behavior under control conditions best identifies these mechanisms. A change in material handling method or any operating variable, such as heating rate, may effect mass transfer.

Drying Operations
When drying at ambient temperature and 50% relative humidity, the vapor pressure of water outside a plastic is greater than within. Moisture migrates into the plastic, increasing its moisture content until a state of equilibrium exists inside and outside the plastic. But conditions are very different inside a drying hopper (etc.) with controlled environment. At a temperature of 170°C (350°F) and –40°C (–40°F) dew point, the vapor pressure of the water inside the plastic is much greater than the vapor pressure of the water in the surrounding area. Result is moisture migrates out of the plastic and into the surrounding air stream, where it is carried away to the desiccant bed of the dryer.

Target is to keep moisture content at a designated low level, particularly for hygroscopic plastics where moisture is collected internally. They have to be carefully dried prior to processing. Usually the moisture content is >0.02 wt%. In practice, a drying heat 30°C below the softening heat has proved successful in preventing caking of the plastic in a dryer. Drying time varies in the range of 2 to 4 h, depending on moisture content. As a rule of thumb, the drying air should have a dew point of –34°C (–30°F) and the capability of being heated up to 121°C (250°F). It takes about 1 ft^3 min^{-1} of plastic processed when using a desiccant dryer.

The non-hygroscopic plastics collect moisture only on the surface. Drying this surface moisture can be accomplished by simply passing warm air over the material. Moisture leaves the plastic in favor of the warm air resulting in dry air. The amount of water is limited or processing can be destructive.

Determine from the material supplier and/or experience the plastic's moisture content limit. Also important is to determine which procedure will be used in determining water content. They include equipment such as weighing, drying, and/or re-weighing. These procedures have definite limitations based on the plastic to be dried. Fast automatic analyzers, suitable for use with a wide variety of plastic systems, are available that provide quick and accurate data for obtaining the in-plant moisture control of plastics.

Fabricating processes

Designing good products requires some familiarity with processing methods. Until the designer becomes familiar with processing, a qualified fabricator must be taken into the designer's confidence early in development. The fabricator and mold or die designer should advise the product designer on materials behavior and how to simplify the design in order to simplify processing and reducing cost. Understanding only one process and in particular just a certain narrow aspect of it should not restrict the designer.

There are dozens of popular different basic processes with each having many modifications so that there are literally hundreds of processes used. The ways in which plastics can be processed into useful end products tend to be as varied as the plastics themselves. However only a few basic processes are used worldwide for most of the products produced. Extrusion consumes approximately 36wt% of all plastics. IM follows by consuming 32wt%. Consumption by other processes is estimated 10wt% blow molding, 8% calendering, 5% coating, 3% compression molding 3%, and others 3%. Thermoforming, which is the fourth major process used (considered a secondary process, since it begins with extruded sheets and films where extrusion is the primary process), consumes principally about 30% of the extruded sheet and film that principally goes into packaging.

It is estimated that there are in USA about 80,000 injection molding machines (IMMs) and about 18,000 extruders operating. This difference in the amount of machines is due the fact that there is more activity (product design, R&D, fabrication, etc.) required with injection molding (IM).

If an extruder can be used to produce products it has definite operating and economical advantages compared to IM. It requires detailed process control. IM requires more sophisticated process control to fabricate many thousands of different complex and intricate products.

While the processes differ, there are elements common to many of them. In the majority of cases, TP compounds in the form of pellets, granules, flake, and powder, are melted by heat so they can flow.

Pressure is often involved in forcing the molten plastic into a mold cavity or through a die and cooling must be provided to allow the molten plastic to harden. With TSs, heat and pressure also are most often used, only in this case, higher heat (rather than cooling) serves to cure or harden the TS plastic, under pressure, in the mold. When liquid TPs or TSs plastics incorporate certain additives, heat and/or pressure need not necessarily be used.

Understanding, controlling, and measuring the plastic melt flow behavior of plastics during processing is important. It relates to a plastic that can be fabricated into a useful product. The target is to provide the necessary homogeneous-uniformly-heated melt during processing to have the melt operate completely stable and working in equilibrium. Unfortunately the perfect melt does not exist. Fortunately with the passing of time where improvements in the plastics and equipment uniformity continues to occur, melt consistency and melt flow behavior continues to improve, simplifying the art of processing.

An important factor for the processor is obtaining the best processing temperature for the plastics used. A guide is obtained from past experience and/or the material producer. The set-up person determines the best process control conditions (usually requires certain temperature, pressure, and time profiles) for the plastic being processed. Recognize that if the same plastic is used with a different machine (with identical operating specifications) the probability is that new control settings will be required for each machine. The reason is that, like the material, machines have variables that are controllable within certain limits that permit meeting the designed product requirements including costs.

The secondary operations fabricating methods can be divided into three broad categories: the machining of solid shapes; the cutting, sewing, and sealing of film and sheeting; and the forming of film and sheet. The machining techniques used are quite common to metal, wood, and other industries. Plastic shapes can be turned into end products by such methods as grinding, turning on a lathe, sawing, reaming, milling, routing, drilling, and tapping.

The cutting, sewing, and sealing of film and sheet involve turning plastic film and sheeting into finished articles like inflatable toys, garment bags, shower curtains, aprons, raincoat, luggage, and literally thousands of products. In making these products, the film or sheet is first cut to the desired pattern by hand, in die-cutting presses, or by other automatic methods. The pieces are then put together using such assembly techniques such as sewing, heat bonding, welding, high frequency vibration, or ultrasonic sealing.

There are post-finished forming methods. Film and sheet can be post-embossed with textures and letterpress, gravure, or silk screening can print them. Rigid plastic parts can be painted or they can be given a metallic surface by such techniques as metallizing, barrel plating, or electroplating. Another popular method is hot-stamping, in which heat, pressure, and dwell time are used to transfer color or design from a carrier film to the plastic part. Popular is the in-mold decorating that involves the incorporation of a printed foil into a plastic part during molding so that it becomes an integral part of the piece and is actually inside the part under the surface. There are applications, such as with blow molded products, where the foil provides structural integrity reducing the more costly amount of plastic to be used in the products.

Extrusions

Extrusion is the method employed to form TPs into continuous films, sheeting, tubes, rods, profile shapes, filaments, coatings (wire, cable, cord, etc.), etc. In extrusion, plastic material is first loaded into a hopper using upstream equipment, then fed into a long heating chamber through which it is moved by the action of a continuously revolving screw. At the end of this plasticator the molten plastic is forced through an orifice (opening) in a die with the relative shape desired in the finished product. As the extrudate (plastic melt) exits the die, it is fed downstream onto a pulling and cooling device such as multiple rotating rolls, conveyor belt with air blower, or water tank with puller.

The multi-screw extruders are used as well as the more popular single-screw extruders. Multiscrew extruders are primarily used for compounding plastic materials. Each has benefits primarily based on the plastic being processed and the products to be fabricated. At times their benefits can overlap, so the type to be used would depend on cost factors, such as cost to produce a quality product, cost of equipment, cost of maintenance, etc.

Size of the die orifice initially controls the thickness, width, and shape of any extruded product dimension. It is usually oversized to allow for the drawing and shrinkage that occur during conveyor pulling and cooling operations. The rate of takeoff also has significant influences on dimensions and shapes. This action, called drawdown, can also influence keeping the melt extrudate straight and properly shaped, as well as permitting size adjustments. Drawdown ratio is the ratio of orifice die size at the exit to the final product size.

Each of the processes (blown film, sheet, tube, etc.) contains secondary equipment applicable to their specific product lines such as computerized fluid chillers and temperature control systems. Equipment has become more energy-efficient, reliable, and cost-effective. The application of microprocessor and computer compatible controls that can communicate within the extruder line results in the more accurate control of the line.

A major part of film, sheet, coating, pipe, profile, etc. lines involve windup rolls. They include winders, dancer rolls, lip rolls, spreader rolls, textured rolls, engraved rolls, and cooling rolls. All have the common feature that they are required to be extremely precise in all their operations and measurements. Their surface conditions include commercial grade mirror finishes, precision bearings and journals are used, and, most important, controlled variable rotating speed controls to ensure uniform product tension control.

Orientations
Systems have been designed to increase the degree of orientation (stretching) in order to obtain films of improved clarity, strength, heat resistance, etc. Except for special applications, where greater strength in one direction may be needed, films are normally made with balanced properties.

Postformings
Various methods can be used for postforming products after the hot plastic melt leaves the extruder die. Examples are netting products that are flat to round shapes, rotated mandrel die makes perforated tubing, spiral spacer web around a coated wire or tube, varying tube or pipe wall thickness, and different perforated tubing or pipe pattern.

Coextrusions
There is the important variation on extrusion that involves the simultaneous or coextrusion of multiple molten layers of plastic from a single extrusion system. Two or more extruders are basically joined together by a common manifold through which melts flow before entering the die face. The plastics can include the same material but with different colors. There are also systems sometimes used where one material with two melts is made from one plasticator whereby certain advantages develop vs. the usual single melt such as reducing pin holes, and/or strengthening the product.

Many advantages exist in coextrusion. The different materials used in the coextruded structure meets different performance requirements based on their combinations. A single expensive plastic could be used to meet performance requirements such as permeability resistance, however with the proper combinations of plastics cost reductions will occur.

Injection Moldings

The process of IM is used principally for processing unreinforced or glass fiber reinforced TPs however it also processes TSs. Examples of the importance of using different mold design approaches are reviewed in Fig. 1.10 concerning product openings and Fig. 1.11 highlighting different ways with or without a parting line on the threads. These are examples that molds have to be properly designed to meet proper operations of product requirements. Where possible design of product shapes should make use of simplifying the design of molds.

The machines used for molding TSs are basically the same system as in molding TPs. Temperatures differ, as does the design of the screw. Unlike TPs that just melt in the plasticator and solidify in the cooled mold, the TSs melt in the plasticator and cure to a harden state in the mold that operates at a higher temperature than the plasticator.

Coinjections

The review in coextrusion also applies with coinjection providing similar advantages. Two or more injection molding barrels are basically joined together by a common manifold and nozzle through which melts flow before entering the mold cavity by a controlled device such

Figure 1.10 Examples of simplifying mold construction to produce openings without side action movements

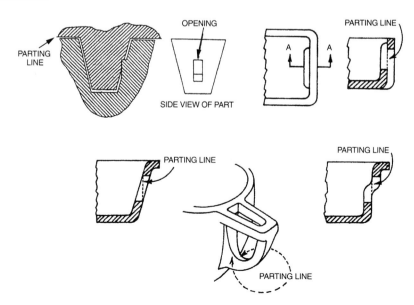

Figure 1.11 Examples of molding with or without parting line on a product.

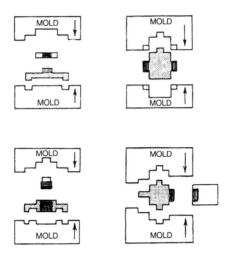

as an open-closed valve system. The plastics can include the same material but with different colors. There are also systems sometimes used where one material with two shots is made from one plasticator whereby certain advantages develop vs. the usual single shot IM such as reducing pin holes, and/or strengthening the product. The nozzle is usually designed with a shutoff feature that allows only one melt to flow through at a controlled time.

The usual coinjection with two or more different plastics is bonded/laminated together. Proper melt flow and compatibility of the plastics is required in order to provided the proper adhesion. Some of the melt flow variable factors can be compensated by the available plasticator and mold process control adjustments.

Gas-Assist Moldings
There are different gas-assist injection molding (GAIM) processes. Other names exist that include injection molding gas-assist (IMGA), gas injection molding (GIM), or injection gas pressure (IGP). Most of the gas-assisted molding systems are patented. This review concerns the use of gas, however there are others such as water-assist injection molding.

The processes use a gas that is usually nitrogen with pressures up to 20 to 30 MPa (2,900 to 4,400 psi). Within the mold cavity the gas in the melt forms channels. Gas pressure is maintained through the cooling cycle. In effect the gas packs the plastic against the cavity wall. Gas can be injected through the center of the IMM nozzle as the melt travels to the cavity or it can be injected separately into the mold cavity. In a properly designed tool run under the proper process conditions, the gas with its much lower viscosity than the melt remains isolated in the gas channels of the part without bleeding out into any thin-walled areas in the mold. The gas produces a balloon-like pressure on the melt.

The gas-assist approach is a solution to many problems associated with conventional IM and structural foam molding. It significantly reduces volume shrinkage that can cause sink marks in injection molding. Products are stiffer in bending and torsion than equivalent conventional IM products of the same weight. The process is very effective in different size and shape products, especially the larger molded products. It offers a way to mold products with only 10 to 15% of the clamp tonnage that would be necessary in conventional injection molding.

Micromoldings
As reviewed, the basic processes have many different fabricating systems. An example for IM is micromolding; precision molding of extremely small products as small as one mm^3. Products usually weigh less than 20 milligrams (0.020g) with some even as low as 0.01g. Products are measured in microns and have tolerances of ±10 microns or less. A micron (μm) is one-millionth of a meter; 25.4 μm make up one-thousandth of an inch. In comparison a human hair is 50 to 100 μm in diameter. A mil, that is about 25 times smaller than a micron, is one-thousandth of an inch.

Molding machines and tooling for small parts are not just smaller versions of their regular larger molding counterparts. Tooling is often created using electrical-discharge machining or diamond turning. It can be created with surface features below the wavelength of light by using lithographic and electrodeposition techniques. Proper venting usually has to include precision venting in the cavity as well as possibly removing air prior to entering the cavity.

Blow Moldings
Generally used only with thermoplastics, this process is applicable to the production of hollow plastic products such as bottles, gas tanks, and complex shaped containers/devices. The two basic systems to melt the TP are extruding (Fig. 1.12) or injection molding (Fig. 1.13). BM involves the melting of the TPs, then forming it into a tube-like or test tube shape (known as a parison when using an extruder or preform when injection molding), sealing the ends of the tube, and injecting air (through a tube or needle inserted in the tube or an opening in the preform core pin). The parison or preform, in a softened state, is inflated inside the mold and forced against the walls of the mold's female cavity. On cooling, the product, now conforming to the shape of the cavity, is solidified, and

Figure 1.12 Schematic of the extrusion BM process

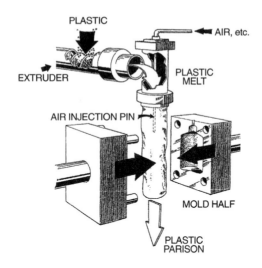

Figure 1.13 Schematic of the injection BM process

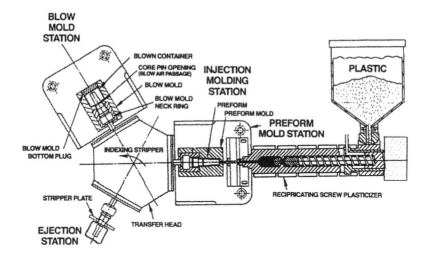

ejected from the mold as a finished piece. The coextrusion and coinjection already reviewed also applies to BM products.

Complex Consolidated Structural Products
BM provides designers with the capability to make products ranging from the simple to rather complex 3-D shapes. Designers should become aware of the potentials BM offers since intricate and complex structural shapes can be fabricated. There are different techniques for BM these shapes (Fig. 1.14). The techniques involve moving the

Figure 1.14 Examples of complex BM products

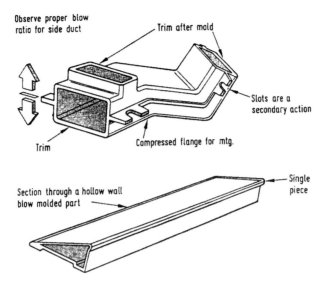

preform or parison, moving the mold, or a combination of moving both the hot melt and mold.

BM permits combining in one product different parts or shapes that are to be assembled when using other processes. Result is simplifying the product design and significantly reducing cost. Some of the consolidating functions include hinges, inserts, fasteners, threads, non-plastic parts, and others somewhat similar to those used in injection molding. Hinges include the different mechanical types as well as integral hinges.

Thermoformings

Thermoforming consists of uniformly heating TP sheet or film to its softening temperature. Next the heated flexible plastic is forced against the contours of a mold. Force is applied by mechanical means (tools, plugs, solid molds, etc.) or by pneumatic means (differentials in air pressure created by pulling a vacuum between plastic and mold or using the pressures of compressed air to force the sheet against the mold).

Almost any TP can be thermoformed. However certain types make it easier to meet certain forming requirements such as deep draws without tearing or excessive thinning in areas such as corners, and/or stabilizing of uniaxial or biaxial deformation stresses. Ease of thermoforming basically depends on stock material's thickness tolerance and forming characteristics. This ease of forming is influenced by factors such as to minimize the variation of the sheet thickness so that a uniform heat occurs in the film or sheet material thicknesswise, ability of the plastic to retain uniform and specific heat gradients across its surface and thickness, elimination or minimizing pinholes in the plastic, and stabilizing of uniaxial or biaxial deformation.

Many forming techniques are used. Each has different capabilities depending on factors such as formed product size, thickness, shape, type plastic, and/or quantity. Mold geometry with their different complex shapes vs. type of plastic material being processed will influence choice of process.

Foams

The manufacture of foam plastic products cuts across most of the processing techniques used. Foams can be fabricated during extrusion, injection molding, blow molding, casting, calendering, coating, rotational molding, etc. Typical requirements in such instances can be the incorporation of blowing agents in the plastic. They can be those that decompose under heat to generate the gasses needed to create the cellular structure. Various controls to accommodate the foaming action are used.

There are, however, some techniques unique to foamed plastics. When working with expanded polystyrene (EPS) beads, for example, to produce cups, picnic dishes, etc., various steam-chest molding methods are used. Based on the blowing agent used (pentane gas, etc.) the application of steam causes the beads to expand and fuse together in a perforated mold.

When working with polyurethane foams, it is possible to use spray guns or mixing metering machines to mix the liquid ingredients together and direct them into a product cavity, mold, etc. The mixed ingredients with their chemical reaction start to foam after leaving the dispensing equipment.

There is a unique technology of molding structural foam, foams with integral solid skins, and a cellular core resulting in a high strength-to-weight ratio. When processing structural foams, several techniques are used with most related to injection molding and extrusion.

Reinforced Plastics

Different fabricating processes and materials of construction are employed to produce RP products that represent about 5wt% of all plastic products produced worldwide. They range in fabricating pressures from zero (contact), through moderate, to relatively high pressure [14 to 207 MPa (2,000 to 30,000 psi)], at temperatures based on the plastic's requirements that range from room temperature and higher. Equipment may be simple/low cost with labor costs high, to rather expensive specialized computer control sophisticated equipment with very low labor costs for the different processes. Each process provides capabilities such as meeting production quantity (small to large), performance requirements, proper ratio of reinforcement to matrix, fiber orientation, reliability/quality control, surface finish, and so forth versus cost (equipment, labor, utilities, etc.).

The plastic may be either reinforced TSs (RTSs) or reinforced TPs (RTPs). The RTSs were the first major plastics to be adapted to this technology. The largest consumption of RTPs are processed by different methods such as injection molding (over 50wt%), rotationally molding, or extruded on conventional equipment. There are even RTP sheets that can be "cold" stamped into shape using matching metal molds that form the products. It is called cold stamping because the molds are kept at or slightly above room temperature. The sheets, however, must be pre-heated.

Calenders

Calenders can be used to process TPs into film and sheeting, and to apply a plastic coating to textiles or other supporting/substraight materials. In calendering film and sheeting, the plastic compound is passed between a series of three or four large, heated, revolving rollers that squeeze the material between them into a sheet or film. An analogy in this case might be flattening out a pasty dough mixture with a rolling pin. The thickness of the finished material is controlled primarily by the space between the rolls. The surface of the plastic film or sheeting may be smooth or matted, depending on the surfacing on the rollers. When large quantities of particularly PVC film and sheet are to be manufactured, this process can provide lower cost products than extrusion.

Castings

Casting may be used with TPs or TS plastics to make products, shapes, rods, tubes, film, sheet, etc. by basically pouring a liquid monomer-polymer solution into an open or closed mold where it finishes polymerizing and/or cooled into a solid. This liquid is often a monomer rather than the polymer used in most molding compounds. In turn the polymer with heat polymerizes into a solid plastic. An essential difference between casting and molding is that pressure need not be used in casting (although large-volume, complex parts can be made by low pressure-casting methods). A variation on casting is known as liquid injection molding (LIM) and involves the proportioning, mixing, and dispensing of liquid components and directly injecting the resultant mix into a mold that is clamped under pressure.

Coatings

TPs or TS plastics may be used as a coating. The materials to be coated may be plastic, metal, wood, paper, fabric, leather, glass, concrete, ceramics, etc. Methods of coating are varied and include knife or spread coating, spraying, roller coating, dipping, brushing, and extrusion. Calendering of a film to a supporting material is also a form of coating.

Special methods can use powdered plastics for coatings. As an example the fluidized bed coating system. The object to be coated is heated and then immersed in a dense-phase fluidized bed of powdered plastic; the plastic adheres to the heated object and subsequent heating provides a smooth, pinhole-free coating. The electrostatic spray system is based on the fact that most plastic powders are insulators with relatively high volume resistivity values. Therefore, they accept a charge (positive or negative polarity) and are attracted to a grounded or oppositely charged objcct (which is the one being coated).

Compression Moldings

CM is the most common method of forming TS plastics. Until the advent of injection molding, it was the most important of plastic processes. CM is the compressing of a material into a desired shape by application of heat and pressure to the material in a mold cavity. Pressure is usually at 7 to 14 MPa (1000 to 2000 psi). Some TSs may require pressures down to 345 kPa (50 psi) or even just contact (zero pressure). The majority of TS compounds are heated to about 150 to 200°C (302 to 392°F) for optimum cure; but can go as high as 650°C (1200°F).

Reaction Injection Moldings

The RIM process predominantly uses TS polyurethane (PUR) plastics. Others include nylon, TS polyester, and epoxy. PUR offers a large range of product performance properties. As an example PUR has a modulus of elasticity in bending of 200 to 1400 MPa (29,000 to 203,000 psi) and heat resistance in the range of 90 to 200°C (122 to 392°F). The higher values are obtained when glass-fiber reinforces the PUR (also with nylon, etc.). The reinforced RIM process is called RRIM or structural RIM (SRIM). Large and very thick RIM products can be molded with or without reinforcements using fast cycles.

When compared to injection molding (IM) that processes a plastic compound (polymer plus additives, etc.), RIM uses two liquid PUR chemical monomer components (polyol and isocyanate) that are mixed to produce the polymer (plastic). Additives such as catalysts, surfactants, fillers, reinforcements, and/or blowing agents are also incorporated. Their purpose is to propagate the reaction and form a finished product possessing the desired properties

Mixing is by a rapid impingement in a chamber (under high pressure in a specially designed mixing head) at relatively low temperatures before being injected into a closed mold cavity at low pressure. An exothermic chemical reaction occurs during mixing and in the cavity requiring less energy than the conventional IM system. Polymerization of the monomer mixture in the mold allows for the custom formulation of material properties and kinetics to suit a particular product application.

RIM is the logical process to consider at least for molding large and/or thick products. With RIM technology, cycle times of 2 min and less have been achieved in production for molding large and thick [10 cm (3.9 in.)] products. It is less competitive for small products. Capital requirements for RIM processing equipment are rather low when compared with injection molding equipment (includes mold) that would be necessary to mold products of similar large size.

Rotational Moldings

This method, like blow molding, is used to make hollow one-piece TP parts. RM consists of charging a measured amount of TP into a warm mold cavity that is rotated in an oven about two axes. In the oven, the heat penetrates the mold, causing the plastic, if it is in powder form, to become tacky and adhere to the mold female cavity surface, or if it is in liquid form, to start to gel on the mold cavity surface. Since the molds continue to rotate during the heating cycle, the plastic will gradually become distributed on the mold cavity walls through gravitational force. As the cycle continues, the plastic melts completely, forming a homogeneous layer of molten plastic. After cooling, the molds are opened and the parts removed.

RM can produce quite uniform wall thicknesses even when the product has a deep draw of the parting line or small radii. The liquid or powdered plastic used in this process flows freely into corners or other deep draws upon the mold being rotated and is fused/melted by heat passing through the mold's wall.

This process is particularly cost-effective for small production runs and large product sizes. The molds are not subjected to pressure during molding, so they can be made relatively inexpensively out of thin sheet metal. The molds may also be made from lightweight cast aluminum and electroformed nickel, both of them light in weight and low in cost. Large rotational machines can be built economically because they use inexpensive gas-fired or hot air ovens with the lightweight mold-rotating equipment.

Variables

Even though equipment operations and plastic compounded materials have understandable and controllable variables that influence processing, the usual most uncontrollable variable in the process can be the plastic material. The degree of properly compounding or blending by the plastic manufacturer, converter, or in-house by the fabricator is important. With the passing of time and looking ahead, existing material and equipment variabilities are continually reduced due to improvement in their manufacturing and process control capabilities. However they still exist.

FALLO Approach

Conditions that are important in making plastic products the success it has worldwide are summarized in Fig. 1.15. All designs, processes, and materials fit into this overall FALLO (Follow ALL Opportunities) approach flow chart that produces products meeting required performance and cost requirements.

Designers and processors, needing to produce qualified products at the lowest cost have used the basic concept of the FALLO approach. This approach makes one aware that many steps are involved to be successful, all of which must be coordinated and interrelated. It starts with the design that involves specifying the plastic and specifying the manufacturing process. The specific process (injection, extrusion, blow molding, thermoforming, and so forth) is an important part of the overall scheme.

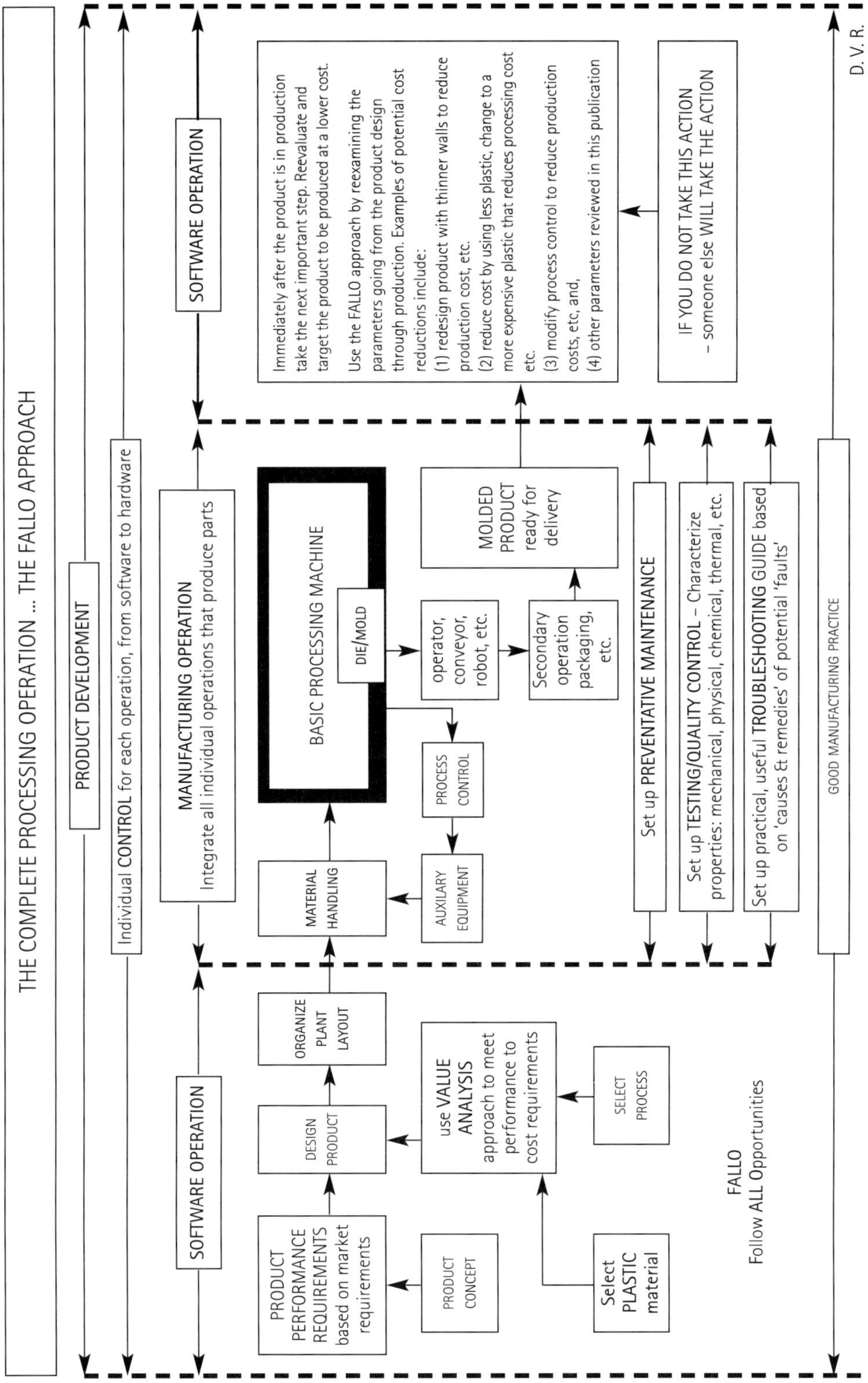

Figure 1.15 The FALLO approach: Follow ALL Opportunities

THE COMPLETE PROCESSING OPERATION ... THE FALLO APPROACH

PRODUCT DEVELOPMENT

SOFTWARE OPERATION

Individual CONTROL for each operation, from software to hardware

MANUFACTURING OPERATION
Integrate all individual operations that produce parts

SOFTWARE OPERATION

BASIC PROCESSING MACHINE

DIE/MOLD

MATERIAL HANDLING

AUXILARY EQUIPMENT

PROCESS CONTROL

operator, conveyor, robot, etc.

Secondary operation packaging, etc.

MOLDED PRODUCT ready for delivery

Immediately after the product is in production take the next important step. Reevaluate and target the product to be produced at a lower cost.

Use the FALLO approach by reexamining the parameters going from the product design through production. Examples of potential cost reductions include:

(1) redesign product with thinner walls to reduce production cost, etc.

(2) reduce cost by using less plastic, change to a more expensive plastic that reduces processing cost etc.

(3) modify process control to reduce production costs, etc and,

(4) other parameters reviewed in this publication

IF YOU DO NOT TAKE THIS ACTION
– someone else WILL TAKE THE ACTION

D. V. R.

Set up PREVENTATIVE MAINTENANCE

Set up TESTING/QUALITY CONTROL – Characterize properties: mechanical, physical, chemical, thermal, etc.

Set up practical, useful TROUBLESHOOTING GUIDE based on 'causes & remedies' of potential 'faults'

GOOD MANUFACTURING PRACTICE

ORGANIZE PLANT LAYOUT

DESIGN PRODUCT

use VALUE ANALYSIS approach to meet performance to cost requirements

SELECT PROCESS

PRODUCT PERFORMANCE REQUIREMENTS based on market requirements

PRODUCT CONCEPT

Select PLASTIC material

FALLO
Follow ALL Opportunities

2 DESIGN OPTIMIZATION

Introduction

To design successful plastic products meeting factors such as quality requirements, consistency, designated life, and profitability, what is needed is understanding and applying the behavior of plastics such as service temperature, load, and time in optimizing the design. Similar action is required for other materials (steel, glass, wood, etc.).

When compared to other materials such as steel and certain other metals their data are rather constant, at least in the temperature range in which plastics are used. When the design engineer is accustomed to working with metals, the same computations are used in order to obtain a plastic product with sufficient strength and deformation under a given load that must not exceed a definite limit for proper performance. One will probably include safety factors of 1.5 to 2 or even more if not too familiar when designing with plastics. That means the designer initially does not utilize the full strength of the material and/or significantly increases product cost.

Terminology

In this book different terms are used. To provide an introduction and ensure a better understanding of these terms, descriptions are presented here with details to follow.

Strength It represents the stress required to break, rupture, or cause a failure of a substance or the property of a material that resists deformation induced by external forces where maximum stress can resist without failure for a given type of loading.

Stress The intensity, at a point in a product, material, etc. of the internal forces (or components of forces) that act on a given plane through the point causing deformation of the body. It is the internal force per unit area that resists a change in size or shape of a body. Stress is expressed in force per unit area and reported in MPa, psi, etc. As used in tension, compression, or shear, stress is normally calculated on the basis of the original dimensions of the appropriate cross section of the test specimen. This stress is sometimes called engineering stress; it is different from true stress.

Stress, alternating amplitude A stress varying between two maximum values that are equal but with opposite signs, according to a law determined in terms of the time.

Stress amplitude Ratio of the maximum applied force, measured from the mean force to the cross sectional area of the unstressed test specimen.

Stress amplitude, alternating A test parameter of a dynamic fatigue test. One half the algebraic difference between the maximum and minimum stress in one cycle.

Stress, component It is the stress that is perpendicular to the plane on which the forces act.

Stress crack Appearance of external and/or internal cracks in the material as a result of stress that is lower than its short-term mechanical strength frequently accelerated by the environment to which the plastic is exposed.

Stress-cracking failure Failure of a material by cracking or crazing some time after it has been placed under load. Time-to-failure can range from minutes to many years.

Stress, elastic limit The greatest stress that a material is capable of sustaining without any permanent strain remaining upon complete release of stress. A material passes its elastic limit when the load is sufficient to initiate non-recoverable deformation.

Stress, engineering The stress calculated on the basis of the original cross sectional dimensions of a test specimen. See *Stress, true.*

Stress, frozen-in Undesirable frozen-in or residual stresses developed during processing.

Stress, initial Also called instantaneous stress. The stress produced by strain in a specimen before stress relaxation occurs.

Stress, nominal The stress at a point calculated on the net cross section without taking into consideration the effect on stress of geometric discontinuities, such as holes, grooves, slots, fillets, etc.

Stress, normal, principal The principal normal stress is the maximum or minimum value of the normal stress at a point in a plane considered with respect to all possible orientations of the considered plane. On such principal planes the shear stresses are zero. There are three principal stresses on three (x-, y-, and z-axes) mutually perpendicular planes. The states of stress at a point may be, (1) uniaxial – a state of stress in which two of the three principal stresses are zero; (2) biaxial – a state of stress in which only one of the three principal stresses is zero, and (3) triaxial– a state of stress in which none of the principal stresses is zero. There is also a multiaxial condition that refers to either biaxial or triaxial.

Stress, offset yield Also called engineering yield strength. The stress at which the strain exceeds by a specified amount (the offset, such as 0.1% of strain) or extension of the initial proportional part of the stress-strain curve. It is the force per unit area (MPa, psi, etc.). This measurement is useful for materials whose S-S curve (stress-strain) in the yield range is of gradual curvature.

Stress relaxation Also called stress relieving or stress decay. It is the decrease in stress after a given time at constant strain that can cause warpage, dimensional changes, or complete damage to the part. The result of changes in internal and/or external conditions relates to the time-dependent decrease in stress in a solid material/product.

Stress relieving Also called annealing. Heating a plastic to a suitable temperature, holding it long enough to reduce residual stresses, and then cooling slow enough to minimize the development of new stresses.

Stress, residual It is the stress existing in a body at rest, in equilibrium, at uniform temperature, and not subjected to external forces. Often caused by the stresses remaining in a plastic part as a result of thermal and/or mechanical treatment during product fabrication. Usually they are not a problem in the finished product. However, with excess stresses, the product could be damaged quickly or afterwards in service from a short to a long time depending on amount of stress and the environmental (such as temperature) conditions around the product.

Stress rupture strength It is the unit stress at which a material breaks or ruptures.

Stress softening The smaller stress required straining a material to a certain strain, after a prior cycle of stressing to the same strain followed by removal of the stress. Primarily observed in filled elastomers (when it is known as the Mullen effect), where it results from the detachment of some plastic molecules from filler particles in the first cycle and which therefore cannot support the stress on subsequent straining to the same strain.

Stress, specific The load divided by the mass per unit length of the test specimen.

Stress, static Stress in which the force is constant or slowly increasing with time such as a test without shock.

Stress-strain Stiffness at a given strain.

Stress-strain curve Also called a S-S curve or stress-strain diagram. Simultaneous readings of load and deformation converted to stress and strain, plotted as ordinates and abscissas, respectively. The S-S relationship applies under test conditions such as tension, compression, or torsion. The area under this curve provides valuable information regarding the characteristics of a plastic such as (usually) toughness where the larger the area the tougher the plastic. An exception is with reinforced plastics (RPs).

Stress, true Stress along the axis calculated on the actual cross section at the time of the observed failure instead of the original cross sectional area. Applicable to tension and compression testing.

Stress whitening Also called crazing. It is the appearance of white regions in a plastic when it is stressed. Stress whitening or crazing is damage that can occur when a TP is stretched near its yield point. The surface takes on a whitish appearance in regions that are under high stress. For practical purposes, stress whiting is the result of the formation of microcracks or crazes that is a form of damage. Crazes are not basically true fractures because they contain supporting string-like supports of highly oriented plastic that connect the two flat surfaces of the crack. These fibrils are surrounded by air voids. Because they are filled with highly oriented fibrils, crazes are capable of carrying stress, unlike true fractures. As a result, a heavily crazed part can carry significant stress even though the part may appear fractured. It is important to note that crazes, microcracking, and stress whitening represent irreversible first damage to a material that could ultimately cause failure. This damage usually lowers the impact strength and other properties. In the total design evaluation, the formation of stress cracking or crazing damage should be a criterion for failure based on the stress applied.

Stress yield point It is the lowest stress at which strain increases without increase in stress. Only a few materials exhibit a true yield point. For other materials the term is sometimes used as synonymous with yield strength.

Stress yield strength It is the unit stress at which a material exhibits a specified permanent deformation. It is a measure of the useful limit of materials, particularly of those whose stress-strain curve in the region of yield is smooth and gradually curved.

Stress, proportional limit It is the greatest stress that a material is capable of sustaining without deviation from the proportionality of the stress-strain straight line.

Strain It is the per unit change, due to force, in the size or shape of a body referred to its original size and shape. Strain is non-dimensional but is usually expressed in unit of length per unit of length or percent. It is the natural logarithm of the ratio of gauge length at the moment of observation instead of the original cross-sectional area. Applicable to tension and compression tests.

Strain amplitude Ratio of the maximum deformation measured from the mean deformation to the free length of the unstrained test specimen. Strain amplitude is measured from zero to peak on one side only.

Strain and elasticity A plastic where its elasticity permits recovery of its shape and size after being subjected to deformation exhibits a Hookean or ideal elasticity.

Strain, critical The strain at the yield point.

Strain extensometer A device for determining elongation of a test specimen as it is strained when conducting tests.

Strain, flexure of fiber The maximum strain in the outer fiber occurring mid-span in a reinforced plastic (RP) test specimen.

Strain hardening An increase in hardness and strength caused by plastic deformation shear strain at temperatures lower than the crystallization range of the plastic (Chapter 1).

Strain, initial The strain produced in a plastic by given loading condition before creep occurs.

Strain, linear During testing (tensile, compression, etc.) change per unit length or percent deformation due to an applied force in an original linear dimension to the applied load.

Strain, logarithm decrement The natural logarithm of the ratio of the amplitude of strain in one cycle to that in the succeeding cycle during a free vibration.

Strain, nominal The strain at a point calculated in the net cross section by simple elastic theory without taking into account the effect on strain produced by geometric discontinuities such as holes, grooves, filters, etc.

Strain rate It is the rate of change with time.

Strain, residual The strain associated with residual stress.

Strain, true It is defined as a function of the original diameter to the instantaneous diameter of the test specimen.

Modulus The constant denoting the relationship (ratio) between a physical effect and the force producing it.

Modulus and stiffness It is a measure of modulus with its relationship of load to deformation or the ratio between the applied stress and resulting strain. It is identified as stiffness (EI) = modulus (E) times moment of inertia (I). The term stiffness is often used when the relationship of stress-strain does not follow the modulus of elasticity's straight-line ratio.

Modulus, apparent The concept of apparent modulus is a convenient method of expressing creep because it takes into account initial strain for an applied stress plus the amount of deformation or strain that occurs with time.

Modulus, dissipation factor Also called loss tangent. It is the ratio of the loss modulus to static modulus of a material under dynamic load. It is proportional to damping capacity.

Modulus, dynamic Ratio of stress to strain under cyclic conditions that is calculated from either free or forced vibration tests in shear, compression, or tension.

Modulus, flexural The ratio, within the elastic limit, of the applied stress in flexure to the corresponding strain in the outermost area of the specimen.

Modulus, initial The slope of the initial straight portion of a stress-strain or load-elongation curve.

Modulus, loss A damping term describing the dissipation of energy into heat when a material is deformed. It is a quantitative measure of energy dissipation defined as the ratio of stress at 90° out of phase with oscillating strain to the magnitude of strain. It can be measured in tension, compression, flexure, or shear.

Modulus of elasticity Also called modulus, Young's modulus, coefficient of elasticity, or E. It is the ratio of normal stress to corresponding strain (straight line) for stresses below the proportional limit of the material (Hooke's law); the ratio of stress to strain in a test specimen (tensile, compression, etc.) that is elastically deformed.

Modulus of resilience The energy that can be absorbed per unit volume without creating a permanent distortion. Calculated by integrating the stress-strain diagram from zero to the elastic limit and dividing by the original volume of the test specimen.

Modulus of rigidity Also called shear modulus or torsional modulus. It is the ratio of stress to strain within the elastic region for shear or torsional stress.

Modulus of rupture Also called torsional strength. Modulus of rupture (MOR) is the strength of a material as determined by flexural or torsional test.

Modulus, secant The slope of a line drawn from the original to a point on the stress-strain curve for a material that corresponds to a particular strain. Used in designing parts subjected to short term, infrequent, intermittent stress of plastics in which the stress-strain is nonlinear.

Modulus, static Ratio of stress-to-strain under static conditions. It is calculated from static S-S tests in shear, tension, or compression. Expressed in force per unit area.

Modulus, storage A quantitative measure of elastic properties defined as the ratio of the stress, in-phase with strain, to the magnitude of the strain. It can be measured in tension, flexure, compression, or shear.

Modulus, stress relaxation Ratio of the time-dependent stress to a fixed strain during stress relaxation of a viscoelastic (plastic) material.

Modulus, tangent The slope of the line at a predefined point on a static-strain curve, expressed in force per unit area per unit strain. This tangent modulus is at the point of shear, tension, or compression.

Modulus, tension The ratio of tensile stress to the strain in the material over the range for which this value is constant.

Mohr circle A graphical representation of the stresses acting on the various planes at a given point.

Mohr circle of stress A graphical representation of the components of stress, and strain, acting across various planes at a given point, drawn with reference to axes of normal stress (strain) and shear stress (strain).

Shear An action of stress resulting from applied forces that causes or tends to cause two contiguous parts of a body to slide relative to each other in a direction parallel to their plane of contact. It is the stress developed because of the action of layers in a material attempting to glide against or separate in a parallel direction.

Shear failure Also called failure by rupture. It is the movement cause by shearing stresses that is sufficient to destroy or seriously endanger a structure.

Shear heating Heat produced within the plastic melt as the plastic layers slide along each other or along the surfaces in a plasticating chamber of the processing machine (Chapter 1).

Shear modulus The ratio of shearing stress to shearing strain within the proportional limit of the material.

Shear strain Also called angular strain. The tangent of the angular change, caused by a force between two lines originally perpendicular to each other through a point in a body. With this strain, there is a change in shape.

Shear strength It is the ability of a material to withstand shear stress or the stress at which a plastic fails in shear. It is calculated from the maximum load during a shear or torsional test and is based on the original cross sectional area of the test specimen.

Shear stress It is the component of stress tangent to the plane on which the forces act. Shear stress is equal to the force divided by the area sheared, yielding MPa (or psi). During processing it is the stress developed in a plastic melt when the layers in a cross section are sliding along each other or along the wall of the channel or cavity in laminar flow.

Shear stress-strain The shear mode involves the application of load to a specimen in such a way that cubic volume elements of the material comprising the specimen become distorted, their volume remaining constant, but with the opposite faces sliding sideways with respect to each other. Shear deformation occurs in structural elements subjected to torsional loads and in short beams subjected to transverse loads. Shear stress-strain data can be generated by twisting a material specimen at a specified rate while measuring the angle of twist between the ends of the specimen and the torque exerted by the specimen on the testing machine. Maximum shear stress at the surface of the specimen can be computed from the measured torque and the maximum shear strain from the measured angle of twist.

Toughness Property of a material indicating its ability to absorb energy by plastic deformation rather than crack or fracture. Toughness tends to relate to the area under the stress-strain curve for thermoplastic (TP) materials. The ability of a TP to absorb energy is a function of strength and ductility, which tends to be inversely related. For high toughness, a plastic needs both the ability to withstand load and the ability to elongate substantially without failing. An exception is in the case of reinforced thermoset plastics that have high strength and low elongation.

Toughness deformation See *Deformation and toughness*

Deformation It is any part of the total deformation of a body that occurs immediately when the load is applied but that remains permanently when the load is removed.

Deformation and toughness Deformation or elasticity is an important attribute in most plastics providing toughness. However, there are some plastics with no deformation that are extremely tough. This is particular true for reinforced plastics (RPs). For designs requiring such capabilities as toughness or elasticity, this characteristic has its advantages, but for other designs it is a disadvantage. However, there are materials that are normally tough but may become embrittled due to processing conditions, chemical attack, prolonged exposure to constant stress, and so on. A high modulus and high strength, with ductility, is a desired combination of performances. However, the inherent nature of most plastics is such that their having a high modulus tends to associate them with low ductility, and the steps taken to improve one will cause the other to deteriorate – except RPs.

Deformation, anelastic It is any portion of the total deformation of a body that occurs as a function of time when load is applied, and which disappears completely after a period of time when the load is removed. In practice the term also describes viscous deformation.

Deformation, elastic A deformation in which a material returns to its original dimensions on release of the deforming load/stress.

Deformation, immediate set It is determined by the measurement immediately after the removal of the load causing the deformation.

Deformation, inelastic The portion of deformation under stress that is not annulled by removal of the stress.

Deformation, permanent set The deformation remaining after a specimen has been stressed a prescribed amount in tension, compression, or shear for a specified time period and released for a specified time period.

Deformation, plastic The change in dimensions of a part under load that is not recovered when the load is removed.

Torque It is a force, energy, or moment, that produces or tends to produce rotation or torsion. The twisting force consists of the load applied multiplied by the distance perpendicular to the center of rotational vertical axis. Torque is very useful to designers where some type of torque action occurs in a product design such as center roller supports, etc.

Torsion Twisting a material causing it to be stressed.

Torsional deformation The angular twist of a specimen produced by a specific torque in the torsion test. This deformation as calculated (radian/in.) by dividing observed total angular twist, the twist of one end of the gauge length with respect to the other, by the original gauge length.

Torsional modulus of elasticity Also called modulus of rigidity. It is approximately equal to the shear modulus.

Torsional strength Also called modulus of rupture in torsion and sometimes in shear strength. It is a measure of the ability of a material to withstand a twisting load.

Brittleness It is that property of a material that permits it to be only slightly deformed without rupture. Brittleness is relative, no material being perfectly brittle, that is, capable of no deformation before rupture. Many materials are brittle to a greater or less degree, glass being one of the most brittle of materials. Brittle materials have relatively short stress-strain curves.

Damping capacity/hysteresis Observations show that when a tensile load is applied to a specimen, it does not produce the complete elongation immediately; there is a definite time lapse that depends on the nature of the material and the magnitude of the stresses involved. Upon unloading, complete recovery of energy does not occur. This phenomenon is called elastic hysteresis or, for vibratory stresses, damping. The area of this hysteresis loop, representing the energy dissipated per cycle, is a measure of the damping properties of the material. Under vibratory conditions the energy dissipated varies approximately as the cube of the stress.

Ductility It is the ability of a material to sustain large permanent deformations in tension.

Dynamic stress It occurs where the dimensions of time is necessary in defining the loads. They include creep, fatigue, and impact stresses.

Creep stress It occurs when either the load or deformation progressively varies with time. They are usually associated with noncyclic phenomena.

Fatigue stress It occurs when type cycle variation of either load or strain is coincident with respect to time.

Impact stress It occurs from loads that are transient with time. The duration of the load application is of the same order of magnitude as the natural period of vibration of the test specimen.

Hardness It is the ability to resist very small indentations, abrasion, and plastic deformation. There is no single measure of hardness, as it is not a single property but a combination of several properties.

Moment of inertia Also called rotational inertia or EI. EI is the sum of the products formed by multiplying the mass (or sometimes the area) of each element of a figure by the square of its distance from a specified line. See *Modulus and stiffness.*

Residual deflection Permanent deformation after complete or partial removal of applied force on a product, component, structure, etc.

Resonance Mechanically it is the reinforced vibration of a body exposed to the vibration at about the frequency of another.

Rupture A break or cleavage resulting from physical stress. Work of rupture is the integral of the stress-strain curve between the origin and the point of rupture.

Rupture strength The true value of rupture strength is the stress of a material at failure based on the original ruptures cross-sectional area.

Stiffness It is the ability to resist deformation under stress. As reviewed the modulus of elasticity is the criterion of the stiffness of the material.

Hysteresis See *Damping capacity/hysteresis*

Viscoelasticity It is the plastics respond to stress with elastic strain. In the material, strain increases with longer loading times and higher temperatures. A material having this property is considered to combine the features of a perfectly elastic solid and a perfect fluid; representing the combination of elastic and viscous behavior of plastics.

Viscoelastic creep When a plastic material is subjected to a constant stress, it undergoes a time-dependent increase in strain. This behavior is called creep. It is a plastic for which at long times of applied stress, such as in creep, a steady flow is eventually achieved. Thus in a generalized Maxwell model, all the dashpot viscosities must have finite values and in generalized models must have zero stiffness.

Maxwell model Also called Maxwell fluid model. A mechanical model for simple linear viscoelastic behavior that consists of a spring of Young's modulus (E) in series with a dashpot of coefficient of viscosity (η). It is an isostress model (with stress δ), the strain (ϵ) being the sum of the individual strains in the spring and dashpot. This leads to a differential representation of linear viscoelasticity as

$d\varepsilon/dt = (1/E) \, d\delta/dt + (\delta/\eta)$. This model is useful for the representation of stress relaxation and creep with Newtonian flow analysis.

Boltzmann superposition principle This principle provides a basis for the description of all linear viscoelastic phenomena. Unfortunately, no such theory is available to serve as a basis for the interpretation of nonlinear phenomena, i.e., to describe flows in which neither the strain or the strain rate is small. As a result, there is no general valid formula for calculating values for one material function on the basis of experimental data from another. However, limited theories have been developed.

Test Testing yields basic information about plastic materials and products, its properties relative to another material, its quality with reference to standards, and applied to designing with plastics. They are usually destructive tests but there are also nondestructive tests (NTD). Most of all, it is essential for determining the performance of plastic materials to be processed and of the finished products. Testing refers to the determination by technical means properties and performances. This action, when possible, should involve application of established scientific principles and procedures. It requires specifying what requirements are to be met. There are many different tests that can be conducted that relate to practically any requirement. Many different tests are provided and explained in different specifications and standards.

Test destructive/nondestructive Any test performed on a part in an attempt to destroy it; often performed to determine how much abuse the part can tolerate without failing. There are also non-destructive tests where a part does not change or is not destroyed.

x-axis The axis in the plane of a material used as 0^0 reference; thus the y-axes is the axes in the plane of the material perpendicular to the x-axis; thus the z-axes is the reference axis normal to the x-y plane. The term plane or direction is also used in place of axis.

y-axis A line perpendicular to two opposite parallel faces.

z-axis The reference axis perpendicular to $\times$ and y axes.

Engineering Optimization

The physical and mechanical properties of plastics, including reinforced plastics (RPs) have some significant difference from those of familiar metallic materials. Consequently in the past those not familiar with designing plastic products may have had less confidence in plastics and in their own ability to design with them. Thus, plastic material selection and optimization was confined to the familiar steel materials approach resulting in overdesign, or failures that may have occurred in service.

Engineering-wise for plastics there are the basic approaches of theory of elasticity and strength of materials with the value of one approach over the other depending on the particular application. In most engineering problems, both methods assume homogeneous, isotropic solid, linearly elastic material. A very important requirement for both approaches is that equilibrium of loads/forces be satisfied. These conditions being met to a reasonable degree, one would expect the elasticity solution to be superior to the strength of materials. Theories exist that provide a unifying principle that explains a body of facts and the laws that is based on those facts.

Strength of material refers to the structural engineering analysis of a part to determine its strength properties. There is also the important empirical approach that is based on experience and observations rather than theory. The basic optimization design theory can be related to the systematic activity necessary, from the identification of the market/user need, to the selling of the successful product to satisfy that need. It is an activity that encompasses product, process, people and organization.

Design Foundation

The target of the integration of technological or non-technological subject material in an effective and efficient manner is greatly enhanced by having a visible operational structure ranging from field service studies to analysis by computers. Some type of visibility is a crucial factor in bringing about integration. Visibility helps everyone find out what people are doing and why. With this approach, design may be construed as having a central foundation of activities, all of which are imperative for any design.

This foundation includes product conceptual design, design specification, detail design, manufacture, and sales. All design starts, or should start, with a need that, when satisfied, will fit into an existing market or create a market of its own. From the statement of the need a specification, or equivalent, must be formulated of the product to be designed. Once this is established, it acts as the envelope that includes all the subsequent stages in the design. It becomes the theoretical control for the total design activity, because it places the boundaries on the subsequent design approaches. Fig. 2.1 provides an example of an

Figure 2.1 Flow chart for designing a product from concept to fabricating the product

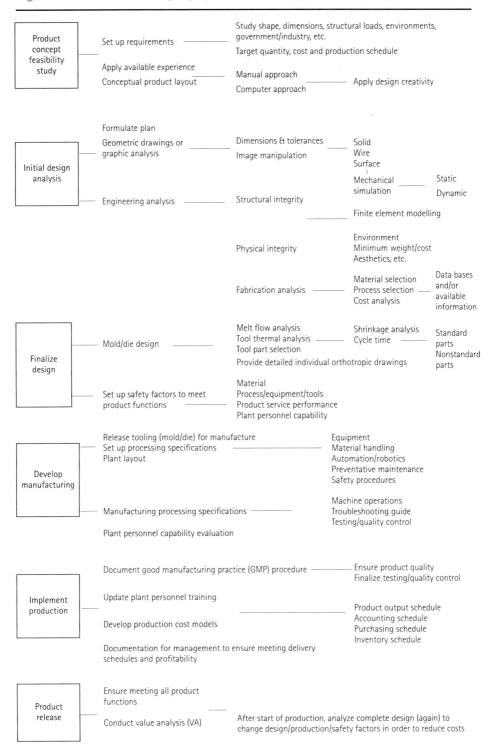

overall flow chart that goes from the product concept to product release.

Use is made of the optimization theory and its application to problems arising in engineering that follows by determining the material and fabricating process to be used. The theory is a body of mathematical results and numerical methods for finding and identifying the best candidate from a collection of alternatives without having to specify and evaluate all possible alternatives. The process of optimization lies at the root of engineering, since the classical function of the engineer is to design new, better, more efficient, and less expensive products, as well as to devise plans and procedures for the improved operation of existing products.

To optimize this approach the boundaries of the engineering system are necessary in order to apply the mathematical results and numerical techniques of the optimization theory to engineering problems. For purposes of analysis they serve to isolate the system from its surroundings, because all interactions between the system and its surroundings are assumed to be fixed/frozen at selected, representative levels. However, since interactions and complications always exist, the act of defining the system boundaries is required in the process of approximating the real system. It also requires defining the quantitative criterion on the basis of which candidates will be ranked to determine the best approach. Included will be the selection system variables that will be used to characterize or identify candidates, and to define a model that will express the manner in which the variables are related.

Use is made of the optimization methods to determine the best condition without actually testing all possible conditions, comes through the use of a modest level of mathematics and at the cost of performing repetitive numerical calculations using clearly defined logical procedures or algorithms implemented on computers. This composite activity constitutes the process of formulating the engineering optimization problem. Good problem formulation is the key to the success of an optimization study and is to a large degree an art. This knowledge is gained through practice and the study of successful applications. It is based on the knowledge and experience of the strengths, weaknesses, and peculiarities of the techniques provided by optimization theory.

Unfortunately at times this approach may result in that the initial choice of performance boundary/requirements is too restrictive. In order to analyze a given engineering system fully it may be necessary to expand the performance boundaries to include other sub-performance systems that strongly affect the operation of the model under study. As an example, a manufacturer finishes products that are mounted on an assembly line and decorates. In an initial study of the secondary decorating operation one may consider it separate from the rest of the assembly line. However, one may find that the optimal batch size and method of attachment sequence are strongly influenced by the operation of the plastic fabrication department that produces the fabricated products (as an example problems of frozen stresses, contaminated surface, and other detriments in the product could interfere with applying the decoration).

Required is selecting an approach to determine a criterion on the basis of which the performance requirements or design of the system can be evaluated resulting in the most appropriate design or set of operating conditions being identified. In many engineering applications this criterion concerns economics. In turn one has to define economics such as total capital cost, annual cost, annual net profit, return on investment, cost to benefit ratio, or net present worth. There are criterions that involve some technology factors such as plastic material to be used, fabricating process to be used, minimum production time, number of products, maximum production rate, minimum energy utilization, minimum weight, and safety.

Problem/Solution Concept

In the art of the design concept there is the generation of solutions to meet the product requirements. It represents the sum of all of the subsystems and of the component parts that go to make up the whole system. During this phase, one is concerned with ideas and the generation of solutions.

In practice, even with the simplest product design, one will probably have ideas as to how you might ultimately approach the problem(s). Record these ideas as they occur; however avoid the temptation to start engineering and developing the ideas further. This tendency is as common with designers as it is with other professionals in their respective areas of interest. So, record the ideas but resist the temptation to proceed.

Target as many ideas as you can possibly generate where single solutions are usually a disaster. While it is recognized that you may have limited experience and knowledge, both of technological and non-technological things, you must work within limits since design is not an excuse for trying to do impossible things outside the limits. Notwithstanding these facts, you need to use what you know and what you can discover. You will need to engineer your concepts to a level where each is complete and recognizable, and technically in balance within the limits and is feasible in meeting product requirements.

Design Approach

The acquisition of analytical techniques and practical skills in the engineering sciences is important to the design system. Through a study of engineering of any label based on mathematics and physics applied through elemental studies, one acquires an all-round engineering competence. This enables, for example, one to calculate fatigue life, creep behavior, inertia forces, torsion and shaft stresses, vibration characteristics, etc.

The list of calculations is limitless if one considers all the engineering disciplines and is therefore generally acceptable as the basis for any engineering review. However, the application of such skills and knowledge to engineering elements is partial design. To include the highly optimized, best material and/or shape in any design when it is not essential to the design may involve engineering analysis of the highest order that is expensive and usually not required.

Limitations, shortcomings, or deficiencies have to be recognized otherwise potentially misdirected engineering analysis give rise to a poor design. What has been helpful in many design teams is to include non-engineers or non-technologists (Fig. 2.1). However, this needs a disciplined, structured approach, so that everyone has a common view of total design and therefore subscribes to a common objective with a minimum of misconceptions. Participants should be able to see how their differing partial design contributions fit into the whole project.

Model Less Costly

When possible the ideal approach is to design products that rely on the formulation and analysis of mathematical models of static and/or dynamic physical systems. This is of interest because a model is more accessible to study than the physical system the model represents. Models typically are less costly and less time-consuming to construct and test. Changes in the structure of a model are easier to implement, and changes in the behavior of a model are easier to isolate and understand in a computer system (Chapter 5).

A model often provides an insight when the corresponding physical system cannot, because experimentation with the actual system could be too dangerous, costly, or too demanding. A model can be used to answer questions about a product that has not yet been finalized or realized. Potential problems can provide an immediate solution.

A mathematical model is a description of a system in terms of the available equations that are available from the engineering books. The desired model used will depend upon: (1) the nature of the system the product represents, (2) the objectives of the designer in developing the model, and (3) the tools available for developing and analyzing the model.

Because the physical systems of primary interest are static and/or dynamic in nature, the mathematical models used to represent these systems most often include difference or differential equations. Such equations, based on physical laws and observations, are statements of the fundamental relationships among the important variables that describe the system. Difference and differential equation models are expressions of the way in which the current values assumed by the variables combine to determine the future values of these variables. As reviewed later it is important to relate static and/or dynamic loads on plastic products to operating temperatures.

Model Type
A variety of models are available that can meet the requirements for any given product. The choice of a particular model always represents a compromise between the accuracy in details of the model, the effort required in model formulation and analysis, and usually the time frame that has to be met in fabricating the product. This compromise is reflected in the nature and extent of simplifying assumptions used to develop the model.

Generally the more faithful or complete the model is as a description of the physical system modeled, the more difficult it is to obtain useful general solutions. Recognize that the best engineering model is not necessarily the most accurate or precise. It is, instead, the simplest model that yields the information needed to support a decision and meet performance requirements for the product. This approach of simplicity also involves the product's shape to the fabricating method used. Most designed products do not complicate fabricating them, however there are those that can complicate the fabrication resulting in extra cost not initially included and the possibility of defective parts.

Recognize that simpler models frequently can be justified, particularly during the initial stages of a product study. In particular, systems that can be described by linear difference or differential equations permit the use of powerful analysis and design techniques. These include the transform methods of classical theory and the state-variable methods of modem theory.

Target is to have more than one model in the evaluation. Simple models that can be solved analytically are used to gain insight into the behavior of the system and to suggest candidate designs. These designs are then verified and refined in more complex models, using computer simulation. If physical components are developed during the course of a study, it is often practical to incorporate these components directly into the simulation, replacing the corresponding model components.

Computer Software
Mathematical models are particularly useful because of the large body of mathematical and computational theory that exists for the study and solution of equations. Based on this theory, a wide range of techniques has been developed. In recent years, computer programs have been written that implement virtually all of these techniques. Computer software packages are now widely available for both simulation and computational assistance in the analysis and design of control systems (Chapter 5).

Design Analysis Approach

Plastics have some design approaches that differ significantly from those of the familiar metals. As an example, the wide choice available in plastics makes it necessary to select not only between TPs, TSs, reinforced plastics (RPs), and elastomers, but also between individual materials within each family of plastic types (Chapter 1). This selection requires having data suitable for making comparisons which, apart from the availability of data, depends on defining and recognizing the relevant plastics behavior characteristics. There can be, for instance, isotropic (homogeneous) plastics and plastics that can have different directional properties that run from the isotropic to anisotropic. As an example, certain engineering plastics and RPs that are injection molded can be used advantageously to provide extra stiffness and strength in predesigned directions.

It can generally be claimed that fiber based RPs offer good potential for achieving high structural efficiency coupled with a weight saving in products, fuel efficiency in manufacturing, and cost effectiveness during service life. Conversely, special problems can arise from the use of RPs, due to the extreme anisotropy of some of them, the fact that the strength of certain constituent fibers is intrinsically variable, and because the test methods for measuring RPs' performance need special consideration if they are to provide meaningful values.

Some of the advantages, in terms of high strength-to-weight ratios and high stiffness-to-weight ratios, can be seen in Figs. 2.2 and 2.3, which show that some RPs can outperform steel and aluminum in their ordinary forms. If bonding to the matrix is good, then fibers augment mechanical strength by accepting strain transferred from the matrix,

Figure 2.2 Tensile stress-strain curves for different materials

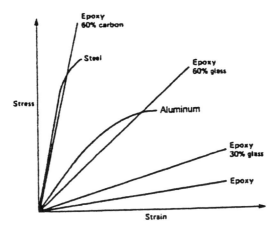

Figure 2.3 Past to future tensile properties of RPs, steel, and aluminium

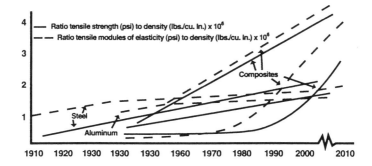

which otherwise would break. This occurs until catastrophic debonding occurs. Particularly effective here are combinations of fibers with plastic matrices, which often complement one another's properties, yielding products with acceptable toughness, reduced thermal expansion, low ductility, and a high modulus.

Viscoelasticity

Viscoelasticity is a very important behavior to understand for the designer. It is the relationship of stress with elastic strain in a plastic. The response to stress of all plastic structures is viscoelastic, meaning that it takes time for the strain to accommodate the applied stress field. Viscoelasticity can be viewed as a mechanical behavior in which the relationships between stress and strain are time dependent that may be extremely short or long, as opposed to the classical elastic behavior in which deformation and recovery both occur instantaneously on application and removal of stress, respectively.

The time constants for this response will vary with the specific characteristics of a type plastic and processing technique. In the rigid section of a plastic the response time is usually on the order of microseconds to milliseconds. With resilient, rubber sections of the structure the response time can be long such as from tenths of a second to seconds. This difference in response time is the cause of failure under rapid loading for certain plastics.

By stressing a viscoelastic plastic material there are three deformation behaviors to be observed. They are an initial elastic response, followed by a time-dependent delayed elasticity that may also be fully recoverable, and the last observation is a viscous, non-recoverable, flow component. Most plastic containing systems (solid plastics, melts, gels, dilute, and concentrated solutions) exhibit viscoelastic behavior due to the long-chain nature of the constituent basic polymer molecules (Chapter 1).

This viscoelastic behavior influences different properties such as brittleness. To understand why the possibility for brittle failure does exist for certain plastics when the response under high-speed stressing is transferred from resilient regions of a plastic, an analysis of the response of the two types of components in the structure is necessary. The elastomeric regions, which stay soft and rubbery at room temperature, will have a very low elastomeric modulus and a very large extension to failure. The rigid, virtually crosslinked regions, which harden together into a crystalline region on cooling, will be brittle and have very high moduli and very low extension to failure, usually from 1 to 10%.

If the stress rate is a small fraction of the normal response time for the rubbery regions, they will not be able to strain quickly enough to accommodate the applied stress. As a consequence for the brittle type plastics, virtually crosslinked regions take a large amount of the stress, and since they have limited elongation, they fail. The apparent effect is that of a high stretch, rubbery material undergoing brittle failure at an elongation that is a small fraction of the possible values.

A fluid, which although exhibits predominantly viscous flow behavior, also exhibits some elastic recovery of the deformation on release of the stress. To emphasize that viscous effects predominate, the term elastico-viscous is sometimes preferred; the term viscoelastic is reserved for solids showing both elastic and viscous behavior. Most plastic systems, both melts and solutions, are viscoelastic due to the molecules becoming oriented due to the shear action of the fluid, but regaining their equilibrium randomly coiled configuration on release of the stress. Elastic effects are developed during processing such as in die swell, melt fracture, and frozen-in orientation.

Polymer Structure

The viscoelastic deviations from ideal elasticity or purely viscous flow depend on both the experimental conditions (particularly temperature with its five temperature regions and magnitudes and rates of application of stress or strain). They also depend on the basic polymer structure particularly molecular weight (MW), molecular weight distribution (MWD), crystallinity, crosslinking, and branching (Chapter 1).

High MW glassy polymer [an amorphous polymer well below its glass transition temperature (T_g) value (Chapter 1)] with its very few chain motions are possible so the material tends to behave elastically, with a very low value for the creep compliance of about 10^{-9} Pa^{-1}. When well above the T_g value (for an elastomer polymer) the creep compliance is about 10^{-4} Pa^{-1}, since considerable segmental rotation can occur.

The intermediate temperature region that corresponds to the region of the T_g value, is referred to as the viscoelastic region, the leathery region, or the transition zone. Well above the T_g value is the region of rubbery flow followed by the region of viscous flow. In this last region flow occurs owing to the possibility of slippage of whole polymer molecular chains occurring by means of coordinated segmental jumps.

These five temperature regions give rise to the five regions of viscoelastic behavior. Light crosslinking of a polymer will have little effect on the glassy and transition zones, but will considerably modify the flow regions.

Viscoelasticity Behavior

There is linear and nonlinear viscoelasticity. The simplest type of viscoelastic behavior is linear viscoelasticity. This type of rheology behavior occurs when the deformation is sufficiently mild that the molecules of a plastic are disturbed from their equilibrium configuration and entanglement state to a negligible extent. Since the deformations that occur during plastic processing are neither very small nor very slow, any theory of linear viscoelasticity to date is of very little use in processing modeling. Its principal utility is as a method for characterizing the molecules in their equilibrium state. An example is in the comparison of different plastics during quality control.

In the case of oscillatory shear experiments, for example, the strain amplitude must usually be low. For large and more rapid deformations, the linear theory has not been validated. The response to an imposed deformation depends on (1) the size of the deformation, (2) the rate of deformation, and (3) the kinematics of the deformation.

Nonlinear viscoelasticity is the behavior in which the relationship of stress, strain, and time are not linear so that the ratios of stress to strain are dependent on the value of stress. The Boltzmann superposition principle does not hold (Appendix B). Such behavior is very common in plastic systems, non-linearity being found especially at high strains or in crystalline plastics.

Relaxation/Creep Analysis
Theories have been developed regarding linear viscoelasticity as it applies to static stress relaxation. This theory is not valid in nonlinear regions. It is applicable when plastic is stressed below some limiting stress (about half the short-time yield stress for unreinforced plastics); small strains are at any time almost linearly proportional to the imposed stresses. When the assumption is made that a timewise linear relationship exists between stress and strain, using models it can be shown that the stress at any time t in a plastic held at a constant strain (relaxation test), is given by:

$$\sigma = \sigma_0 e^{-t/\gamma} \qquad (2\text{-}1)$$

where: σ = stress at any time t
σ_0 = initial stress
e = natural logarithmic base number
γ = relaxation time

The total deformation experienced during creep loading (with the sample under constant stress σ) is given by:

$$\varepsilon = (\sigma/E_0) + (\sigma/E)(1 - e^{-t/\eta}) + (\sigma t/\eta) \qquad (2\text{-}2)$$

where: ε = total deformation,
E_0 = initial modulus of the sample
E = modulus after time t,
η = viscosity of the plastic

Excluding the permanent set or deformation and considering only the creep involved, equation (2-2) becomes:

$$\varepsilon = (\sigma/E) + (\sigma/E)(1 - e^{-t/\eta}) \qquad (2\text{-}3)$$

The term γ in Eqs. (2-1) and (2-2) has a different significance than that in equation (2-3). In the first equations it is based on static relaxation and the other on creep. A major accomplishment of this viscoelastic theory is the correlation of these quantities analytically so that creep deformation can be predicted from relaxation data and relaxation data from creep deformation data as shown in the following equation:

$$(\sigma_0/\sigma) \text{ relaxation} = (\varepsilon/\varepsilon_0) \text{ creep} \tag{2-4}$$

Creep strains can be calculated using equation (2-4) in the form of:

$$\varepsilon = \varepsilon_0 \, (\sigma_0/\sigma) = (\sigma_0/\sigma) \, (\sigma_0/E_0) \, (\sigma_0/\sigma) \tag{2-5}$$

where $(1/E_0) \, (\sigma_0/\sigma)$ may be thought of as a time-modified modulus, i.e., equal to $1/E$, from which the modulus at any time t, is:

$$E = E_0 \, (\sigma_0/\sigma) \tag{2-6}$$

that is the value to replace E in the conventional elastic solutions to mechanical problems.

Where Poisson's ratio, γ, appears in the elastic solution, it is replaced in the viscoelastic solution by:

$$\gamma = (3B - E) \, 6B \tag{2-7}$$

where B is the bulk modulus, a value that remains almost constant throughout deformations.

Stress relaxation and creep behavior for plastics are closely related to each other so that one can be predicted from knowledge of the other. Therefore, such deformations in plastics can be predicted by the use of standard engineering elastic stress analysis formulas where the elastic constants E and γ can be replaced by their viscoelastic equivalents given in equations (2-2) and (2-3).

If data are not available on the effects of time, temperature, and strain rate on modulus, creep tests can be performed at various stress levels as a function of temperature over a prescribed period of time. As an example, for rocket and missile stress relaxations data obtained over a time period of 4 to 5 sec to an hour provide the essential information. For structural applications, such as pipelines, data over a period of years are required. Data from relatively short-term tests can be extrapolated by means of theory to long-term problems. However this approach can have its inherent limitations.

Another method used involves the use of the rate theory based on the Arrhenius equation. In the Arrhenius equation the ordinate is the log of the material life. The abscissa is the reciprocal of the absolute temperature. The linear curves obtained with the Arrhenius plot overcome the deficiency of most of the standard tests, which provide only one point and indicate no direction in which to extrapolate. Moreover, any change in any aspect of the material or the environment could alter the slopes of their curves. Therein lies the value of this method.

This method requires extensive test data but considerably more latitude is obtained and more materials obey the rate theory. The method can also be used to predict stress-rupture of plastics as well as the creep characteristics of a material.

The assumption is made that the physical and chemical properties of the material are the same before and after failure (so that the concentration of material undergoing deformation is related to the rate constant, K, by $x = Kt$, where t is time) then it can be shown, as in the following equation, that for plastics:

$$A/R \quad = K^1 = [TT_0/(T_0 - T)]\,(20+\log t) \tag{2-8}$$

where: A = activation energy for the process
  R = gas constant
  K^1 = constant
  T = absolute temperature of the process
  T_0 = absolute temperature at which the material has no strength
  t = time

Failure curves can be computed for all values of T related to the magnitude of the stress applied. For design purposes, if the required time and operating temperature are specified, K^1 can be computed and the value of stress required to cause rupture at that time and temperature can be obtained.

Stress divided by the modulus of the material results in the creep deformation. The deformation observed in a short-term tensile test at an elevated temperature is related to the deformation that takes place at a lower temperature over a longer period of time. The short-term data obtained can be used to obtain long-term modulus data through the development of a master modulus curve. Being able to determine the modulus at any time t and knowing the constant value of stress to which a material is subjected, it is possible to predict the creep which will have been experienced at time t by simply dividing the stress by the modulus using conventional elastic stress analysis relationships.

Summary

A combination of viscous and elastic properties in a plastic exists with the relative contribution of each being dependent on time, temperature, stress, and strain rate. It relates to the mechanical behavior of plastics in which there is a time and temperature dependent relationship between stress and strain. A material having this property is considered to combine the features of a perfectly elastic solid and a perfect fluid; representing the combination of elastic and viscous behavior of plastics.

In the plastic, strain increases with longer loading times and higher temperatures. It is a phenomenon of time-dependent, in addition to elastic, deformation (or recovery) in response to load. This property possessed by all plastics to some degree, dictates that while plastics have solid-like characteristics such as elasticity, strength, and form-stability, they also have liquid-like characteristics such as flow depending on time, temperature, rate, and amount of loading. These basic characteristics highlight: (a) simplified deformation vs. time behavior, (b) stress-strain deformation vs. time, and (c) stress-strain deformation vs. time (stress-relaxation).

A constitutive relationship between stress and strain describing viscoelastic behavior will have terms involving strain rate as well as stress and strain. If there is direct proportionality between the terms then the behavior is that of linear viscoelasticity described by a linear differential equation. Plastics may exhibit linearity but usually only at low strains. More commonly complex non-linear viscoelastic behavior is observed.

Thus viscoelasticity is characterized by dependencies on temperature and time, the complexities of which may be considerably simplified by the time-temperature superposition principle. Similarly the response to successively loadings can be simply represented using the applied Boltzmann superposition principle. Experimentally viscoelasticity is characterized by creep compliance quantified by creep compliance (for example), stress relaxation (quantified by stress relaxation modulus), and by dynamic mechanical response.

The general design criteria applicable to plastics are the same as those for metals at elevated temperature; that is, design is based on (1) a deformation limit, and (2) a stress limit (for stress-rupture failure). There are cases where weight is a limiting factor and other cases where short-term properties are important. In computing ordinary short-term characteristics of plastics, the standard stress analysis formulas may be used. For predicting creep and stress-rupture behavior, the method will vary according to circumstances. In viscoelastic materials, relaxation data can be used to predict creep deformations. In other cases the rate theory may be used.

Viscosity

In addition to its behavior in viscoelastic behavior in plastic products, viscosity of plastics during processing provides another important relationship to product performances (Chapter 1). Different terms are used to identify viscosity characteristics that include methods to determine viscosity such as absolute viscosity, inherent viscosity, relative viscosity, apparent viscosity, intrinsic viscosity, specific viscosity, stoke viscosity, and coefficient viscosity. Other terms are reduced viscosity, specific viscosity, melt index, rheometer, Bingham body, capillary viscometer, capillary rheometer, dilatancy, extrusion rheometer, flow properties, kinematic viscosity, laminar flow, thixotropic, viscometer, viscosity coefficient, viscosity number, viscosity ratio, viscous flow, and yield value.

The absolute viscosity is the ratio of shear stress to shear rate. It is the property of internal resistance of a fluid that opposes the relative motion of adjacent layers. Basically it is the tangential force on a unit area of either of two parallel planes at a unit distance apart, when the space between the planes is filled and one of the planes moves with unit velocity in its own plane relative to the other. The Bingham body is a substance that behaves somewhat like a Newtonian fluid in that there is a linear relation between rate of shear and shearing forces, but also has a yield value.

Inherent viscosity refers to a dilute solution viscosity measurement where it is the ratio of the natural logarithm of the relative viscosity (sometimes called viscosity ratio) to the concentration of the plastic in grams per 100 ml of a solvent solution.

Relative viscosity (RV) is the ratio of the absolute solution viscosity (of known concentration) and of the absolute viscosity of the pure solvent at the same temperature. IUPAC uses the term viscosity ratio.

Apparent viscosity is defined as the ratio between shear stress and shear rate over a narrow range for a plastic melt. It is a constant for Newtonian materials but a variable for plastics that are non-Newtonian materials (Chapter 1).

Intrinsic viscosity (IV) data is used in processing plastics. It is the limiting value at an infinite dilution of the ratio of the specific viscosity of the plastic solution to the plastic's concentration in moles per liter; it is a measure of the capability of a plastic in solution to enhance the viscosity of the solution. IV increases with increasing plastic molecular weight that in turn influences processability. An example is the higher IV of injection-grade PET (polyethylene terephthalate) plastic can be extruded blow molded; similar to PETG (PET glycol) plastic that can be easily blow molded but is more expensive than injection molded grade PET and PVC for blow molding.

Specific viscosity is the relative viscosity of a solution of known concentration of the plastic minus one. It is usually determined for a low concentration of plastics such as 0.5g/100 ml of solution or less.

Stoke viscosity is the unit of kinetic viscosity. It is obtained by dividing the melt's absolute viscosity by its density. A centipoise is 0.01 of a stoke.

Coefficient viscosity is the shearing stress necessary to induce a unit velocity gradient in a material. In actual measurement, the viscosity coefficient of a material is obtained from the ratio of shearing stress to shearing rate. This assumes the ratio to be constant and independent of the shearing stress, a condition satisfied only by Newtonian fluids. With non-Newtonian plastics, values obtained are apparent and represent one point in the flow chart.

Rheology and mechanical properties

Rheological knowledge combined with laboratory data can be used to predict stresses developed in plastics undergoing strains at different rates and at different temperature; rheology is the science of the deformation and flow of matter under force. The procedure of using laboratory experimental data for the prediction of mechanical behavior under a prescribed use condition involves two rheological principles. There is the Boltzmann's superposition principle that enables one to utilize basic experimental data such as a stress relaxation modulus in predicting stresses under any strain history. The second is the principle of reduced variables, which by a temperature-log time shift allows the time scale of such a prediction to be extended substantially beyond the limits of the time scale of the original experiment.

Regarded as one of the cornerstones of physical science, is the Boltzmann's Law and Principle that developed the kinetic theory of gases and rules governing their viscosity and diffusion. This important work in chemistry is very important in plastics (Ludwig Boltzmann born in Vienna, Austria, 1844–1906). It relates to the mechanical properties of plastics that are time-dependent.

The rheology of solid plastics within a range of small strains and within a range of linear viscoelasticity, has shown that mechanical behavior has often been successfully related to molecular structure. It shows the mechanical characterization of a plastic in order to predict its behavior in practical applications and how such behavior is affected by temperature. It also provides rheological experimentation as a means for obtaining a greater structural understanding of the material that has provided knowledge about the effect of molecular structure on the properties of plastics, particularly in the case of amorphous plastics in a rubbery state as well as extending knowledge concerning the complex behavior of crystalline plastics. Studies illustrate how experimental data can be applied to a practical example of the long-time mechanical stability.

As reviewed, a plastic when subjected to an external force part of the work done is elastically stored and the rest is irreversibly dissipated. Result is a viscoelastic material. The relative magnitudes of such elastic and viscous responses depend, among other things, on how fast the material is being deformed. It can be seen from tensile stress-strain (S-S) curves that the faster the material is deformed, the greater will be the stress developed since less of the work done can be dissipated in the shorter time.

Hooke's Law

When the magnitude of deformation is not too great viscoelastic behavior of plastics is often observed to be linear, that is the elastic part of the response is Hookean and the viscous part is Newtonian. Hookean response relates to the modulus of elasticity where the ratio of normal stress is proportional to its corresponding strain. This action occurs below the proportional limit of the material where it follows Hooke's Law (Robert Hooks 1678). Result is a Newtonian response where the stress-strain curve is a straight-line.

From such curves, however, it would not be possible to determine whether the viscoelasticity is in fact linear. An evaluation is needed where the time effect can be isolated. Typical of such evaluation is stress relaxation. In this test, the specimen is strained to a specified magnitude at the beginning of the test and held unchanged throughout the experiment, while the monotonically decaying stress is recorded against time. The condition of linear viscoelasticity is fulfilled here if the relaxation modulus is independent of the magnitude of the strain. It follows that a relaxation modulus is a function of time only.

There are several other comparable rheological experimental methods involving linear viscoelastic behavior. Among them are creep tests (constant stress), dynamic mechanical fatigue tests (forced periodic oscillation), and torsion pendulum tests (free oscillation). Viscoelastic data obtained from any of these techniques must be consistent data from the others.

If a body were subjected to a number of varying deformation cycles, a complex time dependent stress would result. If the viscoelastic behavior is linear, this complex stress-strain-time relation is reduced to a simple scheme by the superposition principle proposed by Boltzmann. This principle states in effect that the stress at any instant can be broken up into many parts, each of which has a corresponding part in the strain that the body is experienced. This is illustrated where the stress is shown to consist of two parts, each of which corresponds to the time axis as the temperature is changed.

It implies that all viscoelastic functions, such as the relaxation modulus, can be shifted along the logarithmic time axis in the same manner by a suitable temperature change. Thus, it is possible to reduce two independent variables (temperature and time) to a single variable (reduced time at a given temperature). Through the use of this principle of reduced variables, it is thus possible to expand enormously the time range of a viscoelastic function to many years.

The relaxation modulus (or any other viscoelastic function) thus obtained is a means of characterizing a material. In fact relaxation spectra have been found very useful in understanding molecular motions of plastics. Much of the relation between the molecular structure and the overall behavior of amorphous plastics is now known.

Mechanical properties of crystalline plastics are much more complex than those of amorphous plastics. Viscoelastic data, at least in theory, can be utilized to predict mechanical performance of a material under any use conditions. However it is seldom practical to carry out the necessarily large number of tests for the long time periods involved. Such limitations can be largely overcome by utilizing the principle of reduced variables embodying a time-temperature shift. Plastic usually exhibits not one but many relaxation times with each relaxation affected by the temperature.

Static stress

The mechanical properties of plastics enable them to perform in a wide variety of end uses and environments, often at lower cost than other design materials such as metal or wood. This section reviews the static property aspects that relate to short term loads.

As reviewed thermoplastics (TPs) being viscoelastic respond to induced stress by two mechanisms: viscous flow and elastic deformation. Viscous flow ultimately dissipates the applied mechanical energy as frictional heat and results in permanent material deformation. Elastic deformation stores the applied mechanical energy as completely recoverable material deformation. The extent to which one or the other of these mechanisms dominates the overall response of the material is determined by the temperature and by the duration and magnitude of the stress or strain. The higher the temperature, the most freedom of movement of the individual plastic molecules that comprise the TP and the more easily viscous flow can occur with lower mechanical performances.

With the longer duration of material stress or strain, the more time for viscous flow to occur that results in the likelihood of viscous flow and significant permanent deformation. As an example when a TP product is loaded or deformed beyond a certain point, it yields and immediate or eventually fails. Conversely, as the temperature or the duration or magnitude of material stress or strain decreases, viscous flow becomes less likely and less significant as a contributor to the overall response of the material; and the essentially instantaneous elastic deformation mechanism becomes predominant.

Changing the temperature or the strain rate of a TP may have a considerable effect on its observed stress-strain behavior. At lower temperatures or higher strain rates, the stress-strain curve of a TP may exhibit a steeper initial slope and a higher yield stress. In the extreme, the stress-strain curve may show the minor deviation from initial linearity and the lower failure strain characteristic of a brittle material.

At higher temperatures or lower strain rates, the stress-strain curve of the same material may exhibit a more gradual initial slope and a lower yield stress, as well as the drastic deviation from initial linearity and the higher failure strain characteristic of a ductile material.

There are a number of different modes of stress-strain that must be taken into account by the designer. They include tensile stress-strain, flexural stress-strain, compression stress-strain, and shear stress-strain.

Tensile Stress–Strain

In obtaining tensile stress-strain (S-S) engineering data, as well as other data, the rate of testing directly influence results. The test rate or the speed at which the movable cross-member of a testing machine moves in relation to the fixed cross-member influences the property of material. The speed of such tests is typically reported in cm/min. (in./min.). An increase in strain rate typically results in an increase yield point and ultimate strength.

An extensively used and important performance of any material in mechanical engineering is its tensile stress-strain curve (ASTM D 638). It is obtained by measuring the continuous elongation (strain) in a test sample as it is stretched by an increasing pull (stress) resulting in a stress-strain (S-S) curve. Several useful qualities include the tensile strength, modulus (modulus of elasticity) or stiffness (initial straight-line slope of the curve following Hooke's law and reported as Young's modulus), yield stress, and the length of the elongation at the break point.

Stress is defined as the force on a material divided by the cross sectional area over which it initially acts (engineering stress). When stress is calculated on the actual cross section at the time of the observed failure instead of the original cross sectional area it is called true stress. The engineering stress is reported and used practically all the time.

Strain is defined as the deformation of a material divided by a corresponding original cross section dimensions. The units of strain are meter per meter (m/m) or inch per inch (in./in.). Since strain is often regarded as dimensionless, strain measurements are typically expressed as a percentage.

Tensile strength is the maximum tensile stress sustained by a specimen during a tension test. When a maximum stress occurs at its yield point it is designated as tensile strength at yield. When the maximum stress occurs at a break, it is its tensile strength at break. In practice these differences are frequently ignored.

The ultimate tensile strength is usually measured in megapascals (MPa) or pounds per square inch (psi). Tensile strength for plastics range from under 20 MPa (3000 psi) to 75 MPa (11,000 psi) or just above, to more than 350 MPa (50,000 psi) for reinforced thermoset plastics (RTPs).

The area under the stress-strain curve is usually proportional to the energy required to break the specimen that in turn can be related to the toughness of a plastic. There are types, particularly among the many fiber-reinforced TSs, that are very hard, strong, and tough, even though their area under the stress-strain curve is extremely small.

Tensile elongation is the stretch that a material will exhibit before break or deformation. It is usually identified as a percentage. There are plastics that elongate (stretch) very little before break, while others such as elastomers have extensive elongation.

On a stress-strain curve there can be a location at which an increase in strain occurs without any increase in stress. This represents the yield point that is also called yield strength or tensile strength at yield. Some materials may not have a yield point. Yield strength can in such cases be established by choosing a stress level beyond the material's elastic limit. The yield strength is generally established by constructing a line to the curve where stress and strain is proportional at a specific offset strain, usually at 0.2%. Per ASTM testing the stress at the point of intersection of the line with the stress-strain curve is its yield strength at 0.2% offset.

Another important stress-strain identification is the proportional limit. It is the greatest stress at which the plastic is capable of sustaining an applied load without deviating from the straight line of an S-S curve.

The elastic limit identifies a material at its greatest stress at which it is capable of sustaining an applied load without any permanent strain remaining, once stress is completely released.

With rigid plastics the modulus that is the initial tangent to the S-S curve does not change significantly with the strain rate. The softer TPs, such as general purpose polyolefins, the initial modulus is independent of the strain rate. The significant time-dependent effects associated with such materials, and the practical difficulties of obtaining a true initial tangent modulus near the origin of a nonlinear S-S curve, render it difficult to resolve the true elastic modulus of the softer TPs in respect to actual data.

The observed effect of increasing strain is to increase the slope of the early portions of the S-S curve, which differs from that at the origin. The elastic modulus and strength of both the rigid and the softer plastics each decrease with an increase in temperature. Even though the effects of a change in temperature are similar to those resulting from a change in the strain rate, the effects of temperature are much greater.

Modulus of Elasticity
Many unreinforced and reinforced plastics have a definite tensile modulus of elasticity where deformation is directly proportional to their loads below the proportional limits. Since stress is proportional to load and strain to deformation, stress is proportional to strain. Fig. 2.4 shows this relationship. The top curve is where the S-S straight line identifies a modulus and a secant modulus based at a specific strain rate at point C' that could be the usual 1% strain. Bottom curve secant moduli of different plastics are based on a 85% of the initial tangent modulus.

There are unreinforced commodities TPs that have no straight region on the S-S curve or the straight region of this curve is too difficult to locate. The secant modulus is used. It is the ratio of stress to the corresponding strain at any specific point on the S-S curve. It is the line from the initial S-S curve to a selected point C on the stress-strain curve based on an angle such as 85% or a vertical line such as at the usual 1% strain.

Hooke's Law highlights that the straight line of proportionality is calculated as a constant that is called the modulus of elasticity (E). It is

Figure 2.4 Examples of tangent moduli and secant moduli

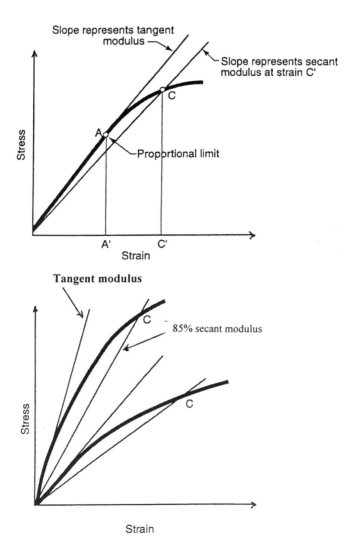

the straight-line slope of the initial portion of the stress-strain curve:

Stress/Strain = Constant (2-9)

The modulus of elasticity is also called Young's modulus, elastic modulus, or just modulus. E was defined by Thomas Young in 1807 although others used the concept that included the Roman Empire and Chinese-BC. It is expressed in terms such as MPa or GPa (psi or Msi). A plastic with a proportional limit and not loaded past its proportional limit will return to its original shape once the load is removed.

With certain plastics, particularly high performance RPs, there can be two or three moduli. Their stress-strain curve starts with a straight line that results in its highest E, followed by another straight line with a lower S, and so forth. To be conservative providing a high safety factor the lowest E is used in a design, however the highest E is used in certain designs where experience has proved success.

Standard ASTM D 638 states that it is correct to apply the term modulus of elasticity to describe the stiffness or rigidity of a plastic where its S-S characteristics depend on such factors as the stress or strain rate, the temperature, and its previous history as a specimen. However, D 638 still suggests that the modulus of elasticity can be a useful measure of the S-S relationship, if its arbitrary nature and dependence on load duration, temperature, and other factors are taken into account.

Interesting straight-line correlations exist of the tensile modulus of elasticity to specific gravity of different materials (Fig. 2.5). In this figure, the modulus/specific gravity of reinforced plastics with its high performing fibers (graphite, aramid, carbon, etc.) continues to increase in the upward direction.

Flexural Stress–Strain

Flexural stress-strain testing according to ASTM D 790 determines the load necessary to generate a given level of strain on a specimen typically using a three-point load (Fig. 2.6). Testing is performed at specified

Figure 2.5 Modulus vs. specific gravity ratio for different materials follows a straight line

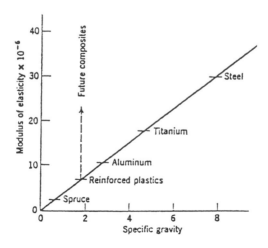

Figure 2.6 Example of a three-point ASTM D 790 flexural test specimen

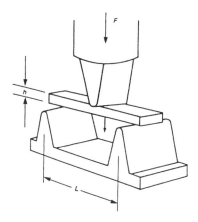

constant rate of crosshead movement based on material being tested. A solid plastic is usually at 0.05 in./min., foamed plastic at 0. 1 in./ min., etc.

Simple beam equations are used to determine the stresses on specimens at different levels of crosshead displacement. Using traditional beam equations and section properties in Fig. 2.5, the following relationships can be derived where Y is the deflection at the load point:

Bending stress where $\sigma = 3FL/2bh^2$ (2-10)

Bending or flexural modulus where $E = FL^3/4bh^3Y$ (2-11)

Using these relationships, the flexural strength (also called the modulus of rupture) and the flexural modulus of elasticity can be determined. Table 2.1 provides examples of the flexural modulus of elasticity for polypropylene with nothing added (NEAT/Chapter 1) and reinforced with glass fibers and talc.

Table 2.1 Polypropylene NEAT and filled flexural modulus of elasticity data

NEAT plastic	180,000 psi (1,240 MPa)
40wt% glass fiber	1,100,000 psi (7,600 MPa)
40wt% talc	575,000 psi

A flexural specimen is not in a state of uniform stress on the specimen. When a simply supported specimen is loaded, the side of the material opposite the loading undergoes the greatest tensile loading. The side of the material being loaded experiences compressive stress (Fig. 2.7). These stresses decrease linearly toward the center of the sample.

Figure 2.7 Tensile-compressive loading occurs on a flexural specimen

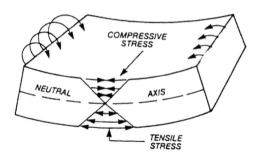

Theoretically the center is a plane, called the neutral axis, that experiences no stress.

In the flexural test the tensile and compressive yield stresses of a plastic may cause the stress distribution within the test specimen to become very asymmetric at high strain levels. This change causes the neutral axis to move from the center of the specimen toward the surface that is in compression. This effect, along with specimen anisotropy due to processing, may cause the shape of the stress-strain curve obtained in flexure to differ significantly from that of the normal S-S curve.

The S-S behavior of plastics in flexure generally follows that of tension and compression tests for either unreinforced or reinforced plastics. The flexural E tends to be the average between the tension and compression Es. The flexural yield point follows that observed in tension.

For the standard ASTM flexural strengths most plastics are higher than their ultimate tensile strengths, but may be either higher or lower than compressive strengths. Since most plastics exhibit some yielding or nonlinearity in their tensile S-S curve, there is a shift from triangular stress distribution toward rectangular distribution when the product is subject to bending. This behavior with plastics is similar to that when designing in steel and also for ultimate design strength in concrete. Shifts in the neutral axis resulting from differences in the yield strain and post-yield behavior in tension and compression usually affect the correlation between the modulus of rupture and the uniaxial strength results. The modulus of rupture reflects in part nonlinearities in stress distribution caused by plastification or viscoelastic nonlinearities in the cross-section.

Plastics such as short-fiber reinforced plastics with fairly linear stress-strain curves to failure usually display moduli of rupture values that are higher than the tensile strength obtained in uniaxial tests; wood behaves much the same way. Qualitatively, this can be explained from statistically considering flaws and fractures and the fracture energy available in flexural samples under a constant rate of deflection as compared to tensile samples under the same load conditions. These differences become less as the thickness of the bending specimen increases, as would be expected by examining statistical considerations.

The cantilever beam is another flexural test that is used to evaluate different plastics and structures such as beam designs. It is used in creep and fatigue testing and for conducting testing in different environments where the cantilever test specimen under load is exposed to chemicals, moisture, etc.

Compressive Stress–Strain

A test specimen under loading conditions located between the two flat, parallel faces of a testing machine is compressed at a specified rate (ASTM D 695). Stress and strain are computed from the measured compression test, and these are plotted as a compressive stress-strain curve for the material at the temperature and strain rate employed for the test.

Procedures in compression testing are similar to those in tensile testing. However in compression testing particular care must be taken to specify the specimen's dimensions and relate test results to these dimensions. If a sample is too long and narrow buckling may cause premature failure resulting in inaccurate compression test results. Buckling can be avoided by examining different size specimens. Consider a test specimen with a square cross-section and a longitudinal dimension twice as long as a side of the cross-section.

At high stress levels, compressive strain is usually less than tensile strain. Unlike tensile loading, which usually results in failure, stressing in compression produces a slow, indefinite yielding that seldom leads to failure. Where a compressive failure does occur, the designer should determine the material's strength by dividing the maximum load the sample supported by its initial cross-sectional area. When the material does not exhibit a distinct maximum load prior to failure, the designer should report the strength at a given level of strain that is usually at 10%.

The compression specimen's ends usually do not remain rigid. They tend to spread out or flower at its ends. Test results are usually very scattered requiring close examination as to what the results mean in reference to the behavior of the test specimens. Different clamping devices (support plates on the specimen sides, etc.) are used to

Figure 2.8 Stress-strain tensile and compressive response tends to be similar

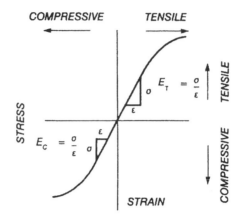

eliminate the flowering action that could provide inaccurate readings that in turn influence results by usually making them stronger.

The majority of tests to evaluate the characteristics of plastics are performed in tension and/or flexure because compression data could be misleading. The result is that compressive stress-strain behavior of many plastics is not well described. Generally, the behavior in compression is different from that in tension, but the S-S response in compression is usually close enough to that of tension so that possible differences can be neglected. Fig. 2.8 compares tensile and compression data of TPs.

The compressive strength of a non-reinforced plastic or a mat-based RP laminate is usually greater than its tensile strength (unidirectional fiber-reinforced plastic is usually slightly lower than its tensile strength). However, this is not generally true for reinforced TSs (RTSs). Different results are obtained with different plastics. As an example the compression testing of foamed plastics provides the designer with the useful recovery rate. A compression test result for rigid foamed insulating polyurethane (3.9 lb/ft^3) resulted in almost one-half of its total strain recovered in one week.

Shear Stress–Strain

Shear deformation occurs in structural elements such as those subjected to torsional loads and in short beams subjected to transverse loads. Shear S-S data can be generated by twisting (applying torque) a specimen at a specified rate while measuring the angle of twist between the ends of the specimen and the torque load exerted by the specimen on the testing machine (ASTM D 732). Maximum shear stress at the surface of the specimen can be computed from the measured torque that is the maximum shear strain from the measured angle of twist.

The shear mode involves the application of a load to a material specimen in such a way that cubic volume elements of the material comprising the specimen become distorted, their volume remaining constant, but with opposite faces sliding sideways with respect to each other. Basically, shearing stresses are tangential stresses that act parallel to the planes they stress. The shearing force in a beam provides shearing stresses on both the vertical and horizontal planes within the beam. The two vertical stresses must be equal in magnitude and opposite in direction to ensure vertical equilibrium. However, under the action of those two stresses alone the element would rotate.

Another couple must counter these two stresses. If the small element is taken as a differential one, the magnitude of the horizontal stresses must have the value of the two vertical stresses. This principle is sometimes phrased as cross-shears are equal that refers to a shearing stress that cannot exist on an element without a like stress being located 90 degrees around the corner.

Fig. 2.9 schematic diagram is subjected to a set of equal and opposite shearing force/load (Q). The top view represents a material with equal and opposite shearing forces and the bottom view is a schematic of a theoretical infinitesimally thin layers subject to shear stress. As shown at the bottom with the infinitesimally thin layers there is a tendency for one layer of the material to slide over another to produce a shear form of deformation or failure if the force is great enough. The shear stress will always be tangential to the area upon which it acts. The shearing strain is the angle of deformation as measured in radians. For materials that behave according to Hooke's Law, shear strain is proportional to the shear stress. The constant G is called the shear modulus, modulus of rigidity, or torsion modulus.

G is directly comparable to the modulus of elasticity used in direct-stress applications. Only two material constants are required to characterize a material if one assumes the material to be linearly elastic, homogeneous, and isotropic. However, three material constants exist: the tensile modulus of elasticity (E), Poisson's ratio (v), and the shear modulus (G). An equation relating these three constants, based on engineering's elasticity principles is as follows:

$$E/G = 2 \left(1 + v\right) \tag{2-12}$$

This calculation that is true for most metals, is generally applicable to

Figure 2.9 Theoretical approaches to shear stress behavior

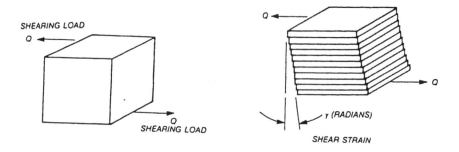

plastics. However, this calculation does not apply with the inherently nonlinear, anisotropic nature of most plastics, particularly the fiber-reinforced and liquid crystal ones.

Torsion Stress-Strain
Shear modulus can be determined by a static torsion test or by a dynamic test using primarily a torsional pendulum (ASTM D 2236). Also used is an oscillatory rheometer test. The torsional pendulum is applicable to virtually all plastics and uses a simple specimen readily fabricated by all commercial fabricating processes or easily cut from fabricated part. The moduli of elasticity, G for shear and E for tension, are ratios of stress to strain as measured within the proportional limits of the material. Thus the modulus is really a measure of the rigidity for shear of a material or its stiffness in tension and compression. For shear or torsion, the modulus analogous to that for tension is called the shear modulus or the modulus of rigidity, or sometimes the transverse modulus.

Direct Load Shear Strength
Unlike the methods for tensile, flexural, or compressive testing, the typical procedure used for determining shear properties is intended only to determine the shear strength. It is not the shear modulus of a material that will be subjected to the usual type of direct loading (ASTM D 732). When analyzing plastics in a pure shear situation or when the maximum shear stress is being calculated in a complex stress environment, a shear strength equal to half the tensile strength or that from shear tests is generally used, whichever is less.

The shear strength values are obtained by such simple tests using single or double shear actions. In these tests the specimen to be tested is sheared between the hardened edges of the supporting block and the block to which the load is applied. The shearing strength is calculated as the load at separation divided by the total cross-sectional area being sheared. The maximum short-term shear stress (strength) of a material can also be determined from a punch shear test.

The test fixtures in these test devices indicates that bending stresses exist and the stress cannot be considered as being purely that of shear. Therefore, the shearing stress calculated must be regarded as an average stress. This type of calculation is justified in analyzing bolts, rivets, and any other mechanical member whose bending moments are considered negligible. Note that tests of bolts and rivets have shown that their strength in double shear can at times be as much as 20% below that for single shear. The values for the shear yield point (MPa or psi) are generally not available; however, the values that are listed are usually obtained by the torsional testing of round test specimens.

The data obtained using the test method reviewed should be reported as direct shear strength. These data can only be compared to data determined by the same direct-shear methods. This test cannot be used to develop shear S-S curves or determine a shear modulus, because bending or compression rather than pure shear transfer a considerable portion of the load. The test results depend on the susceptibility of the material to the sharpness of load faces.

It is important to note material such as certain plastics, concrete, or wood that are weak in either tension or compression will also be basically weak in shear. For example, concrete is weak in shear because of its lack of strength in tension. Reinforced bars in the concrete are incorporated to prevent diagonal tension cracking and strengthen concrete beams. Similar action occurs with RPs using fiber filament structures.

Although no one has ever been able to determine accurately the resistance of concrete to pure shearing stress, the matter is not very important, because pure shearing stress is probably never encountered in concrete structures. Furthermore, according to engineering mechanics, if pure shear is produced in one member, a principal tensile stress of an equal magnitude will be produced on another plane. Because the tensile strength of concrete is less than its shearing strength, the concrete will fail in tension before reaching its shearing strength. This action also occurs with plastics.

Residual Stress

It is the stress existing in a body at rest, in equilibrium, at uniform temperature, and not subjected to external forces. Often caused by the stresses remaining in a plastic part as a result of thermal and/or mechanical treatment in fabricating parts. Usually they are not a problem in the finished product. However, with excess stresses, the product could be damaged quickly or after in service from a short to long time depending on amount of stress and the environmental conditions around the product.

Dynamic stress

Knowledge in such behavioral responses of plastics as those ranging from short time static (tensile, flexural, etc.) to long time dynamic (creep, fatigue, impact, etc.) mechanical load performances in different environments are important to product designs subjected to those loads. Dynamic loading in the present context is taken to include deformation rates above those achieved on the standard laboratory-testing machine (commonly designated as static or quasi-static). These slower tests may encounter minimal time-dependent effects, such as creep and stress-relaxation, and therefore are in a sense dynamic. Thus the terms static and dynamic can be overlapping.

Long time dynamic load involves behaviors such as creep, fatigue, and impact. Two of the most important types of long-term material behavior are more specifically viscoelastic creep and stress relaxation. Whereas stress-strain behavior usually occurs in less than one or two hours, creep and stress relaxation may continue over the entire life of the structure such as 100,000 hours or more.

In many applications, intermittent or dynamic loads arise over much shorter time scales. Examples of such products include chair seats, panels that vibrate and transmit noise, engine mounts and other antivibration products, and road surface-induced loads carried to wheels and suspension systems. Plastics' relevant properties in this regard are material stiffness and internal damping, the latter of which can often be used to advantage in design. Both properties depend on the frequency of the applied loads or vibrations, a dependence that must be allowed for in the design analysis. The possibility of fatigue damage and failure must also be considered.

Mechanical loads on a structure induce stresses within the material such as those shown in Fig. 2.10. The magnitudes of these stresses depends on many factors, including forces, angle of loads, rate and point of application of each load, geometry of the structure, manner in which that structure is supported, and time at temperature. The behavior of the material in response to these induced stresses determines the performance of the structure.

Figure 2.10 Static and dynamic loads (courtesy of Plastics FALLO)

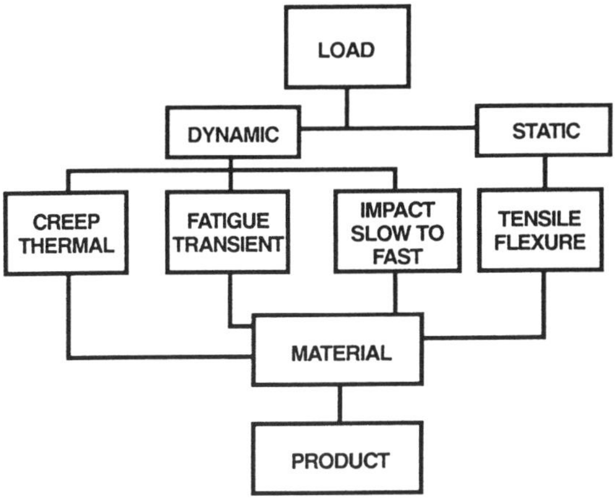

Table 2.2 on p. 90 provides examples of reinforced thermoplastics flexural data.

Dynamic/Static Mechanical Behavior

Mechanical tests measure the response or deformation of a material to periodic or varying forces. Generally an applied force and its resulting deformation both vary sinusoidally with time. From such tests it is possible to obtain simultaneously an elastic modulus and mechanical damping, the latter of which gives the amount of energy dissipated as heat during the deformation of the material.

Description of material behavior is basic to all designing applications. Many of the problems that develop may be treated entirely within the framework of plastic's viscoelastic material response. While even these problems may become quite complex because of geometrical and loading conditions, linearity, reversibility, and rate independence generally applicable to elastic material description certainly eases the task of the analyst for dynamic and static loads that include conditions such as creep, fatigue, and impact.

Many plastic products seen in everyday life are not required to undergo sophisticated design analysis because they are not required to withstand high static and dynamic loads. However, we are increasingly confronted with practical problems that involve material response that is inelastic, hysteretic, and rate dependent combined with loading which is transient in nature. These problems include structural response to moving or impulsive loads, all the areas of ballistics (internal, external, and terminal), contact stresses under high speed operations, high speed

Table 2.2 Examples of reinforced thermoplastics flexural creep data.

Plastic	Reinforcement and % content	Stress, psi	Strain (%) hours 10	100	1,000	Apparent modulus (10^3 psi) hours 10	100	1,000
Nylon 6/6	Glass fiber 15, mineral 25	2,500	0.555	0.623	0.709	450	401	353
		5,000	0.823	0.967	1.140	607	517	439
Polyester (PBT)	Glass fiber 15, mineral 25	2,500	0.452	0.470	0.482	553	532	519
		5,000	0.693	0.742	0.819	721	674	610
Nylon 6/10	Ferrite 83	2,500	0.463	0.507	0.568	540	493	440
		5,000	0.638	0.732	0.952	784	683	525
Polypropylene	Carbon powder	2,500	1.100	1.140	1.970	114	87	63
		5,000	6.230	6.920	8.660	40	36	29
Nylon 6/6	Glass fiber 15, carbon powder	2,500	2.160	2.400	2.510	116	104	100
		5,000						
Nylon 6	Glass beads 30	1,250	0.140	0.320	0.368	893	391	340
		5,000	0.290	0.650	0.750	862	385	333

fabricating processes (injection molding, extrusion, blow molding, thermoforming, etc.), shock attenuation structures, seismic wave propagation, and many others of equal importance.

From past problems it became evident that the physical or mathematical description of the behavior of materials necessary to produce realistic solutions did not exist. Since at least the 1940s, there has been considerable effort expended toward the generation of both experimental data on the dynamic and static mechanical response of materials (steel, plastic, etc.) as well as the formulation of realistic constitutive theories. As a plastic is subjected to a fixed stress or strain, the deformation versus time curve shows an initial rapid deformation followed by a continuous action.

As reviewed dynamic loading is taken to include deformation rates above those achieved on the standard laboratory-testing machine that are designated as static or quasi-static. These slower tests may encounter minimal time-dependent effects, such as creep and stress-relaxation, and therefore are in a sense dynamic. This situation shows that the terms static and dynamic can be overlapping. The behavior of materials under dynamic load is of considerable importance and interest in most mechanical analyses of design problems where these loads exist. The complex workings of the dynamic behavior problem can best be appreciated by summarizing the range of interactions of dynamic loads that exist for all the different types of materials. Dynamic loads involve the interactions of creep and relaxation loads, vibratory and transient fatigue loads, low-velocity impacts measurable sometimes in milliseconds, high-velocity impacts measurable in microseconds, and hypervelocity impacts.

An interesting point is that metals are unique under both dynamic and static loads that can be cited as outstanding cases. The mechanical engineer and the metallurgical engineer have both found these materials to be most attractive to study. When compared to plastics, they are easier to handle for analysis. However there is a great deal that is still not understood about metals, even in the voluminous scientific literature available. The importance of plastics and reinforced plastics (RPs) has been growing steadily, resulting in more dynamic mechanical behavior data becoming available.

Summarization of all material behaviors can be by classifications. They include (1) creep, and relaxation behavior with a primary load environment of high or moderate temperatures; (2) fatigue, viscoelastic, and elastic range vibration or impact; (3) fluidlike flow, as a solid to a gas, which is a very high velocity or hypervelocity impact; and (4) crack propagation and environmental embrittlement, as well as ductile and brittle fractures.

Energy and Motion Control

Elastomers are frequently subjected to dynamic loads where heat energy and motion control systems are required. One of the serious dynamic loading problems frequently encountered in machines, vehicles, moving belts, and other products is vibration-induced deflection. Such effects can be highly destructive, particularly if a product resonates at one of the driving vibration frequencies.

One of the best ways to reduce and in many cases eliminate vibration problems is by the use of viscoelastic plastics. Some materials such as polyurethane plastics, silicone elastomers, flexible vinyl compounds of specific formulations, and a number of others have very large hysteresis effects (Chapter 3). By designing them into the structure it is possible to have the viscoelastic material absorb enough of the vibration inducing energy and convert it to heat so that the structure is highly damped and will not vibrate.

In each case the plastic is arranged in such a way that movement or flexing of the product results in large deflections of the viscoelastic materials so that a large hysteresis curve is generated with a large amount of energy dissipated per cycle. By calculating the energy to heat it is possible to determine the vibration levels to which the structure can be exposed and still exhibit critical damping.

Plastics exhibit a spectrum of response to stress and there are certain straining rates that the material will react to almost elastically. If this characteristic response corresponds to a frequency to which the structure is exposed the damping effect is minimal and the structure may be destroyed. In order to avoid the possibility of this occurring, it is desirable to have a curve of energy absorption vs. frequency for the material that will be used.

The same approach can be used in designing power transmitting units such as moving belts. In most applications it is desirable that the belts be elastic and stiff enough to minimize heat buildup and to minimize power loss in the belts. In the case of a driver which might be called noisy in that there are a lot of erratic pulse driven forces present, such as an impulse operated drive, it is desirable to remove this noise by the available damping action of plastics and obtain a smooth power curve. This is easily done using a viscoelastic belt that will absorb the high rate load pulses.

Making one gear in a gear train or one link in a linear drive mechanism an energy absorber can use the same approach as the belt. The viscoelastic damping is a valuable tool for the designer to handle impulse loading that is undesirable and potentially destructive to the product.

There is another type of application that has a long history where the damping effect of plastic structures can be used to advantage. The early airplanes used doped fabric as the covering for wings and other aerodynamic surfaces. The cellulose nitrate and later cellulose acetate is a damping type of plastic. Consequently, surface flutter was a rare occurrence. It became a serious problem when aluminum replaced the fabric because of the high elasticity of the metal surfaces. The aerodynamic forces acting on the thin metal coverings easily induce flutter and this was a difficult design problem that was eventually corrected for minimizing the effect.

Isolator

When products are subjected to dynamic loads where energy and motion controls are required use is made of thermoplastic elastomer (TPE) components. These products involve buildings (Fig. 2.11), bridges, highways, sporting goods, home appliances, automobiles, boats, aircraft, and spacecraft.

Figure 2.11 Schematic of a building isolator

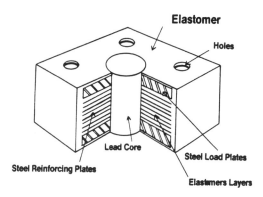

In a building, TPE controls vibration and noise from motors and engines that is generated to the building itself. For rapid transit TPE supports the rail and the vehicle reducing noise and vibration to adjacent buildings. For ships TPE absorb their berthing energies used in vibration units that are as large as 3 m (9.9 ft.) high and weighing up to 19 tons. For all these and other applications, TPEs are used either in shear, compression, tension, torsion, buckling, or a combination of two or more load conditions depending on the needs of the specific application. The particular application will dictate which would be best.

When berthing a vessel the structure has to be designed to withstand the energy developed by the vessel. The more rigid the system, the higher the reactive forces must be to absorb the vessel's kinetic energy. The area under the structure's load as against its deflection response curve is typical to that shown in Fig. 2.12.

Figure 2.12 Load-deflection energy absorbed behavior in these type isolators

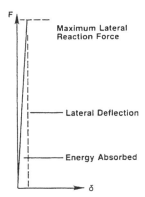

Figure 2.13 Shear energy curves   Figure 2.14 Compression energy curves

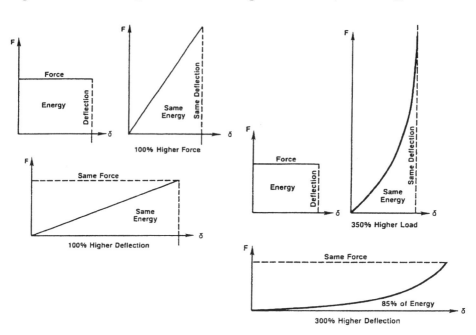

Figure 2.15 Buckling energy curves

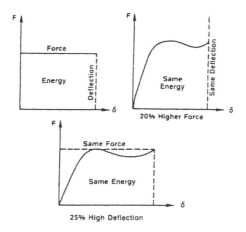

An elastomer is ideal in the vessel environment because it will not corrode. Metal components are protected by being totally encapsulated in an elastomer and then bonded to all-metal surfaces. In examining the load-deflection characteristics of shear, compression, and buckling systems, the one that results in the lowest reaction force generally also produces the lowest-cost structure. Figs 2.13 to 2.15 show six results that could be obtained compared to an ideal hydraulic system with 100% energy efficiency.

In Fig. 2.13 the energy capacity is approximately 50% efficient, requiring 100% more deflection or load, if the deflection or load of a 100% efficient curve is required. Fig. 2.14 shows an energy capacity approximately 35% efficient, requiring 300% higher deflection or a 350% higher load if the deflection or load of a 100% efficient curve is required. The energy capacity shown in Fig. 2.15 is approximately 75% efficient, requiring a 25% higher deflection or a 20% higher load if the deflection or load of a 100% efficient curve is required.

The buckling column was selected because it produces the lowest reaction load and the lowest deflection (deflection controls the projection of the berthing system out from the structure). When designing to support structures and allow the horizontal input of an earthquake or to allow the structural movement needed on a structure such as a bridge pier, the vertical and horizontal stiffnesses must be calculated, then a system can be designed. Take, for example, a rectangular elastomeric section with a length of 762 cm, a width of 508 cm, and a thickness of 508 cm ($305 \times 203 \times 203$ in.). Table 2.3 lists the formulas used to calculate the respective compression and shear stiffnesses for these data. For the elastomeric section of the example use the following formula:

$$K_c = \frac{(k_c) \times 508)(E_c)}{508} \qquad K_s = \frac{(k_s(762 \times 508)(G_s)}{508} \qquad (2\text{-}13)$$
$$= 762 k_c E_c \qquad\qquad = 762 k_s G_s$$

Table 2.3 Data required for formulas.

Direction	Compression (K_c)	Shear (K_s)
Formula	$\dfrac{(k_c)(\text{L.A.})(Ec)}{t}$	$\dfrac{(k_c)(\text{L.A.})(G_s)}{t}$
Variable k	Factor of geometry	Factor of geometry
Variable L.A.	Load area	Load area
Variable E_c	Compression modulus	—
Variable G_s	—	
Variable t	Elastomer thickness	Elastomer thickness

The calculations for k_c and k_s adjust for such design parameters as strain, bulk compression, and bending by the elastomer section, as developed over many years of sample testing. A way to forego the pain of getting there is to let $k_s = 0.98$ and $k_c = 1.0$, using a 0.69 MPa shear-modulus elastomer as follows:

$K_c = 762E_c$ and $K_s = 747G_s$.

For compression, the shape factor (SF), which is the projected load area of the product divided by the elastomer area that is free to move (known in the industry as the bulge area, or BA), is the major design parameter. For this example the shape factor is calculated thus:

Load area/Bulge area = (726)(508)/(762 + 508)(2)(508) = 0.3   (2-14)

Using SF = 0.3 and a G_s = 0.69MPa, and E_{c+} = 2.42-MPa is obtained, Therefore K_c = 1. 8-MN/m and K_s = 0.5-MN/m.

Applying a 20% maximum compression strain to the product results in a maximum compression deflection of 102 cm (41 in.). This allows a maximum compression load of only 19 kg (42 lb.), hardly sufficient to support a building or a bridge.

Given that shear stiffness K_s cannot change, the sole remaining option is to change the compression stiffness, by adjusting the shape factor. Designing the product as two units each 762 × 508 × 254 cm (305 × 203 × 102 in.) thick will not change the shear characteristic, but it does change the shape factor to 0.060. In this instance, E_c = 3.03 MPa and K_c becomes 4.6 MN/m per section. With two sections in series the spring rate is 2.3 MN/m, which now allows for 24.3 kg of compression load. Dividing each of these sections into a total of four sections each 762 × 508 × 127 cm (305 × 203 × 51 in.) thick yields a shape factor of 1.2 and an E_c equal to 151.7 MPa. An individual section K_c will be 46 MN/m, with a series of four being 115 MN/m, allowing a compression load of about 120 kg. It always maintains a shear spring rate of about 0.5 MN/m. This is the basic design philosophy for obtaining high-compression loads while maintaining the soft shear stiffness needed for seismic considerations or for thermal expansion and contraction. Continued thinning of the individual elastomer sections will drive the compression load-carrying capacity upward.

In the rapid transit industry, wherever there is an elevated structure or subway tunnel, noise and vibration caused by the vehicles can generate unrest among those living along the route. There are three areas that can be adjusted to reduce annoying frequencies: the vehicles' suspension system; the trackbed, including the tie-to-rail interface and the floating slab; and various acoustic barriers. For the vehicle suspension system and trackbed techniques, good elastomeric product design and application are generally sufficient. The product design requires including the design considerations mentioned previously of the compression and shear curve and shape factor, but also introduces new combinations of compression and shear. These occur either by shear and compression in planes 90 degrees to each other or shear with compression, as in seismic and thermo designs, only installed at angles to the horizontal.

We have now looked at the more common types of single-axis load-deflection characteristics possible with elastomeric products, as well as various methods to change the stiffness in one direction while maintaining an initial stiffness in a plane that is 90 degrees from the reference plane. The angular considerations in different directions can now yield an unlimited number of design-configuration options.

Torsion Load

The next six typical concepts will allow motion through an elastomer in three or more directions. First, Fig. 2.16 would be a typical design for seismic concerns and bridge-bearing pads. It is capable of supporting high compressive loads while allowing for soft lateral (shear) and torsional characteristics. Although the part shown is a circular unit, square and rectangular products are more common in the construction industry.

Fig. 2.17 shows another high shape factor design, which in this case takes the elastomer in compression (that is, radially), with the shear modes being axial and torsional.

Fig. 2.18 is similar to Fig. 2.17 except that its metal components have a spherical contour, allowing for torsion about the center line and radial axis. For each of these directions the radial-load deflection characteristic

Figure 2.16 Axial compression concept with two shear modes

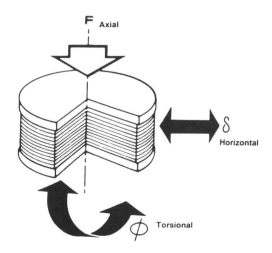

Figure 2.17 Radial compression concept with two shear modes

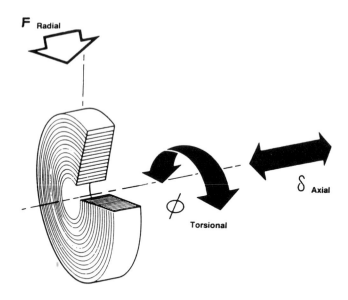

is quite stiff. The axial characteristic is indeed stiffer than the shear in the previous configuration. The torsion about the axial axis is also similar, but the torsion about the radial axial axis is soft.

Figure 2.19 shows a product similar to one used in the rapid-transit and

Figure 2.18 Radial compression concept with three shear modes

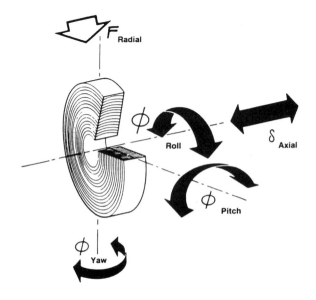

Figure 2.19 Radial and axial compression concept with one shear mode

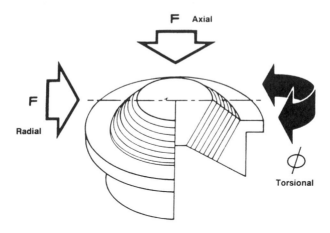

railroad industry that is, however, of a conical configuration as opposed to a chevron configuration. This product exhibits high axial and radial characteristics while maintaining soft torsional characteristics.

Fig. 2.20 shows another variation of Fig. 2.19, which produces high radial and axial load-deflection characteristics but maintains a soft torsional nature. This design is basically a combination of the flat, high shape-factor part described in Fig. 2.18, and the circular part described in Fig. 2.19.

Fig. 2.21 shows a spherical bearing, a configuration that allows for high compressive loads in the axial direction while maintaining soft torsion

Figure 2.20 Radial and axial compression with one shear mode and high deflection

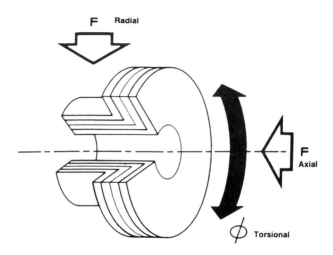

Figure 2.21 Axial compression with three shear modes

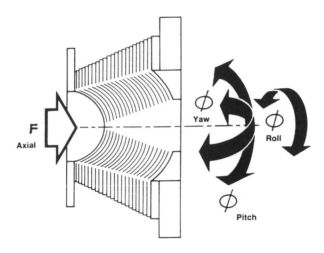

about the axial axis and torsion perpendicular to that axis. This configuration is a combination of Figs. 2.20 and 2.21.

For auxiliary generators and compressors any of these configurations would be viable. However, each individual application has its own design requirements.

One of the parameters to consider when applying an elastic suspension system to an energy-producing device is the degree of motion that will be acceptable to the installation.

The performance of elastomers is of major interest and concern to the design engineer. The readily available data concern the tensile-elongation factor, the compression set, results from durometer tests, and information on oil resistance, heat aging, and the static modulus. In designing for a given environment, certain information makes the designer's job easier and the actual results closer to that predicted. These types of data are normally generated at the designer's facility with in-house-developed test equipment and procedures. They include: (1) dynamic modulus at various strains, frequencies, and temperatures; (2) ozone resistance at different concentration levels; (3) loss factor at various strains, frequencies, and temperatures; (4) fatigue of various shape factors and cyclic strains and temperatures; (5) effects of different ingredients such as carbon black; (6) drift and set characteristics at various initial strains and temperatures; and (7) electrical resistance.

Rapid loading

Different behavioral characteristics for a wide range of loading rates have been reviewed. This review concerns load or strain duration that are much shorter than those reviewed that are usually referred to as being rapid impact loading. They range from a second or less (Fig. 2.22). There are a number of basic forms of rapid impact loading or impingement on products to which plastics react in a manner different from other materials. These dynamic stresses include loading due to direct impact, impulse, puncture, frictional, hydrostatic, and erosion. They have a difference in response and degree of response to other forms of stress.

The concept of a ductile-to-brittle transition temperature in plastics is well known in metals where notched metal parts cause brittle failure when compared to unnotched specimens. There are differences such as the short time moduli of many plastics compared with those in metals that may be 200 MPa (29×10^6 psi). Although the ductile metals often undergo local necking during a tensile test, followed by failure in the neck, many ductile plastics exhibit the phenomenon called a propagating neck.

Figure 2.22 Rapid loading velocity (Courtesy of Plastics FALLO)

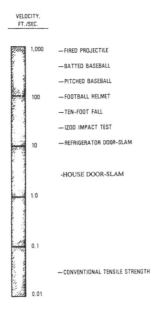

Impact

Impact loading analysis may take the form of design against impact damage requiring an analysis under high-rate loading or design for acceptable energy absorption, or a combination of the two. Impact resistance of a structure is defined as its ability to absorb and dissipate the energy delivered to it during relatively high speed collisions with other objects without sustaining damage that would damage its intended performance.

To determine whether failure will occur the acceptable energy absorption case requires an analysis of the stress and strain distribution during the impact loading followed by comparison with materials impact failure data. Whenever a product is loaded rapidly, it is subjected to impact loading. Any product that is moving has kinetic energy. When this motion is somehow stopped because of a collision, its energy must be dissipated. The ability of a plastic product to absorb energy is determined by such factors as its shape, size, thickness, type of material, method of processing, and environmental conditions of temperature, moisture, and/or others.

Temperature conditions effect impact strength. The impact strength of plastics is reduced drastically at low temperatures with the exception of fibrous filled materials that improve in impact strength at low temperature. The reduction in impact strength is especially severe if the material undergoes a glass transition where the reduction in impact strength is usually an order of magnitude.

From a design approach several design features affect impact resistance. For example, rigidizing elements such as ribs may decrease a part's impact resistance, while less-rigid sections may absorb more impact energy without damage by deflecting elastically. Dead sharp corners or notches subjected to tensile loads during impact may decrease the impact resistance of a product by acting as stress concentrators, whereas generous radii in these areas may distribute the tensile load and enhance the impact resistance. This factor is particularly important for products comprised of materials whose intrinsic impact resistance is a strong function of a notch radius. An impact resistance that decreases drastically with notch radius characterizes such notch sensitive materials. Wall thickness may also affect impact resistance. Some materials have a critical thickness above which the intrinsic impact resistance decreases dramatically.

There are different methods used to determine the impact resistance of plastics. They include pendulum methods (Izod, Charpy, tensile impact, falling dart, Gardner, Dynatup, etc.) and instrumented techniques. In the case of the Izod test, what is measured is the energy required to break a test specimen transversely struck (the test can be done either with the specimen notched or unnotched). The tensile impact test has a bar loaded in tension and the striking force tends to elongate the bar.

Impact strengths of plastics are widely reported, these properties have no particular design value. However, they are important, because they can be used to provide an initial comparison of the relative responses of materials. With limitations, the impact value of a material can broadly separate those that can withstand shock loading from those that are poorly in this response. The results provide guidelines that will be more meaningful and empirical to the designer. To eliminate broad general-izations, the target is to conduct impact tests on the final product or, if possible, at least on its components.

An impact test on products requires setting up an approach on how it should be conducted. The real test is after the product has been in service and field reports are returned for evaluation. Regardless, the usual impact tests conducted on test samples can be useful if they are properly related with product requirements.

Test and service data with PVC both rate low in notched Izod impact tests and performs well in normal service applications that involve impact loading. Another example is with some grades of rubber-modified high impact PSs that show up well in the Izod test fail on impact under field test conditions. These results have led to continual reexamination of the tests used to determine the toughness of plastics.

There are thermoplastics that tend to be very notch sensitive on impact. This is apparent from the molecular structure of the TP that consist of random arrangements of plastic chains (Chapter 1). If the material exists in the glassy state at room temperature the notch effect is to cut the chains locally and increase the stress on the adjacent molecular chains which will scission and propagate the effect through the material. At the high loading rate encountered in impact loading the only form of molecular response is the chain bending reaction which is limited in extent and generally low in magnitude compared to the viscoelastic response which responds at longer loading times.

TPs impact properties can be improved if the material selected does not have sufficient impact strength. One method is by altering the com-position of the material so that it is no longer a glassy plastic at the operating temperature of the product. In the case of PVC this is done by the addition of an impact modifier which can be a compatible plastic such as an acrylic or a nitrile rubber. The addition of such a material lowers the T_g (glass transition temperature) and the material becomes a rubbery viscoelastic plastic with improved impact properties (Chapter 1).

Molecular orientation can improve impact TP properties. As an example nylon has a fair impact strength but oriented nylon has a very high transverse impact strength. The intrinsic impact strength of the nylon comes from the polar structure of the material and the fact that the polymer is crystalline. The substantial increase in impact strength as a result of the orientation results from the molecular chains being aligned. This makes them very difficult to break and, in addition, the alignment improves the polar interaction between the chains so that even when there is a chain break the adjacent chains hold the broken chain and resist parting of the structure. The crystalline nature of the nylon material also means that there is a larger stress capability at rapid loading since the crystalline areas react much more elastically than the amorphous glassy materials.

Other methods in which impact strength can be substantially improved are by the use of fibrous reinforcing fillers and product design. With reinforcements materials act as a stress transfer agent around the region that is highly stressed by the impact load. Since most of the fibrous fillers such as glass have high elastic moduli, they are capable of responding elastically at the high loading rates encountered in impact loading. Designwise prevent the formation of notched areas that act as stress risers. Especially under impact conditions the possibility of localized stress intensification can lead to product failure. In almost every case the notched strength is substantially less than the unnotched strength.

Impulse

Impulse loading differs from impact loading. The load of two billiard balls striking is an impact condition. The load applied to an automobile brake shoe when the brake load is applied or the load applied to a fishing line when a strike is made is an impulse load. The time constants are short but not as short as the impact load and the entire structural element is subjected to the stress.

It is difficult to generalize as to whether a plastic is stronger under impulse loading than under impact loading. Since the entire load is applied to the elastic elements in the structure the plastic will exhibit a high elastic modulus and much lower strain to rupture. For example acrylic and rigid PVC (polyvinyl chloride) that appear to be brittle under normal loading conditions, exhibit high strength under impulse loading conditions. Rubbery materials such as TP polyurethane elastomers and other elastomers behave like brittle materials under impulse loading. This is an apparently unexpected result that upon analysis is obvious because the elastomeric rubbery response is a long time constant response and the rigid connecting polymer segments that are brittle are the ones that respond at high loading rates.

Impact loading implies striking the object and consequently there is a severe surface stress condition present before the stress is transferred to the bulk of the material. The impact load is applied instantly limiting the straining rate only by the elastic constants of the material being struck. A significant portion of the energy of impact is converted to heat at the point of impact and complicates any analytically exact treatment of the mechanics of impact. With impulse loading the load is applied at very high rates of speed limited by the member applying the load. However, the loading is not generally localized and the heat effects are similar to conventional dynamic loading in that the hysteresis characteristics of the material determines the extent of heating and the effects can be analyzed with reasonable accuracy.

Plastics generally behave in a much different manner under impulse loading than they do under loading at normal straining rates. Some of the same conditions occur as under impact loading where the primary response to load is an elastic one because there is not sufficient time for the viscoelastic elements to operate. The primary structural response in thermoplastic is by chain bending and by stressing of the crystalline areas of crystalline polymers. The response to loading is almost completely elastic for most materials, particularly when the time of loading is of the order of milliseconds.

Improvements made with respect to impact loading for structures such as fibers and orientations apply equally to impulse loading conditions. Crystalline polymers generally perform well under impulse loading, especially polar materials with high interchain coupling.

To design products subjected to impulse loading requires obtaining applicable data. High-speed testing machines are used to determine the response of materials at millisecond loading rates. If this type data is not available evaluation can be done from the results of the tensile impact test. The test should be done with a series of loads below break load, through the break load, and then estimating the energy of impact under the non-break conditions as well as the tensile impact break energy. Recognize that brittle plastics perform well and rubbery materials that would seem to be a natural for impulse loading are brittle.

Puncture

Puncture loading is very applicable in applications with sheet and film as well as thin-walled tubing or molding, surface skins of sandwich panels, and other membrane type loaded structures. The test involves a localized force that is applied by a relatively sharp object perpendicular to the plane of the plastic being stressed. In the case of a thin sheet or film the stresses cause the material to be (1) displaced completely away from the plane of the sheet (compressive stress under the point of the puncturing member) and (2) the restraint is by tensile stress in the sheet and by hoop stress around the puncturing member (part of the hoop stress is compressive adjacent to the point which changes to tensile stress to contain the displacing forces). Most cases fall somewhere between these extremes, but the most important conditions in practice involve the second condition to a larger degree than the first condition.

If the plastic is thick compared to the area of application of the stress, it is effectively a localized compression stress with some shear effects as the material is deformed below the surface of the sheet.

Plastics that are biaxially oriented have good puncture resistance. Highly polar polymers would be resistant to puncture failure because of their tendency to increase in strength when stretched. The addition of randomly dispersed fibrous filler will also add resistance to puncture loads.

Anisotropic materials will have a more complicated force pattern. Uniaxially oriented materials will split rather than puncture under \puncturing loading. To improve the puncture resistance materials are needed with high tensile strength. In addition, the material should have a high compression modulus to resist the point penetration into the material. Resistance to notch loading is also important.

Friction

Friction is the opposing force that develops when two surfaces move relative to each other. Basically there are two frictional properties exhibited by any surface; static friction and kinetic friction. The ranges of friction properties are rather extensive. Frictional properties of plastics are important in applications such as machine products and in sliding applications such as belting and structural units such as sliding doors. In friction applications suggested as well as in many others, there are important areas that concern their design approach.

It starts in plastic selection and modification to provide either high or low friction as required by the application. There is also determining the required geometry to supply the frictional force level needed by controlling contact area and surface quality to provide friction level. A controlling factor limiting any particular friction force application is heat dissipation. This is true if the application of the friction loads is either a continuous process or a repetitive process with a high duty cycle. The use of cooling structures either incorporated into the products or by the use of external cooling devices such as coolants or airflow should be a design consideration. For successful design the heat generated by the friction must be dissipated as fast as it is generated to avoid overheating and failure.

The relationship between the normal force and the friction force is used to define the coefficient of static friction. Coefficient of friction is the ratio of the force that is required to start the friction motion of one surface against another to the force acting perpendicular to the two surfaces in contact. Friction coefficients will vary for a particular plastic from the value just as motion starts to the value it attains in motion. The coefficient depends on the surface of the material, whether rough or smooth. These variations and others make it necessary to do careful testing for an application which relies on the friction characteristics of plastics. Once the friction characteristics are defined, however, they are stable for a particular material fabricated in a prescribed method.

The molecular level characteristics that create friction forces are the intermolecular attraction forces of adhesion. If the two materials that make up the sliding surfaces in contact have a high degree of attraction for each other, the coefficient of friction is high. This effect is modified by surface conditions and the mechanical properties of the materials. If the material is rough there is a mechanical locking interaction that adds to the friction effect. Sliding under these conditions actually breaks off material and the shear strength of the material is an important factor in the friction properties. If the surface is polished smooth the governing factor induced by the surface conditions is the amount of area in contact between the surfaces. In a condition of large area contact and good adhesion, the coefficient of friction is high since there is intimate surface contact. It is possible by the addition of surface materials that have high adhesion to increase the coefficient of friction.

If one or both of the contacting surfaces have a low compression modulus it is possible to make intimate contact between the surfaces which will lead to high friction forces in the case of plastics having good adhesion. It can add to the friction forces in another way. The displacement of material in front of the moving object adds a mechanical element to the friction forces.

In regard to surface contamination, if the surface is covered with a material that prevents the adhesive forces from acting, the coefficient is reduced. If the material is a liquid, which has low shear viscosity, the condition exists of lubricated sliding where the characteristics of the liquid control the friction rather than the surface friction characteristics of the plastics.

The use of plastics for gears and bearings is the area in which friction characteristics have been examined most carefully. As an example highly polar plastic such as nylons and the TP polyesters have, as a result of the surface forces on the material, relatively low adhesion for themselves and such sliding surfaces as steel. Laminated plastics make excellent gears and bearings. The typical coefficient of friction for such materials is 0.1 to 0.2. When they are injection molded (IM) the skin formed when the plastic cools against the mold tends to be harder and smoother than a cut surface so that the molded product exhibit lower sliding friction and are excellent for this type of application. Good design for this type of application is to make the surfaces as smooth as possible without making them glass smooth which tends to increase the intimacy of contact and to increase the friction above that of a fine surface.

To reduce friction, lubricants are available that will lower the friction and help to remove heat. Mixing of slightly incompatible additive materials such as silicone oil into an IM plastic are used. After IM the additive migrates to the surface of the product and acts as a renewable source of lubricant for the product. In the case of bearings it is carried still further by making the bearing plastic porous and filling it with a lubricating material in a manner similar to sintered metal bearings, graphite, and molybdenum sulfide are also incorporated as solid lubricants.

Fillers can be used to increase the thermal conductivity of the material such as glass and metal fibers. The filter can be a material like PTFE (polytetrafluoroethylene) plastic that has a much lower coefficient of friction and the surface exposed material will reduce the friction.

With sliding doors or conveyor belts sliding on support surfaces different type of low friction or low drag application is encountered. The normal forces are generally small and the friction load problems are of the adhering type. Some plastics exhibit excellent surfaces for this type of application. PTFEs (tetrafluoroethylene) have the lowest coefficient of any solid material and represent one of the most slippery surfaces known. The major problem with PTFE is that its abrasion resistance is low so that most of the applications utilize filled compositions with ceramic filler materials to improve the abrasion resistance.

In addition to PTFE in reducing friction using solid materials as well as films and coatings there are other materials with excellent properties for surface sliding. Polyethylene and the polyolefins in general have low surface friction, especially against metallic surfaces. UHMWPE (ultra high molecular weight polyethylene) has an added advantage in that it has much better abrasion resistance and is preferred for conveyor applications and applications involving materials sliding over the product. In the textile industry loom products also use this material extensively because it can handle the effects of the thread and fiber passing over the surface with low friction and relatively low wear.

There are applications where high frictions have applications such as in torque surfaces in clutches and brakes. Some plastics such as poly-urethanes and plasticized vinyl compositions have very high friction coefficients. These materials make excellent traction surfaces for products ranging from power belts to drive rollers where the plastics either drives or is driven by another member. Conveyor belts made of oriented nylon and woven fabrics are coated with polyurethane elastomer compounds to supply both the driving traction and to move the objects being conveyed up fairly steep inclines because of the high friction generated. Drive rollers for moving paper through printing presses, copy machines, and business machines are frequently covered with either urethane or vinyl to act as the driver members with minimum slippage.

Erosion

Friction in basically the effect of erosion forces such as wind driven sand or water, underwater flows of solids past plastic surfaces, and even the effects of high velocity flows causing cavitation effects on material surfaces. One major area for the utilization of plastics is on the outside of moving objects that range from the front of automobiles to boats, aircraft, missiles, and submarine craft. In each case the impact effects of the velocity driven particulate matter can cause surface damage to plastics. Stationary objects such as radomes and buildings exposed to the weather in regions with high and frequent winds are also exposed to this type of effect.

Hydrostatic
In applications where water is involved if the water does not wet the surface, the tendency will be to have the droplets that do not impact close to the perpendicular direction bounce off the surface with considerably less energy transfer to the surface. Non-wetting coatings reduce the effect of wind and rain erosion. Impact of air-carried solid particulate matter is more closely analogous to straight impact loading since the particles do not become disrupted by the impact. The main characteristic required of the material, in addition to not becoming brittle under high rate loading is resistance to notch fracture.

The ability to absorb energy by hysteresis effects is important, as is the case with the water. In many cases the best type of surface is an elastomer with good damping properties and good surface abrasion resistance. An example is polyurethane coatings and products that are excellent for both water and particulate matter that is air-driven. Besides such applications as vehicles, these materials are used in the interior of sand and shot blast cabinets where they are constantly exposed to this type of stress. These materials are fabricated into liners in hoses for carrying pneumatically conveyed materials such as sand blasting hoses and for conveyor hose for a wide variety of materials such as sand, grain, and plastics pellets.

The method of minimizing the effects of erosion produced when the surface impact loading by fluid-borne particulate matter, liquid or solid, or cavitation loading is encountered, relates to material selection and modification. The plastics used should be ductile at impulse loading rates and capable of absorbing the impulse energy and dissipating it as heat by hysteresis effects. The surface characteristics of the materials in terms of wettability by the fluid and frictional interaction with the solids also play a role. In this type of application the general data available for materials should be supplemented by that obtained under simulated use conditions since the properties needed to perform are not readily predictable.

Cavitation
Another rapid loading condition in underwater applications is the application of external hydrostatic stress to plastic structures (also steel, etc.). Internal pressure applications such as those encountered in pipe and tubing or in pressure vessels such as aerosol containers are easily treated using tensile stress and creep properties of the plastic with the appropriate relationships for hoop and membrane stresses. The application of external pressure, especially high static pressure, has a rather unique effect on plastics. The stress analysis for thick walled spherical and tubular structures under external pressure is available.

The interesting aspect that plastics have in this situation is that the relatively high compressive stresses increase the resistance of plastic materials to failure. Glassy plastics under conditions of very high hydrostatic stress behave in some ways like a compressible fluid. The density of the material increases and the compressive strength are increased. In addition, the material undergoes sufficient internal flow to distribute the stresses uniformly throughout the product. As a consequence, the plastic products produced from such materials as acrylic and polycarbonate make excellent view windows for undersea vehicles that operate at extreme depths where the external pressures are 7MPa (1000 psi) and more.

With increasing ship speeds, the development of high-speed hydraulic equipment, and the variety of modem fluid-flow applications to which metal materials are being subjected, the problem of cavitation erosion becomes more important since it was first reported during 1873 (Chapter 8). Erosion may occur in either internal-flow systems, such as piping, pumps, and turbines, or in external ones like ships' propellers.

This erosion action occurs in a rapidly moving fluid when there is a decrease in pressure in the fluid below its vapor pressure and the presence of such nucleating sources as minute foreign particles or definite gas bubbles. Result is the formation of vapor bubble that continues to grow until it reaches a region of pressure higher than its own vapor pressure at which time it collapses. When these bubbles collapse near a boundary, the high-intensity shock waves (rapid loading) that are produced radiate to the boundary, resulting in mechanical damage to the material. The force of the shock wave or of the impinging may still be sufficient to cause a plastic flow or fatigue failure in a material after a number of cycles.

Materials, particularly steel, in cavitating fluids results in an erosion mechanism that includes mechanical erosion and electrochemical corrosion. Protection against cavitation is to use hardened materials, chromium, chrome-nickel compounds, or elastomeric plastics. Also used are methods to reduce the vapor pressure with additives, add air to act as a cushion for the collapsing bubbles, reduce the turbulence, and/ or change the liquid's temperature.

Rain
As it has been reported since the 1940s as one walks through a gentle spring rain one seldom considers that raindrops can be small destructive "bullets" when they strike high-speed aircraft. These rapid loaded bullet-like raindrops can erode paint coatings, plastic products, and even steel, magnesium or aluminum leading edges to such an extent that the surfaces may appear to have been sandblasted. Even the structural integrity of the aircraft may be affected after several hours of flight through rain. Also affected are commercial aircraft, missiles, high-speed vehicles on the ground, spacecraft before and after a flight when rain is encountered, and even buildings or structures that encounter high-speed rainstorms. Critical situations can exist in flight vehicles, since flight performance can be affected to the extent that a vehicle can be destroyed.

First reports on rain erosion on aircraft were first reported during WW II when the B-29 bomber was flying over the Pacific Ocean. Aerodynamic RP radar wing-type shaped structure on the B-29 was flying at a so called (at that time) high-speed was completely destroyed by rain erosion (DVR was a flight engineer on B-29). The "Eagle Wing" radome all-weather bomber airplanes were then capable of only flying at 400 mph. The aluminum aerodynamic leading edges of wings and particularly of the glass-fiber-reinforced TP polyester-nose radomes were particularly susceptible to this form of degradation. The problem continues to exist, as can be seen on the front of commercial and military airplanes with their neoprene protective coated RP radomes; the paint coating over the rain erosion elastomeric plastic erodes and then is repainted prior to the catastrophic damage of the rain erosion elastomeric coating.

Extensive flight tests conducted to determine the severity of the rain erosion were carried out in 1944. They established that aluminum and RP leading edges of airfoil shapes exhibited serious erosion after exposure to rainfall of only moderate intensity. Inasmuch as this problem originally arose with military aircraft, the U.S. Air Force initiated research studies at the Wright-Patterson Development Center's Materials Laboratory in Dayton, Ohio (DVR department involved; young lady physicist actually developed the theory of rain erosion that still applies). It resulted in applying an elastomeric neoprene coating adhesively bonded to RP radomes. The usual 5 mil coating of elastomeric material used literally bounces off raindrops, even from a supersonic airplane traveling through rain. Even though a slight loss (1%/mil of coating) of radar transmission occurred it was better than losing 100% when the radome was destroyed.

To determine the type of physical properties materials used in this environment should have, it is necessary to examine the mechanics of the impact of the particulate matter on the surfaces. The high kinetic energy of the droplet is dissipated by shattering the drop, by indenting the surface, and by frictional heating effects. The loading rate is high as in impact and impulse loading, but it is neither as localized as the impact load nor as generalized as the impulse load. Material that can dissipate the locally high stresses through the bulk of the material will respond well under this type of load. The plastic should not exhibit brittle behavior at high loading rates.

In addition, it should exhibit a fairly high hysteresis level that would have the effect of dissipating the sharp mechanical impulse loads as heat. The material will develop heat due to the stress under cyclical load. Materials used are the elastomeric plastics used in the products or as a coating on products.

High performance

As reviewed throughout this book the high performance materials are engineering plastics such as polycarbonate, nylon, acetal, and reinforced plastic (RP). Data on these plastics are provided throughout this book. In this section information on RPs is presented since they can provide a special form of high performance material that provides a designer with different innovative latitudes of performances than usually reviewed in textbooks.

Reinforced Plastic

They are strong, usually inert materials bound into a plastic to improve its properties such as strength, stiffness/modulus of elasticity, impact resistance, reduce dimensional shrinkage, etc. (Figs 2.2, 2.23, & 2.24). They include fiber and other forms of material. There are inorganic and organic fibers that have the usual diameters ranging from about one to over 100 micrometers. Properties differ for the different types, diameters, shapes, and lengths. To be effective, reinforcement must form a strong adhesive bond with the plastic; for certain reinforcements special cleaning, sizing, etc. treatments are used to improve bonds. A microscopic view of an RP reveals groups of fibers surrounded by the matrix.

In general adding reinforcing fibers significantly increases mechanical properties. Particulate fillers of various types usually increase the modulus, plasticizers generally decrease the modulus but enhance flexibility, and so on. These reinforced plastics (RPs) can also be called composites. However the name composites literally identifies thousands of different combinations with very few that include the use of plastics. In using the term composites when plastics are involved the more appropriate term is plastic composite.

Figure 2.23 RPs tensile S-S data (Courtesy of Plastics FALLO)

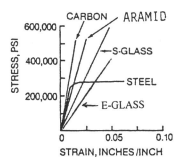

Figure 2.24 Properties of RPs and other materials (Courtesy of Plastics FALLO)

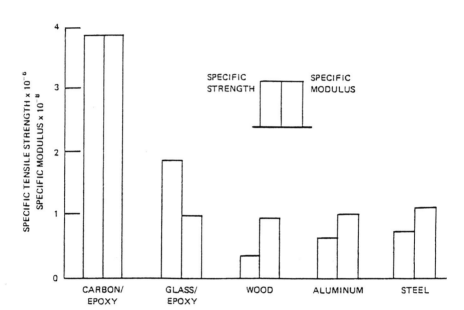

Types of reinforcements include fibers of glass, carbon, graphite, boron, nylon, polyethylene, polypropylene, cotton, sisal, asbestos, metals, whiskers, etc. Other types and forms of reinforcements include bamboo, burlap, carbon black, platelet forms (includes mica, glass, and aluminum), fabric, and hemp. There are whiskers that are metallic or nonmetallic single crystals (micrometer size diameters) of ultrahigh strength and modulus. Their extremely high performances (high modulus of elasticity, high melting points, resistance to oxidation, low weights, etc.) are attributed to their near perfect crystal structure, chemically pure nature, and fine diameters that minimize defects. They exhibit a much higher resistance to fracture (toughness) than other types of reinforcing fibers (Chapter 1).

The advanced RP (ARP) refers to a plastic matrix reinforced with very high strength, high modulus fibers that include carbon, graphite, aramid, boron, and S-glass. They can be at least 50 times stronger and 25 to 150 times stiffer than the matrix. ARPs can have a low density (1 to 3 g/cm^3), high strength (3 to 7 GPa) and high modulus (60 to 600 GPa).

It can generally be claimed that fiber based RPs offer good potential for achieving high structural efficiency coupled with a weight saving in products, fuel efficiency in manufacturing, and cost effectiveness during service life. Conversely, special problems can arise from the use of RPs, due to the extreme anisotropy of some of them, the fact that the strength of certain constituent fibers is intrinsically variable, and because the test methods for measuring RPs' performance need special consideration if they are to provide meaningful values.

Orientation of Reinforcement

RPs behavior is dominated by the arrangement and the interaction of the stiff, strong reinforcing fibers with the less stiff, weaker plastic matrix. The fiber arrangement determines the behavior of RPs where a major advantage is that directional properties can be maximized. Arrangements include the use of woven (with different weaves) and nonwoven (with different lengths and forms) fabrics.

Design theories of combining actions of plastics and reinforcement arrangements have been developed and used successfully. Theories are available to predict overall behavior based on the properties of fiber and matrix. In a practical design approach, the behavior can use the original approach analogous to that used in wood for centuries where individual fiber properties are neglected; only the gross properties, measured at various directions relative to the grain, are considered. This was the initial design evaluation approach used during the 1940s.

Orientation Terms

Orientation terms of RP directional properties include the following:

Anisotropic construction RP properties are different in different directions along the laminate flat plane.

Balanced construction RP in which properties are symmetrical along the laminate flat plane.

Bidirectional construction RP with the fibers oriented in various directions in the plane of the laminate usually identifying a cross laminate with the direction 90° apart.

Heterogeneous construction RP material's composition varies from point to point in a heterogeneous mass.

Homogeneous construction Uniform RP.

Isotropic construction RPs having uniform properties in all directions along the laminate flat plane.

Nonisotropic construction RP does not have uniform properties in all directions.

Orthotropic construction RP having mutually perpendicular planes of elastic symmetry along the laminate flat plane.

Unidirectional, construction Refers to fibers that are oriented in the same direction (parallel alignment) such as filament-winding, pultrusion, unidirectional fabric laminate, and tape.

RPs can be constructed from a single layer or built up from multiple layers using fiber preforms, nonwoven fabrics, and woven fabrics. In many products woven fabrics are very practical since they drape better over 3-D molds than constructions that contain predominantly straight fibers. However they include kinks where fibers cross. Kinks produce repetitive variations and induce local stresses in the direction of reinforcement with some sacrifice in properties. Regardless, extensive use of fabrics is made based on their advantages.

The glass content of a part has a direct influence on its mechanical properties where the more glass results in more strength. This relates to the ability to pack the reinforcement. Fiber content can be measured in percent by weight of the fiber portion (wt%) or percent by volume (vol%). (Fig. 2.25) When content is only in percent, it usually refers to wt%. Depending on how glass fibers are arranged content can range from 65 to 95.6 wt% or up to 90.8 vol%. When one-half of the strands are placed at right angles to each half, glass loadings range from 55 to 88.8 wt% or up to 78.5 vol% (Fig. 2.26).

Basic Design Theory

In designing RPs, certain important assumptions are made so that two materials act together and the stretching, compression, twisting of fibers and of plastics under load is the same; that is, the strains in fiber and plastic are equal. Another assumption is that the RP is elastic, that is, strains are directly proportional to the stress applied, and when a load is removed the deformation disappears. In engineering terms, the material obeys Hooke's Law. This assumption is a close approximation to the actual behavior in direct stress below the proportional limit, particularly in tension, where the fibers carry essentially all the stress. The assumption is possibly less valid in shear where the plastic carries a substantial portion of the stress.

In this analysis it is assumed that all the glass fibers are straight; however, it is unlikely that this is true, particularly with fabrics. In practice, the load is increased with fibers not necessarily failing at the same time. Values of a number of elastic constants must be known in addition to strength properties of the resins, fibers, and combinations. In this analysis, arbitrary values are used that are low for elastic constants and strength values. Any values can be used; here the theory is illustrated.

Figure 2.25 Weight to volume relation example for filament wound fabricated products

Figure 2.26 Fiber arrangement influences properties

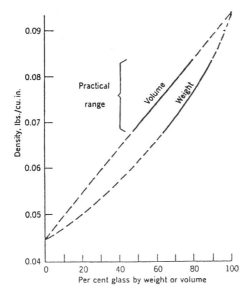

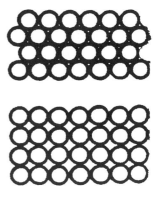

Any material, when stressed, stretches or is otherwise deformed. If the plastic and fiber are firmly bonded together, the deformation is the same. Since the fiber is more unyielding, a higher stress is developed in the glass than the plastic. If the stress-strain relationships of fiber and plastic are known, the stresses developed in each for a given strain can be computed and their combined action determined. Fig. 2.27 stress-strain (S-S) diagrams provide the basis for this analysis; it provides related data such as strengths and modulus.

These S-S diagrams may be applied to investigate a rod in which half of the volume is glass and the other half is plastic. If the fibers are parallel to the axis of the rod, at any cross-section, half of the total is fiber with half plastic. If the rod is stretched 0.5%, the S-S diagrams show that the glass is stressed to 50,000 psi (345 MPa), resin B at 7,500 psi (52 MPa), and resin C at 2,500 psi (17 MPa). If the rod has a total cross-section of $\frac{1}{2}$ in^2, the glass is $\frac{1}{4}$ in^2. The total load on the glass is $\frac{1}{4}$ × 50,000 or 12,500 lb. Similarly resin B is 1,875 lb and resin C is 625 lb.

The load required to stretch the rod made of resin B becomes the sum of glass and resin load or 14,375 lb. With resin C the load is 13,125 lb.

The foregoing can be put into the form of an equation:

$$\sigma A = \sigma_f A_f + \sigma_r A_r \tag{2-15}$$

Figure 2.27 Analysis of RPs stress-strain curves (Courtesy of Plastics FALLO)

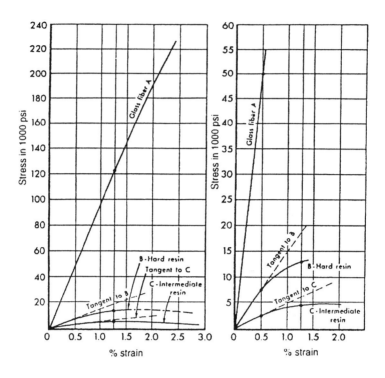

σ	=	mean stress in tensity on entire cross-section
σ_f	=	stress intensity in fiber
σ_r	=	stress intensity in resin
A	=	total cross-sectional area
A_f	=	cross-sectional area of fiber
A_r	=	cross-sectional area of resin

If the moduli of elasticity, as measured by the tangents to the S-S diagrams, are known the following equations are obtained:

$$\frac{\sigma_r}{\sigma_f} = \frac{E_r}{E_f}, \quad \text{or} \quad \sigma_r = \frac{E_r}{E_f}\sigma_f \tag{2-16}$$

E_r	=	modulus of elasticity of resin
E_f	=	modulus of elasticity of fiber

Substituting (2-16) in (2-15) results in:

$$\sigma A = \sigma_f \left(A_f + \frac{E_r}{E_f}A_r \right) \tag{2-17}$$

Referring to Fig. 2.27, the tangent to the S-S curve for glass gives a value of $E_f = 10 \times 10^6$ psi. The resin tangents are given for B and C at 1.5×10^6 psi and 0.5×10^6 psi, respectively. Substituting these values in

Eq. (2-17) results in:

$$\text{Resin B} \qquad \sigma A = 50{,}000 \left(0.25 + \frac{1.5}{10} . 0.25 \right) \qquad \text{(2-18)}$$

$$= 14{,}375 \text{ lb}$$
$$\text{or } \sigma = 28{,}750 \text{ psi}$$

$$\text{Resin C} \qquad \sigma A = 50{,}000 \left(0.25 + \frac{0.5}{10} . 0.25 \right) \qquad \text{(2-19)}$$

$$= 13{,}125 \text{ lb}$$
$$\sigma = 26.250 \text{ psi}$$

Average values of modulus of elasticity of the entire cross-section may be computed by dividing σ by the strain. The strain is 0.5%, therefore the two average values of E of the rod, incorporating resins B and C, are 5.75×10^6 psi and $5.35 \times 10^{6-}$ psi, respectively.

For a cross-section made up of a number of different materials, Eq. (2-15) may be generalized to:

$$\sigma A = \sum_{i=1}^{i=n} \sigma_i A_i \qquad \text{(2-20)}$$

in which σ is the tensile strength and A_i the cross-sectional area of any component of the cross-section. This equation can be still further generalized to include tension, compression, and shear:

$$SA = \sum_{i=1}^{i=n} S_i A_i \qquad \text{(2-21)}$$

in which S_i is the strength property of the cross-sectional area A_i, and S is the mean strength property over the entire cross-section A.

Similar to finding the overall modulus of a cross-section, the equation becomes:

$$EA = \sum_{i=1}^{i=n} E_i A_i \qquad \text{(2-22)}$$

in which E is the overall modulus of elasticity, A the total cross-section, and E_i the modulus of elasticity corresponding to the partial cross-sectional area A_i. For shear modulus G the equation becomes:

$$GA = \sum_{i=1}^{i=n} G_i A_i \qquad \text{(2-23)}$$

Fiber Strength Theory

The deformation and strength of filamentary structures subjected to combined loading can be theoretically predicted using experimentally-determined intrinsic stiffnesses and strength of the individual constituent layers. In order to have an integrated material and structure design, the gross properties as functions of the micromechanical parameters represent an important issue on the continuing and expanding use of RPs. It has been established, both in theory and experiment, that four principal elastic moduli and three principal strengths govern the deformation and strength of unidirectional fiber RPs. With the aid of a yield condition, the initial failure of filamentary structures can be predicted. After the initial failure, the structure may carry additional loads. An analysis of a partially failed or degraded structure can be used to predict the ultimate deformation and strength.

With an understanding of the gross behavior of a filamentary structure, a proper assessment of the mechanical and geometric properties of the constituent materials is possible. In particular, the use of fiber strength, the binding resin matrix, and the interface may be placed in a perspective based on the results of a mathematical analysis. They provide accurate guidelines for the design of RPs.

A better understanding exists of the elastic stiffness of filamentary materials than of the strengths. The generalized Hooke's law is usually accepted as the governing equation of the linear elastic deformation of RP materials. The simultaneous or sequential modes of deformation and fracture are difficult to describe from the phenomenological standpoint. In general, a strength theory on one criterion will not be sufficient to cover the entire range of failure modes of RP. In addition, fabrication variables and test methods are also known to introduce uncertainties in strength data that makes the verification of theories more difficult.

A macroscopic theory of strength is based on a phenomenological approach. No direct reference to the mode of deformation and fracture is made. Essentially, this approach employs the mathematical theories of elasticity and tries to establish a yield or failure criterion. Among the most popular strength theories are those based on maximum stress, maximum strain, and maximum work. The maximum stress theory states that, relative to the material symmetry axes x-y, failure of the RP will occur if one of three ultimate strengths is reached. There are three inequalities, as follows:

$$\sigma_x \leq X \qquad\qquad\qquad (2\text{-}24)$$

$$\sigma_y \leq Y \qquad\qquad\qquad (2\text{-}25)$$

$$\sigma_s \leq S \qquad\qquad\qquad (2\text{-}26)$$

With negative normal stress components, compressive strengths designated by X' and Y' must be used:

$$\sigma_z \leq X' \tag{2-27}$$

$$\sigma_y \leq Y' \tag{2-28}$$

Shear strength S has no directional property and it retains the same value for both positive and negative shear stress components.

The maximum strain theory is similar to the maximum stress theory. Associated with each strain component, relative to the material symmetry axes, e_x, e_y, or e_s, there is an ultimate strain or an arbitrary proportional limit, X_e, Y_e, or S_e, respectively. The maximum strain theory can be expressed in terms of the following inequalities:

$$e_x \leq X_e \tag{2-29}$$

$$e_y \leq Y_e \tag{2-30}$$

$$e_x \leq S_e \tag{2-31}$$

Where e_x and y_y are negative, use the following inequalities:

$$e_x \leq X'_e \tag{2-32}$$

$$e_y \leq Y'_e \tag{2-33}$$

The maximum work theory in plane stress takes the following form:

$$\left(\frac{\sigma_x}{X}\right)^2 - \left(\frac{\sigma_z}{X}\right)\left(\frac{\sigma_y}{X}\right) + \left(\frac{\sigma_y}{Y}\right)^2 + \left(\frac{\sigma_s}{S}\right)^2 = 1 \tag{2-34}$$

If σ_x and σ_y are negative, compressive strengths X' and Y' should be used in Eq. 2-34, respectively.

In the following reviews, the tensile and compressive strengths of unidirectional and laminated RPs, based on the three theories, is computed and compared with available data obtained from glass fiber-epoxy RPs. The uniaxial strength of unidirectional RPs with fiber orientation θ can be determined according to the maximum theory. Strength is determined by the magnitude of each stress component according to Eqs. 2-24, 2-25, and 2-26 or Eqs. 2-27 and 2-28. As fiber orientation varies from 0° to 90°, it is only necessary to calculate the variation of the stress components as a function of θ. This is done by using the usual transformation equations of a second rank tensor, thus:

$$\sigma_x = \sigma_1 \cos^2 \theta \tag{2-35}$$

$$\sigma_x = \sigma_1 \sin^2 \theta \tag{2-36}$$

$$\sigma_x = \sigma_1 \sin \theta \cos \theta \tag{2-37}$$

where σ_x, σ_y, σ_s are the stress components relative to the material symmetry axes, i.e., σ_x is the normal stress along the fibers, σ_y, transverse to the fibers, σ_s, the shear stress; σ_1 = uniaxial stress along to the test specimen. Angle θ is measured between the 1-axis and the fiber axis. By combining Eqs. 2-35, 2-36, and 2-37 with 2-24, 2-25, and 2-26, the uniaxial strength is determined by:

$$\sigma_1 \leq X/\cos^2 \theta \tag{2-38}$$

$$\leq Y/\sin^2 \theta \tag{2-39}$$

$$\leq S/(\sin \theta \cos^2 \theta) \tag{2-40}$$

The maximum strain theory can be determined by assuming that the material is linearly elastic up to the ultimate failure. The ultimate strains in Eqs. 2-29, 2-30, and 2-31 as well as 2-32 and 2-33 can be related directly to the strengths as follows:

$$X_e = X/E_{11} \tag{2-41}$$

$$Y_e = X/E_{22} \tag{2-42}$$

$$S_e = S/G \tag{2-43}$$

The usual stress-strain relations of orthotropic materials is:

$$e_x = \frac{1}{E_{11}} (\sigma_x - v_{12}\sigma_y) \tag{2-44}$$

$$e_y = \frac{1}{E_{22}} (\sigma_y - v_{12}\sigma_x) \tag{2-45}$$

$$e_s = \frac{1}{G} \sigma_s \tag{2-46}$$

Substituting Eq. 2-35, 2-36, and 2-37 into 2-44, 2-45, and 2-46 results in,

$$e_x = \frac{1}{E_{11}} (\cos^2 \theta - v_{12} \sin^2 \theta)\sigma_1 \tag{2-47}$$

$$e_y = \frac{1}{E_{22}} (\sin^2 \theta - v_{21} \cos^2 \theta)\sigma_1 \tag{2-48}$$

$$e_s = \frac{1}{G} (\sin \theta \cos \theta)\sigma_1 \tag{2-49}$$

Finally, substituting Eqs. 2-47, 2-48, and 2-49 and 2-41, 2-42, and 2-43 into Eqs. 2-29, 2-30, and 2-31, and after rearranging, one obtains the uniaxial strength based on the maximum theory:

$$\sigma_1 \leq X/(\cos^2 \theta - v_{12}\sin^2 \theta) \tag{2-50}$$

$$\leq Y/(\sin^2 \theta - v_{21}\cos^2 \theta) \tag{2-51}$$

$$\leq X/(\sin \theta - \cos \theta) \tag{2-52}$$

The maximum work theory can be obtained directly by substituting Eq. 2-35, 2-36, and 2-37 into Eq. 2-34:

$$\frac{1}{\sigma_1^2} = \frac{\cos^4\theta}{X^2} + \left(\frac{1}{S^2} - \frac{1}{X^2}\right)\cos^2\theta \sin^2\theta + \frac{\sin^4\theta}{Y^2} \tag{2-53}$$

Determining the strength of laminated RPs is no more difficult conceptually than determining the strength of unidirectional RPs. It is only necessary to determine the stress and strain components that exist in each constituent layer. Strength theories can then be applied to ascertain which layer of the laminated composite has failed. Stress and strain data is obtained for E-glass-epoxy, and cross-ply and angle-ply RPs. Under uniaxial loading, only N, is the nonzero stress resultant and when temperature effect is neglected, the calculations become:

$$e_i = (A'_{i1} + zB'_{i1})N_1 \tag{2-54}$$

$$\sigma_i^{(k)} = c_{ij}^{(k)} [A'_{j1} + zB^j_{j1}]N_1 \tag{2-55}$$

where A' and B' matrices are the in-plane and coupling matrices of a laminated anisotropic composite.

The stress and strain components can be computed from Eqs. 2-54 and 2-55. They can then be substituted into the strength theories, from which the maximum Ni, the uniaxial stress resultant can be determined. Uniaxial tensile strengths of unidirectional and laminated composites made of E-glass-epoxy systems are obtained. Also, uniaxial axial-compressive strengths are obtained. The three strength theories can be applied to the glass-epoxy RP by using the following material coefficients:

$$
\begin{aligned}
E_{11} &= 7.8 \times 10^6 \text{ psi} \\
E_{22} &= 2.6 \times 10^6 \text{ psi} \\
v_{12} &= 0.25 \\
G &= 1.25 \times 10^6 \text{ psi}
\end{aligned}
\tag{2-56}
$$

$$
\begin{aligned}
X &= 150 \text{ ksi} \\
X' &= 150 \text{ ksi} \\
Y &= 4 \text{ ksi} \\
Y' &= 20 \text{ ksi} \\
S &= 8 \text{ ksi}
\end{aligned}
$$

The maximum stress theory is shown as solid lines in Fig. 2.28. On the right-hand side of the figure is the uniaxial strength of directional RPs with fiber orientation θ from 0° to 90°; on the left-hand side, laminated RPs with helical angle α from 0° to 90°. Both tensile and compressive loadings are shown. The tensile data are the solid circles and the com-

Figure 2.28 Maximum stress theory

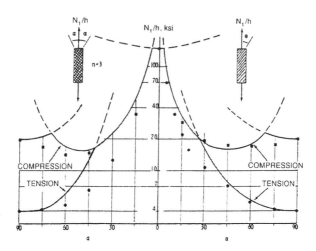

pressive are squares. Tensile data are obtained from dog-bone specimens. Compressive data are from specimens with uniform rectangular cross-sections.

Figure 2.29 shows the comparison between the maximum strain theory and the same experimental data shown in Fig. 2.28. The formats are similar. Fig. 2.28 shows a comparison between the maximum work theory of the same experimental data as shown in Figs 2.29 and 2.30.

Based on a Tsai review, it shows that the maximum work theory is more accurate than the maximum stress and strain theories. The maximum work theory encompasses the following additional features.

1. There is a continuous variation, rather than segmented variation, of the strength as a function of either the fiber orientation θ or helical angle α.

2. There is a continuous decrease as the angles θ and α deviate from 0°. There is no rise in axial strength, as indicated by the maximum stress and strain theories.

3. The uniaxial strength is plotted on a logarithmic scale and an error of a factor of 2 exists in the strength prediction of the maximum stress and strain theories in the range of 30°.

4. A fundamental difference between the maximum work and the other theories lies in the question of interaction among the failure modes. The maximum stress and strain theories assume that there is no interaction among the three failure modes (axial, transverse, and shear failures).

Figure 2.29 Maximum strain theory

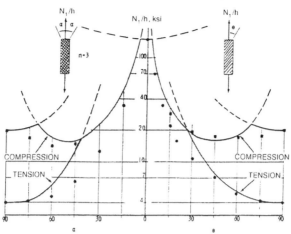

Figure 2.30 Maximum work theory

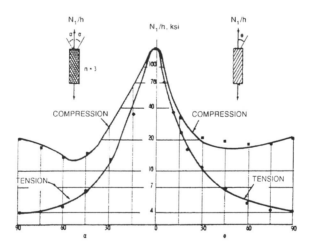

5. In the limit, when:

$$X = Y = \sqrt{3}S \qquad (2\text{-}57)$$

which corresponds to isotropic materials, equation 2-53 becomes,

$$\sigma_1 = X \qquad (2\text{-}58)$$

This means that the axial strength is invariant. If equation 2-57 is substituted into equation 2-38, 2-39, and 2-40 thus:

$$\sigma_1 \leq X/\cos^2\theta \qquad (2\text{-}59)$$
$$\leq X/\sin^2\theta$$
$$\leq X/\sqrt{3}\sin\theta\cos\theta$$

The angular dependence does not vanish, that should not be the case for an isotropic material.

Fiber Geometry on Strength

Various investigators have developed mathematical means for determining the efficiency of glass-fiber RPs. In order to analyze the effect of fiber geometry on strength, the fundamental mechanics of RP theory is reviewed. Relationships have been derived to relate the load distribution in an RP to the properties of the individual materials. The derivations are based on the following: (1) stress is proportional to the strain in both materials; (2) resin-fiber bond is efficient, so that resin and fiber are strained an equal amount under load; (3) fibers are straight, continuous, and aligned with the axis of the applied load; and (4) material components are isotropic and homogeneous. The nomenclature used is as follows:

A_c	Area of comosite, inches^2
A_{ft}	Total fiber area, inches^2
A_f	Area of fiber in load direction, inches^2
A_m	Area of matrix, inches^2
D	Fiber diameter, inches
E_c	Modulus of elasticity of composite, psi (tension)
E_f	Modulus of elasticity of fiber, psi (tension)
E_m	Modulus of elasticity of matrix, psi (tension)
F_f	Base strength of the fiber, psi
F_m	Base strength of the matrix, psi
F_c	Theoretical composite strength, psi
h	Height of shear plane, inches
L_s	Length of shear plane and required overlap of fibers
P_c	Load of composite, pounds
P_f	Load on fiber, pounds
P_m	Load on matrix, pounds
s_c	Unit stress in composite, psi
s_f	Unit stress in fiber, psi
s_m	Unit stress in matrix, psi
e_c	Unit strain of composite, inches/inches
e_f	Unit strain of fiber, inches/inches
e_m	Unit strain of matrix, inches/inches

The derivations are as follows:

$$s = Ee \qquad (2\text{-}60)$$

where E is the proportionality constant or the modulus of elasticity. For RPs

$$s_c = E_c e_c \qquad (2\text{-}61)$$

and the stress in the fiber and matrix is

$$s_f = E_f e_f \qquad (2\text{-}62)$$

$$s_m = E_m e_m \qquad (2\text{-}63)$$

By assumption

$$e_f = e_e = e_m \qquad (2\text{-}64)$$

Thus Eq. 2-62 and 2-63 can be written as

$$s_f = E_f e_c \qquad (2\text{-}65)$$

$$s_f = E_m e_c \qquad (2\text{-}66)$$

Since the load is equal to unit stress times area, the load on the fiber is:

$$P_f = s_f A_f = E_f e_c A_f \qquad (2\text{-}67)$$

and the load on the matrix is:

$$P_m = s_m A_m = E_m e_c A_m \qquad (2\text{-}68)$$

The load applied to the composite, P_o is resisted by the resisting loads in the fiber and matrix; therefore the following equation exists:

$$P_c = P_f + P_m \qquad (2\text{-}69)$$

The ratio of the load carried by the fibers to the applied load is

$$\frac{P_f}{P_c} = \frac{P_f}{P_f + P_m} \qquad (2\text{-}70)$$

and substituting for P_f and P_m

$$\frac{P_f}{P_c} = \frac{E_f A_f}{E_f Af + E_m A_m} \qquad (2\text{-}71)$$

Equation 2-71 can be further simplified by assuming the composite to have an area of one square inch. Thus:

$$A_f + A_m = A_c = 1 \qquad (2\text{-}72)$$

Equation 2-71 can now be written as:

$$\frac{P_f}{P_c} = \frac{1}{1 + \dfrac{E_m}{E_f}\left(\dfrac{1}{A_f} - 1\right)} \qquad (2\text{-}73)$$

The ratio of fiber stress to composite stress can be determined by dividing the fiber and composite loads by their respective area, thus

$$\frac{S_f}{S_c} = \frac{P_f/A_f}{P_c/A_c} = \frac{A_c}{A_f} \frac{E_f A_f}{E_f A_f + E_m (1 - A_f)} \tag{2-74}$$

and since $A_c = 1$

$$\frac{S_f}{S_c} = \frac{1}{A_f + \frac{E_m}{E_f}(1 - A_f)} \tag{2-75}$$

It can be concluded from equation 2-73 that the percentage of the applied load carried by the fiber is a function of the relative moduli of matrix and fiber and also a function of the area fiber resisting the applied load. The same statement is true for the ratio of the stress and the fibers to the stress in the composite. By equation 2-75, it is determined that the stress in the fiber increases as E_m/E_f decreases and A_f decreases.

Continuous fibers, such as those in filament winding, cross laminates, and cloth laminates, can transmit the applied load or stress from the point of application to the reaction via a continuous load path. If the fibers are not continuous between the load and the reaction, the matrix must transfer the load from one fiber to the next at the points of discontinuity. Fiber continuity also affects the type of failure of the composite.

With continuous fibers, it can be assumed that the failure will ultimately occur by fracture of the fibers. Discontinuous fibers, on the other hand, can have three other types of failures: (1) fracture of the resin at a weak net section; (2) shear failure in the matrix at the points of discontinuity of the fiber; and (3) failure of the bond between the fibers and the matrix.

The theoretical composite strength is defined as the sum of the strengths of the fiber and matrix materials. This can be written as

$$P_c = A_{ft}F_f + A_m F_m \tag{2-76}$$

$$F_c = A_{ft}F_f + A_m F_m \tag{2-77}$$

(where A_{ft} and A_m are part of a unit area) when the composite is assumed to have a unit area. The composite efficiency is the ratio of the composite strength as tested to the simple composite theoretical strength expressed in percent. Thus,

$$\text{composite efficiency} = \frac{\text{test strength of the composite}}{\text{theoretical composite strength}} \times 100 \tag{2-78}$$

The effective fiber stress can be determined from the load in the fiber and the fiber area. The percentage of the applied load that is carried by the fiber is dependent on E_m/E_f and A_f. This load divided by the fiber area is the effective fiber stress. Thus,

$$s_f(\text{effective}) = \frac{P_f}{P_c} \times \frac{\text{test strength composite}}{A_f} \qquad (2\text{-}79)$$

Fiber efficiency can now be defined as the ratio of the developed fiber stress to the base strength of the fibers. Thus,

$$\text{fiber efficiency} = \frac{\text{developed fiber stress}}{\text{basic fiber strength}} \times 100 \qquad (2\text{-}80)$$

RP efficiency is based on the total glass content plus the total resin matrix content, and fiber efficiency is based on the glass area oriented in the load direction. The average tensile strength of glass fibers in their several common forms are about 500 000 psi for a virgin single filament, 40 000 psi for single glass roving, and 250 000 psi for glass strands (as woven into cloth). As a basis for comparing fiber geometries, the base glass strength will be assumed as 400 000 psi for glass roving. Single-glass filaments are not practical to handle, and glass strands that are used in cloth have undergone the first phase of fabrication.

Stiffness-Viscoelasticity

The stiffness response of RPs can be identified as viscoelasticity. RPs are nearly elastic in behavior and tend to reduce the importance of the time-dependent component of viscoelastic behavior. Also, the stiffness of fiber reinforcements and the usual TS resin matrices are less sensitive to temperature change than most unreinforced plastics. The stiffness of both the fibers and the matrices are frequently more stable on exposure to solvents, oils, and greases than TPs although for certain composites water, acids, bases, and some strong solvents still may alter stiffness properties significantly.

Stiffness properties of RPs are used (as with other materials) for the usual purposes of estimating stresses and strains in a structural design, and to predict buckling capacity under compressive loads. Also, stiffness properties of individual plies of a layered "flat plate approach" may be used for the calculation of overall stiffness and strength properties. The relationship between stress and strain of unreinforced or reinforced plastics varies from viscous to elastic. Most RPs, particularly RTSs are intermediate between viscous and elastic. The type of plastic, stress, strain, time, temperature, and environment all influence the degree of their viscoelasticity.

Creep and Stress Relaxation
Properties of unreinforced plastics are strongly dependent on temperature and time. This is also true, to a lesser degree, for RPs, particularly RTSs,

compared with other materials, such as steel. This strong dependence of properties on temperature and how fast the material is deformed, based on a time scale, is a result of the viscoelastic nature of plastics. Consequently, it is important in practice to know how the product is likely to be loaded with respect to time.

In structural design, it is important to distinguish between various modes in the product. The behavior of any material in tension, for example, is different from its behavior in shear, as with plastics, metals, concrete, etc. For viscoelastic materials such as plastics, the history of deformation also has an effect on the response of the material, since viscoelastic materials have time- and temperature-dependent material properties.

Conceptual design approach

A skilled designer blends knowledge of materials, an understanding of manufacturing processes, and imagination of new or innovative designs. It is the prediction of performance in its broadest sense, including all the characteristics and properties of materials that are essential and relate to the processing of the plastic. To the designer, an example of a strict definition of a design property could be one that permits calculating of product dimensions from a stress analysis. Such properties obviously are the most desirable upon which to base material selections. These correlative properties, together with those that can be used in design equations, generally are called engineering properties. They encompass a variety of stress situations over and above the basic static strength and rigidity, such as impact, fatigue, high and low temperature capability, flammability, chemical resistance, and arc resistance.

Recognize that there are many stresses that cannot be accurately analyzed in plastics, metals, aluminum, etc. Thus one relies on properties that correlate with performance requirements. Where the product has critical performance requirements, such as ensuring safety to people, production prototypes will have to be exposed to the requirements it is to meet in service.

Design Analysis

The designer starts by one visualizing a certain family of material, makes approximate calculations to see if the contemplated idea is practical to meet requirements that includes cost, and, if the answer is favorable, proceeds to collect detailed data on a range of materials that may be considered for the new product. When plastics are the candidate materials, it must be recognized from the beginning that the available test data require understanding and proper interpretation before an attempt can be made to apply them to the initial product design. For this reason, an explanation of data sheets is required in order to avoid anticipating product characteristics that may not exist when merely applying data sheet information without knowing how such information was derived. The application of appropriate data to product design can mean the difference between the success and failure of manufactured products made from any material (plastic, steel, etc.).

In structural applications for plastics, which generally include those in which the product has to resist substantial static and/or dynamic loads, it may appear that one of the problem design areas for many plastics is their low modulus of elasticity. The moduli of unfilled plastics are usually under 7×10^3 MPa (1×10^6 psi) as compared to materials such as metals and ceramics where the range is usually 7 to 28×10^4 MPa (10 to 40×10^6 psi). However with reinforced plastics (RPs) the high moduli of metals are reached and even surpassed as summarized in Fig. 2.3.

Since shape integrity under load is a major consideration for structural products, low modulus plastic products are designed shapewise for efficient use of the material to afford maximum stiffness and overcome their low modulus. These type plastics and products represent most of the plastic products produced worldwide.

With the structural analysis reviewed characteristics or behaviors of plastics are included. These characteristics or behaviors are reviewed throughout this book. The following information provides examples of what could apply in a design.

The value of heat insulation is fully appreciated in the use of plastic drinking cups and of plastic handles on cooking utensils, electric irons, and other devices where heat can cause discomfort or burning. In electrical devices the plastic material's application is extended to provide not only voltage insulation where needed, but also the housing that would protect the user against accidental electrical grounding. In industry the thermal and electrical uses of plastics are many, and these uses usually combine additional features that prove to be of overall benefit.

Corrosion resistance and color are extremely important in many products. Protective coatings for most plastics are not required owing to their inherent corrosion-resistant characteristics. The eroding effects of rust are well known with certain materials, and materials such as certain plastics that do not deteriorate offers distinct advantages. Colors for esthetic appearance are incorporated in the material compound and become an integral part of the plastic for the life of the product.

Those with transparency capabilities provide many different products that include transportation vehicle lighting, camera lenses, protective shields (high heat resistance, gunfire, etc.), etc. When transparency is needed in conjunction with toughness and safety, plastic materials are the preferred candidates. Add to the capability of providing simple to very complex shapes.

Other important properties for certain products include coefficient of friction, chemical resistance, and others. Many plastic materials inherently have a low coefficient of friction. Other plastic materials can incorporate this property by compounding a suitable ingredient such as graphite powder into the base material. It is an important feature for moving products, which provides for self-lubrication. Chemical resistance is another characteristic that is inherent in most plastic materials; the range of this resistance varies among materials.

Materials that have all these favorable properties also have their limitations. As with other materials, every designer of plastic products has to be familiar with their advantages and limitations. It requires being cautious and providing attention to all details – nothing new since this is what designers have been doing for centuries with all kinds of materials if they want to be successful.

Pseudo–Elastic Method

As reviewed viscoelastic behavior relates to deformations that are dependent on time under load and the temperature. Therefore, when structural components are to be designed using plastics it must be remembered that the standard engineering equations that are available (Figs 2.31 and 2.32) have been derived under the assumptions that (1) the strains are small, (2) the modulus is constant, (3) the strains are independent of the loading rate or history and are immediately reversible, (4) the material is isotropic, and (5) the material behaves in the same way in tension and compression.

These equations cannot be used indiscriminately. Each case must be considered on its merits, with account being taken of the plastic behavior in time under load, mode of deformation, static and/or dynamic loads, service temperature, fabrication method, environment, and others. The traditional engineering equations are derived using the relationship that stress equals modulus times strain, where the modulus is a constant. The moduli of many plastics are generally not a constant. Several approaches have been reviewed permitting use of these type

Figure 2.31 Engineering equations (na = neutral axis).

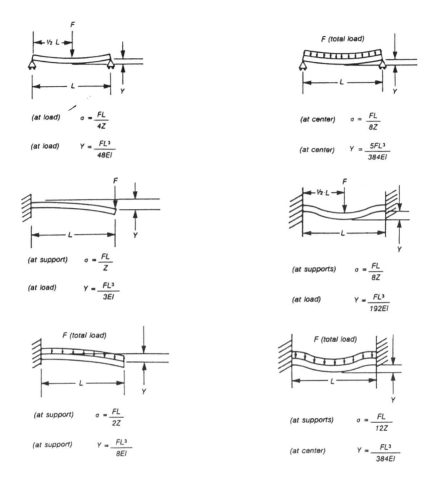

(at load) $\sigma = \dfrac{FL}{4Z}$

(at load) $Y = \dfrac{FL^3}{48EI}$

(at center) $\sigma = \dfrac{FL}{8Z}$

(at center) $Y = \dfrac{5FL^3}{384EI}$

(at support) $\sigma = \dfrac{FL}{Z}$

(at load) $Y = \dfrac{FL^3}{3EI}$

(at supports) $\sigma = \dfrac{FL}{8Z}$

(at load) $Y = \dfrac{FL^3}{192EI}$

(at support) $\sigma = \dfrac{FL}{2Z}$

(at support) $Y = \dfrac{FL^3}{8EI}$

(at supports) $\sigma = \dfrac{FL}{12Z}$

(at center) $Y = \dfrac{FL^3}{384EI}$

plastics. The drawback is that these methods can be quite complex, involving numerical techniques that may not be attractive to designers. However, one method has been widely accepted, the so-called pseudo-elastic (PE) design method.

In the PE method time-dependent property values for the modulus (include secant modulus) are selected and substituted into the standard equations. This approach is sufficiently accurate if the value chosen for the modulus takes into account the projected service life of the product and/or the limiting strain of the plastic. This approach is not a straightforward solution applicable to all plastics or even to one plastic in all its applications. This type of evaluation takes into consideration the value to use as a safety factor (SF). If no history exists a high value will be required. In time with service condition inputs, the SF can be reduced if justified (Chapter 7).

Figure 2.32 Beam equations

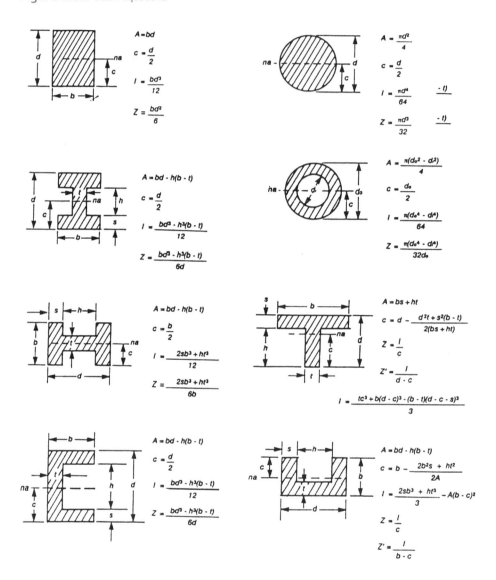

Determining a secant modulus is usually based on 1% strain or that is 0.85% of the initial tangent modulus (Fig. 2.4). However, for many plastics, particularly the crystalline TPs, this method is too restrictive, so in most practical situations the limiting strain is decided in consultation between the designer and the plastic material's manufacturer. Once the limiting strain is known, design methods based on its static and/or dynamic load becomes rather straightforward.

Theory of Combined Action

Overview

The following sections attempt only to set forth the elements of design theory of reinforced plastics (RPs). Fibrous RPs differ from most other engineering materials because they combine two essentially different materials, fibers and synthetic resin, into a single composite. In this they are somewhat analogous to reinforced concrete which combines concrete and steel, but in RPs the fibers are generally much more evenly distributed throughout the mass and the ratio of fibers to resin is much higher than the ratio of steel to concrete.

In their design it is necessary to take into account the combined action of fiber and resin. Sometimes the combination can be considered to be homogeneous and, therefore, to be similar to engineering materials like metal but in other cases, homogeneity cannot be assumed and it is necessary to take into account the fact that two widely dissimilar materials have been combined into a single unit.

In designing these reinforced plastics, certain important assumptions are made. The first and most fundamental is that the two materials act together and that the stretching, compression, and twisting of fibers and of resin under load is the same, that is, the strains in fiber and resin are equal. This assumption implies that a good bond exists between resin and fiber to prevent slippage between them and to prevent wrinkling of the fiber.

The second major assumption is that the material is elastic, that is, strains are directly proportional to the stresses applied, and when a load is removed the deformation disappears. In engineering terms the material is assumed to obey Hooke's Law. This assumption is probably a close approximation of the actual behavior in direct stress below the proportional limit, particularly in tension, if the fibers are stiff and elastic in the Hookean sense and carry essentially all of the stress. The assumption is probably less valid in shear where the resin carries a substantial portion of the stress. The resin may undergo plastic flow leading to creep or to relaxation of stress, especially when stresses are high.

More or less implicit in the theory of materials of this type is the assumption that all of the fibers are straight and unstressed or that the initial stresses in the individual fibers are essentially equal. In practice it is quite unlikely that this is true. It is to be expected, therefore, that as the load is increased some fibers reach their breaking points first. As they fail, their loads are transferred to other as yet unbroken fibers, with the consequence that failure is caused by the successive breaking of fibers rather than by the simultaneous breaking of all of them. The effect is to reduce the overall strength and to reduce the allowable working stresses accordingly, but the design theory is otherwise largely unaffected as long as essentially elastic behavior occurs. The development of higher working stresses is, therefore, largely a question of devising fabrication techniques to make the fibers work together to obtain maximum strength.

Design theory shows that the values of a number of elastic constants must be known in addition to the strength properties of the resin, fibers, and combination. Reasonable assumptions are made in carrying out designs. In the examples used, more or less arbitrary values of elastic constants and strength values have been chosen to illustrate the theory. Any other values could be used.

As more experience is gained in the design of these materials, and as more complete experimental data are forthcoming, the design procedures will no doubt be modified. This review can be related to the effects of environment.

Stress-Strain Analysis
Any material when stressed stretches or is deformed. If the resin and the fiber in RPs are firmly bonded together, the deformation is the same in both. For efficient structural behavior high strength fibers are employed, but these must be more unyielding than the resin, therefore for a given deformation or strain, a higher stress is developed in the fiber than in the resin. If the stress to strain relationships of fiber and resin are known from their stress-strain diagrams, the stresses developed in each for a given strain can be computed, and their combined action determined.

In Fig. 2.27 stress-strain diagrams for glass fiber and for two resins are shown. Curve A, typical of glass, shows that stress and strain are very nearly directly proportional to each other to the breaking point. Stiffness, or modulus of elasticity, as measured by the ratio of stress to strain, is high. Curve B represents a hard resin. Stress is directly proportional to strain when both are low, but stress gradually levels off as strain increases. Stiffness, or modulus of elasticity, is much lower than that of glass. The tangent measures it to the curve, usually at the origin. Curve C represents a softer resin intermediate between the hard resin and the very soft plastics. Stress and strain are again directly proportional at low levels, but not when the strains become large. Modulus of elasticity, as measured by the tangent to the curve, is lower than for the hard resin.

These stress-strain diagrams may be applied, for example, in the investigation of a rod in which half the total volume is glass fiber and half is resin. If the glass fibers are laid parallel to the axis of the rod, at any cross section, half of the total cross-sectional area is glass and half is resin. If the rod is stretched 0.5%, the glass is stressed at an intensity of 345 MPa (50,000 psi) and the resin, if resin B, at 52 MPa (7500 psi), or if resin C, at 17 MPa (2500 psi). If, for example, the rod has a total cross section of one-half square inch, the glass is one-quarter square inch, and the total stress in the glass is ¼ times 50,000 or 5,675 kg (12,500 lb). Similarly, the stress in the resin, if resin B, is 850 kg (1875 lb), and in resin C is 280 kg (625 lb). The load required to stretch the rod made with resin B is therefore the sum of the stresses in glass and resin, or 6,526 kg (14,375 lb). Similarly, for a rod utilizing resin C, the load is 5,960 kg (13,125 lb). The average stress on the one-half square inch cross section is therefore 198 MPa (28,750 psi) or 180 MPa (26,250 psi), respectively.

An analogous line of reasoning shows that at a strain of 1.25% the stress intensity in the glass is 860 MPa (125,000 psi), and in resins B and C it is 87 and 31 MPa (12,600 and 4,500 psi), respectively. The corresponding loads on rods made with resins B and C are 237 and 223 MPa (34,400 and 32,375 lb), respectively.

Table 2.4 Examples of loading conditions

Loading	Beam ends	Deflections at	K_m	K_s
Uniformly distributed	Both simply supported	Midspan	5/384	1/8
Uniformly distributed	Both clamped	Midspan	1/384	1/8
Concentrated at midspan	Both simply supported	Midspan	1/48	1/4
Concentrated at midspan	Both clamped	Midspan	1/192	1/4
Concentrated at outer quarter points	Both simply supported	Midspan	11/768	1/8
Concentrated at outer quarter points	Both simply supported	Load point	1/96	1/8
Uniformly distributed	Cantilever, 1 free, 1 clamped	Free end	1/8	1/2
Concentrated at free end	Cantilever, 1 free, 1 clamped	Free end	1/3	1

The foregoing can be put into the form of an equation

$$\sigma A = \sigma_f A_f + \sigma_r A_r \tag{2-81}$$

where

σ = mean stress intensity on entire cross section
σ_f = stress intensity in fiber
σ_r = stress intensity in resin
A = total cross-sectional area

A_f = cross-sectional area of fiber
A_r = cross-sectional area of resin

If the moduli of elasticity, as measured by the tangents to the stress-strain diagrams are known, the following relationships hold:

$$\sigma_r/\sigma_f = E_r/E_f \text{ or } \sigma_r = (E_r/E_f) \sigma_f \tag{2-82}$$

E_r = modulus of elasticity of resin
E_f = modulus of elasticity of fiber

Substituting (2-82) in (2-81)

$$\sigma A = \sigma_f \left(A_f + \frac{E_r}{E_f} A_r \right) \tag{2-83}$$

Referring to Fig. 2.27, the tangent to the stress-strain curve for glass gives a value of modulus of elasticity $E_f = 10 \times 10^6$ psi. The tangents to the two resin curves give values of E_r equal to 1.5×10^6 psi and 0.5×10^6 psi, respectively. Substituting these values in Eq. 2-83 and solving for the stresses in the one-half square inch rod of the previous example, gives

$$\text{Resin B} \quad \sigma A = 50,000 \left(0.25 + \frac{1.5}{10} 0.25 \right)$$
$$= 14,375 \text{ lb}$$
$$\sigma = 28,750 \text{ psi}$$

$$\text{Resin C} \quad \sigma A = 50,000 \left(0.25 + \frac{0.5}{10} 0.25 \right)$$
$$= 13,125 \text{ lb}$$
$$\sigma = 26,250 \text{ psi}$$

Average values of modulus of elasticity of the entire cross section may be computed by dividing σ by the strain. The strain is 0.5%, therefore the two average values of E of the rod, incorporating resins B and C, are 5.75×10^6 psi and 5.35×10^6 psi, respectively.

For a cross section made up of a number of different materials, Eq. 2-81 may be generalized to

$$\sigma A = \sum_{i=1}^{i=n} \sigma_i A_i \tag{2-84}$$

in which σ_i is the tensile strength and A_i the cross-sectional area of any component of the cross section. This equation can be still further generalized to include tension, compression, and shear

$$SA = \sum_{i=1}^{i=n} S_i A_i \tag{2-85}$$

in which S_i is the strength property of the cross-sectional area A_i of component i, and S is the mean strength property over the entire cross section A.

Similarly, to find the overall modulus of elasticity of a cross-section, the equation becomes

$$EA = \sum_{i=1}^{i=n} E_i A_i \qquad (2\text{-}86)$$

in which E is the overall modulus of elasticity, A the total cross section, and E_i the modulus of elasticity corresponding to the partial cross-sectional area A_i. For shear modulus G the equation becomes

$$GA = \sum_{i=1}^{i=n} G_i A_i \qquad (2\text{-}87)$$

Plain Reinforced Plates

Fibrous reinforced plates, flat or curved, are commonly made with mat, fabrics, and parallel filaments, either alone or in combination. Mat is usually used for good strength at minimum cost, fabrics for high strength, and parallel filaments for maximum strength in some particular direction.

Because the fibers in mat are randomly oriented, mat-reinforced materials have essentially the same strength and elastic properties in all directions in the plane of the plate, that is, they are essentially isotropic in the plane. Consequently, the usual engineering theories and design methods employed for isotropic engineering materials may be applied. It is only necessary to know strength, modulus of elasticity, shearing modulus, and Poisson's ratio of the combined mat and resin. These can be obtained from standard stress-strain measurements made on specimens of the particular combination of fiber and resin under consideration.

In fabric and roving-reinforced materials the strength and elastic properties are different in different directions, that is, they are not isotropic, and the usual engineering equations must accordingly be modified. Because fabrics are woven with yarns at right angles (warp and fill directions), a single layer of fabric-reinforced material has two principal directions or natural axes, longitudinal (warp) and transverse (fill) at right angles to each other. This structure is called orthotropic (right-angled directions). Parallel strands of fiber, as in a single layer of roving-reinforced or unidirectional fabric-reinforced plates, also result in orthotropic materials, with one direction parallel, and one at right angles to the fibers. Multilayer plates, in which layers of fabric or of roving are laid up parallel or perpendicular to each other, are also orthotropic. If the same number of strands or yarns is found in each principal direction (balanced construction), the strength and elastic properties are the same in those directions but not at intermediate angles; if the number of strands or yarns is different in the two principal directions (unbalanced construction), the strength and elastic properties are different in those directions as well as at all intermediate angles.

In the foregoing discussion the direction perpendicular to the plane of the plate has been neglected because the plate is assumed to be thin and the stresses are assumed to be applied in the plane of the plate rather than perpendicular to it. This assumption, which considerably simplifies the theory, carries through all of the following discussion. It is true, of course, that properties perpendicular to the plane of the plate are undoubtedly different than in the plane of the plate, and in thick plates this difference has to be taken into account, particularly when stresses are not planar.

For isotropic materials, such as mat-reinforced construction, if E is the modulus of elasticity in any reference direction, the modulus E_1 at any angle to this direction is the same, and the ratio E_1/E is therefore unity. Poisson's ratio v is similarly a constant in all directions, and the shearing modulus $G = E/2 (1 + v)$. If v, for example, is 0.3, $G/E = 0.385$ at all angles. These relationships are shown in Fig. 2.33.

The following familiar relationships between direct stress σ and strain, ε, and shearing stress τ and strain γ hold:

$$\varepsilon = \sigma/E \tag{2-88}$$

$$\gamma = \tau/G \tag{2-89}$$

A transverse strain (contraction or dilation) ε_T is caused by σ equal to

$$\varepsilon_T = -v\varepsilon \tag{2-90}$$

For orthotropic materials, such as fabric and roving-reinforced construction, E_L and E_T are the elastic moduli in the longitudinal (L) and transverse (T) directions, G_{LT} is the shearing modulus associated with these directions, v_{LT} is the Poisson's ratio giving the transverse strain caused by a strett in the longitudinal direction, and v_{LT} is Poisson's ratio giving the longitudinal strain caused by a stress in the transverse direction. The modulus at any intermediate angle is E_1, and if σ_1 is a stress applied in the 1-direction at an angle α with a longitudinal direction (Fig. 2.34, top), the stress σ_1 causes a strain ε_1

Figure 2.33 Modulus of elasticity, shear modulus, and Poisson's ratio for isotropic material such as mat-reinforced plastics

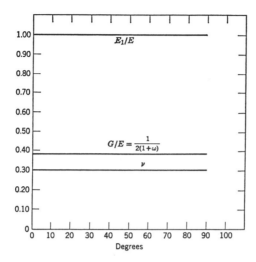

Figure 2.34 Elastic constants of unbalanced orthotropic material E_1, G_{12}, and v_{12} are all functions of the angle between direction of stress and the longitudinal axis (warp direction) of the material. Factors m_1 and m_2 account for direct and shear strains caused by shear and direct stresses, respectively. Angle 0° is longitudinal direction, and angle 90° is transverse direction

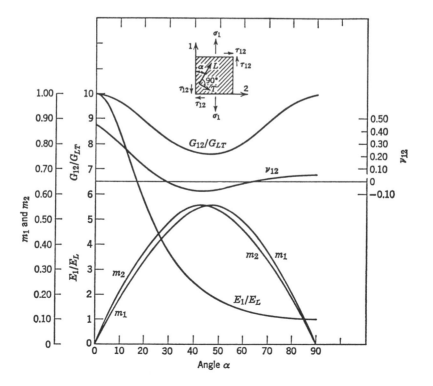

$$\varepsilon_1 = \sigma_1/E_1 \tag{2-91}$$

in which E_1 may be found from

$$\frac{E_L}{E_1} = \cos^4 \alpha + \frac{E_L}{E_T} \sin^4 \alpha + \frac{1}{4}\left(\frac{E_L}{G_{LT}} - 2v_{LT}\right)\sin^2 2\alpha \tag{2-92}$$

This relationship is plotted as E_1/E_L in Fig. 2.34, in which 0° corresponds to the longitudinal direction and 90° to the transverse direction.

A transverse strain ε_2 is caused by σ_1

$$\varepsilon_2 = v_{12}\varepsilon_1 \tag{2-93}$$

In which (Fig. 2.34)

$$v_{12} = \frac{E_1}{E_L}\left\{v_{LT} - \frac{1}{4}\left(1 + 2v_{LT} + \frac{E_L}{E_T} - \frac{E_L}{G_{LT}}\right)\sin^2 2\alpha\right\} \tag{2-94}$$

Unlike isotropic materials, stress σ_1, when applied at any angle except 0° and 90°, causes shear distortion and the shear strain γ_{12} is found from

$$\gamma_{12} = m_1\sigma_1/E_L \tag{2-95}$$

in which (Fig. 2.34)

$$m_1 = \sin 2\alpha\left\{v_{LT} + \frac{E_L}{E_T} - \frac{1}{2}\frac{E_L}{G_{LT}} - \cos^2 \alpha\left(1 + 2v_{LT} + \frac{E_L}{E_T} - \frac{E_L}{G_{LT}}\right)\right\} \tag{2-96}$$

A shearing stress π_{12} applied in the 1–2 directions causes a shear strain γ_{12}

$$\gamma_{12} = \tau_{12}/G_{12} \tag{2-97}$$

in which (Fig. 2.34)

$$\frac{G_{LT}}{G_{12}} = \frac{G_{LT}}{E_L}\left\{\left(1 + 2v_{LT} + \frac{E_L}{E_T}\right) - \left(1 + 2v_{LT} + \frac{E_L}{E_T} - \frac{E_L}{G_{LT}}\right)\cos^2 2\alpha\right\} \tag{2-98}$$

This relationship is plotted as G_{12}/G_{LT} in Fig. 2.34.

Unlike isotropic materials, stress τ_{12} causes a strain ε_1 in the 1-direction

$$\varepsilon_1 = -m_1\tau_{12}/E_L \tag{2-99}$$

and a strain ε_2 in the 2-direction

$$\varepsilon_2 = -m_2\tau_{12}/E_L \tag{2-100}$$

in which (Fig. 2.34)

$$m_2 = \sin 2\alpha\left\{v_{LT} + \frac{E_L}{E_T} - \frac{1}{2}\frac{E_L}{G_{LT}} - \sin^2 \alpha\left(1 + 2v_{LT} + \frac{E_L}{E_T} - \frac{E_L}{G_{LT}}\right)\right\} \tag{2-101}$$

Figure 2.35 Elastic constants of balanced orthotropic material. Constants and angles have same meaning as previous figure

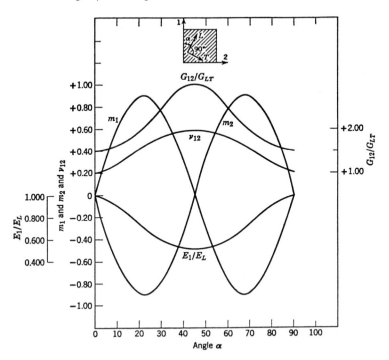

The two values of Poisson's ratio are related :

$$v_{LT}/v_{LT} = E_L/E_T \tag{2-102}$$

In plotting Fig. 2.34 the following values were used:

$$E_L = 5,000,000 \text{ psi}$$
$$E_T = 500,000 \text{ psi}$$
$$G_{LT} = 550,000 \text{ psi}$$
$$v_{LT} = v_{0°} = 0.450$$
$$v_{TL} = v_{90°} = 0.045$$

These values, for example, might correspond to a parallel glass filament reinforced panel employing an intermediate polyester resin.

When the orthotropic material is balanced, the longitudinal and transverse properties are the same, that is, $E_L = E_T$ and $v_{LT} = v_{LT} = v_{TL}$. The properties are symmetrical about the 45° angle, as shown in Fig. 2.35, in which the following values were used:

$$E_L = E_T = 3,000,000 \text{ psi}$$
$$G_{LT} = 500,000 \text{ psi}$$
$$v_{LT} = v_{TL} = 0.20$$

These values might correspond, for example, to a square-weave or symmetrical satin-weave fabric-reinforced construction.

As an example of the application of the foregoing equations, the tensile stress σ_1 acting on the small plate at the top of Fig. 2.34 is 10,000 psi, the shear stress τ_{12} is 4000 psi, and the angle α is 30°. Then from Fig. 2.34,

$$E_1/E_L = 0.367 \text{ or } E_1 = 0.367 \times 5,000,000 = 1,830,000 \text{ psi}$$
$$G_{12}/G_{LT} = 0.81 \text{ or } G_{12} = 0.81 \times 550,000 = 445,000 \text{ psi}$$
$$v_{12} = -0.0286 \quad m_1 = 4.66 \quad m_2 = 4.98$$

Then, strains caused by σ_1 are

$$\varepsilon_1 = 10,000/1,830,000 = 5.45 \times 10^{-3} \tag{2-103}$$

$$\varepsilon_2 = -(-0.0286)\,5.45 \times 10^{-3} = 0.16 \times 10^{-3} \tag{2-104}$$

$$\gamma_{12} = -4.66 \times 10,000/5,000,000 = -9.32 \times 10^{-3} \tag{2-105}$$

and strains caused by τ_{12} are

$$\gamma_{12} = 4,000/550,000 = 7.28 \times 10^{-3} \tag{2-106}$$

$$\varepsilon_1 = -4.66 \times 4,000/5,000,000 = -3.73 \times 10^{-3} \tag{2-107}$$

$$\varepsilon_2 = -4.98 \times 4,000/5,000,000 = -3.98 \times 10^{-3} \tag{2-108}$$

Total strains, therefore, are

$$\gamma_{12} = -2.04 \times 10^{-3}$$
$$\varepsilon_1 = 1.72 \times 10^{-3}$$
$$\varepsilon_2 = -3.82 \times 10^{-3}$$

Problems involving Fig. 2.35 can be solved in an analogous manner.

It must be kept in mind that Eqs. 2–92, 2–94, 2–96, 2–98, and 2–101 are valid and useful if the fibers and the resin behave together in accordance with the assumptions upon which their derivation is based. If only the values of E_L, E_T, G_{LT}, and v_{LT} are available, the intermediate values of E_1, G_{12}, v_{12}, and the values of m_1 and m_2 can be estimated by means of these equations.

Composite Plates

Fibrous reinforced plates in practice are often made up of several layers, and the individual layers may be of different construction, such as mat, fabric, or roving. Furthermore, the various layers may be oriented at different angles with respect to each other in order to provide the best combination to resist some particular loading condition. Outside loads or stresses applied to a composite plate of this type result in internal stresses which are different in the individual layers. External direct

Figure 2.36 Composite panel with layers a and b of different orthotropic materials oriented at arbitrary angles α and β with respect to applied stresses σ_1, σ_2 and γ_{12}

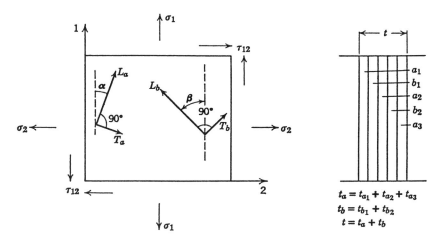

$$t_a = t_{a1} + t_{a2} + t_{a3}$$
$$t_b = t_{b1} + t_{b2}$$
$$t = t_a + t_b$$

stresses may result not only in internal direct stresses but in internal shear stresses, and external shear stresses may result in internal direct stresses as well as internal shear stresses.

Fig. 2.36 depicts a small composite plate made up of materials a and b having principal longitudinal and transverse directions L_a and T_a, and L_b and T_b, respectively. Several layers of each are present but their total thicknesses are t_a and t_b, respectively, and the overall thickness is t. Outside stresses σ_1, σ_2, and τ_{12} are applied in the 1 and 2 directions, as shown. The 1-direction makes an angle α with L_{a3} and a reverse angle β with L_b. The angle α is considered to be positive and the angle β negative.

The internal stresses σ_{1a}, σ_{2a}, τ_{12a}, and σ_{1b}, σ_{2b}, τ_{12b} in the individual layers can be found by observing that the sums of the internal stresses in the 1 and 2 directions must equal the external stresses in these directions, and that the strains must be the same in all layers. These relationships may be written in the following forms:

$$\sigma_{1a}t_a + \sigma_{1b}t_b = \sigma_1 t; \qquad \sigma_{1b} = \frac{\sigma_1 t - \sigma_{1a}t_a}{t_b} \tag{2-109}$$

$$\sigma_{2a}t_a + \sigma_{2b}t_b = \sigma_2 t; \qquad \sigma_{2b} = \frac{\sigma_2 t - \sigma_{2a}t_a}{t_b} \tag{2-110}$$

$$\tau_{12a}t_a + \tau_{12b}t_b = \tau_{12}t; \qquad \tau_{12b} = \frac{\tau_{12}t - \tau_{12a}t_a}{t_b} \tag{2-111}$$

$$\varepsilon_{1a} = \varepsilon_{1b} = \varepsilon_1 \tag{2-112}$$

$$\varepsilon_{2a} = \varepsilon_{2b} = \varepsilon_2 \tag{2-113}$$

$$\gamma_{12a} = \gamma_{12b} = \gamma_{12} \tag{2-114}$$

Strains and stresses are induced in each layer. Because the layers are firmly bonded together the strains are the same in the a and b layers, and are equal to the strains in the whole plate:

$$\varepsilon_1 = \begin{cases} \varepsilon_{1a} = \dfrac{\sigma_{1a}}{E_{1a}} - v_{21a}\dfrac{\sigma_{2a}}{E_{2a}} - m_{1a}\dfrac{\tau_{12a}}{E_{La}} \\[3mm] \varepsilon_{1b} = \dfrac{\sigma_{1b}}{E_{1b}} - v_{21b}\dfrac{\sigma_{2b}}{E_{2b}} - m_{1b}\dfrac{\tau_{12b}}{E_{Lb}} \end{cases} \tag{2-115}$$

$$\varepsilon_2 = \begin{cases} \varepsilon_{2a} = -\,v_{21a}\dfrac{\sigma_{1a}}{E_{1a}} + \dfrac{\sigma_{2a}}{E_{2a}} - m_{2a}\dfrac{\tau_{12a}}{E_{La}} \\[3mm] \varepsilon_{2b} = -\,v_{21b}\dfrac{\sigma_{1b}}{E_{1b}} + \dfrac{\sigma_{2b}}{E_{2b}} - m_{2b}\dfrac{\tau_{12b}}{E_{Lb}} \end{cases} \tag{2-116}$$

$$\gamma_{12} = \begin{cases} \gamma_{12a} = -\,m_{1a}\dfrac{\sigma_{1a}}{E_{La}} - m_{2a}\dfrac{\sigma_{2a}}{E_{La}} + \dfrac{\tau_{12a}}{G_{12a}} \\[3mm] \gamma_{12a} = -\,m_{1b}\dfrac{\sigma_{1b}}{E_{Lb}} - m_{2b}\dfrac{\sigma_{2b}}{E_{Lb}} + \dfrac{\tau_{12b}}{G_{12b}} \end{cases} \tag{2-117}$$

Solution of the foregoing Eqs. 2–109 to 2–117 leads to the following simultaneous equations:

$$A_{11}\sigma_{1a} + A_{12}\sigma_{2a} + A_{13}\tau_{12a} = \dfrac{t}{t_a t_b}\left(\dfrac{\sigma_1}{E_{1b}} - v_{21b}\dfrac{\sigma_2}{E_{2b}} - m_{1b}\dfrac{\tau_{12}}{E_{Lb}}\right) \tag{2-118}$$

$$A_{21}\sigma_{1a} + A_{22}\sigma_{2a} + A_{23}\tau_{12a} = \dfrac{t}{t_a t_b}\left(-v_{12b}\dfrac{\sigma_1}{E_{1b}} + \dfrac{\sigma_2}{E_{2b}} - m_{2b}\dfrac{\tau_{12}}{E_{Lb}}\right) \tag{2-119}$$

$$A_{31}\sigma_{1a} + A_{32}\sigma_{2a} + A_{33}\tau_{12a} = \dfrac{t}{t_a t_b}\left(-m_{1b}\dfrac{\sigma_1}{E_{Lb}} - m_{2b}\dfrac{\sigma_2}{E_{Lb}} + \dfrac{\tau_{12}}{G_{12b}}\right) \tag{2-120}$$

in which:

$$A_{11} = \dfrac{1}{E_{1a}t_a} + \dfrac{1}{E_{1b}t_b} \qquad A_{12} = -\dfrac{v_{21a}}{E_{2a}t_a} - \dfrac{v_{21b}}{E_{2b}t_b} \qquad A_{13} = -\dfrac{m_{1a}}{E_{La}t_a} - \dfrac{m_{1b}}{E_{Lb}t_b}$$

$$A_{21} = -\dfrac{v_{12a}}{E_{1a}t_a} - \dfrac{v_{12b}}{E_{1b}t_b} \qquad A_{22} = \dfrac{1}{E_{2a}t_a} + \dfrac{1}{E_{2b}t_b} \qquad A_{23} = -\dfrac{m_{2a}}{E_{La}t_a} - \dfrac{m_{2b}}{E_{Lb}t_b}$$

where $A_{21} = A_{12}$, numerically.

$$A_{31} = A_{13} \qquad\qquad A_{32} = A_{23} \qquad\qquad A_{33} = \dfrac{1}{G_{12a}t_a} + \dfrac{1}{G_{12b}t_b}$$

Figure 2.37 Fibrous glass-reinforced plastic thin-wall cylinder. (a) internal pressure alone and (b) internal pressure plus twisting moment

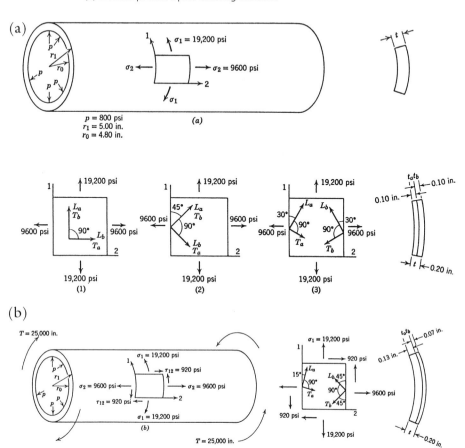

Example:

The application of the foregoing expressions may be illustrated by a cylindrical pressure vessel as shown in Fig. 2.37a. The wall of this vessel, having an external radius of 5 in., and wall thickness of 0.20 in., may be considered to be a thin plate. It is subjected to an internal pressure of 800 psi. the circumferential stress σ_1 and the longitudinal stress σ_2 in the wall are calculated

$$\sigma_1 = \frac{pr_0}{t} = 19{,}200 \text{ psi}$$

$$\sigma_2 = \frac{pr_0}{2t} = 9{,}600 \text{ psi}$$

The stresses acting on a small part of the wall are therefore as shown in Fig. 2.37a.

Three types of construction will be investigated as shown in Fig. 2.37 (1), (2), (3). All three employ the balanced fabric having the characteristics shown in Fig. 2.35. In (1) the fabric is simply wrapped in layers a and b with the L and T directions laid in the circumferential and axial directions. In (2) the layers are laid at 45° to the axis of the cylinder, and in (3) they are laid at alternate 30° angles in left-hand and right-hand spirals as shown. In each instance $t_a = t_b = 0.10$ in.

Referring to Fig. 2.35, it is seen that for Case 1

$$E_{1a} = E_{1b} = E_{2a} = E_{2b} = 3 \times 10^6 \text{ psi}$$
$$V_{18a} = V_{21a} = V_{12b} = 0.20$$
$$m_{1a} = m_{1b} = m_{2a} = m_{2b} = 0$$
$$A_{11} = A_{22,} \ A_{12} = A_{21}$$
$$A_{18} = A_{31} = A_{32} = A_{23} = 0$$

Eqs. 2–118 to 2–120 therefore become

$$A_{11}\sigma_{1a} + A_{12}\sigma_{2a} + 0 = \frac{t}{t_a t_b}\left(\frac{\sigma_1}{E_{1b}} - V_{21b}\frac{\sigma_2}{E_{2b}} + 0\right)$$

$$A_{21}\sigma_{1a} + A_{22}\sigma_{2a} + 0 = \frac{t}{t_a t_b}\left(-V_{12}\frac{\sigma_1}{E_{1b}} + \frac{\sigma_2}{E_{2b}} + 0\right)$$

$$A_{33}\tau_{12a} = 0$$

Solution of these equations and reference to Eqs. 2–109 to 2–111 show that

$$\sigma_{1a} = \sigma_{1b} = \sigma_1 = 19{,}200 \text{ psi}$$
$$\sigma_{2a} = \sigma_{2b} = \sigma_2 = 9{,}600 \text{ psi}$$
$$\tau_{12a} = \tau_{12b} = 0$$

This proves what might have been expected intuitively; because of symmetry with respect to the 1–2 directions chosen; the internal direct stresses σ_{1a}, σ_{1b}, σ_{2a}, and σ_{2b} are equal to the imposed stresses σ_1 and σ_2, and there is no internal shear stress.

The same result is found for Case 2. In this balanced fabric $m_1 = m_2 = 0$ at 45°, there is no shear distortion caused by direct stress, and shear therefore is zero. In Case 3:

$$E_{1a} = E_{1b} = E_{30°} = E_{60°} = E_{2a} = E_{2b}$$
$$= 0.597 \times 3 \times 10^6 = 1.78 \times 10^6 \text{ psi}$$
$$G_{12a} = G_{12b} = 1.82 \times 0.5 \times 10^6 = 0.91 \times 10^6$$
$$V_{12a} = V_{12b} = V_{80°} = V_{60°} = V_{21a} = V_{12b} = 0.523$$
$$m_{1a} = m_{80°} = 0.775, \ = m_{1b} = -m_{1a} = -0.775$$
$$m_{2a} = m_{60°} = 0.775, \ = m_{2b} = -m_{2a} = -0.775$$

The values of m_{1b} and m_{2b} are negative because the 30° angles of orientation of the longitudinal direction L_b of layers b is measured in the negative direction whereas it is positive for the a-layers.

Equations 22 become

$$A_{11}\sigma_{1a} + A_{12}\sigma_{2a} + 0 = \frac{t}{t_a t_b}\left(\frac{\sigma_1}{E_{1b}} - v_{21b}\frac{\sigma_2}{E_{2b}} + 0\right)$$

$$A_{21}\sigma_{1a} + A_{22}\sigma_{2a} + 0 = \frac{t}{t_a t_b}\left(-v_{12b}\frac{\sigma_1}{E_{1b}} + \frac{\sigma_2}{E_{2b}} + 0\right)$$

$$0 + 0 + A_{33}\tau_{12a} = \frac{t}{t_a t_b}\left(-m_{1b}\frac{\sigma_1}{E_{Lb}} - m_{2b}\frac{\sigma_2}{E_{Lb}} + 0\right)$$

The first two of these equations are exactly like the first two equations for Cases 1 and 2 and show that the internal direct stresses are equal to the imposed, that is

$\sigma_{1a} = \sigma_{1b} = \sigma_1 = 19{,}200$ psi
$\sigma_{2a} = \sigma_{2b} = \sigma_2 = 9{,}600$ psi

The third equation, however, is not equal to zero, and its solution, together with equation 19c, shows that

$\tau_{12a} = 6750$ psi
$\tau_{12b} = 6750$ psi

Appreciable shear stresses are set up within the body of the cylinder wall when layers are oriented as in Case 3, even though no shear forces are applied to the cylinder itself. The shear stresses in layers b are oriented in the direction opposite to the shear stresses in layers a.

The difference in the shear stresses between the two layers must be taken up by shear in the adhesive bond between them, that is, in the layer of resin that holds the fiber-reinforced layers together. The difference is

$6750 - (-6750) = 13{,}500$ psi

This shear stress in the resin bonding the layers together is therefore seen to be high.

In Cases 1 and 2 the orientation of the fibers with respect to the 1–2 directions chosen resulted in zero shear stresses associated with those directions, whereas in Case 3 the shear stresses were not zero. In all three cases, symmetry of the fiber orientations with respect to the stress directions resulted in internal stresses equal to the external stresses. These are special cases. In the more general case the internal direct stresses in the individual layers are not necessarily equal to the external direct stresses, nor are they the same in the various layers. Furthermore, even symmetrical Case 3 leads to internal shear stresses when external shear stresses are absent. In the more general case it is still more true that internal shear stresses may be appreciable, or they may be absent, depending upon the magnitude of the external stresses and the orientation of the 1–2 directions with respect to the external stresses.

A more general case in shown in Fig. 2.37b in which the same cylinder is chosen as in Fig. 2.37a except that torsional effect equal to a twisting couple of 25,000 in.-lb has been added. The construction of the wall has also been changed. Layers *a* of unbalanced material having the properties of Fig. 2.34 are a total of 0.13 in. thick, and are oriented at 15° to the circumferential direction as shown. Layers *b*, of balanced material having the properties of Fig. 2.35, are a total of 0.07 in. thick and are oriented at 45° as shown. Referring to Fig. 2.35, the properties are found to be

a-layers	b-layers
t_a = 0.13 in.	t_b = 0.07 in.
a = 15°	a = 45°
E_{1a} = 0.703 × 5 × 10^6	E_{1b} = E_{2b} = 0.526 × 3 × 10^6
= 3.515 × 10^6 psi	= 1.578 × 10^6 psi
E_{2a} = 0.109 × 5 × 10^6	v_{12b} = v_{21b} = 0.579
= 0.545 × 10^6 psi	m_{1b} = m_{2b} = 0
v_{12a} = 0.193	G_{12b} = 2.5 × 0.5 × 10^6
m_{1a} = 2.63	=1.250 × 10^6 psi
m_{2a} = 2.94	
G_{12a} = 0.93 × 0.5 × 10^6	
= 0.465 × 10^6 psi	

Solving for the various constants and substituting in Eqs. 2-118 to 2-120

$$11.2412\sigma_{1a} - 5.6641\sigma_{2a} - 4.0461t_{12a} = +190,180$$
$$-5.6641\sigma_{1a} + 23.2030\sigma_{2a} - 4.5156\tau_{12a} = -21,150$$
$$-4.0461\sigma_{1b} - 4.5156\sigma_{2a} + 26.4668\tau_{12a} = +16,190$$

The solution of the forgoing simultaneous equations leads to the following results for stresses in the *a*-layers:

$$\sigma_{1a} = 21,100 \text{ psi}$$
$$\sigma_{2a} = 5,200 \text{ psi}$$
$$\tau_{12a} = 4,740 \text{ psi}$$

When these results are employed with Eqs. 2-109 to 2-111 it is found that stresses in the *b*-layers are:

$$\sigma_{1b} = 15,700 \text{ psi}$$
$$\sigma_{2b} = 17,800 \text{ psi}$$
$$\tau_{12b} = -6,150 \text{ psi}$$

Bending of Beams and Plates

Plates and beams of fibrous glass reinforced plastics may be homogeneous and isotropic or composite and nonisotropic depending upon their structure. Mat-reinforced plates may be considered to be essentially isotropic and the usual engineering formulas may be applied. Composite structures require suitably modified formulas but otherwise the procedures for computing bending stresses, stiffness, and bending shear stresses are essentially the same as for isotropic materials. The differences and similarities may be brought out by considering two beams of identical overall dimensions, one isotropic and the other composite. Two such cross sections are shown in Fig. 2.38. For each cross section it is necessary to know the stiffness factor EI to compute deflection, the section modulus to compute bending stresses, and the statical moments of portions of the cross section to compute shear stresses. For isotropic materials (a) the neutral axis of a rectangular cross section is at middepth, and the familiar formulas are

$$\text{Moment of inertia } I = \frac{bd^3}{12} \text{, stiffness factor} = EI \qquad (2\text{-}121)$$

$$\text{Section modulus } = \frac{I}{y} = \frac{bd^2}{6} \text{ for outermost fiber} \qquad (2\text{-}122)$$

$$\text{Bending stress } = \sigma = M\frac{y}{I} = \frac{6M}{bd^2} \text{ for outermost fiber} \qquad (2\text{-}123)$$

$$\text{Shear stres } = \frac{VQ}{bI} = \frac{3}{2}\frac{V}{bd} \text{ for maximum shear at the neutral axis} \qquad (2\text{-}124)$$

For composite materials the neutral axis is not necessarily at middepth of a rectangular section, and it must first be found.

$$\text{Neutral axis } x = \Sigma E_i A_i x_i / \Sigma E_i A_i \qquad (2\text{-}125)$$

in which E_i, A_i, x_i are the modulus of elasticity; cross-sectional area (bd_i); and distance from some reference line, such as the bottom of the cross section, to the center of gravity of any particular layer.

$$\text{Stiffness factor } = EI = \Sigma E_i I_i \qquad (2\text{-}126)$$

in which E_i and I_i are, for any particular layer, the modulus of elasticity and the moment of inertia about the neutral axis.

$$\text{Bending stress } \sigma = M E_y y / E_i \qquad (2\text{-}127)$$

in which y is the distance from the neutral axis to any point, and E_y is the modulus of elasticity of the layer at that point. The maximum bending stress does not necessarily occur at the outermost (top or bottom) fiber, as it does in isotropic materials.

Shear stress $\tau = VQ'/bEI$ (2-128)

in which V is the total shear on the cross section, t is the shear stress intensity along some horizontal plane, and Q' is the weighted statical moment, E_iA_iy about the beam's neutral axis, of the portion of the cross section between the horizontal plane in question and the outer edge (top or bottom) of the cross section.

An example of the forgoing is illustrated in Fig. 2.38c in which a composite beam is made up of five layers having three different moduli of elasticity, and three different strengths, as shown.

The neutral axis, found by applying Eq. 2-125 is 0.415 in. from the bottom of the cross section. Distance from the neutral axis to the centers of the individual layers are computed, and the stiffness factor EI calculated by means of Eq. 2-126. This is found to be

$$EI = \Sigma E_i I_i = 0.174 \times 10^6 \text{ lb in.}^2$$

Bending stresses are next computed for the top and bottom edges of the cross section and for the outer edge of each layer, that is, the edge of each layer farther from the neutral axis. From these, the bending moment the cross section is capable of carrying can be computed. This may be done for example by applying a bending moment M of one in.–lb and computing the unit bending stresses. These unit bending stresses divided into the strengths of the individual layers give a series of calculated resisting moments, the smallest of which is the maximum bending moment the beam is capable of carrying without exceeding the strength of any portion of the cross section.

For a unit bending moment $M = 1$ in.–lb,

$$\sigma_y = \frac{E_y y}{EI} \text{ from Eq. 2-127}$$

Plane	y	E_y	σ_y/in.lb	$\sigma = \sigma_y$/in.lb	=	M
a-a	0.385 in.	5×10^6	11.1 psi	40,000/11.1	=	3,600 in.-lb
b-b	0.185 in.	3×10^6	3.19 psi	25,000/3.19	=	7,800 in.-lb
c-c	0.085 in.	1×10^6	0.49 psi	5,000/0.49	=	10,200 in.-lb
d-d	0.115 in.	1×10^6	0.66 psi	5,000/0.66	=	7,600 in.-lb
e-e	0.315 in.	5×10^6	9.07 psi	40,000/9.07	=	4,400 in.-lb
f-f	0.415 in.	3×10^6	7.16 psi	25,000/7.16	=	3,500 in.-lb

If, for example, the beam were a simple beam carrying a load W on a 10-in. span, as shown in Fig. 2.38, the bending moment at the center

Figure 2.38 Cross section of: (a) isotropic beam (b) composite beam made in layers of different materials, and (c) composite beam having properties (where all E are × 10⁶ psi) $E_1 = 5$, $E_2 = 3$, $E_3 = 1$, $E_4 = 5$, and E5 = 3; also (where σ are 10³ psi) $\sigma_1 = 40$, σ_2 25, $\sigma_3 = 5$, $\sigma_4 = 40$, and $\sigma_5 = 25$

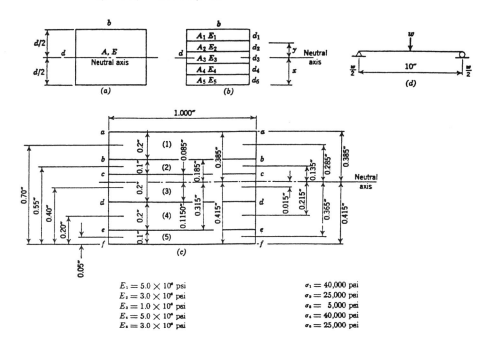

$E_1 = 5.0 \times 10^6$ psi
$E_2 = 3.0 \times 10^6$ psi
$E_3 = 1.0 \times 10^6$ psi
$E_4 = 5.0 \times 10^6$ psi
$E_5 = 3.0 \times 10^6$ psi

$\sigma_1 = 40{,}000$ psi
$\sigma_2 = 25{,}000$ psi
$\sigma_3 = 5{,}000$ psi
$\sigma_4 = 40{,}000$ psi
$\sigma_5 = 25{,}000$ psi

of the span would be $WL/4$. Setting this equal to 3,500 in.-lb gives the load W as 1400 lb. Shear V is $W/2$ or 700 lb. Using this value, the shear stress intensity at various horizontal planes in the beam may be computed by means of Eq. 2-128.

For planes *b-b*, *c-c*, and *d-d*, for example:

Plane	Layers	E_iA_i	y'	Q'	τ
b-b	1	$0.2 \times 5 \times 10^6$	0.285''	0.285×10^6	1150 psi
c-c	1	$0.2 \times 5 \times 10^6$	0.285	0.326×10^6	1315 psi
	2	$+ 0.1 \times 3 \times 10^6$	0.135		
d-d	4	$0.2 \times 5 \times 10^6$	0.215	0.324×10^6	1310 psi
	5	$+ 0.1 \times 3 \times 10^6$	0.365		

These would be the critical planes because they represent planes between layers of different materials, and consequently the resin alone would largely carry the stress. The shear stress at the neutral axis would be slightly higher and might or might not represent the critical plane, depending upon the structure of the material in layer 3.

Structural Sandwiches

In usual construction practice, a structural sandwich is a special case of a laminate in which two thin facings of relatively stiff, hard, dense, strong material are bonded to a thick core of relatively lightweight material considerably less dense, stiff, and strong than the facings.

With this geometry and relationship of mechanical properties, the facings are subjected to almost all of the stresses in transverse bending or in axial loading, and the geometry of the arrangement provides high stiffness combined with lightness because the stiff facings are at maximum distance from the neutral axis, similar to the flanges of an I-beam. The continuous core takes the place of the web of an I-beam or box beam, it absorbs most of the shear, and it also stabilizes the thin facings against buckling or wrinkling under compressive stresses. The bond between core and facings must resist shear and any transverse tensile stresses set up as the facings tend to wrinkle or pull away from the core.

Stiffness

For an isotropic material with modulus of elasticity E, the bending stiffness factor EI of a rectangular beam b wide and h deep is

$$EI = E(bh^3/12) \tag{2-129}$$

In a rectangular structural sandwich of the same dimensions as above whose facings and core have moduli of elasticity E_f and E_c, respectively, and a core thickness C, the bending stiffness factor EI is

$$EI = \frac{E_f b}{12}(h^3 - c^3) + \frac{E_c b}{12}c^3 \tag{2-130}$$

This equation is exact if the facings are of equal thickness, and approximate if they are not, but the approximation is close if facings are thin relative to the core.

If, as is usually the case, E_c is much smaller than E_f, the last term in the equation can be ignored.

For unsymmetrical sandwiches with different materials or different thicknesses in the facings, or both, the more general equation for ΣEI given in the previous section may be used.

In many isotropic materials the shear modulus G is high compared to the elastic modulus E, and shear distortion of a transversely loaded beam is so small that it can be neglected in calculating deflection. In a structural sandwich the core shear modulus G_c is usually so much smaller than E_f of the facings that shear distortion of the core may be large and therefore contribute significantly to the deflection of a transversely loaded beam. The total deflection of a beam is therefore composed of two factors: the deflection caused by bending moment alone, and the deflection caused by shear, that is

$$\delta = \delta_m + \delta_8 \qquad (2\text{-}131)$$

where

δ = total deflection
δ_m = moment deflection
δ_8 = shear deflection

Under transverse loading, bending moment deflection is proportional to the load and the cube of the span and inversely proportional to the stiffness factor EI. Shear deflection is proportional to the load and span and inversely proportional to a shear stiffness factor N whose value for symmetrical sandwiches is

$$N = \frac{(h+c)^b}{2} G_c \qquad (2\text{-}132)$$

where

G_c = core shear modulus

The total deflection may therefore be written

$$\delta = \frac{K_m W L^3}{EI} + \frac{K_8 W L}{N} \qquad (2\text{-}133)$$

The values of K_m and K_8 depend on the type of load. Values for several typical loading conditions are given as shown in Table 2.4.

Stresses in Sandwich Beams
The familiar equation for stresses in an isotropic beam subjected to bending

$$\sigma_y = \frac{M_y}{I} \qquad (2\text{-}134)$$

must be modified for sandwiches to the form

$$\sigma_y = \frac{M E_y y}{EI} \qquad (2\text{-}135)$$

where y = distance from neutral axis to fiber at y

E_y = elastic modulus of fiber at y
EI = stiffness factor

For a symmetrical sandwich the stress in the outermost facing fiber is found by setting

$$y = h/2$$
$$E_y = E_f$$

and the stress in the outermost core fiber by setting

$$y = c/2$$
$$E_y = E_c$$

The mean stress in the facings of a symmetrical sandwich can be found from

$$\sigma = \frac{2M}{bt(h+c)} \qquad (2\text{-}136)$$

where t = facing thickness

Similarly the general equation for the shear stresses in a laminate (see preceding section)

$$\tau = \frac{VQ}{bEI} \qquad (2\text{-}137)$$

can be used for any sandwich. For the symmetrical sandwich the value of τ can be closely approximated by

$$\tau = \frac{2V}{b(h+c)} \qquad (2\text{-}138)$$

Axially-Loaded Sandwich
Edge-loaded sandwiches such as columns and walls are subject to failure by overstressing the facings or core, or by buckling of the member as a whole. Direct stresses in facings and core can be calculated by assuming that their strains are equal, so that

$$P = \sigma_f A_f + \sigma_c A_c \qquad (2\text{-}139)$$

$$= \sigma_f\left(A_f + A_c \frac{E_c}{E_f}\right) \qquad (2\text{-}139a)$$

where

$$
\begin{aligned}
P &= \text{total load} \\
\sigma_f &= \text{facing stress} \\
\sigma_c &= \text{core stress} \\
A_f &= \text{cross-sectional area of facings} \\
A_c &= \text{cross-sectional area of core}
\end{aligned}
$$

Usually the elastic modulus E_c of the core is so small that the core carries little of the total load, and the equation can be simplified by ignoring the last term, so that for a sandwich b wide with facings t thick,

$$P = 2\sigma_f bt \qquad (2\text{-}140)$$

The column buckling load of a sandwich L long simply supported at the ends is given by

$$P = \frac{\pi^2 EI}{L^2 \left(1 + \frac{\pi^2 EI}{L^2 N}\right)^2} \qquad (2\text{-}141)$$

This variation of the Euler equation takes into account the low shear stiffness of the core.

For wall panels held in line along their vertical edges an approximate buckling formula is

$$P = \frac{4\pi^2 EI}{b^2 \left(1 + \frac{\pi^2 EI}{b^2 N}\right)^2} \qquad (2\text{-}142)$$

provided the length L of the panel is at least as great as the width b and provided the second term in the bracket of the denominator is not greater than unity.

Filament–Wound Shells, Internal Hydrostatic Pressure

Basic Equations
Cylindrically symmetric shells are considered which are of the form $r=r$ (z) in a system of cylindrical polar coordinates (r, θ, z). Inextensible fibers are wound on and bonded to this shell in such a way that at any point on it equal numbers of fibers are inclined at angles α and π-α to the line of latitude (z=constant) passing through that point. At a point (r, θ, z) of the shell there are $n_1, n_{12} \dots n_p$ fibers, per unit length measured perpendicular to the length of the fibers, with positive inclinations $\alpha_1, \alpha_2, \dots \alpha_p$; and an equal number of fibers with inclinations $\pi - \alpha_1, \pi - \alpha \dots \pi$-$\alpha_p$, to the line of latitude passing through (r, θ, z). The number $n_1, n_2 \dots n_p$, and angles $\alpha_1, \alpha_2 \dots \alpha_p$ are independent of θ and, since r is a function of z, they may be regarded as functions of z only.

The shell is subjected to internal hydrostatic pressure P and all resulting forces are carried by the fibers, which are considered to constitute an undeformable membrane. T_1 and T_2 are the normal components of stress in the latitudinal and longitudinal directions at a point (r, θ, z). Because of cylindrical symmetry, T_1 and T_2 are independent of θ; and because there are equal numbers of fibers inclined at α_i and $\pi - \alpha_i$ ($i = 1,2, \dots p$), the shearing components of stress are zero.

Let τ_1, τ_2, ... τ_p be the tensions at (r, θ, z) in the fibers inclined at α_1, α_2 ... α_p. Then because of symmetry, τ_i (i=1, 2, ...p) must also be the tensions in the fibers inclined at $\pi-\alpha_1$. The number of fibers with inclination α_i per unit length measured along a line of longitude (θ = constant) is $n_i \cos \alpha_i$. Resolving the tensions in the fibers in the latitudinal direction the latitudinal tension T_1 is obtained

$$T_1 = 2 \sum_{i=1}^{p} \tau_i n_i \cos^2\alpha_i \qquad (2\text{-}143)$$

Similarly, the number of fibers inclined at α_i per unit length measured along a line of latitude (z = const) in $n_i \sin \alpha_i$, and resolving the tensions in fibers parallel to the longitudinal direction, the longitudinal tension T_2 is obtained

$$T_2 = 2 \sum_{i=1}^{p} \tau_i \sin 2\alpha_i \qquad (2\text{-}144)$$

If k_1 and k_2 are the curvatures of the membrane in the latitudinal and longitudinal directions

$$k_1 = \frac{1}{r}\frac{dz}{d} \quad \text{and} \quad k_2 = \frac{d}{dr}\left(\frac{dz}{d\sigma}\right) \qquad (2\text{-}145)$$

where τ is the distance from the equator, measured along a line of longitude.

The equations of equilibrium for the membrane are

$k_1 T_2 = {}^1/_2 P$ equilibrium in direction normal to membrane
$k_1 T_1 + k_2 T_2 = P$ equilibrium along a line of longitude $\qquad (2\text{-}146)$

Solving for T_1 and T_2 and combining

$$\sum_{i=1}^{p} \tau_i n_i \cos^2\alpha_i = \frac{P}{4k_1}\left(2 - \frac{k_2}{k_1}\right)$$

$$\sum_{i=1}^{p} \tau_i n_i \sin^2\alpha_i = P/4k_1 \qquad (2\text{-}147)$$

Since τ_i and n_i must be positive, and P and k_i are inherently positive

$2k_1 \geq k_2$

Weight of Fiber
Consider a ring lying between z and $z + dz$. A single fiber with inclination α to the latitudinal direction has a length $d\sigma/\sin \alpha$, where $d\sigma$ is the width of the ring measured along a line of longitude. The number of fibers at inclination a_i per unit length of a line of latitude is $n_i \sin \alpha_i$. The total length of fiber in the ring is therefore $4\pi_r$.

$(n_1 + n_2 \ldots n_p) \, d\sigma$

If the mass per unit length of fiber is π, the total mass of fiber covering the surface is

$$M = 4\pi p \int r \, (n_1 + n_2 + \ldots n_p) \frac{d\sigma}{dz} \, dz \tag{2-148}$$

Minimum Weight
If each fiber supports its maximum tensile force τ_1 the minimum weight of fiber required to withstand the internal pressure P is simply related to the volume of the vessel V by the following relationship

$$M = (3\pi P/\tau) \, V \tag{2-149}$$

For example, for an ellipsoid of revolution in which Z is the semi-axis of revolution, and R is the semiaxis at right angles to Z, $V = (4/3) \, \pi R^2 Z$, and

$$M = 4\pi\pi \, PR^2 Z/\tau \tag{2-150}$$

Isotensoid Design
In isotensoid design every fiber is at the same tension τ and if τ is at the same time the maximum stress the fiber is permitted to carry, this also becomes the minimum weight design. For this to be true, the inequalities must become equalities in which $\tau_i = \tau$.

Rewriting:

$$\sum_{i=1}^{P} n_i \cos^2 \alpha_i = \frac{P}{4\tau k_1} \left(2 - \frac{k_2}{k_1} \right)$$

$$\sum_{i=1}^{p} n_i \sin^2 \alpha_i = P/4\tau k_1 \tag{2-151}$$

Adding these together

$$\sum_{i=1}^{P} n_i = \frac{P}{4\tau k_1} \left(3 - \frac{k_2}{k_1} \right) \tag{2-152}$$

Any choice of $n_i \cos^2 \alpha_i$ and $n_i \sin^2 \alpha_i$, both functions of z which satisfy the equation, provides an isotensoid design of minimum mass M.

Geodesic-Isotensoid Design
On a surface of revolution, a geodesic satisfies the following equation

$$r \cos \alpha = R \cos \beta \tag{2-153}$$

in which α is the inclination of the geodesic to the line of latitude that has a radial distance r from the axis, and β is the inclination of the geodesic to the line of latitude of radius R. Attention here is restricted to shells of revolution in which r decreases with increasing z^2. an equator occurs at $z = 0$, all geodesics cross the equator, and all geodesics have an equation with R the radius at the equator.

If $r > R \cos \beta$, then $\cos \alpha > 1$ and α is imaginary. Therefore $r = R \cos \beta$ gives the extreme lines of latitude on the shell reached by the geodesic.

If $N(\beta) \sin \beta d\beta$ is the number of fibers per unit length of the equator with inclinations to it lying between β and $\beta + d\beta$, it can be shown that for a sphere

$$N(\beta) = PR/\tau\pi \tag{2-154}$$

The fiber distribution is independent of the angle β. For a cone with half-vertex angle γ

$$N(\beta) = \frac{3PR \cos \beta}{4\tau \cos \gamma} \tag{2-155}$$

For an ellipsoid of revolution

$$N(\beta) = \frac{PR^2}{\tau\pi Z} \left[\frac{1 + (3/2)\, v \cos^2\beta}{1 + v \cos^2\beta} + \frac{3}{4}\, (-v)^{1/2} \cos \beta \, 1n \frac{1 - (-v)^{1/2} \cos \beta}{1 + (-v)^{1/2} \cos \beta} \right] \tag{2-156}$$

in which $v + = (Z^2 - R^2)/R^2$ and provided that $v \geq 0$.

3 DESIGN PARAMETER

Load Determination

Loads on a fabricated product can produce different types of stresses within the material. There are basically static and dynamic stresses (Fig. 3.1). The magnitude of these stresses depends on many factors such as applied forces/loads, angle of loads, rate and point of application of each load, geometry of the structure, manner in which the structure is supported, and time at temperature. The behavior of the material in response to these induced stresses determines the performance of the structure.

Figure 3.1 Examples of stresses due to loads (Courtesy of Plastics FALLO)

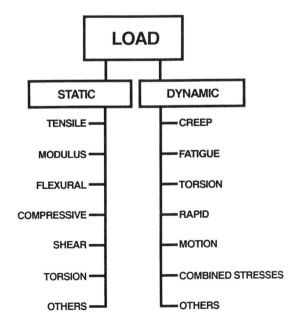

The behavior of materials (plastics, steels, etc.) under dynamic loads is important in certain mechanical analyses of design problems. Unfortunately, sometimes the engineering design is based on the static loading properties of the material rather than dynamic properties. Quite often this means over-design at best or incorrect design resulting in failure of the product in the worst case.

The complex nature of the dynamic behavior problem can be seen from Fig. 3.1, which depicts a wide range of interaction of dynamic loads that occurs with various materials (metals, plastics, etc.). Ideally, it would be desirable to know the mechanical response to the full range of dynamic loads for each material under all types of conditions. However, certain load-material interactions have more relative importance for engineering design, and significant as well as sufficient work on them exists already. The mechanical engineers, civil engineers, and metallurgical engineers have always found materials (includes plastic, steel, aluminum, etc.) to be most attractive to study. Even so, there is a great deal that we do not understand about these materials in spite of voluminous scientific literature existing worldwide. Each type of load response, e.g., creep, fatigue/vibratory, or impact, is a major field in itself. Data on each response is available. However there is always a desire to obtain more data.

The nature and complexity of applied loads as well as the shape requires the usual engineering calculations. For a simple engineering form like a plate, beam, or box structure the standard design formulas can be used with appropriate parameters relating to the factors of short- and long-time loadings, creep, fatigue, impact, and applying the viscoelastic plastic material behavior (Chapter 2). The term engineering formulas refers to those equations in engineering handbooks by which the stress analysis can be accomplished.

In a product load analysis the structure as a whole and each of its elements together are in a state of equilibrium. There are no unbalanced forces of tension, compression, flexure, or shear acting on the structure at any point. All the forces counteract one another, which results in equilibrium. When all the forces acting on a given element in the same direction are summed up algebraically, the net effect is no load. However the product does respond to the various forces internally.

These forces could deform the product due to internal stresses of varying types and magnitudes. This action could be immediate or to some time-temperature period based on its viscoelastic behavior and underestimating potential internal stresses. To overcome this situation different approaches are used, as explained in the engineering books.

An example is when the cross-sectional area of a product increases for a given load, the internal stresses are reduced, so make it thicker. Design is concerned with determining the stresses for a given shape and subsequently adjusting the shape until the stresses are neither high enough to risk fracture nor low enough to suggest that material is being wasted (costly).

The stress analysis design involves various factors. It requires the descriptions of the product's geometry, the applied loads and displacements, and the material's properties including its viscoelastic behavior. The result is to obtain numerical expressions for internal stresses as a function of the stress's position within the product and as a function of time-temperature as well.

With the more complex shapes the component's geometry complicates the design analysis for plastics (and other materials) and may make it necessary to carry out a direct analysis, possibly using finite element analysis (FEA) followed with prototype testing (Chapter 5).

Loads applied on products induce tension, compression, flexure, torsion, and/or shear, as well as distributing the loading modes. The product's particular shape will control the type of materials data required for analyzing it. The location and magnitude of the applied loads in regard to the position and nature of such other constraints as holes, attachment points, and ribs are important considerations that influences its shape. Also influencing the design decision will be the method of fabricating the product (Chapter 1).

Loads will generally fall into one of two categories, directly applied loads and strain-induced loads (Chapter 2. Isolator). Directly applied loads are usually easy to understand. They are defined loads that are applied to defined areas of the product, whether they are concentrated at a point, line, or boundary or distributed over an area. The magnitude and direction of these loads are known or can easily be determined. An example of a strain-induced load is when it is required that a product be deflected. The load developed is directly related to the strain that occurs. Unlike directly applied loads, strain-induced loads are dependent on the modulus of elasticity; when comparing TPs with TSs, the TPs will generally decrease quicker in magnitude over time. Many assembly and thermal stresses could be the result of these strain-induced loads.

Time-dependent applied loading effects the materials viscoelasticity (Chapter 2). Loads applied for short times and at normal rate cause material response that is essentially elastic in character. However, under sustained load plastics, particularly TPs, tend to creep, a factor that is

Figure 3.2 Example of intermittent loading

Figure 3.3 Loading and unloading examples of creep changes for engineering TPs

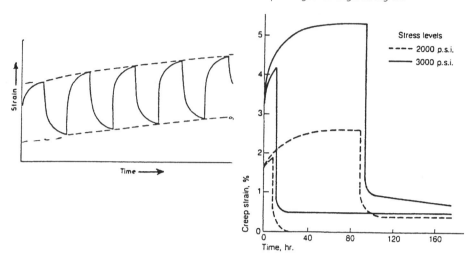

included in the design analysis.

Intermittent loading can involve creep and recovery over relatively long time periods. Creep deformation during one loading can be partly recovered in the unloading cycle, leading to a progressive accumulation of creep strain as the continuous intermittent load action continues (Figs. 3.2 and 3.3). This action in an improperly designed product will probably result in creep rupture. An analogue of creep behavior is the stress-relaxation cycle that can occur under constant strain. This behavior is particularly relevant with push-fit assemblies and bolted joints that rely on maintaining their load under constant strain. Special design features or analysis may be required to counteract excessive stress-relaxation.

There are intermittent or dynamic loads that occur over short time periods that can cause failure due to creep and possibly fatigue. This type loading condition applies to products such as motion control isolators, engine mounts, and other antivibration products; panels that vibrate and transmit noise; chairs; and road surface-induced loads carried to vehicle wheels and suspension systems. Plastics' relevant properties in this regard are material stiffness and internal damping, the latter of which can often be used to advantage in design (Chapter 2). Both properties depend on the frequency of the applied loads or vibrations, a dependence that must be allowed for in the design analysis. Design engineers unfamiliar with plastics' behavior will be able to apply the information contained in this book to applicable equations that involve such analysis as multiple and complex stress concentrations. The various machine-design texts and mechanical engineering handbooks review this subject.

Products can be stressed in a manner that is more complex than simple tension, compression, flexure, or shear. Because yielding will also occur under complex stress conditions, a yield criterion can be specified that will apply in all stress states. Any complex stress state can be resolved into the usual engineering three normal components acting along three mutually perpendicular (X, Y, Z) axes and into three shear components along the three planes of those axes. By making a proper choice it is possible to find a set of three axes along which the shear stresses will be zero. These are the principal axes, with the normal stresses along them being called the principal stresses.

Design analysis process

The nature of design analysis obviously depends on having product-performance requirements. The product's level of technical sophistication and the consequent level of analysis that can be justified costwise basically control these requirements. The analysis also depends on the design criteria for a particular product. If the design is strength limited, to avoid component failure or damage, or to satisfy safety requirements, it is possible to confine the design analysis simply to a stress analysis. However, if a plastic product is stiffness limited, to avoid excessive deformation from buckling, a full stress-strain analysis will likely be required. Even though many potential factors can influence a design analysis, each application fortunately usually involves only a few factors. For example, TPs' properties are dominated by the viscoelasticity relevant to the applied load. Anisotropy usually dominates the behavior of long-fiber RPs and so on.

The design analysis processes for metals, plastics, and RPs are essentially the same, However due to a certain degree of differences, they some-times appear to be drastically different. Experience of design analysis can be misleading if applied without consideration to plastics and RPs behaviors. The design analysis process is composed essentially of the three main steps: (a) assessment of stress and strain levels in the proposed design; (b) comparison of critical stress and/or deformation values with design criteria to ensure that the proposed design will satisfy product requirements and materials limitations; and (c) modification of the pro-posed design to obtain optimum satisfaction of product requirement.

For metallic materials, component design is usually strength limited so that the design criteria in step (b) are often defined in terms of materials strength values, that is, in terms of a maximum permissible stress. Even when the design criterion is avoidance of plastic flow, rather than avoidance of material failure, the criterion is specified by the limiting yield stress. In these cases, step (a) is only required to provide an analysis of the stress distribution in the component, and the strain and deformation distributions are of little practical interest. These conclusions are a consequence of the relatively high stiffness of metals, and the principal exception is the deformation of thin sections that may lead to buckling.

A further simplification can often arise if the stress analysis problem required in step (a) is statically determinate. In particular, this requires that the externally applied constraints (or boundary conditions) can all be expressed in the form of applied loads and not in terms of imposed relative displacements. The stress distribution depends on the applied loads and on the component geometry, but not on the material stiffness properties. Thus, it is identical for all materials, whether they be elastic, rigid, or any other form, provided only that the material is sufficiently stiff for satisfaction of the assumption that the applied loads can be considered to be applied to the undeformed, rather than deformed, component geometry.

Thus, for metallic materials in many idealized practical situations, the design process is simplified to a stress (but not strain or displacement) analysis followed by comparison and optimization with critical stress values. When the problem is not statically determinate, the stress analysis requires specification of material stiffness values, but the associated strain and deformation values are usually not required. Since the material behavior is usually represented adequately by linear isotropic elasticity, the stress analysis can be limited to that form, and there are many standard formulae available to aid the designer.

For plastics (unreinforced), the emphasis is somewhat different. Due to their relatively low stiffness, component deformations under load may be much higher than for metals, and the design criteria in step (b) are often defined in terms of maximum acceptable deflections. Thus, for example, a metal panel subjected to a transverse load may be limited by the stresses leading to yield and to a permanent dent. Whereas a plastics panel may be limited by a maximum acceptable transverse deflection even though the panel may recover without permanent damage upon removal of the loads. Even when the design is limited by material failure, it is usual to specify the materials criterion in terms of a critical failure strain rather than a failure stress. Thus, it is evident that strain and deformation play a much more important role for plastics than they do for metals. As a consequence, step (a) is usually required to provide a full stress/strain/deformation analysis and, because of the viscoelastic nature of plastics, this can pose a more difficult problem than for metals.

A particular distinction between the mechanical behaviors of metals and plastics is explained in order to avoid a possible confusion that could have arisen from the preliminary review. A typical stress/strain curve for a metal, exhibits a linear elastic region followed by yield at the yield stress, plastic flow, and ultimately failure at the failure stress. Yield and failure occur at corresponding strains, and one could define yield and failure in terms of these critical strains. This is not common practice because it is simpler in many cases to restrict step (a) to a stress analysis alone. By comparison, it may appear strange that it was stated above that plastics failure criteria are usually defined in terms of a critical strain (rather than stress) and, by comparison with the metals case, switching back from strain to stress may appear to be a minor operation.

Explanation of this apparent fallacy depends on recognition of the fact that stress and strain are not as intimately related for plastics as they are for metals. This is demonstrated by a set of stress/strain curves for a typical plastic where their loading rates increase. This emphasizes that the stress/strain curve for a plastic is not unique, but depends on the loading type, that is, also on time, frequency, or rate. For example, the stress/strain curves obtained at different loading rates and for metals these curves would essentially coincident. However, the behavior of plastics can be very different at low and high rates, and there is no unique relation between stress and strain since this depends on the loading rate too. It is evident that characterization of failure through a unique failure strain cannot be valid in general, but it can be a good approximation in certain classes of situations such as, for example, at high rates or under creep conditions.

Reinforced Plastic Analysis

For RPs, the emphasis and difficulty in the design analysis depends on the nature of the RPs. For a thermoplastic reinforced with short fibers, the viscoelastic nature of the matrix remains an important factor, and the discussion given above for unreinforced plastics is relevant. In addition, there may be a significant degree of anisotropy and/or inhomogeneity due to processing that could further complicate the analysis (Chapter 2). For thermosets reinforced with short fibers (for example, BMC; Chapter 1) there may be only a low level of visco-elasticity, anisotropy, and inhomogeneity, and metals-type design analysis may be a reasonable approximation. However, thermosets reinforced with long fibers can have a high degree of anisotropy (depends on lay up of reinforcement), and this must be taken into account in the design analysis. When thermoplastics are reinforced with long fibers there may be significant anisotropy and viscoelasticity, and this creates a potentially complex design analysis situation. In all cases, RPs failure characteristics may be specified in terms of a critical strain, and this requires the design analysis to be performed for stress and strain.

Long-fiber materials can often be tailored to the product requirements, and therefore materials design analysis and component design analysis interact strongly. If the component design analysis is statically determinate (stresses independent of materials properties) then this can be carried out first, and then the material can be designed to carry the stresses in the most efficient manner. However, if the analysis is not statically determinate, then the component stresses depend on material anisotropy, and material and product design have to be carried out and optimized at the same time. This is also the case if component shape is regarded as one of the variable design parameters.

In summary, it can be seen that plastics and RPs design analysis follows the same three steps (a) to (c) as that for metals, but there are some differences of emphasis and difficulty. In particular, step (a) is usually more substantial for the newer materials, partly because a full stress/strain/deformation analysis is required and partly because of the need to take account of viscoelasticity, inhomogeneity, and/or anisotropy. For long fiber materials, the component design analysis may need to contain the associated material design analysis.

Stress Analysis

Different nondestructive techniques are used to evaluate the stress level in products. They can predict or relate to potential problems. There is the popular electrical resistance strain gauges bonded on the surface of the product. This method identifies external and internal stresses. The various configurations of gauges are made to identify stresses in different directions. This technique has been extensively used for over half a century on very small to very large products such as toys to airplanes and missiles.

There is the optical strain measurement system that is based on the principles of optical interference. It uses Moire, laser, or holographic interferometry. Another very popular method is using solvents that actually attack the product. It works only with those plastics that can be attacked by a specific solvent. Immersed products in a temperature controlled solvent for a specific time period identifies external and internal stresses. After longer time periods products will self-destruct. Stress and crack formations can be calibrated using different samples subjected to different loads.

With the brittle coating system that is applied on the surface of a product one identifies conditions such as stressed levels, cracks, etc. A lacquer coating is applied, usually sprayed on the surface of the product. It provides experimental quantitative stress-strain measurement data. As the product is subjected to a load simulating the load that would be encountered in service, cracks begin to appear in the coating. The extent of cracks is noted for each increment of load. Prior to this action, the coating is calibrated by applying the coating on a simple beam and observing the strain at which cracks appear and relating them to the stress behavior of the beam.

Photoelastic measurement is a popular and useful method for identifying stress in transparent plastics. Quantitative stress measurement is possible with a polarimeter equipped with a calibrated compensator. It makes stresses visible. The optical property of the index of refraction will change with the level of stress (strain). When the photoelastic material is stressed, the plastic becomes birefringent identifying the different levels of stress via color patterns.

This photoelastic stress analysis is a technique for the nondestructive determination of stress and strain components at any point in a stressed product by viewing a transparent plastic product. If not transparent, a plastic coating is used such as certain epoxy, polycarbonate, or acrylic plastics. This test method measures residual strains using an automated electro-optical system.

The photoelastic technique relates to the Brewster's Constant law. It states that the index of refraction in a strained material becomes directional, and the change of the index is proportional to the magnitude of the strain present. Thus a polarized beam in a clear plastic splits into two wave fronts (X and Y directions) that contain vibrations oriented along the directions of principal strains. The index of refraction in these directions is different and the difference (or birefringence) is proportional to the stress level. Result is the colorful patterns seen when stressed plastic are placed between two polarized filters providing qualitative analysis. Observed colors correspond to different levels of retardation at that point, which in turn correspond to stress levels.

Stress-strain behavior

The information presented throughout this book is used in different loading equations. As an example stress-strain data may guide the designer in the initial selection of a material. Such data also permit a designer to specify either design stresses or strains safely within the proportional/elastic limit of the material. However for certain products such as a vessel that is being designed to fail at a specified internal pressure, the designer may choose to use the tensile yield stress of the material in the design calculations.

Designers of most structures specify material stresses and strains well within the proportional/elastic limit. Where required (with no or limited experience on a particular type product materialwise and/or processwise) this practice builds in a margin of safety to accommodate the effects of improper material processing conditions and/or unforeseen loads and environmental factors. This practice also allows the designer to use design equations based on the assumptions of small deformation and purely elastic material behavior. Other important properties derived from stress-strain data that are used include modulus of elasticity and tensile strength.

Rigidity (EI)

Tensile modulus of elasticity (E) is one of the two factors that determine the stiffness or rigidity (EI) of structures comprised of a material. The other is the moment of inertia (I) of the appropriate cross section, a purely geometric property of the structure. In identical products, the higher the modulus of elasticity of the material, the greater the rigidity; doubling the modulus of elasticity doubles the rigidity of the product. The greater the rigidity of a structure, the more force must be applied to produce a given deformation.

It is appropriate to use E to determine the short-term rigidity of structures subjected to elongation, bending, or compression. It may be more appropriate to use the flexural modulus to determine the short-term rigidity of structures subjected to bending, particularly if the material comprising the structure is non-homogeneous, as foamed or fiber-reinforced materials tend to be. Also, if a reliable compressive modulus of elasticity is available, it can be used to determine short-term compressive rigidity, particularly if the material comprising a structure is fiber-reinforced. The room temperature E for several plastics and some other materials are presented in Chapter 2.

Hysteresis Effect

Hysteresis relates to the relation of the initial load applied to a material and its recovery rate when the load is released. There can be a time lapse that depends on the nature of the material and the magnitude of the stresses involved. This plastic behavior is typically nonlinear and history dependent. This incomplete recovery of strain in a material subjected to a stress during its unloading cycle is due to energy consumption. Upon unloading, complete recovery of energy does not occur. During a static test this phenomenon is called elastic hysteresis; for vibratory stresses it is called damping. The area of this hysteresis loop, representing the energy dissipated per cycle, is a measure of the damping properties of the material. Under vibratory conditions the energy dissipated varies approximately as the cube of the stress.

This energy is converted from mechanical to frictional energy (heat). It can represent the difference in a measurement signal for a given process property value when approached first from a zero load and then from a full scale as shown in Figs. 3.4 and 3.5. They provide examples of recovery to near zero strain. It shows that material can withstand stress beyond its proportional limit for a short time, resulting in different degrees of the hysteresis effect.

The hysteresis heating failure occurs more commonly in plastic members subject to dynamic loads. An example is a plastic gear. With the gear teeth under load once per revolution, it is subjected to a bending load that transmits the power from one gear to another. Another example is a link that is used to move a paper sheet in a copier or in an accounting machine

Figure 3.4 Hysteresis recovery effects

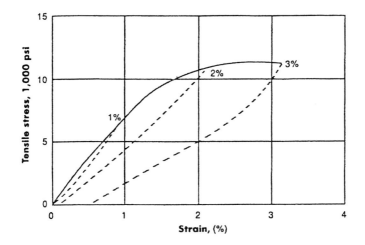

Figure 3.5 Hysteresis loop related to cyclic loading

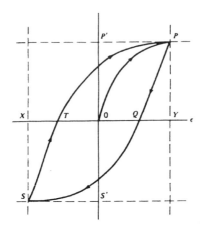

from one operation to the next. The load may be simple tensile or com-pressive stresses, but more commonly it is a bending load.

Poisson's Ratio

Poisson's ratio is a required constant in engineering analysis for deter-mining the stress and deflection properties of materials (plastics, metals, etc.). It is a constant for determining the stress and deflection properties of structures such as beams, plates, shells, and rotating discs. With plastics when temperature changes, the magnitude of stresses and strains, and the direction of loading all have their effects on Poisson's ratio. However, these factors usually do not alter the typical range of values enough to affect most practical calculations, where this constant is frequently of only secondary importance. The application of Poisson's ratio is frequently required in the design of structures that are markedly 2-D or 3-D, rather than 1-D like a beam. For example, it is needed to calculate the so-called plate constant for flat plates that will be subjected to bending loads in use. The higher Poisson's ratio, the greater the plate constant and the more rigid the plate.

When a material is stretched, it's cross-sectional area changes as well as its length. Poisson's ratio (v) is the constant relating these changes in dimensions. It is defined as the ratio of the change in lateral width per unit width to change in axial length per unit length caused by the axial stretching or stressing of a material. The ratio of transverse strain to the corresponding axial strain below the tensile proportional limits.

For plastics the ratio falls within the range of 0 to 0.5. With a 0 ration there is no reduction in diameter or contraction laterally during the elongation but would undergo a reduction in density. A value of 0.5 would indicate that the specimen's volume would remain constant during elongation or as the diameter decreases such as with elastomeric or rubbery material. Plastic range is usually from about 0.2 to 0.4; natural rubber is at 0.5 and reinforced TPs at 0.1 to 0.4. In mathematical terms, Poisson's ratio is the diameter of the test specimen before and after elongation divided by the length of the specimen before and after elongation. Poisson's ratio will have more than one value if the material is not isotropic. (Table 3.1)

Table 3.1 Poisson's ratios (and shear data) for different thermoplastics

Plastic	Poisson's ratio	Shear modulus MPa	Shear stress MPa
ABS	0.35	965	51.2
	0.36	660	30.0
Acetal homopolymer	0.35	1340	65.5
Acetal copolymer	0.35	1000	53.0
Nylon (0.2 wt%)	0.34–0.43		66.4
Polycarbonate	0.37	785	41.5
Polymethyl methacrylate	0.35		44.6

Brittleness

Brittleness identifies material easily broken, damaged, disrupted, cracked, and/or snapped. Brittleness can result from different conditions such as from drying, plasticizer migration, etc. Brittle materials exhibit tensile S-S behaviors different from the usual S-S curves. Specimens of such materials fracture without appreciable material yielding. They lack toughness. Their brittle point is the highest temperature at which a plastic or elastomer fractures in a prescribed impact test procedure.

Plastics that are brittle frequently have lower impact strength and higher stiffness properties. A major exception is reinforced plastics. The tensile S-S curves of brittle materials often show relatively little deviation from the initial linearity, relatively low strain at failure, and no point of zero slope. Different materials may exhibit significantly different tensile S-S behavior when exposed to different factors such as the same temperature and strain rate or at different temperatures.

A brittleness temperature value is used. It is the temperature statistically calculated where 50% of the specimens would probably fail 95% of the time when a stated minimum number are tested. The 50% failure temperature may be determined by statistical calculations.

There is a Griffith design failure theory. It expresses the strength of a material in terms of crack length and fracture surface energy. Brittle fracture is based on the idea that the presence of cracks determines the brittle strength and crack propagation occurs. It results in fracture rate of decreased elastically stored energy that at least equals the rate of formation of the fracture surface energy due to the creation of new surfaces.

Ductile

Ductility is the amount of strain that a material can withstand before fracture. In turn the fracture behavior of plastics, especially microscopically brittle plastics, is governed by the microscopic mechanisms operating in a heterogeneous zone at their crack or stress tip because of internal or external forces. In TPs, craze zones can develop that are important microscopic features around a crack tip governing strength behavior. Fracture is preceded by the formation of a craze zone, which is a wedge shaped region spanned by oriented microfilms. Methods of craze zone measurements include optical emission spectroscopy, diffraction techniques, scanning electron beam microscopy, and transmission electron microscopy.

Fig. 3.6 is an example of the ductile plastic tensile stress-strain curve. This curve identifies behavior so that as the strain increases, stress initially increases approximately proportionately (from point 0 to point A). Point A is called the proportional limit. From point 0 to point B, the behavior of the material is purely elastic/stretches; but beyond point B, the material exhibits an increasing degree of permanent deformation/stretch. Point B is the elastic limit of the material. At point C the material is yielding and so its coordinates are called the yield strain and stress (strength) of the material. Point D relates to the S-S elongation at break/failure. Table 3.2 provides these type data at room temperature for different materials.

Temperature influences the S-S curve. With a decrease in temperature the yield stress and strain usually decreases or the strain rate decreases. Point D corresponds to specimen fracture/failure. It represents the maximum elongation of the material specimen; its coordinates are called the ultimate, or failure strain and stress. As temperature decreases the ultimate elongation usually decreases or the strain rate increases.

Crazing

Crazing is also called hairline craze. They can be fine, thin, tiny type cracks that may extend in an unreinforced or reinforced plastic network

Figure 3.6 Tensile stress-strain behavior of ductile plastics

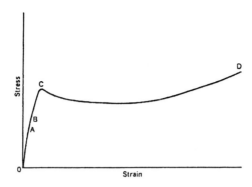

Table 3.2 Tensile data

Plastic	Modulus MPa	Yield stress MPa	Elongation at yield, %	Elongation at break, %
ABS	2,700	55	2.5	75
Acetal homopolymer	3,100	69	12	75
Acetal copolymer	2,800	61	12	60
Acrylic	3,000	72		5.4
Nylon	2,400	82	5	60
Phenolic	19,300	62	8	90
Polyethylene	1,200	30	20	600
Polypropylene	1,400	35	12	400
Polystyrene	3,100	25	8	60
Polysulfone	2,500	70	6	100

on or under the surface or through a layer of a plastic material. Different conditions and effects occur depending on the type plastic, load conditions, and environment. The formation of crazes are like cracks in that they are wedge shaped and formed perpendicular to the applied stress. They differ from cracks by containing plastic that is stretched in a highly oriented manner perpendicular to the plane of the craze. They are parallel to the applied stress direction. Another major distinguishing feature is that unlike cracks, crazes are able to support stress.

With the application of static loading, the strain at which crazes start to form decreases as the applied stress decreases. In constant strain-rate testing crazes always start to form at a well-defined stress level. Crazes start sooner under high stress levels. When tensile stress is applied to an amorphous (Chapter 1) plastic such as acrylics, PVCs, PS, and PCs, crazing may occur before fracturing. Crazing occurs in crystalline plastics, but in those its onset is not readily visible. It also occurs in most fiber-reinforced plastics, at the time-dependent knee in the stress-strain curve.

Environmental stress cracking is the cracking of certain plastic products that becomes exposed to a chemical agent while it is under stress. This effect may be caused by exposure to such agents as cleaners or solvents. The susceptibility of affected plastics to stress cracking by a particular chemical agent varies considerably among plastics, particularly the TPs.

The resistance of a given plastic to attack may be evaluated by using either constant-deflection or constant-stress tests in which specimens are usually coated with the chemical or immersed in the chemical agent. After a specified time the degree of chemical attack is assessed by measuring such properties as those of tensile, flexural, and impacts. The results are then compared to specimens not yet exposed to the chemical. In addition to chemical agents and the environment for testing may also require such other factors as thermal or other energy-intensive conditions.

It is possible with solvents of a particular composition to determine quantitatively the level of stress existing in certain TP products where undesirable or limited fabricated-in stresses exist. The stresses can be residual (internal) stresses resulting from the molding, extrusion, or other process that was used to fabricate the plastic product. Stresses can also be applied such as bending the product. As it has been done for over a half century, the product is immersed in the solution that attacks the plastic for various time periods. Any initial cracks or surface imperfections provide information that stresses exist. Other tests conducted can be related to the stress-time information. Information on the solvent mixtures suitable for this type of test and how to interrupt them are available from plastic material suppliers or can be determined from industry test data which show solvents that effect the specific plastic to be evaluated.

TP cracking develops under certain conditions of stress and environment sometimes on a microscale. Because there are no fibrils to connect surfaces in the fracture plane (except possibly at the crack tip), cracks do not transmit stress across their plane. Cracks result from embrittlement, which is promoted by sustained elevated temperatures and ultraviolet, thermal, chemical, and other environments.

For the designer it is not important whether cracking develops upon exposure to a benign or an aggressive medium. The important considerations are the embrittlement itself and the fact that apparently benign environments can cause serious brittle fractures when imposed on a product that is under sustained stress and strain, which is true of certain plastics.

Crazing or stress whiting is damage that can occur when a TP is stretched near its yield point. The surface takes on a whitish appearance in regions that are under high stress. Crazing is usually associated with yielding. For practical purposes stress whiting is the result of the formation of microcracks or crazes, which is another form of damage. Crazes are not true fractures, because they contain strings of highly oriented plastic that connect the two flat faces of the crack. These fibrils are surrounded by air voids. Because they are filled with highly oriented fibrils, crazes are capable of carrying stress, unlike true fractures. As a result, a heavily crazed product can still carry significant stress, even though it may appear to be fractured.

It is important to note that crazes, microcracking, and stress whiting represent irreversible first damage to a material, which could ultimately cause failure. This damage usually lowers the impact strength and other properties of a material compared to those of undamaged plastics. One reason is that it exposes the interior of the plastic to attack and subsequent deterioration by aggressive fluids. In the total design evaluation, the formation of stress cracking or crazing damage should be a criterion for failure, based on the stress applied.

Stress Whitening

It is the appearance of white regions in a TP when it is stressed. A stress-whitening zone may be a sign of crazing in some plastics where individual fine crazes may be difficult to detect. Stress whitening occurs fairly late in the rupture stage, just prior to yielding. The surface takes on a whitish appearance in regions that are under high stress. It is usually associated with yielding. For practical purposes, stress whiting is the result of the formation of microcracks or crazes that is a form of damage.

Combined stresses

In the direct design procedure the assumption is made that no abrupt changes occur in cross-section, discontinuities in the surface, or holes through the member. This is not the case in most structural parts. The stresses produced at these discontinuities are different in magnitude from those calculated by various design methods. The effect of the localized increase in stress, such as that caused by a notch, fillet, hole, or similar stress raiser, depends mainly on the type of loading, the geometry of the product, and the material. As a result, it is necessary to consider a stress-concentration factor. In general it will have to be determined by the methods of experimental stress analysis or the theory of elasticity, and by a simple theory without taking into account the variations in stress conditions caused by geometrical discontinuities such as holes, grooves, and fillets. For ductile materials it is not customary to apply stress-concentration factors to members under static loading. For brittle materials, however, stress concentration is serious and should be considered.

There are conditions of loading a product that is subjected to a combination of tensile, compressive, and/or shear stresses. For example, a shaft that is simultaneously bent and twisted is subjected to combined stresses, namely, longitudinal tension and compression, and torsional shear. For the purposes of analysis it is convenient to reduce such systems of combined stresses to a basic system of stress coordinates known as principal stresses. These stresses act on axes that differ in general from the axes along which the applied stresses are acting and represent the maximum and minimum values of the normal stresses for the particular point considered. There are different theories that relate to these stresses. They include Mohr's Circle, Rankine's, Saint Venant, Guest, Hencky-Von Mises, and Strain-Energy.

Surface Stresses and Deformations

It can be said that the design of a product involves analytical, empirical, and/or experimental techniques to predict and thus control mechanical stresses. Strength is the ability of a material to bear both static (sustained) and dynamic (time-varying) loads without significant permanent deformation. Many non-ferrous materials suffer permanent deformation under sustained loads (creep). Ductile materials withstand dynamic loads better than brittle materials that may fracture under sudden load application. As reviewed, materials such as plastics often exhibit significant changes in material properties over the temperature range encountered by a product.

There are examples where control of deflection or deformation during service may be required. Such structural elements are designed for stiffness to control deflection but must be checked to assure that strength criteria are reached. A product can be viewed as a collection of individual elements interconnected to achieve an overall systems function. Each element may be individually modeled; however, the model becomes complex when the elements are interconnected.

The static or dynamic response of one element becomes the input or forcing function for elements adjacent or mounted to it. An example is the concept of mechanical impedance that applies to dynamic environments and refers to the reaction between a structural element or component and its mounting points over a range of excitation frequencies. The reaction force at the structural interface or mounting point is a function of the resonance response of an element and may have an amplifying or damping effect on the mounting structure, depending on the spectrum of the excitation. Mechanical impedance design involves control of element resonance and structure resonance, providing compatible impedance for interconnected structural and component elements.

As an example view a 3-D product that has a balanced system of forces acting on it, F_1 through F_5 in Fig. 3.7, such that the product is at rest. A product subjected to external forces develops internal forces to transfer and distribute the external load. Imagine that the product in

Figure 3.7 Example of stresses in a product

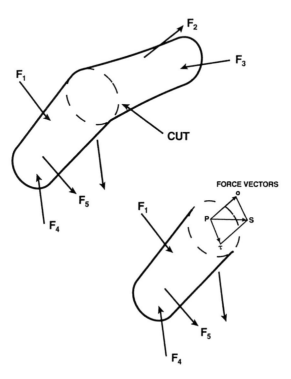

Fig. 3.7 is cut at an arbitrary cross-section and one part removed. To keep the body at rest there must be a system of forces acting on the cut surface to balance the external forces. These same systems of forces exist within the uncut body and are called stresses. Stresses must be described with both a magnitude and a direction. Consider an arbitrary point, P, on the cut surface in the figure where the stress, S, is as indicated. For analysis, it is more convenient to resolve the stress, S, into two stress components. One acts perpendicular to the surface and is called a normal or direct stress, σ. The second stress acts parallel to the surface and is called a shear stress, τ.

Creep

Plastic materials subjected to a constant stress can deform continuously with time and the behavior under different conditions such as temperature. This continuous deformation with time is called creep or cold flow. In some applications the permissible creep deformations are critical, in others of no significance. But the existence of creep necessitates information on the creep deformations that may occur during the expected life of the product. Materials such as plastic, RP, zinc, and tin creep at room temperature. Aluminum and magnesium alloys start to creep at around 300°F. Steels above 650°F must be checked for creep.

There are three typical stages. The initial strain takes place almost immediately, consisting of the elastic strain plus a plastic strain near its end, if the deformation extends beyond the yield point. This initial action in the first stage shows a decreasing rate of elongation that can be called strain hardening (as in metals). The action most important to the designer's working area concerns the second stage that is at a minimum strain rate and remains rather constant. In the third stage a rapid increase in the creep rate occurs with severe specimen necking/ thickness reduction and ultimately rupture. It is important for the designer to work in the second stage and not enter the third stage. Thus, after plotting the creep vs. time data of a 1,000 h test, the second stage can be extrapolated out to the number of hours of desired product life.

These test specimens may be loaded in tension or flexure (with some in compression) in a constant temperature environment. With the load kept constant, deflection or strain is recorded at regular intervals of hours, days, weeks, months, or years. Generally, results are obtained at different stress levels.

In conducting a conventional creep test, curves of strain as a function of time are obtained for groups of specimens; each specimen in one group is subjected to a different constant stress, while all of the specimens in the group are tested at one temperature. In this manner families of curves are obtained. Important are the several methods that have been proposed for the interpretation of such data.

The rate of viscoelastic creep and stress relaxation at a given temperature may vary significantly from one TP to another because of differences in the chemical structure and shape of the plastic molecules (Chapter 1). These differences affect the way the plastic molecules interact with each other. Viscoelastic creep and stress relaxation tests are generally conducted up to 1,000 hours. Time-temperature super-positioning is often used to extrapolate this 1,000 hours of data to approximately 100,000 hours ($\approx$ 12 years). Basically with TPs subjected to heat there is an increase in the rate of creep and stress relaxation. The TSs and particularly reinforced thermosets (RTSs) remains relatively unaffected until a high temperature is encountered.

Usually the strain readings of a creep test can be more accessible if they are presented as a creep modulus that equals stress divided by strain. In the viscoelastic plastic, the strain continues to increase with time while the stress level remains constant. Result is an appearance of a changing modulus. This creep modulus also called the apparent modulus or viscous modulus when graphed on log-log paper, is a straight line and lends itself to extrapolation for longer periods of time.

Plastic viscoelastic nature reacts to a constant creep load over a long period of time by an ever-increasing strain. With the stress being constant, while the strain is increasing, result is a decreasing modulus. This apparent modulus and the data for it are collected from test observations for the purpose of predicting long-term behavior of plastics subjected to a constant stress at selected temperatures.

The creep test method of loading and material constituents influences creep data. Increasing the load on a part increases its creep rate. Particulate fillers provide better creep resistance than unfilled plastics but are less effective than fibrous reinforcements. Additives influence data such as the effect of a flame-retardant additive on the flexural modulus provides an indication of its effect on long-time creep. Increasing the level of reinforcement in a composite increases its resistance to creep. Glass-fiber-reinforced amorphous TP RPs generally has greater creep resistance than glass fiber-reinforced crystalline TP RPs containing the same amount of glass fiber. Carbon-fiber reinforcement is more effective in resisting creep than glass-fiber reinforcement.

Figure 3.8 Mechanical Maxwell model

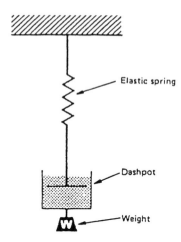

Elastic spring

Dashpot

Weight

For the designer there is generally a less-pronounced curvature when creep and relaxation data are plotted log-log. Predictions can be made on creep behavior based on creep and relaxation data. This usual approach makes it easier to extrapolate, particularly with creep modulus and creep-rupture data.

To relate the viscoelastic behavior of plastics with an S-S curve the popular Maxwell model is used, this mechanical model is shown in Fig. 3.8. This model is useful for the representation of stress relaxation and creep with Newtonian flow analysis that can be related to plastic's non-Newtonian flow behavior. It consists of a spring [simulating modulus of elasticity (E)] in series with a dashpot of coefficient of viscosity (η). It is an isostress model (with stress δ), the strain (ε) being the sum of the individual strains in the spring and dashpot.

Based on this mechanical loading system a differential representation of linear viscoelasticity is produced as:

$$d\varepsilon/dt = (1/E) \, d\delta/dt + (\delta/\eta) \tag{3-1}$$

When a load is applied to the system the spring will deform. The dashpot will remain stationary under the applied load, but if the same load continues to be applied, the viscous fluid in the dashpot will slowly leak past the piston, causing the dashpot to move. Its movement corresponds to the strain or deformation of the plastic material.

When the stress is removed, the dashpot will not return to its original position, as the spring will return to its original position. The result is a viscoelastic material behavior as having dual actions where one is of an elastic material (spring), and the other like the viscous liquid in the dashpot. The properties of the elastic phase is independent of time, but the properties of the viscous phase are very much a function of time, temperature, and stress (load). A thinner fluid resulting from increased temperature under a higher pressure (stress) will have a higher rate of leakage around the piston of the dashpot during the time period. A greater creep occurs at this higher temperature that caused higher stress levels and strain.

The Maxwell model relates to a viscoelastic plastic's S-S curve. The viscoelasticity of the plastic causes an initial deformation at a specific load and temperature. It is followed by a continuous increase in strain under identical test conditions until the product is either dimensionally out of tolerance or fails in rupture as a result of excessive deformation.

Test data using the apparent creep modulus approach is used as a method for expressing creep. It is a convenient method of expressing creep because it takes into account initial strain for an applied stress plus the deformation or strain that occurs with time. Because parts tend to deform in time at a decreasing rate, the acceptable strain based on service life of the part must be determined. The shorter the duration of load, the higher the apparent modulus and the higher the allowable stress.

When plotted against time, they provide a simplified means of predicting creep at various stress levels. It takes into account the initial strain for an applied stress plus the amount of deformation or strain that occurs over time. Fig. 3.9 shows curves of deformation versus time. Beyond a certain point, creep is small and may safely be neglected for many applications.

Figure 3.9 Apparent creep modulus vs. log time with increased load (Courtesy of Mobay/Bayer)

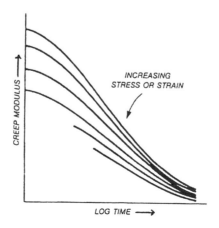

The acceptable strain based on the desired service life of a product can be determined since they deform under load in time at a decreasing rate. Short duration results in the higher apparent modulus and in turn a higher allowable stress. The apparent modulus is most easily explained with an example. The apparent modulus E_a is calculated in a very simplified approach as:

$$E_a = Stress/Initial\ strain + Creep \tag{3-2}$$

As long as the stress level is below the elastic limit of the material, its E can be obtained from the usual equation:

$$E = Stress/Strain \tag{3-3}$$

If a compressive stress of 10,000 psi (69 MPa) is used, the result is a strain of 0.015 in./in. (0.038 cm/cm) for FEP plastic at 63°F (17°C). Thus:

$$E = 10,000/0.015 = 667,000\ psi\ (4,600\ MPa) \tag{3-4}$$

If this stress level remains for 200 hours, the total strain will be the sum of the initial strain plus the strain due to time. This total strain can be obtained from a creep-data curve. With a total deformation under a tension load for 200 hours of 0.02 in./in., the result is:

$$E = 10,000/0.02 = 5,000,000\ psi\ (3,500\ MPa) \tag{3-5}$$

An E can then be determined for one year. Extrapolating from the straight-line creep-data curve gives a deformation of 0.025 in./in. the E becomes:

$$E = 10,000/0.025 = 400,000\ psi\ (2,800\ MPa) \tag{3-6}$$

Different attempts have been used to create meaningful formulas for the apparent modulus change with respect to time. However the factors in the formulas that would fit all conditions are more complicated to use than presenting test data in a graph form and using it as the means for predicting the strain (elongation) at some distant point in time. Log-log test data usually form a straight line and lend themselves to easy extrapolation by the designer. The slope of the straight line depends on the material being tested such as its rigidity and temperature of heat deflection with the amount of stress in relation to tensile strength.

Long term behavior of plastics involves plastic exposure to conditions that include continuous stresses, environment, excessive heat, abrasion, and/or continuous contact with liquids. Tests such as those outlined by ASTM D 2990 that describe in detail the specimen preparations and testing procedure are intended to produce consistency in observations and records by various manufacturers, so that they can be correlated to provide meaningful information to product designers. The procedure under this heading is intended as a recommendation for uniformity of making setup conditions for the test, as well as recording the resulting data. The reason for this move is the time consuming nature of the test (many years' duration), which does not lend itself to routine testing. The test specimen can be round, square, or rectangular and manufactured in any suitable manner meeting certain dimensions. The test is conducted under controlled temperature and atmospheric conditions.

The requirements for consistent results are outlined in detail as far as accuracy of time interval, of readings, etc., in the procedure. Each report of test results should indicate the exact grade of material and its supplier, the specimen's method of manufacture, its original dimensions, type of test (tension, compression, or flexure), temperature of test, stress level, and interval of readings. When a load is initially applied to a specimen, there is an instantaneous strain or elongation. Subsequent to this, there is the time-dependent part of the strain (creep), which results from the continuation of the constant stress at a constant temperature. In terms of design, creep means changing dimensions and deterioration of product strength when the product is subjected to a steady load over a prolonged period of time.

All the mechanical properties described in tests for the conventional data sheet properties represented values of short-term application of forces. In most cases, the data obtained from such tests are used for comparative evaluation or as controlling specifications for quality determination of materials along with short-duration and intermittent-use design requirements. The visualization of the reaction to a load by the dual component interpretation of a material is valuable to the understanding of the creep process, but meaningless for design purposes. For this reason, the designer is interested in actual deformation or part failure over a specific time span. The time segment of the creep test is common to all materials, strains are recorded until the specimen ruptures or the specimen is no longer useful because of yielding. In either case, a point of failure of the test specimen has been reached, this means making observations of the amount of strain at certain time intervals which will make it possible to construct curves that could be extrapolated to longer time periods. The initial readings are 1, 2, 3, 5, 7, 10, and 20 h, followed by readings every 24 h up to 500 h and then readings every 48 h up to 1,000 h.

The strain readings of a creep test can be more convenient to a designer if they are presented as a creep modulus. In a viscoelastic material, strain continues to increase with time while the stress level remains constant.

Since the modulus equals stress divided by strain, there is the appearance of a changing modulus.

The method of obtaining creep data and their presentation have been described; however, their application is limited to the exact same material, temperature use, stress level, atmospheric conditions, and type of test (tensile, compression, flexure) with a tolerance of ±10%. Only rarely do product requirement conditions coincide with those of the test or, for that matter, are creep data available for all grades of material. In those cases a creep test of relatively short duration such as 1,000 h can be instigated, and the information can be extrapolated to the long-term needs. It should be noted that reinforced thermoplastics and thermosets display much higher resistance to creep (Chapter 4).

The stress-strain-time data can be plotted as creep curves of strain vs. log time (Fig. 3.10 top view). Different methods are also used to meet specific design requirements. Examples of methods include creep curves at constant times to yield isochronous stress versus strain curves or at a constant strain, giving isometric stress versus log-time curves, as shown in the bottom views in Fig. 3.10.

To date the expected operating life of most plastic products designed to

Figure 3.10 Examples of different formatted creep vs. log time curves (Courtesy of Mobay/Bayer)

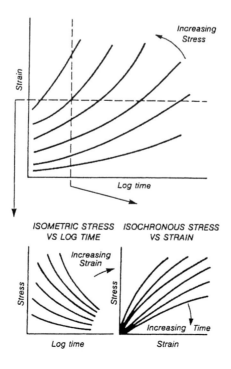

withstand creep is usually at least ten to twenty years. Available data at the time of designing will not be available so one uses available creep test-data based on at least 1,000 hours that is the recommended time specified in the ASTM standard. These long-time data have been developed and put to use in designs for over a half-century in designing plastic materials. An example is the engineering design and fabrication of the first all-plastic airplane.

Creep information is not as readily available as that from short-term property data sheets. From a designer's viewpoint, it is important to have creep data available for products subjected to a constant load for prolonged periods of time. The cost of performing or obtaining the test in comparison with other expenditures related to product design would be insignificant when considering the element of safety and confidence it would provide. Furthermore, the proving of product performance could be carried out with a higher degree of favorable expectations as far as plastic material is concerned. Progressive material manufacturers can be expected to supply the needed creep and stress-strain data under specified use conditions when requested by the designer; but, if that is not the case, other means should be utilized to obtain required information.

In conclusion regarding this subject, it can be stated that creep data and a stress-strain diagram indicate whether plain plastic properties can lead to practical product dimensions or whether a RP has to be substituted to keep the design within the desired proportions. For long-term product use under continuous load, plastic materials have to be considered with much greater care than would be the case with metals.

Preparing the important creep rupture data for the designer is similar to that for creep except that higher stresses are used and the time is measured to failure. It is not necessary to record strain. The data are plotted as the log stress vs. log time to failure. In creep-rupture tests it is the material's behavior just prior to the rupture that is of primary interest. In these tests a number of samples are subjected to different levels of constant stress, with the time to failure being determined for each stress level.

The overall behavior is the time-dependent strain at which crazing, stress whitening, and rupture decreases with a decreasing level of sustained stress. The time to develop these defects increases with a decreasing stress level.

Thermoplastic fiber RPs display a degree of creep, and creep rupture compared to RPs with thermoset plastics. TS plastic RPs reinforced with carbon and boron is very resistant to deformation (creep) and failure (creep rupture) under sustained static load when they are loaded in a fiber-dominated direction. The creep and creep rupture behavior of aramid fiber is not as good but still rather high. Creep and creep rupture with RPs has to take into consideration the stresses in matrix-dominated directions. That is fiber oriented directional properties influence the data.

In service products may be subjected to a complex pattern of loading and unloading cycles that is represented by stress relaxation. This variability of intermittent loading can cause design problems in that it would clearly not be feasible to obtain experimental data to cover all possible loading situations, yet to design on the basis of constant loading at maximum stress would not make efficient use of materials or be economical. In such cases it is useful to have methods for predicting the extent of the accumulated strain that will be recovered during the no load periods after cyclic loading.

Tests have been conducted that provide useful stress relaxation data. Plastic products with excessive fixed strains imposed on them for extended periods of time could fail. Data is required in applications such as press fits, bolted assemblies, and some plastic springs. In time, with the strain kept constant the stress level will decrease, from the same internal molecular movement that produces creep. This gradual decay in stress at a constant strain (stress-relaxation) becomes important in these type applications in order to retain preloaded conditions in bolts and springs where there is concern for retaining the load.

The amount of relaxation can be measured by applying a fixed strain to a sample and then measuring the load with time. The resulting data can be presented as a series of curves. A relaxation modulus similar to the creep modulus can also be derived from the relaxation data, it has been shown that using the creep modulus calculated from creep curves can approximate the decrease in load from stress relaxation. From a practical standpoint, creep measurements are generally considered more important than stress-relaxation tests and are also easier to conduct.

The TPs are temperature dependent, especially in the region of the plastics' glass transition temperature (Tg). Many unreinforced amorphous types of plastics at temperatures well below the T_g have a tensile modulus of elasticity of about 3×10^{10} dynes/cm^2 [300 Pa (0.04 psi)] at the beginning of a stress-relaxation test. The modulus decreases gradually with time, but it may take years for the stress to decrease to a value near zero. Crystalline plastics broaden the distribution of the relaxation times and extend the relaxation stress to much longer periods. This pattern holds true at both the higher and low extremes of crystallinity. With some plastics, their degree of crystallinity can change during the course of a stress-relaxation test.

Stress-relaxation test data has been generated for the designer. Plastic is deformed by a fixed amount and the stress required maintaining this deformation is measured over a period of time. The maximum stress occurs as soon as the deformation takes place and decreases gradually with time from this value.

Creep data in designing products has been used for over a century; particularly since the 1940s. Unfortunately there is never enough data especially with the new plastics that are produced. However, relationships of the old and new are made successfully with a minor amount of testing.

Fatigue

When reviewing fatigue one studies their behaviors of having materials under cyclic loads at levels of stress below their static yield strength. Fatigue test, analogous to static creep tests, provides information on the failure of materials under repeated stresses. The more conventional short-term tests give little indication about the lifetime of an object subjected to vibrations or repeated deformations. When sizing products so that they can be modeled on a computer, the designer needs a starting point until feedback is received from the modeling. The stress level to be obtained should be less than the yield strength. A starting point is to estimate the static load to be carried, to find the level of vibration testing in G levels, to assume that the part vibrates with a magnification of 10, and to multiply these together to get an equivalent static load. The computer design model will permit making design changes within the required limits.

If the loading were applied only once the magnitude of the stresses and strains induced would be so low that they would not be expected to cause failure. With repeated constant load amplitude tests, maximum material stress is fixed, regardless of any decay in the modulus of elasticity of the material. Constant deflection amplitude fatigue testing is less demanding, because any decay in the modulus of elasticity of the material due to hysteretic heating would lead to lower material stress at the fixed maximum specimen deflection.

Material fatigue data are normally presented in constant stress (S) amplitude or constant (s) strain amplitude plotted vs. the number of cycles (N) to specimen failure to produce a fatigue endurance S-N

Figure 3.11 Typical S-N curve

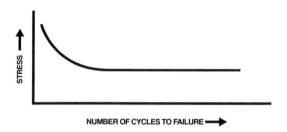

NUMBER OF CYCLES TO FAILURE ➡

curve for the material (Fig. 3.11). The test frequency for plastics is typically 30 Hz, and test temperature is typically conditioned and tested in an environment of 23°C (73°F). The behavior of viscoelastic materials is very temperature and strain rate dependent. Consequently, both test frequency and test temperature has a significant effect upon the observed fatigue behavior. The fatigue testing of TPs is normally terminated at 10^7 cycles.

S-N curve provides information on the higher the applied material stresses or strains, the fewer cycles the specimen can survive. It also provides the curve that gradually approaches a stress or strain level called the fatigue endurance limit below which the material is much less susceptible to fatigue failure. A curve of stress to failure vs. the number of cycles to this stress level to cause failure is made by testing a large number of representative samples of the material under cyclical stress. Each test made at a progressively lowered stress level. This S-N curve is used in designing for fatigue failure by determining the allowable stress level for a number of stress cycles anticipated for the product. In the case of materials such as metals, this approach is relatively uncompli- cated. Unfortunately, in the case of plastics the loading rate, the repetition rate, and the temperature all have a substantial effect on the S-N curve, and it is important that the appropriate tests be conducted.

There is the potential for having a large amount of internal friction generated within the plastics when exposed to fatigue. This action involves the accumulation of hysteretic energy generated during each loading cycle. Because this energy is dissipated mainly in the form of heat, the material experiences an associated temperature increase. When heating takes place the dynamic modulus decreases, which results in a greater degree of heat generation under conditions of constant stress. The greater the loss modulus of the material, the greater the amount of heat generated that can be dissipated. TPs, particularly the crystalline type that are above their glass-transition temperatures (T_g), will be more sensitive to this heating and highly cross-linked plastics or glass-reinforced TS plastics (GRTSs) are less sensitive to the frequency of load.

If the TP's surface area of a product is insufficient to permit the heat to be dissipated, the plastic will become hot enough to soften and melt. The possibility of adversely affecting its mechanical properties by heat generation during cyclic loading must therefore always be considered. The heat generated during cyclic loading can be calculated from the loss modulus or loss tangent of the plastics.

Damping is the loss of energy usually as dissipated heat that results when a material or material system is subjected to fatigue, oscillatory load, or displacement. Perfectly elastic materials have no mechanical damping. Damping reduces vibrations (mechanical and acoustical) and prevents resonance vibrations from building up to dangerous amplitudes. However, high damping is generally an indication of reduced dimensional stability, which can be very undesirable in structures carrying loads for long time periods. Many other mechanical properties are intimately related to damping; these include fatigue life, toughness and impact, wear and coefficient of friction, etc. Measuring damping capacity is equal to the area of the elastic hysteresis loop divided by the deformation energy of a vibrating material. It can be calculated by measuring the rate of decay of vibrations induced in a material.

This dynamic mechanical behavior of plastics is important. The role of mechanical damping is not as well known. Damping is often the most sensitive indicator of all kinds of molecular motions going on in a material. Aside from the purely scientific interest in understanding the molecular motions that can occur, analyzing these motions is of great practical importance in determining the mechanical behavior of plastics. For this reason, the absolute value of a given damping and the temperature and frequency at which the damping peaks occur can be of considerable interest and use.

High damping is sometimes an advantage, sometimes a disadvantage. For instance, in a car tire high damping tends to give better friction with the road surface, but at the same time it causes heat buildup, which makes a tire degrade more rapidly. Damping reduces mechanical and acoustical vibrations and prevents resonance vibrations from building up to dangerous amplitudes. However, the existence of high damping is generally an indication of reduced dimensional stability, which can be undesirable in structures carrying loads for long periods of time.

To improve fatigue performance, as with other properties of other properties use is made of reinforcements. RPs are susceptible to fatigue.

Figure 3.12 High-performance fatigue properties of RPs and other materials

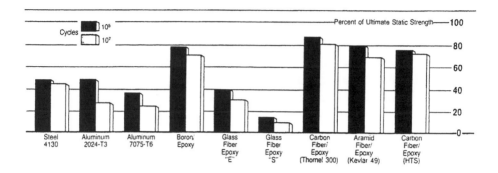

However, they provide high performance when compared to unreinforced plastics and many other materials (Fig. 3.12). With a TP there is a possibility of thermal softening failures at high stresses or high frequencies. However, in general the presence of fibers reduces the hysteretic heating effect, with a reduced tendency toward thermal softening failures. When conditions are chosen to avoid thermal softening, the normal fatigue process takes places as a progressive weakening of the material from crack initiation and propagation.

Plastics reinforced with carbon, graphite, boron, and aramid are stiffer than the glass-reinforced plastics (GRP) and are less vulnerable to fatigue. (E-glass is the most popular type used; S-glass improves both short- and long-term properties.) In short-fiber GRPs cracks tend to develop easily in the matrix, particularly at the interface close to the ends of the fibers. It is not uncommon for cracks to propagate through a TS matrix and destroy the material's integrity before fracturing of the fabricated product occurs. With short-fiber composites fatigue life can be prolonged if the fiber aspect ratio of its length to its diameter is large, such as at least a factor of five, with ten or better for maximum performance.

In most GRPs debonding can occur after even a small number of cycles, even at modest load levels. If the material is translucent, the buildup of fatigue damage can be observed. The first signs (for example, with glass-fiber TS polyester) are that the material becomes opaque each time the load is applied. Subsequently, the opacity becomes permanent and more pronounced, as can occur in corrugated RP translucent roofing panels. Eventually, plastic cracks will become visible, but the product will still be capable of bearing the applied load until localized intense damage causes separation in the components. However, the first appearance of matrix cracks may cause sufficient concern, whether for safety or aesthetic reasons, to limit the useful life of the product. Unlike most other materials, GRPs give visual warning of their fatigue failure.

Since GRPs can tend not to exhibit a fatigue limit, it is necessary to design for a specific endurance, with safety factors in the region of 3 to 4 being commonly used. Higher fatigue performance is achieved when the data are for tensile loading, with zero mean stress. In other modes of loading, such as flexural, compression, or torsion, the fatigue behavior can be more unfavorable than that in tension due to potential abrasion action between fibers if debonding of fiber and matrix occurs. This is generally thought to be caused by the setting up of shear stresses in sections of the matrix that are unprotected by some method such as having properly aligned fibers that can be applied in certain designs. An approach that has been used successfully in products such as high-performance RP aircraft wing structures, incorporates a very thin, high-heat-resistant film such as Mylar between layers of glass fibers. With GRPs this construction significantly reduces the self-destructive action of glass-to-glass abrasion and significantly increases the fatigue endurance limit.

Fatigue data provides the means to design and fabricate products that are susceptible to fatigue. Ranking fatigue behavior among various plastics should be conducted after an analysis is made of the application and the testing method to be used or being considered. It is necessary to also identify whether the product will be subjected to stress or strain loads. Plastics that exhibit considerable damping may possess low fatigue strength under constant stress amplitude but exhibit a considerably higher ranking in constant deflection amplitude and strain testing. Also needing consideration is the volume of material under stress in the product and its surface area-to-volume ratio. Because plastics are viscoelastic, this ratio is critical in that it influences the temperature that will be reached. At the same stress level, the ratio of stressed volume to area may well be the difference between a thermal short-life failure and a brittle long-life failure, particularly with TPs.

Like in metal and other material in any design books, factors should be eliminated or reduced such as sharp corners or abrupt changes in their cross-sectional geometry or wall thickness should be avoided because they can result in weakened, high-stress areas. The areas of high loading where fatigue requirements are high need more generous radii, combined with optimal material distribution. Radii of ten to twenty times are suggested for extruded parts, and one quarter to one half the wall thickness may be necessary for moldings to distribute stress more uniformly over a large area.

Figure 3.13 Carbon fiber-epoxy RPs fatigue data

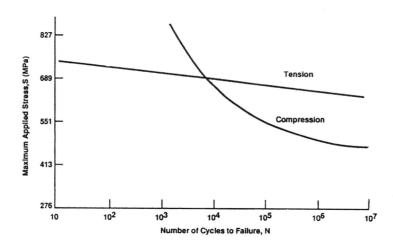

In evaluating plastics for a particular cyclic loading condition, the type of material and the fabrication variables are important. As an example, the tension fatigue behavior of unidirectional RPs is one of their great advantages over other plastics and other materials. In general the tension S-N curves (curves of maximum stressed plotted as a function of cycles to failure) of RPs with carbon, boron, and aramid fibers are relatively flat. Glass fiber RPs show a greater reduction in strength with increasing number of cycles. However, RPs with high strength glass fiber are widely used in applications for which fatigue resistance is a critical design consideration, such as helicopter blades.

Fig. 3.13 shows the cycles to failure as a function of maximum stress for carbon fiber-reinforced epoxy laminates subjected to tension and compression fatigue. The laminates have 60% of their layers oriented at 0°, 20% at +45°, and 20% at −45°. They are subjected to a fluctuating load in the 0° direction. The ratios of minimum stress-to-maximum stress for tensile and compressive fatigue are 0.1 and 10, respectively. One observes that the reduction in strength is much greater for compression fatigue. However as an example, the RPs compressive fatigue strength at 10^7 cycles is still considerably greater than the corresponding tensile value for aluminum.

Metals are more likely to fail in fatigue when subjected to fluctuating tensile rather than compressive load. This is because they tend to fail by crack propagation under fatigue loading. However, the failure modes in RPs are very different and more complex. One consequence is that RPs tend to be more susceptible to fatigue failure when loaded in compression.

Fiber reinforcement provides significant improvements in fatigue with carbon fibers and graphite and aramid fibers being higher than glass fibers. The effects of moisture in the service environment should also be considered, whenever hygroscopic plastics such as nylon, PCs, and others are to be used. For service involving a large number of fatigue cycles in TPs, crystalline-types offer the potential of more predictable results than those based on amorphous types, because the crystalline ones usually have definite fatigue endurance. Also, for optimum fatigue life in service involving both high-stress and fatigue loading, the reinforced high-temperature performance plastics like PEEK, PES, and Pi are recommended.

Reinforcement performance

Reinforcements can significantly improve the structural characteristics of a TP or TS plastic. They are available in continuous forms (fibers, filaments, woven or non-woven fabrics, tapes, etc.), chopped forms having different lengths (Fig. 3.14), or discontinuous in form (whiskers, flakes, spheres, etc.) to meet different properties and/or processing methods. Glass fiber represents the major material used in RPs worldwide. There are others that provide much higher structural performances, etc. The reinforcements can allow the RP materials to be tailored to the design, or the design tailored to the material.

To be effective, the reinforcement must form a strong adhesive bond with the plastics; for certain reinforcements special cleaning, sizing, finishing, etc. treatments are used to improve bond. Also used alone or in conjunction with fiber surface treatments are bonding additives in the plastic to promote good adhesion of the fiber to the plastic.

Figure 3.14 Fiber strength vs. fiber length (Courtesy of Plastics FALLO)

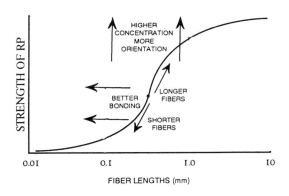

Applicable to RPs is the aspect ratio of fibers. It is the ratio of length to diameter (L/D) of a fiber. In RP fiber L/D will have a direct influence on the reinforced plastic performance. High values of 5 to 10 provide for good reinforcements. Theoretically, with proper lay-up the highest performance plastics could be obtained when compared to other materials. To maximize strength and modulus of RPs the long fiber approach is used.

Different types of reinforcement construction are used to meet different RP properties and/or simplify reinforcement layup for certain fabricating processes to meet design performance requirements. They include woven, nonwoven, rovings, and others (Table 3.3). These different constructions are used to provide different processing and directional properties.

Table 3.3 Example of E-glass constructions used in TS polyester RPs

	Bulk Molding Compound	Sheet Molding Compound	Chopped Strand Mat	Woven Roving	Unidirectional Axial	Unidirectional Transverse
Glass content (wt %)	20	30	30	50	70	70
Tensile modulus GPa (Msi)	9 (1.3)	13 (1.9)	7.7 (1.1)	16 (2.3)	42 (6.1)	12 (1.7)
Tensile strength MPa (Ksi)	45 (6.5)	85 (12)	95 (14)	250 (36)	750 (110)	50 (7)

There are certain types of so-called nonwoven fabric that are directly formed from short or chopped fiber as well as continuous filaments. They are produced by loosely compressing together fibers, yarns, rovings, etc. with or without a scrim cloth carrier; assembled by mechanical, chemical, thermal, or solvent methods. Products of this type include melted and spun-bonded fabrics. The nonwoven spun-bonded integrates the spinning, lay-down, consolidation, and bonding of continuous filaments to form fabrics. Felt is the term used to describe nonwoven compressed fabrics, mats, and bats prepared from staple fibers without spinning, weaving, or knitting; made up of fibers interlocked mechanically.

A fibrous material extensively used in RPs are the mat constructions. They consist of different randomly and uniformly oriented products: (1) chopped fibers with or without carrier fibers or binder plastics; (2) short fibers with or without a carrier fabric; (3) swirled filaments loosely held together with a plastic binder; (4) chopped or short fiber with long fibers included in any desired pattern to provided addition mechanical properties in specific directions; (5) and so on.

There are reinforcement preform constructions. A preform is a method of making chopped fiber mats of complex shapes that are to be used as reinforcements in different RP molding fabricating processes (injection, etc.). Oriented patterns can be incorporated in the preforms.

When conventional flat mats are used, they may tear, wrinkle, or give uneven glass distribution when producing complex shapes. To alleviate this problem, it is necessary to take great care in tailoring the mat and in placing it properly in the mold cavity. Otherwise, mats may cause poor products or poor production rates. Preforms are used to overcome these problems. They are slightly more expensive for short production runs. However they are used when mats are considered impractical, or a relatively high production run exists that offsets the higher cost.

Fiber-reinforced plastics differ from many other materials because they combine two essentially different materials of fibers and a plastic into a single plastic composite. In this way they are somewhat analogous to reinforced concrete, that combines concrete and steel. However, in the RPs the fibers are generally much more evenly distributed throughout the mass and the ratio of fibers to plastic is much higher.

In designing fibrous-reinforced plastics it is necessary to take into account the combined actions of the fiber and the plastic. At times the combination can be considered homogeneous, but in most cases homogeneity cannot be assumed (Chapter 2).

4 PRODUCT DESIGN

Introduction

Plastics offer the opportunity to optimize designs by focusing on material composition as well as product structural geometry to meet different product requirements. In structural applications for plastics, which generally include those in which the product has to resist substantial static and/or dynamic loads, it may appear that one of the problem design areas for many plastics is their low modulus of elasticity. Since shape integrity under load is a major consideration for structural products, low modulus type plastic products are designed shapewise and/or thicknesswise for efficient use of the material to afford maximum stiffness and overcome their low modulus. This type of plastics and products represent most of the plastic products produced worldwide.

Throughout this book as the viscoelastic behavior of plastics has been described, it has been shown that deformations are dependent on such factors as the time under load and the temperature. Therefore, when structural components are to be designed using plastics it must be remembered that the extensive amount of standard equations that are available (Figs 2.31 and 2.32) for designing springs, beams, plates, and cylinders, and so on have all been derived under certain assumptions. They are that (1) the strains are small, (2) the modulus is constant, (3) the strains are independent of the loading rate or history and are immediately reversible, (4) the material is isotropic, and (5) the material behaves in the same way in tension and compression.

Since these assumptions are not always justifiable when applied to plastics, the classic equations cannot be used indiscriminately. Each case must be considered on its merits, with account being taken of such factors as the time under load, the mode of deformation, the service temperature, the fabrication method, the environment, and others. In particular, it should be noted that the traditional equations are derived using the relationship that stress equals modulus times strain, where the modulus is a constant. As reviewed in Chapters 2 and 3 the modulus of a plastic may not be a constant.

There are different design approaches to consider as reviewed in this book and different engineering textbooks concerning specific products. They range from designing a drinking cup to the roof of a house. As an example consider a house to stand up to the forces of a catastrophic hurricane. Low pitch roofs are less vulnerable than steeper roofs because the same aerodynamic factors that make an airplane fly can lift the roof off the house. The roof also requires being properly attached to the building structure.

Example of a product design program approach follows:

1. Define the function of the product with performance requirements.
2. Identify space and load limitations of the product if they exist.
3. Define all of the environmental stresses that the product will be exposed to in its intended function.
4. Select several materials that appear to meet the required environmental requirements and strength behaviors.
5. Do several trial designs using different materials and geometries to perform the required function.
6. Evaluate the trial designs on a cost effectiveness basis. Determine several levels of performance and the specific costs associated with each to the extent that it can be done with available data.
7. Determine the appropriate fabricating process for the design.
8. Based on the preliminary evaluation select the best apparent choices and do a detailed design of the product.
9. Based on the detailed design select the probable final product design, material, and process.
10. Make a model if necessary to test the effectiveness of the product.
11. Build prototype tooling.
12. Make prototype products and test products to determine if they meet the required function.
13. Redesign the product if necessary based on the prototype testing.
14. Retest.
15. Make field tests.
16. Add instructions for use.

Reinforced Plastic

More extensively used are the conventional engineering plastics that are not reinforced to maximize mechanical performances of plastics. However there are reinforced plastics (RPs), as reviewed throughout this handbook, that offer certain important structural and other performance requirements. These requirements provides the designer great flexibility and provides freedom practically not possible with most other materials. However, it requires a greater understanding of the interrelations to take full advantage of RPs. It is important to understand that RPs has an extremely wide range of properties, structural responses, product performance characteristics, product shapes, manufacturing processes, and influence on product performances.

The usual approach is that the designer is involved in "making the material." RP designed products have often performed better than expected, despite the use of less sophisticated fabricating tools in their design. Depending on construction and orientation of stress relative to reinforcement, it may not be necessary to provide extensive data on time-dependent stiffness properties since their effects may be small and can frequently be considered by rule of thumb using established practical design approaches. When time dependent strength properties are required, creep, fatigue, and other data are used most effectively. These type data are available.

The arrangement and the interaction of the usual stiff, strong fibers dominate the behavior of RPs with the less stiff, weaker plastic matrix [thermoset (TS) or thermoplastic (TP)]. A major advantage is that directional properties can be maximized in products by locating fibers that maximize mechanical performances in different directions.

When compared to unreinforced plastics, the analysis and design of RPs is simpler in some respects and more complicated in others. Simplifications are possible since the stress-strain behavior of RPs is frequently fairly linear to failure and they are less time-dependent. For high performance applications, they have their first damage occurring at stresses just below their high ultimate strength properties. They are also much less temperature-dependent, particularly RTSs (reinforced TSs).

When constructed from any number and arrangement of RP plies, the stiffness and strength property variations may become much more complex for the novice. Like other materials, there are similarities in that the first damage that occurs at stresses just below ultimate strength. Any review that these types of complications cause unsolvable problems is incorrect. Reason being that an RP can be properly designed, fabricated and evaluated to take into account any possible variations; just as with other materials. The variations may be insignificant or significant. In either case, the designer will use the required values and apply them to an appropriate safety factor; similar approach is used with other materials. The designer has a variety of alternatives to choose from regarding the kind, form, amount of reinforcement to use, and the process versus requirements.

With the many different fiber types and forms available, practically any performance requirement can be met and molded into any shape. However they have to be understood regarding their advantages and limitations. As an example there are fiber bundles in lower cost woven rovings that are convoluted or kinked as the bulky rovings conform to the square weave pattern. Kinks produce repetitive variations in the direction of reinforcement with some sacrifice in properties. Kinks can also induce high local stresses and early failure as the fibers try to straighten within the matrix under a tensile load. Kinks also encourage local buckling of fiber bundles in compression and reduce compressive strength. These effects are particularly noticeable in tests with woven roving in which the weave results in large-scale reinforcement.

Fiber content can be measured in percent by weight of the fiber portion (wt%). However, it is also reported in percent by volume (vol%) to reflect better the structural role of the fiber that is related to volume (or area) rather than to weight. When content is only in percent, it usually refers to wt%.

Basic behaviors of combining actions of plastics and reinforcements have been developed and used successfully. As an example, conventional plain woven fabrics that are generally directional in the 0° and 90° angles contribute to the highest mechanical strength at those angles. The rotation of alternate layers of fabric to a layup of 0°, +45°, 90°, and –45° alignment reduces maximum properties in the primary directions, but increases in the +45° and –45° directions. Different fabric and/or individual fiber patterns are used to develop different property performances in the plain of the molded RPs. These woven fabric patterns CAN include basket, bias, cowoven, crowfoot, knitted, leno, satin (four-harness satin, eight-harness satin, etc.), and twill.

For almost a century many different RP products have been designed, fabricated, and successfully operated in service worldwide. They range from small to large products such as small insulators for high voltage cable lines to large 250 ft diameter deep antenna parabolic reflectors. RPs have been used in all types of transportation vehicles, different designed bridges, road surfacing such as aircraft landing strips and roads, mining equipment, water purification and other very corrosive environmental equipment, all types of electrical/electronic devices, etc.

Monocoque Structure

With the flexibility in shaping and fabricating plastics provides an easy approach to designing monocoque structures. This is the type of construction in which the outer skin carries all or a major part of the stresses. Different applications take advantage of this design approach such as automotive body, motor truck, railroad car, aircraft fuselage and wings, toys, houses, rockets, and so on. This load bearing construction can integrate the product body and support/chassis into a single structure.

Geometric shape

Design analysis is required to convert applied loads and other external constraints into stress and strain distributions within the product, and to calculate associated deformations (Chapters 1 and 2). The nature and complexity of these calculations is influenced strongly by the shape of the component. It is most convenient if the component approximates to some simple idealized form, such as a plate or shell (for example a body panel), a beam or tube (for example a chassis member or bumper), or a combination of idealized forms (for example a box structure). In these cases, standard design formulae can be provided into which appropriate parameters can be substituted for a particular application.

However, there are many more cases where the component shape does not approximate to a simple standard form (for example a wheel, pump housing, or manifold) or where a more detailed analysis is required for part of a product (for example the area of a hole, boss, or attachment point). In these cases, the geometry complicates the design analysis and it may be necessary to carry out a direct analysis, possibly using finite element analysis.

One of plastics' design advantages is its formability into almost any conceivable shape. It is important for designers to appreciate this important characteristic. Both the plastic materials and different ways to fabricate products provide this rather endless capability. Shape, which can be almost infinitely varied in the early design stage, is capable for a given volume of materials to provide a whole spectrum of strength properties, especially in the most desirable areas of stiffness and bending resistance. With shell structures, plastics can be either singly or doubly curved via the different fabricating processes.

Modulus of Elasticity

There are different techniques that have been used for over a century to increase the modulus of elasticity of plastics. Orientation or the use of fillers and/or reinforcements such as RPs can modify the plastic. There is also the popular and extensively used approach of using geometrical design shapes that makes the best use of materials to improve stiffness even for those that have a low modulus. Structural shapes that are applicable to all materials include shells, sandwich structures, dimple sheet surfaces, and folded plate structures (Fig. 4.1).

EI theory

In all materials (plastics, metals, wood, etc.) elementary mechanical theory demonstrates that some shapes resist deformation from external loads. This phenomenon stems from the basic physical fact that deformation in beam or sheet sections depends upon the mathematical product of the modulus of elasticity (E) and the moment of inertia (I), commonly expressed as EI (Chapter 3, Stress-strain behavior). It is applied to all types of constructions such as solids, foams, and sandwich structures. In many applications plastics can lend themselves in the form of a sophisticated lightweight stiff structure and the requirements are such that the structure must be of plastics. In other instances, the economics of fabrication and erection of a plastics lightweight structure and the intrinsic appearance and other desirable properties make it preferable to other materials.

This theory has been applied to many different constructions including many plastic products. In each case displacing material from the neutral plane makes the improvement in flexural stiffness. Use of this engineering principle that has been used for many centuries relates to the basic

Figure 4.1 Examples of shapes to increase stiffness

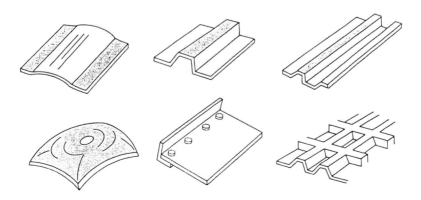

physical fact that deformation in beams or sheets depends upon the mathematical product of *E* and *I*, more commonly expressed as *EI*.

The *EI* principle applies to the basic beam structures as well as hollow channel, I-shape, T-shape, etc. where it imparts increased stiffness in one direction much more than in the other. Result is more efficient strength-to-weight products and so forth. While this construction may not be as efficient as the sandwich panel, it does have the advantage that it can be easily fabricated (molded, extruded, etc.) directly in the required configuration at a low cost and the relative proportions be designed to meet the load requirements.

Plate

Methods for the design analysis in the past for plastics were based on models of material behavior relevant to traditional metals, as for example elasticity and plastic yield. These principles were embodied in design formulas, design sheets and charts, and in the modern techniques such as those of CAD using finite element analysis (FEA). Design analyst was required only to supply appropriate elastic or plastic constants for the material, and not question the validity of the design methods. Traditional design analysis is thus based on accepted methods and familiar materials, and as a result many designers have little, if any, experience with such other materials as plastics, wood, and glass.

Using this approach it is both tempting and common practice for certain designers to treat plastics as though they were traditional materials such as steel and to apply familiar design methods with what seem appropriate materials constants. It must be admitted that this pragmatic approach does often yield acceptable results. However, it should also be recognized that the mechanical characteristics of plastics are different from those of metals, and the validity of this pragmatic approach is often fortuitous and usually uncertain.

It would be more acceptable for the design analysis to be based on methods developed specifically for the materials, but this action will require the designer of metals to accept new ideas. Obviously, this acceptance becomes easier to the degree that the newer methods are presented as far as possible in the form of limitations or modifications to the existing methods discussed in this book.

Table 4.1 provides examples of mechanical property data of different materials ($GN/m^2 = kPa$). A review is presented concerning the four materials in Table 4.1, where the exact values used are unimportant.

Higher performance types could be used for the metals and plastics but those in this table offer a fair comparison for the explanation being presented. This review identifies the need for using design analysis methods appropriate for plastics. It also indicates the uncertainty of using with plastics methods derived from metals, and demonstrates the dangers of making generalized statements about the relative merits of different classes of materials.

Table 4.1 Mechanical properties of materials

Property	Aluminum	Mild steel	Polypropylene (PP)	Glass-fiber reinforced plastics (GRP)
Tensile modulus (E) 10^6 GN/m^2 (psi)	70 (10)	210 (30)	1.5 (0.21)	15 (2.2)
Tensile strength (σ) 10^3 MN/m^2 (psi)	400 (58)	450 (65)	40 (5.8)	280 (40.5)
Specific gravity (S)	2.7	7.8	0.9	1.6

Based on the usual data on metals, they are much stiffer and significantly stronger than plastics. This initial evaluation could eliminate the use of plastics in many potential applications, but in practice it is recognized by those familiar with the behavior of plastics that it is the stiffness and strength of the product that is important, not its material properties.

The proper approach is to consider the application in which a material is used such as in panels with identical dimensions with the service requirements of stiffness and strength in flexure. Their flexural stiffnesses and strengths depend directly on the respective material's modulus and strength. Other factors are similar such as no significantly different Poisson ratios. The different panel properties relative to stiffness and strength are shown in Fig. 4.2. The metal panels are stiffer and stronger than the plastic ones because the panels with equal dimensions use equal volumes of materials.

By using the lower densities of plastics it allows them to be used in thicker sections than metals. This approach significantly influences the panel's stiffnesses and strengths. With equal weights and therefore different thickness (t) the panels are loaded in flexure. Their stiffnesses depend on (Et^3) and their strength on (σt^2) where E and σ are the material's modulus and its strength. For panels of equal weight their relative stiffnesses are governed by (E/s^3) and their relative strengths by ($\sigma/s2$) where s denotes specific gravity. As shown in Fig. 4.3, the plastics now are much more favorable. So depending on how one wants to present data or more important apply data either Figs. 4.2 or 4.3 or

Figure 4.2 Open bar illustrations represent stiffness and shaded illustrations represent strength with panels having the same dimensions

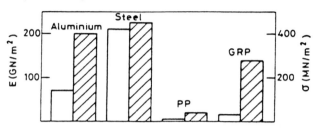

Figure 4.3 Open bar illustrations represent stiffness and shaded illustrations represent strength with panels having the equal weights

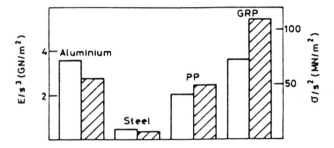

Table 4.1 is used. Thus the designer has the opportunity to balance out the requirements for stiffness, strength, and weight saving.

Recognize that it is easy to misinterpret property data and not properly analyze the merits of plastics. No general conclusions should be drawn on the relative merits of various materials based on this description alone. In comparing materials in Table 4.1 a designer can easily obtain different useful data. As an example the GRP panel has 2.4 times the thickness of a steel panel for the same flexural stiffness. It has 3.6 times its flexural strength and only half its weight. The tensile strength of the GRP panel would be 50% greater than that of the steel panel, but its tensile stiffness is only 17% that of the steel panel.

Similar remarks could be made with respect to various materials' costs and energy contents, which can also be specified per unit of volume or weight. General statements about energy content or cost per unit of stiffness or strength, as well as other factors, should be treated with caution and applied only where relevant. If these factors are to be treated properly, they too must relate to final product values that include the method of fabrication, expected lifetime, repair record, and in-service use.

This review shows what the veteran plastic designer knows; that plastic products are often stiffness critical, whereas metal products are usually strength critical. Consequently, metal products are often made stiffer than required by their service conditions, to avoid failure, whereas plastic products are often made stronger than necessary, for adequate stiffness. In replacing a component in one material with a similar product in another material is not usually necessary to have the same product stiffness and strength.

Folded Plate

Capability in fabricating simple to complex folded products makes designing them easy based on analysis such as beams with rectangular, triangular, spherical, or other shapes. Products vary from those with spring actions to movable contoured walls and ceilings to bottles and outer-space structures. Elementary beam equations are used. When assemblies are plates whose lengths are large relative to their cross-sectional dimensions (thin-wall beam sections, ribbed panels, and so on) and are in large plates whose fold lines deflect identically, such as the interior bays of roofs or bridge structures, they can be analyzed as beams. More elaborate documented procedures are used to determine transverse multi-bending stresses in assemblies.

An example of folded plate technology are bellows-style collapsible plastic containers such as blow molded bottles that are foldable. The foldable shaped containers in contrast to that of the usual shaped bottles provides advantages and conveniences. Examples include reduced storage as the container's content is reduced, transportation volume and weight costs relate to the collapsed size, and disposal space; prolonged product freshness by reducing oxidation and loss of carbon dioxide; and provides extended life via continuous collapsing surface access to foods like mayonnaise and jams.

Figs 4.4 and 4.5 show the bellows of collapsible containers that overlap and fold to retain their folded condition without external assistance, thus providing a self-latching feature. The views show uncollapsed bottle, collapsed bottle, and top view of the bottle. This latching is the result of bringing together under pressure two adjacent conical sections of unequal proportions and different angulations to the bottle axis. On a more technical analytical level the latching is created basically by the swing action of one conical section around a fixed pivot point, from an outer to an inner, resting position. The two symmetrically opposed pivot points and rotating segments keep a near-constant diameter as they travel along the bottle axis. This action explains the bowing action of the smaller, conical section as it approaches the overcentering point.

Figure 4.4 Views of a collapsible bottle (Courtesy of Collapsible Bottle of America Co.)

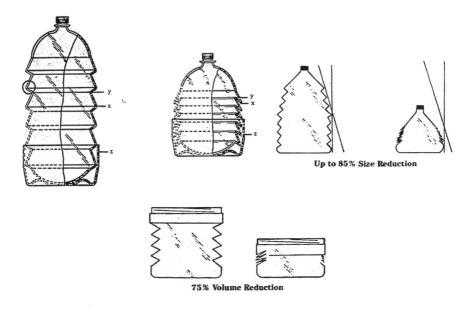

Figure 4.5 Geometric concepts to the collapsible bottle (Courtesy of Collapsible Bottle of America Co.)

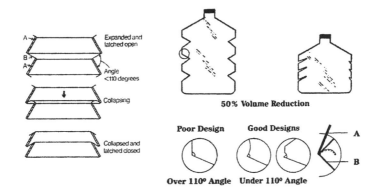

A characteristic in molding some products that are designed to include collapsing, living hinge, etc. action requires them to be subjected to the action as soon as possible after processing. This initial action is used to create permanent fold rings and completely orient the plastic molecules. In most disposable applications these bottles would undergo three changes of volume: an initial collapsing of the container before shipment and storage; expansion of the container at its destination, before or during filling; and finally gradually collapsing the bottle for reuse or disposal.

Beam

As the engineering books explain, a beam is a bar or structural member subjected to transverse loads that tend to bend it. Any structural members act as a beam if external transverse forces induce bending. A simple beam is a horizontal member that rests on two supports at the ends of the beam. All parts between the supports have free movement in a vertical plane under the influence of vertical loads.

There are fixed beams, constrained beams, or restrained beams rigidly fixed at both ends or rigidly fixed at one end and simply supported at the other. A continuous beam is a member resting on more than two supports. A cantilever beam is a member with one end projecting beyond the point of support, free to move in a vertical plane under the influence of vertical loads placed between the free end and the support.

When a simple beam bends under its own weight, the plastic or fibers in a plastic on the upper or concave side is shortened, with the stress acting on them is compression. The fibers on the under or convex side are lengthened, and the stress acting on them is tension. In addition, shear exists along each cross section, the intensity of which is greatest along the sections at the two supports and zero at the middle section. When a cantilever beam bends under its own weight the fibers on the upper or convex side are lengthened under tensile stresses. The fibers on the under or concave side are shortened under compressive stresses, the shear is greatest along the section at the support, and zero at the free end.

The neutral surface is that horizontal section between the concave and convex surfaces of a loaded beam, where there is no change in the length of the fibers and no tensile or compressive stresses acting upon them. The neutral axis is the trace of the neutral surface on any cross section of a beam. The elastic curve of a beam is the curve formed by the intersection of the neutral surface with the side of the beam, it being assumed that the longitudinal stresses on the fibers are within the elastic limit.

The reactions, or upward pressures at the points of support, are computed by applying certain conditions necessary for equilibrium of a system of vertical forces in the same plane. They are the algebraic sum of all vertical forces that must equal zero; that is, the sum of the reactions equals the sum of the downward loads. There is also the algebraic sum of the moments of all the vertical forces that equals zero.

The first condition applies to cantilever beams and to simple beams uniformly loaded, or with equal concentrated loads placed at equal distances from the center of the beam. In the cantilever beam, the reaction is the sum of all the vertical forces acting downward, comprising the weight of the beam and the superposed loads. In the simple beam each reaction is equal to one-half the total load, consisting of the weight of the beam and the superposed loads. The second condition applies to a simple beam not uniformly loaded. The reactions are computed separately, by determining the moment of the several loads about each support. The sum of the moments of the load around one support is equal to the moment of the reaction of the other support around the first support.

The fundamental laws for the stresses at any cross-section of a beam in equilibrium are: (1) sum of the horizontal tensile stresses equal sum of horizontal compressive stresses, (2) resisting shear equal vertical shear, and (3) resisting moment equal bending moment. Bending moment at any cross-section of a beam is the algebraic sum of the moments of the external forces acting on either side of the section. It is positive when it causes the beam to bend convex downward, thus causing compression in upper fibers and tension in lower fibers of the beam. When the bending moment is determined from the forces that lie to the left of the section, it is positive if they act in a clockwise direction; if determined from forces on the right side, it is positive if they act in a counter-clockwise direction. If the moments of upward forces are given positive signs, and the moments of downward forces are given negative signs, the bending moment will always have the correct sign, whether determined from the right or left side. The bending moment should be determined for the side for which the calculation will be the simplest.

The deflection of a beam as computed by the ordinary formulas is that due to flexural stresses only. The deflection in honeycomb and short beams due to vertical shear can be high, and should always be checked. Because of the nonuniform distribution of the shear over the cross section of the beam, computing the deflection due to shear by exact methods is difficult. It may be approximated by:

$$y_s = M/AE_s \tag{4-1}$$

where y_s = deflection, inch, due to shear; M = bending moment, lb-in, at the section where the deflection is calculated; A = area of cross section of beam, square inches; and E_s = modulus of elasticity in shear, psi; For a rectangular section, the ratio of deflection due to shear to the deflection due to bending, will be less than 5% if the depth of the beam is less than one-eighth of the length.

In designing a beam an approach is: (1) compute reactions; (2) determine position of the dangerous section and the bending moment at that section; (3) divide the maximum bending moment (lb-in) by the allowable unit stress (psi) to obtain the minimum value of the section modulus; and (4) select a beam section with a section modulus equal to or slightly greater than the section modulus required.

Assumptions are made in simple beam-bending theory that involve (1) all deflections are small, so that planar cross-sections remain planar before and after bending; (2) the beam is initially straight, unstressed, and symmetrical; (3) its proportional limit is not exceeded; and (4) Young's modulus for the material is the same in both tension and compression. In the analysis maximum stress occurs at the surface of the beam farthest from the neutral surface (Fig. 2.32), as given by the following equation:

$$\sigma = Mc/I = M/Z \tag{4-2}$$

where M = the bending moment in in./lbs., c = the distance from the neutral axis to the outer surface where the maximum stress occurs in in., I = the moment of inertia in in.^{4}, and $Z = I/c$-, the section modulus in in.^{3}. Observe that this is a geometric property, not to be confused with the modulus of the material, which is a material property.

Rib

Plastic

When discussion problems in minimizing or increasing load-bearing requirements in wall thickness of plastics, ribbing is recommended if it is determined that space exist for adding ribs (Figs. 4.6 and 4.7). If there is sufficient space, the use of ribs is a practical, economic means of increasing the structural integrity of plastic products without creating thick walls. Table 4.2 shows how a rib structure rigidity (2 in. rib spacings) compares with non-ribbed panels where each are 1 × 2 ft. The rib height is 0.270 in. with a thickness of 0.065 in. If it is determined after a product is produced that ribbing is required adding ribs after the tool is built is usually simple and relatively inexpensive since it involves removing steel in the fabricating tool. With thinner walls a major cost saving could develop since processing them reduces processing time and provides more heat uniformity during processing.

Handbooks reviewing Stress and Deflections in Beams and Moments of Inertia provide information such as the moment of inertia and resistance to deflection that expresses the resistance to stress by the section modulus. By finding a cross section with the two equivalent

Figure 4.6 Example of a design where thin wall with ribbing supports high edge loading

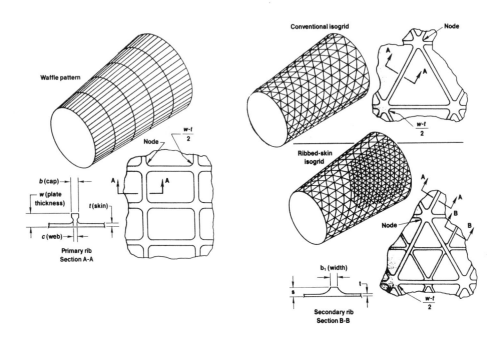

factors, one ensures equal or better performance of the fabricated product. The moment of inertia can be changed substantially by adding ribs and other shapes such as gussets as well as their combinations.

There are available basic engineering rib-design guidelines. The most general approach is to make the rib thickness at its base a minimum of one-half the adjacent wall's thickness. With ribs opposite appearance areas, the width should be kept as thin as possible. In areas where structure is more important than appearance, or with very low shrinkage materials, ribs are often 75 or even 100% of the outside wall's thickness. A goal in rib design is to prevent the formation of a heavy mass of material that can result in a sink, void, distortion, long cycle time, or any combination of these problems.

Table 4.2 Example where ribbing is beneficial

Property	Steel	Solid plastic	Structural foam
Thickness (in.)	0.040	0.182	0.196
E (psi)	3×10^7	3.2×105	2.56×105
I (in.4)	0.000064	0.006	0.0075
E × I (rigidity)	1,920	1,920	1,920
Weight (lbs.)	3.24	1.98	1.78

Figure 4.7 Comparing a rib design strengthwise and weightwise with other materials

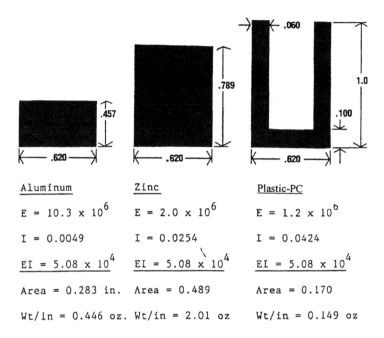

Aluminum	Zinc	Plastic-PC
$E = 10.3 \times 10^6$	$E = 2.0 \times 10^6$	$E = 1.2 \times 10^6$
$I = 0.0049$	$I = 0.0254$	$I = 0.0424$
$EI = 5.08 \times 10^4$	$EI = 5.08 \times 10^4$	$EI = 5.08 \times 10^4$
Area = 0.283 in.	Area = 0.489	Area = 0.170
Wt/in = 0.446 oz.	Wt/in = 2.01 oz	Wt/in = 0.149 oz

Reinforced/Foamed Plastic

Frequently cases arise in which ribs are used to reinforce plastic (RP) plates such as in tanks, boat hulls, bridges, floors, towers, buildings, and so on. The design of ribbed plates such as these is somewhat analogous to reinforced concrete and T-beam design. In view (a) of Fig. 4.8, for example, a construction is shown consisting of a plate composed of balanced fabric RP 0.15 in. thick and mat RP 0.05 in. thick, combined with a rib, making a structure whose overall depth is 1.500 in. The rib is formed of a cellular material such as foamed plastic, plus a cluster of resin-bonded parallel fibers such as roving, at the bottom. The mat is carried around the rib and serves to tie the rib and plate together.

The construction of a plate and rib form a T-beam. The principal design problem is to determine how much of the plate can be considered to be acting as a flange with the rib, that is, the magnitude of b in Fig. 4.8 (a). For purposes of illustration, b is taken as 5 in. If the T-beam is loaded in bending so as to induce compression at the top and tension at the bottom, the balanced fabric and the mat will be in compression at the top, and the roving and mat will be in tension at the bottom. Because the roving is much stronger than the mat, it is evident that the mat adjacent to it will break before the roving reaches its maximum stress. That is if the roving were stressed 50,000 psi the adjacent mat would be stressed 10,000 psi, which is double its strength. Consequently, in finding the neutral axis and computing the strength of the cross-section, the mat is neglected on the tension side. Above the neutral axis the mat is in compression, but in order to simplify the computations, only the mat in the flange of the T is considered.

The foamed plastic has such a low modulus of elasticity and such low strength that it contributes little to either the stiffness or the bending strength of the T-beam. It must, however, be stiff enough to prevent buckling or wrinkling of the mat or the roving.

The active elements of the T-beam are therefore as shown in Fig. 4.8 (b). The flange consists of balanced fabric 5.000 in. wide and 0.150 in. thick plus mat 4.200 in. wide and 0.050 in. thick. The web consists of the bundle of roving 0.800 in. wide and 0.200 in. thick. By the application of equation 4.32, the neutral axis is found to be 1.105 in. from the bottom, or 1.055 in. above the lower edge of the roving.

The basic assumptions discussed in the introduction imply that when this beam is bent, strains at any point in both tension and compression are proportional to the distance from the neutral axis, and that stress is equal to strain multiplied by modulus of elasticity. As an example, the stress in the lower-most roving fiber is 50,000 psi, the stress in the topmost fiber of the flange is $50,000 \ (3 \times 10^6/5) \ (0.395/1.055)$ or 11,250 psi. Similarly, the stresses in the upper edge of the bundle of rovings, at the lower edge of the balanced fabric, and at the upper and lower edges of the balanced fabric, and at the upper and lower edges of the mat in the flange are as shown in Fig. 4.8 (c). These are all less than the corresponding values of σ_1, σ_2, and σ_3 listed.

The internal resisting moment, or resistance to outside bending forces, can be found by computing the total resultant compression C_1 in the balanced fabric, total resultant compression C_2 in the mat, finding the distances α_1 and α_2 between the lines of action of these two resultants and the line of action of the total resultant tension T in the roving, computing the values $C_1\alpha_1$ and $C_2\alpha_2$, and adding.

Resultant C_1 acts at the centroid of the trapezoidal stress area $1a$, resultant C_2 at the centroid of area $2a$, and resultant T at the centroid of area $3a$. Solving for these centroids, the distance α_1 is found to be 1.285 in. and distance α_2 is 1.179 in. These are the internal moment arms of the two resultant compressive forces C_1 and C_2.

$$C_1 = [(11250 + 7000)/2][5.000 \times 0.150] \ = \ 6,840 \text{ lb} \qquad (4\text{-}3)$$
$$C_2 = [(2330 + 1850)/2][4.200 \times 0.05] \ = \ \ \ 440 \text{ lb}$$
$$\text{Total } C \ = \ 7,280$$

Figure 4.8 Cross-section of a rib applied to a plate

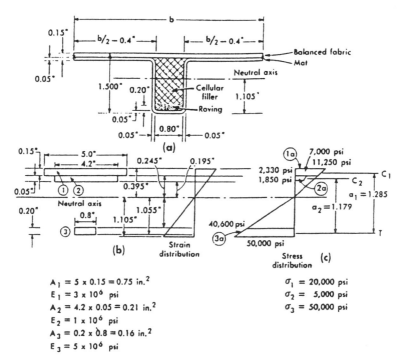

$$A_1 = 5 \times 0.15 = 0.75 \text{ in.}^2$$
$$E_1 = 3 \times 10^6 \text{ psi}$$
$$A_2 = 4.2 \times 0.05 = 0.21 \text{ in.}^2$$
$$E_2 = 1 \times 10^6 \text{ psi}$$
$$A_3 = 0.2 \times 0.8 = 0.16 \text{ in.}^2$$
$$E_3 = 5 \times 10^6 \text{ psi}$$

$$\sigma_1 = 20,000 \text{ psi}$$
$$\sigma_2 = 5,000 \text{ psi}$$
$$\sigma_3 = 50,000 \text{ psi}$$

A check on the accuracy of the computations is afforded by the fact that T must equal C.

$$T = [(40{,}600 + 50{,}000)/2][0.800 \times 0.200] = 7{,}250 \text{ lb} \tag{4-4}$$

The initial resisting moment is $C_1\alpha_1 + C_2\alpha_2$

$$C_1\alpha_1 = 6840 \times 1.285 = 8790 \text{ in-lb} \tag{4-5}$$
$$C_2\alpha_2 = 440 \times 1.179 = \underline{520}$$
$$M_{res.} = 9310 \text{ in-lb}$$

Evidently, if the mat in the flange had been left out of the computation ($C_2\alpha_2$), the error in the calculated result would have been approximately 5%. Omitting the mat in the rib between the flange and the neutral axis affected the result much less.

If a shear force is imposed on the rib, two critical planes of internal shear stress occur, one at the neutral axis and one at the plane between the mat and the fabric in the flange [Fig. 4.8 (a)]. Shear stresses are computed by equation,

$$\tau = \frac{VQ'}{bEI} \tag{4-6}$$

For example, suppose the T-beam is 30 in. long (L) and carries a uniformly distributed load W. Then $M = wL/8 = 9310$ in-lb and $W= 2500$ lb. The maximum shear V is half the total load or 1250 lb.

At the neutral axis the statistical moment Q' is the weighted moment of the flange or of the roving about the neutral axis; El is the stiffness factor of flange plus roving; and b is the total thickness of the mat at the neutral axis plus the effective thickness of the cellular filler. This effective thickness may be computed in accordance with the principles set forth on combined action. If, for example, the shear modulus of the cellular core is $1/15$ the shear modulus of the mat-reinforced material, the effective width of the core is $0.8/15$ or 0.053 in. The total effective width of the mat and the core at the neutral axis is therefore $0.05 + 0.05 + 0.053 = 0.153$ in.

The computed value of Q' is 0.725×10^6 lb-in, and the value of El is 0.968×10^6 psi. The shear stress intensity in the mat at the neutral axis is:

$$\tau_m = \frac{1250 \times 0.725}{0.153 \times 0.968} = 6100 \text{ psi} \tag{4-7}$$

The shear stress intensity in the cellular core is

$$\tau_c = 6100/15 = 410 \text{ psi} \tag{4-8}$$

If either τ_m or τ_r is excessive it is necessary to increase the rib thickness at the neutral axis, probably by increasing the thickness of the mat.

Properties of the cellular core may not be known well enough or may be too low to warrant inclusion with the mat in calculating shear. If the core is not included, the thickness at the neutral axis is that of the mat alone, or 0.10 in. The shear stress τ_m then becomes 9400 psi instead of 6100 psi.

At the interface between mat and fabric in the flange of the T-beam, the value of Q' is that of the fabric alone. This is found to be 0.720×10^6 psi. The width b is 4.2 in. (neglecting the width of the cellular core). Therefore:

$$\tau = \frac{1250 \times 0.720}{4.2 \times 0.968} = 220 \text{ psi} \tag{4-9}$$

In all probability the shear stress intensity is actually higher adjacent to the rib, and lower near the outer ends of the flange, but in any event it is not likely to be excessive.

Column

A column or strut is a bar or structural member under axial compression, which has an unbraced length greater than about eight or ten times the least dimension of its cross section. Because of its length, it is impossible to hold a column in a straight line under a load. A slight sidewise bending always occurs, causing flexural stresses in addition to the compressive stresses induced directly by the load. The lateral deflection will be in a direction perpendicular to that axis of the cross section about which the moment of inertia is the least. Thus in a complex shape such an H-column will bend in a direction perpendicular to its major axis. In a square shape it will bend perpendicular to its two major axes. With a tubular shape it is likely to bend in any direction.

The radius of gyration of a column section with respect to a given axis is equal to the square root of the quotient of the moment of inertia with respect to that axis, divided by the area of the section, that is:

$$k = \sqrt{\frac{I}{A}}; \quad \frac{I}{A} = k^2 \qquad (4\text{-}10)$$

where I is the moment of inertia and A is the sectional area. Unless otherwise mentioned, an axis through the center of gravity of the section is the axis considered. As in beams, the moment of inertia is an important factor in the ability of the column to resist bending, but for purposes of computation it is more convenient to use the radius of gyration. The length of a column is the distance between points unsupported against lateral deflection. The slenderness ratio is the length l divided by the least radius of gyration k.

Various conditions may exist at the ends of columns that usually are divided into four classes. (1) Columns with round ends; the bearing at either end has perfect freedom of motion, as there would be with a ball-and-socket joint at each end. (2) Columns with hinged ends; they have perfect freedom of motion at the ends in one plane, as in compression members in bridge trusses where loads are transmitted through endpins. (3) Columns with flat ends; the bearing surface is normal to the axis of the column and of sufficient area to give at least partial fix to the ends of the columns against lateral deflection. (4) Columns with fixed ends; the ends are rigidly secured, so that under any load the tangent to the elastic curve at the ends will be parallel to the axis in its original position.

Tests prove that columns with fixed ends are stronger than columns with flat, hinged, or round ends, and that columns with round ends are weaker than any of the other types. Columns with hinged ends are equivalent to those with round ends in the plane in which they have movement; columns with flat ends have a value intermediate between those with fixed ends and those with round ends. Usually columns have one end fixed and one end hinged, or some other combination. Their relative values may be taken as intermediate between those represented by the condition at either end. The extent to which strength is increased by fixing the ends depends on the length of column; fixed ends have a greater effect on long columns than on short ones.

There is no exact theoretical formula that gives the strength of a column of any length under an axial load. Different formulas involving the use of empirical coefficients have been deduced, however, and they give results that are consistent with the results of tests. These formulas include the popular Euler's formula, different eccentric formulas, cross-bend formulas, wood and timber column formulas, and general principle formulas.

Euler's Formula

The Euler's formula developed during 1759 by Leonard Euler (Swiss mathematician, 1707 to 1783) is used in product designs and also in designs using columns in molds and dies that processes plastic. Euler's formula assumes that the failure of a column is due solely to the stresses induced by sidewise bending. This assumption is not true for short columns that fail mainly by direct compression, nor is it true for columns of medium length. The failure in such cases is by a combination of direct compression and bending.

Column formulas are found in most machine and tooling hand books as well as strength of materials textbooks. Euler first published this critical-load formula for columns in 1759. For slender columns it is usually expressed in the following form:

$$F = \frac{m\pi^2 EI}{l^2} = \frac{m\pi^2 EA}{(l/k)^2} \qquad (4\text{-}11)$$

where F = Collapsing load on the column in pounds, l = length of the column in inches, A = area of the section in square inches, k = least radius of gyration, which = I/A, E = modulus of elasticity, I = the least moment of inertia of the section, m = a constant depending on the end conditions of the column.

Euler's formula is strictly applicable to long and slender columns, for which the buckling action predominates over the direct compression action and thus makes no allowance for compressive stress. The slenderness ratio is defined as the ratio of length l to the radius of gyration k, represented as l/k.

When the slenderness ratio exceeds a value of 100 for a strong slim column, failure by buckling can be expected. Columns of stiffer and

Table 4.3 Moments of inertia and radii of gyration

	MOMENT OF INERTIA I	RADIUS OF GYRATION $\sqrt{\frac{I}{A}} = k$
	$\dfrac{\pi D^4}{64} = \dfrac{\pi r^4}{4}$	$\dfrac{D}{4} = \dfrac{r}{2}$
	$\dfrac{bh^3}{12}$	$\dfrac{h}{\sqrt{12}} = 0.289h$
	$\dfrac{\pi D^4}{128} = \dfrac{\pi r^4}{8}$	$\dfrac{D}{5.66}$

more brittle materials will buckle at lower slenderness ratios. The constant factor m in Euler's critical-load formula clearly shows that the failure of a column depends on the configuration of the column ends. The basic four types with their respective m are:

1. Both ends pivoted or hinged ($m = 1$)

2. One end fixed and the other free ($m = \frac{1}{4}$)

3. One end fixed and, the other pivoted ($m = 2$)

4. Both ends fixed ($m = 4$)

Table 4.3 shows cross sections of the three common slender column configurations. Formulas for each respective moment of inertia I and radius of gyration k are given. With the above formulas buckling force F can be calculated for a column configuration. Table 4.4 lists values of slim ratios (I/k) for small-nominal-diameter column lengths.

Table 4.4 Slenderness ratio l/k of round columns

Column length (in.)	Diameter (in.)						
	0.031	0.047	0.0625	0.078	0.083	0.125	0.1875
1.0	128	85	64	51	43	32	21
1.5	192	128	96	77	64	48	32
1.75	224	149	112	90	75	56	37
2.0	256	171	128	102	85	64	43
2.25	288	192	144	115	96	72	48
2.5	320	213	160	128	107	80	53
3.0	384	256	192	154	128	96	64
3.25	416	277	206	166	139	104	69

Most failures with the slender columns occur because the slenderness ratio exceeds 100. The prudent designer devises ways to reduce or limit the slenderness ratio.

In the following formula P = axial load; l = length of column; I = least moment of inertia; k = least radius of gyration; E = modulus of elasticity; y = lateral deflection, at any point along a larger column, that is caused by load P. If a column has round ends, so that the bending is not restrained, the equation of its elastic curve is:

$$EI \frac{d^2y}{dx^2} = -Py \tag{4-12}$$

When the origin of the coordinate axes is at the top of the column, the positive direction of x being taken downward and the positive direction of y in the direction of the deflection. Integrating the above expression twice and determining the constants of integration give:

$$P = \Omega\pi^2 \frac{EI}{l^2} \tag{4-13}$$

which is Euler's formula for long columns. The factor Ω is a constant depending on the condition of the ends. For round ends Ω = 1; for fixed ends Ω = 4; for one end round and the other fixed Ω = 2.05. P is the load at which, if a slight deflection is produced, the column will not return to its original position. If P is decreased, the column will approach its original position, but if P is increased, the deflection will increase until the column fails by bending.

For columns with value of l/k less than about 150, Euler's formula gives results distinctly higher than those observed in tests. Euler's formula is used for long members and as a basis for the analysis of the stresses in some types of structural parts. It always gives an ultimate and never an allowable load.

Torsion

A bar is under torsional stress when it is held fast at one end, and a force acts at the other end to twist the bar. In a round bar (Fig. 4.9) with a constant force acting, the straight-line ab becomes the helix ad, and a radial line in the cross-section, ob, moves to the position ad. The angle bad remains constant while the angle bod increases with the length of the bar. Each cross section of the bar tends to shear off the one adjacent to it, and in any cross section the shearing stress at any point is normal to a radial line drawn through the point. Within the shearing proportional limit, a radial line of the cross section remains straight after the twisting force has been applied, and the unit shearing stress at any point is proportional to its distance from the axis.

Figure 4.9 Round bar subject to torsion stress

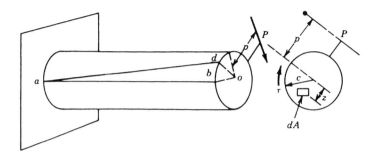

The twisting moment, T, is equal to the product of the resultant, P, of the twisting forces, multiplied by its distance from the axis, p. Resisting moment, T_r, in torsion, is equal to the sum of the moments of the unit shearing stresses acting along a cross section with respect to the axis of the bar. If dA is an elementary area of the section at a distance of z units from the axis of a circular shaft [Fig. 4.9 (b)], and c is the distance from the axis to the outside of the cross section where the unit shearing stress is τ, then the unit shearing stress acting on dA is $(\tau z/c) \, dA$, its moment with respect to the axis is $(\tau z^2/c) \, dA$, and the sum of all the moments of the unit shearing stresses on the cross section is $f \, (\tau z^2/c) \, dA$. In this expression the factor $f z^2 \, dA$ is the polar moment of inertia of the section with respect to the axis. Denoting this by J, the resisting moment may be written $\tau J/c$.

The polar moment of inertia of a surface about an axis through its center of gravity and perpendicular to the surface is the sum of the products obtained by multiplying each elementary area by the square of its distance from the center of gravity of its surface; it is equal to the sum of the moments of inertia taken with respect to two axes in the plane of the surface at right angles to each other passing through the center of gravity section of a round shaft.

The analysis of torsional shearing stress distribution along noncircular cross sections of bars under torsion is complex. By drawing two lines at right angles through the center of gravity of a section before twisting, and observing the angular distortion after twisting, it has been found from many experiments that in noncircular sections the shearing unit stresses are not proportional to their distances from the axis. Thus in a rectangular bar there is no shearing stress at the comers of the sections, and the stress at the middle of the wide side is greater than at the middle of the narrow side. In an elliptical bar the shearing stress is greater along the flat side than at the round side.

It has been found by tests as well as by mathematical analysis that the torsional resistance of a section, made up of a number of rectangular parts, is approximately equal to the sum of the resistances of the separate parts. It is on this basis that nearly all the formulas for noncircular sections have been developed. For example, the torsional resistance of an I-beam is approximately equal to the sum of the torsional resistances of the web and the outstanding flanges. In an I-beam in torsion the maximum shearing stress will occur at the middle of the side of the web, except where the flanges are thicker than the web, and then the maximum stress will be at the midpoint of the width of the flange. Reentrant angles, as those in 1-beams and channels, are always a source of weakness in members subjected to torsion.

The ultimate/failure strength in torsion, the outer fibers of a section are the first to shear, and the rupture extends toward the axis as the twisting is continued. The torsion ula for round shafts has no theoretical basis after the shearing stresses on the outer fibers exceed the proportional limit, as the stresses along the section then are no longer proportional to their distances from the axis. It is convenient, however, to compare the torsional strength of various materials by using the formula to compute values of τ at which rupture takes place.

Sandwich

The same or different materials are combined in the form of sandwich structures (Fig. 4.10). They can be used in products with an irregular distribution of the different materials, and in the form of large structures or sub-structures. A sandwich material composed of two skins and a different core material is similar to RP laminates. Overall load-carrying capabilities depend on average local sandwich properties, but materials failure criteria depend on local detailed stress and strain distributions. Design analysis procedures for sandwich materials composed of linear elastic constituents are well developed. In principle, sandwich materials can be analyzed as composite structures, but incorporation of viscoelastic properties will be subject to the limitations discussed throughout this book.

Structures and sub-structures composed of a number of different components and/or materials, including traditional materials, obey the same principles of design analysis. Stresses, strains, and displacements within individual components must be related through the character-istics (anisotropy, viscoelasticity, and so on) relevant to the particular material, and loads and displacements must be compatible at component

Figure 4.10 Honeycomb core sandwich structure (Courtesy of Plastics FALLO)

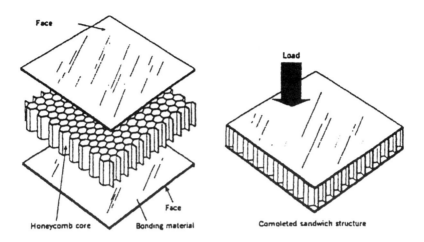

interfaces. Thus, each individual component or sub-component must be treated.

Load and support conditions for individual components depend on the complete structure (or system) analysis, and are unknown to be determined in that analysis. As an example, if a plastic panel is mounted into a much more rigid structure, then its support conditions can be specified with acceptable accuracy. However, if the surrounding structure has comparable flexibility to the panel, then the interface conditions will depend on the flexural analysis of the complete structure. In a more localized context, structural stiffness may be achieved by ribbing and relevant analyses may be carried out using available design formulae (usually for elastic behavior) or finite element analysis, but necessary anisotropy or viscoelasticity complicate the analysis, often beyond the ability of the design analyst.

Design

A structural sandwich is a specially shaped product in which two thin facings of relatively stiff, hard, dense, strong material is bonded to a relatively thick core material. With this geometry and relationship of mechanical properties, facings are subjected to almost all the stresses in transverse bending or axial loading. The geometry of the arrangement provides for high stiffness combined with lightness, because the stiff facings are at a maximum distance from the neutral axis, similar to the flanges of an I-beam. Overall load-carrying capabilities depend on average local sandwich properties, but material failure criteria depend on local detailed stress and strain distributions. Design analysis procedures and fabricating procedures for sandwich materials composed of linear elastic constituents are well developed and reported in the literature. In principle, sandwich materials can be analyzed as RP composites.

The usual objective of a sandwich design is to save weight, increase stiffness, use less expensive materials, or a combination of these factors, in a product. Sometimes, other objectives are also involved such as reducing tooling and other costs, achieving smooth or aerodynamic smoothness, reducing reflected noise, or increasing durability under exposure to acoustic energy. The designers consider factors such as getting the loads in, getting the loads out, and attaching small or large load-carrying members under constraints of deflection, contour, weight, and cost. To design properly, it is important to understand the fabrication sequence and methods, use of the correct materials of construction, the important influence of bond between facing materials and core, and to allow a safety factor that will be required on original, new developments. Use of sandwich panels are extensively used in building and, construction, aircraft, containers, etc.

The primary function of the face sheets is to provide the required bending and in-plane shear stiffness, and to carry the axial, bending, and in-plane shear loading. In high-performance structures, facings most commonly chosen are RPs (usually prepreg), solid plastic, aluminum, titanium, or stainless steel.

The primary function of a core in structural sandwich parts is that of stabilizing the facings and carrying most of the shear loads through the thickness. In order to perform this task efficiently, the core must be as rigid and as light as possible, and must deliver uniformly predictable properties in the environment and meet performance requirements. Several different materials are used such as plastic foam, honeycomb [using RP, film (plastic, steel, aluminum, paper, etc.), balsa wood, etc.].

Different fabricating processes are used. These include bag molding, compression molding, reinforced reaction rejection molding (RRIM), filament winding, corotational molding, etc. There is also the so-called structural foam (SF) that is also called integral skin foaming or reaction injection molding. It can overlap in lower performance use with the significantly larger market of the more conventional sandwich. Up until the 1980s in the U.S., the RIM and SF processes were kept separate. Combining them in the marketplace was to aid in market penetration. During the 1930s to 1960s, liquid injection molding (LIM) was the popular name for what later became RIM and SF (Chapter 1).

These structures are characterized as a plastic structure with nearly uniform density foam core and integral near-solid skins (facings). When these structures are used in load-bearing applications, the foam bulk density is typically 50 to 90% of the plastic's unfoamed bulk density. Most SF products (90wt%) are made from different TPs, principally PS, PE, PVC, and ABS. Polyurethane is the primary TS plastic. Unfilled and reinforced SFs represents about 70% of the products. The principal method of processing (75%) is modified low-pressure injection molding. Extrusion and RIM account for about 10% each.

In a sandwich design, overall proportions of structures can be established to produce an optimization of face thickness and core depth which provides the necessary overall strength and stiffness requirements for minimum cost of materials, weight of components, or other desired objectives. Competing materials should be evaluated on the basis of optimized sandwich section properties that take into account both the structural properties and the relative costs of the core and facing materials in each combination under consideration. For each combination of materials being investigated, thickness of both facings and core should be determined to result in a minimum cost of a sandwich design that provides structural and other functional requirements.

Sandwich configurations are used in small to large shapes. They generally are more efficient for large components that require significant bending strength and/or stiffness. Examples of these include roofs, wall and floor panels, large shell components that are subject to compressive buckling, boat hulls, truck and car bodies, and cargo containers. They also provide an efficient solution for multiple functional requirements such as structural strength and stiffness combined with good thermal insulation, or good buoyancy for flotation.

Sandwich materials can be analyzed as composite structures. Structures and sub-structures composed of a number of different components and/or materials, including traditional materials, obey the same principles of design analysis. Stresses, strains, and displacements within individual components must be related through the characteristics (anisotropy, viscoelasticity, etc.) relevant to the particular material; also loads and displacements must be compatible at component interfaces. Thus, each individual component or sub-component must be treated using the relevant methods.

Load and support conditions for individual components depend on the complete structure (or system) analysis. For example, if a panel is mounted into a much more rigid structure, then its support conditions can be specified with acceptable accuracy. However, if the surrounding structure has comparable flexibility to the panel, then the interface conditions will depend on the flexural analysis of the complete structure.

In a more localized context, structural stiffness may be achieved by ribbing, and relevant analyses may be carried out using available design formulae (usually for elastic behavior) or finite element analysis. But necessary anisotropy or viscoelasticity complicate the analysis, often beyond the ability of the design analyst.

Primary structural role of the face/core interface in sandwich construction is to transfer transverse shear stresses between faces and core. This condition stabilizes the faces against rupture or buckling away from the core. It also carries loads normally applied to the panel surface. They resist transverse shear and normal compressive and tensile stress resultants. For the most part, the faces and core that contain all plastics can be connected during a wet lay-up molding or, thereafter, by adhesive bonding. In some special cases, such as in a truss-core pipe, faces and core are formed together during the extrusion process, resulting in an integral homogeneous bond/connection between the components. Fasteners are seldom used to connect faces and core because they may allow erratic shear slippage between faces and core or buckling of the faces between fasteners. Also, they may compromise other advantages such as waterproofing integrity and appearance.

For RP-faced sandwich structures the design approaches includes both the unique characteristics introduced by sandwich construction and the special behavior introduced by RP materials. The overall stiffness provided by the interaction of the faces, the core, and their interfaces must be sufficient to meet deflection and deformation limits set for the structures. Overall stiffness of the sandwich component is also a key consideration in design for general instability of elements in compression.

In a typical sandwich constructions, the faces provide primary stiffness under in-plane shear stress resultants (N_{xy}), direct stress resultants (N_x, N_y), and bending stress resultants (M_x, M_y). Also as important, the adhesive and the core provide primary stiffness under normal direct stress resultants (N_z), and transverse shear stress resultants (Q_x, Q_y). Resistance to twisting moments (T_{xz}, T_{yz}), which is important in certain plate configurations, is improved by the faces. Capacity of faces is designed not to be limited by either material strength or resistance to local buckling.

The stiffness of the face and core elements of a sandwich composite must be sufficient to preclude local buckling of the faces. Local crippling occurs when the two faces buckle in the same mode (anti-symmetric). Local wrinkling occurs when either or both faces buckle locally and independently of each other. Local buckling can occur under either axial compression or bending compression. Resistance to local buckling is developed by an interaction between face and core that depends upon the stiffness of each.

With the structural foam (SF) construction, large and complicated parts usually require more critical structural evaluation to allow better prediction of their load-bearing capabilities under both static and dynamic conditions. Thus, predictions require careful analysis of the structural foam's cross-section.

The composite cross-section of an SF part contains an ideal distribution of material, with a solid skin and a foamed core. The manufacturing process distributes a thick, almost impervious solid skin that is in the range of 25% of overall wall thickness at the extreme locations from the neutral axis where maximum compressive and tensile stresses occur during bending.

When load is applied flatwise the upper skin is in compression while the lower one is in tension, and a uniform bending curve will develop. However, this happens only if the shear rigidity or shear modulus of the cellular core is sufficiently high. If this is not the case, both skins will deflect as independent members, thus eliminating the load-bearing capability of the composite structure. In this manner of applying a load the core provides resistance against shear and buckling stresses as well as impact (Fig. 4.11). There is an optimum thickness that is critical in designing this structure.

When the SF cross-section is analyzed, its composite nature still results in a twofold increase in rigidity, compared to an equivalent amount of solid plastic, since rigidity is a cubic function of wall thickness. This

Figure 4.11 Core thickness vs. density impact strength

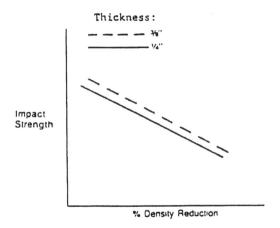

Figure 4.12 Sandwich and solid material construction

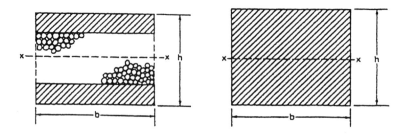

Figure 4.13 Sandwich cross-section with and without a core

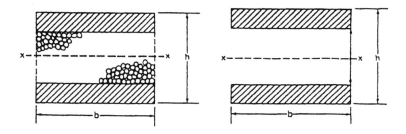

increased rigidity allows large structural parts to be designed with only minimal distortion and deflection when stressed within the recommended values for a particular core material. When analyzing rigidity and the moment of inertia (I) can be evaluated three ways. In the first approach, the cross-section is considered to be solid material (Fig. 4.12).

The moment of inertia (I_x) is then equal to:

$$I_x = bh^3/12 \qquad\qquad (4\text{-}14)$$

where b = width and h = height.

This commonly used approach provides acceptable accuracy when the load-bearing requirements are minimal. An example is the case of simple stresses or when time and cost constraints prevent more exact analysis.

The second approach ignores the strength contribution of the core and assumes that the two outer skins provide all the rigidity (Fig. 4.13).

The equivalent moment of inertia is then equal to:

$$I_x = b(h^3 - h^3/12) \qquad\qquad (4\text{-}15)$$

This formula results in conservative accuracy, since the core does not

Figure 4.14 Sandwich and I-beam Cross-section

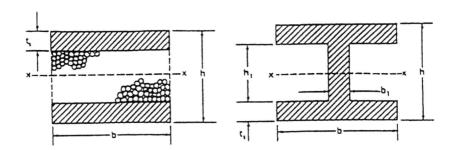

contribute to the stress-absorbing function. It also adds a built-in safety factor to a loaded beam or plate element when safety is a concern.

A third method is to convert the structural foam cross-section to an equivalent I-beam section of solid resin material (Fig. 4.14).

The moment of inertia is then formulated as:

$$I_x = [bh^3 - (b - b_1)(h - 2t_x)^3]/12 \qquad (4\text{-}16)$$

where $b_1 = b(E_c)/(E_s)$, E_c = modulus of core, E_s = modulus of skin, t_s = skin thickness, and h_1 = core height

This approach may be necessary where operating conditions require stringent load-bearing capabilities without resorting to overdesign and thus unnecessary costs. Such an analysis produces maximum accuracy and would, therefore, be suitable for finite element analysis (FEA) on complex parts. However, the one difficulty with this method is that the core modulus and the as-molded variations in skin thicknesses cannot be accurately measured.

The following review relates to the performance of sandwich constructions such as those with RP skins and honeycomb core. For an isotropic material with a modulus of elasticity (E), the bending stiffness factor (EI) of a rectangular beam b wide and h deep is:

$$EI = E(bh^3/12) \qquad (4\text{-}17)$$

A rectangular structural sandwich with the same dimensions whose facings and core have moduli of elasticity E_f and E_c, respectively, and a core thickness c, the bending stiffness factor EI becomes:

$$EI = (E_f b/12)(h^3 - c^3) + (E_c b/12)\, c^3 \qquad (4\text{-}18)$$

This equation is OK if the facings are of equal thickness, and approximate or approximately equal, but the approximation is close if the facings are thin relative to the core. If, as is usually the case, E_c is much smaller than E_f, the last term in the equation is deleted.

Asymmetrical sandwich structures with different materials or different thicknesses in their facings, or both, the more general equation for *El* may be used. With isotropic materials, the shear modulus *G* is high compared to the elastic modulus *E*, and the shear distortion of a transversely loaded beam is so small that it can be neglected in calculating deflection. Sandwich core shear modulus *G* is usually so much smaller than E_f of the facings that the shear distortion of the core may be large and therefore contribute significantly to the deflection of a transversely load. The total deflection of a sandwich beam involves the two factors of the deflection caused by the bending moment alone and the deflection caused by shear, that is;

$$\delta = \delta_m + \delta_s \qquad\qquad (4\text{-}19)$$

where δ = total deflection, δ_m = moment deflection, and δ_s = shear deflection.

Under transverse loading, bending moment deflection is proportional to the load and the cube of the span and inversely proportional to the stiffness factor, *EI*.

Gear

Designing gears can be very complex since many interfacing load factors are involved. There are bending, shear, rolling, tension, and sliding stresses all acting upon a mechanism whose purpose is to transmit uniform motion and power. This situation is well understood by those designing gears. For over a century plastic gears have been extensively used in all industries worldwide and with time passing the plastic industry has provided lightweight and quieter operating gears. They provide a means of cutting cost, weight, and noise without reducing performance.

Information on designing gears is extensive. Knowledge of gear fundamentals is a prerequisite for the understanding of applying appropriate plastic information into the gear formulas so that the application results in favorable operation. Textbooks, technical handbooks, and industrial literature of gear suppliers provide information such as teeth load requirements, transmitting motion and power by means of gears, their construction, and detail performance requirements. Reviews on teeth of heavily loaded gears require tip relief to reduce effects of deflection, and have full fillet radii to reduce stress concentrations. If the pinion in a pair of gears has a small number of teeth, undercutting may result. Undercutting weakens teeth, causes undue wear, and may affect continuity of action.

Designing gears involves ribbing, coring, and shaping material. When going from a metal to a plastic gear it is redesigned to meet the behaviors of plastics. The popular injection molded plastic gear saves material, eliminates stresses from having thick and thin sections, provides uniform shrinkage in teeth and the remainder of the gear, and provides a full load-carrying capacity for the teeth. Plastic gears are dimensioned so that they will provide sufficient backlash at the highest temperatures likely to be encountered in service.

Wear, scoring, material flow, pitting, fracture, creep, and fatigue cause plastic and metal gears to fail. Continuous lubrication can increase the allowable bending stress by a factor of at least 1.5. However there are plastics (acetals, nylons, fluoropolymers, and others) that operate efficiently with no lubrication. There are plastics with wear resistance and durability of plastic gears makes them exceptionally useful.

The bending stress in engineering TPs is based on fatigue tests run at specific pitch-line velocities. A velocity factor should be used if the operating pitch-line velocity exceeds the test speed. Plastic gears should have a full fillet radius at the tooth root, so they are not subjected to stress concentration as are metal gears. If a gear is lubricated, bending stress will be important to evaluate. As with bending stresses, calculating surface-contact stress requires using a number of correction factors. As an example, a velocity factor is used when the pitch-line velocity exceeds the test velocity. A correction factor is also used to account for changes in operating temperature, gear materials, and the pressure angle. Stalled torque, another important factor, could be considerably more than the normal loading torque.

A damaging situation for gears is to operate over a specified temperature for the plastic used. Reducing the rate of heat generation or increasing the rate of heat transfer will stabilize the gear's temperature so that they will run indefinitely until stopped by fatigue failure. Using unfilled engineering plastics usually gives them a fatigue life on an order of magnitude higher than metal gears.

Plastic gears are subject to hysteresis heating, particularly at high speeds (Chapter 3). If the proper plastic is not used to meet the gears service requirements the hysteresis heat may be severe enough that the plastic melts. Avoid this failure by designing the gear drive so that there is favorable thermal balance between the heat that is generated and that which is removed by cooling processes. Hysteresis heating in plastics can be reduced by several methods, the usual one being to reduce the peak stress by increasing the tooth root area available for torque transmission. Another way to reduce stress on the teeth is by increasing the gear's diameter.

Materials used such as stiffer plastics can reduce hysteresis heating. Crystalline TPs for example (the popularly used acetal and nylon) can be stiffened by 25 to 50% with the addition of fillers and reinforcements. Others used include ABS, polycarbonates, polysulfones, phenylene oxides, polyurethanes, and thermoplastic polyesters. Additives, fillers, and reinforcements are used in plastics gears to meet different performance requirements (Chapter 1), Examples include glass fiber for added strength, and fibers, beads, and powders for reduced thermal expansion and improved dimensional stability. Other materials, such as molybdenum disulfide, polytetrafluoroethylene (PTFE), and silicones, may be added as lubricants to improve wear resistance.

Choice of plastics gear material depends on requirements for size and nature of loads to be transmitted, speeds, required life, working environment, type of cooling, lubrication, and operating precision. The strength of these TPs varies with temperature. If the incorrect plastic is used, the higher temperatures can reduce root stress and permit tooth deformation. In calculating power to be transmitted by spur, helical, and straight bevel gearing, the following formulas should be used with the factors given in Table 4.5.

For internal and external spur gears:

$$HP = \frac{S_s FYV}{55(600 + V)PC_s} \qquad (4\text{-}20)$$

For internal and external helical gears,

$$HP = \frac{S_s FYV}{423(78 + \sqrt{V})P_n C_s} \qquad (4\text{-}21)$$

For straight bevel gears,

$$HP = \frac{S_s FYV(C - F)}{55(600 + V)PCC_s} \qquad (4\text{-}22)$$

where S = safe stress in bending (Table 4.5a); F = face width in inches; Y = tooth form factor (Table 4.5b); C = pitch cone distance in inches; C, = service factor (Table 4.5c); P = diarnetral pitch; P. = normal diametral pitch; and V = velocity at pitch circle diameter in ft/min.

The surrounding condition, whether liquid, air, or oil (most efficient) will have substantial cooling effects. A fluid like oil is at least ten times better at cooling than air. Agitating these mediums increases their cooling rates, particularly when employing a cooling heat exchanger.

Methods of fabricating gears involve cutting/hobbing from processed blocks or sheet plastics, compression molding laminated (RP) material, or the most popular injection molding. Use is made of unfilled and

Table 4.5 Plastic gear (a) safe bending stress (psi), (b) tooth form examples of Y factors, and (c) service factors

(a)

Plastics Type	Safe Stress	
	Unfilled	Glass-filled
ABS	3,000	6,000
Acetal	5,000	7,000
Nylon	6,000	12,000
Polycarbonate	6,000	9,000
Polyester	3,500	8,000
Polyurethane	2,500	–

(b)

Number of Teeth	$14^1/_2$-deg Involute or Cycloidal	20-deg Full Depth Involute	20-deg Stub Tooth Involute	20-deg Internal Full Depth Pinion	Gear
12	0.210	0.245	0.311	0.327	...
13	0.220	0.261	0.324	0.327	...
14	0.226	0.276	0.339	0.330	...
15	0.236	0.289	0.348	0.330	...
16	0.242	0.259	0.361	0.333	...
17	0.251	0.302	0.367	0.342	...
18	0.261	0.308	0.377	0.349	...
19	0.273	0.314	0.386	0.358	...
20	0.283	0.320	0.393	0.364	...
21	0.289	0.327	0.399	0.371	...
22	0.292	0.330	0.405	0.374	...
24	0.298	0.336	0.415	0.383	...
26	0.307	0.346	0.424	0.393	...
28	0.314	0.352	0.430	0.399	0.691
30	0.320	0.358	0.437	0.405	0.679
50	0.352	0.480	0.474	0.437	0.613
100	0.371	0.446	0.506	0.462	0.565
150	0.377	0.459	0.518	0.468	0.550
300	0.383	0.471	0.534	0.478	0.534
Rack	0.390	0.484	0.550	...	...

(c)

Type of load	8–10 hr/day	24 hr/day	Intermittent 3 hr/day	Occasional $^1/_2$ hr/day
Steady	1.00	1.25	0.80	0.50
Light shock	1.25	1.5	1.00	0.80
Medium shock	1.5	1.75	1.25	1.00
Heavy shock	1.75	2.00	1.5	1.25

filled or reinforced laminated TPs or TSs. Phenolic laminated gears are in a class of their own. One can make all the perfect calculations and insert the necessary values for plastic gears, but if molding conditions and molding materials are not processed properly one may end up with mediocre or even unsatisfactory results.

Being not as strong as steel, plastics perform far closer to their design limits than do metal gears. Metal and plastic in gear design differ. Designs for metal are based on the strength of a single tooth, but plastic shares the load among the various gear teeth to spread it out. In plastics the allowable stress for a specific number of cycles to failure increases as the tooth size decreases to a pitch of about 48. Very little increase is seen above a 48 pitch, because of the effects of size and other considerations.

Contact Stress

Stresses caused by the pressure between two elastic contacting parts are of importance in design such as gears and bearings. Centuries ago H. Hertz developed the mathematical theory for the surface stresses and the deformation produced by pressure between curved parts, and the results of this analysis are supported by research. Formulas based on this theory give the maximum compressive stresses that occur at the center of the surfaces of contact, but do not consider the maximum subsurface shear stresses or the maximum tensile stresses that occur at the boundary of the contact area.

Bearing

Plastic used in bearings, as in gears, have many success stories because certain plastics have the required properties to meet different product application performances. Similarly to plastic gears, plastic bearings have a long history of successful performance. Small to large bearings are used in applications requiring light to heavy duty. Some TPs have inherent lubricating characteristics and additives such as molybdenum disulfide, polytetrafluoroethylene (PTFE), and others enhance other TPs as well as TSs. Other TPs, as well as TSs, by the addition of PTFE and/or molybdenum disulfide, become excellent candidates for bearing materials.

There are high performance laminated (RP) fabric, bonded with phenolic plastic incorporating antifriction ingredients. They have given excellent service when properly applied in various applications particularly in the past. This group of bearings has a low coefficient of friction, antiscoring properties, and adequate strength for use in steel mills and other heavy-duty applications.

PV Factor

Bearings are designed to keep their frictional heat at a low value and have conditions that lead to dissipation of such generated heat. Major heat contributors are the magnitude pressures P exerted on the projected area of the bearing and the velocity V or the speed of the rotating bearing. Experience has set limits on this PV value. Limits within PV factors have been developed for specific plastics that provide the industries with successful bearings. Other heat contributors are coefficient of friction of mating materials, lubrication, clearances between bearing and shaft, rusted shaft, surrounding temperature, surface finish, hardness of the mating materials, contaminants, and bearing wall thickness that relates to heat dissipation.

The basic rule is that neither the pressure or the velocity should exceed a value of 1000 (psi or ft/min). As an example with a PV limit for acetal of 3000, the PV factor could be 1000 ft/min times 3 pounds or 1000 pounds times 3 ft/min at the extreme, provided heat conditions resulted in uniform rate of wear. The coefficient of friction data, available from suppliers, can also provide guidelines to the efficiency in comparing the different materials.

The limit of the PV factor for each material or the internally lubricated materials for the constant wear of bearing is usually available from the supplier of the plastic. Lubrication whether incorporated in a plastic or provided by feeding the lubricant to the bearing will raise the PV limit 2.5 or more times over the dry system.

Grommet

Damping designed products may be required. As an example, large flat areas may require damping so that they do not act like loudspeakers. The damping action can quiet a cabinet that resonates. Various plastics have helped alleviate problems in all types of noisemakers. Different damping approaches can be used, such as applying plastic foam sound insulator or plastic panels that have low damping characteristics.

A popular approach is to use plastic grommets where applicable. Sound-absorbing grommets are used on equipment such as motors' bolt attachments and trash compactors. Testing and all other types of

Figure 4.15 Grommet replaces a five individual metal assembly (Courtesy of Mobay/Bayer)

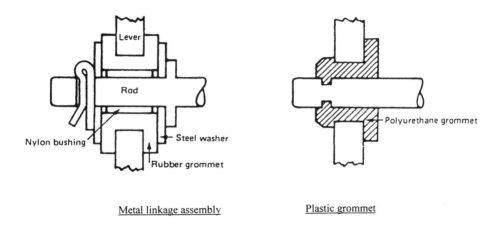

Metal linkage assembly   Plastic grommet

equipment can take advantage of grommets or be redesigned to use plastic. Grommets provide their greatest noise reduction through damping in the octave frequency bands above 500 Hz where the ear is most sensitive and sound most annoying.

Grommets have replaced assembled linkages (Fig. 4.15). In addition to reducing noise, the usual injection molded polyurethane (PUR) grommet eliminated the time consuming/costly metal assembly. During assembly it is snapped into a hole in the steel lever, then a grooved rod is inserted into the grommet. The grommet isolates vibration from the metal parts and eliminates the hardening and cracking that used to shorten the life of the old assembly. The plastic design mechanically is at least as strong as the metal assembly that includes withstanding high load pull-outs and can withstand high cyclic loads applied at different degrees off the rod axis at temperatures up to 300°F (149°C).

To quiet a noise-generating mechanism, the first impulse is often to enclose it. Enclosure can be the best solution, but not always. By determining what is causing the noise, appropriate action can be taken to be more specific and provide a cost-effective fix. A plastic enclosure can be used to suppress noise. Recognize that with a metal enclosure a small noise is transmitted to the metal structure that serves to amplify the sound.

Gasket

Different plastics are used to meet different gasket and seal requirements. An example is PTFE (polytetrafluoroethylene) that provides a virtually inert exposure to all kinds of elements and has outstanding high-temperature performance. Unreinforced PTFE can be vulnerable to creep and stress relaxation. However, reinforced PTFE modifies these limitations.

Gaskets are designed to meet different requirements such as retaining loads or meeting stress relaxation requirements, chemical or heat resistance, severe environments, and containment of liquids, greases, and so on. There are different industry tests and standards to meet many different service requirements. There are tests for applying compressive stress simulating the way many gaskets and seals are stressed in service (ASTM F 38). They measure the effects of such pertinent variables as stress relaxation in regard to time and environment.

Stress analysis is used to determine their capability to seal against leakage resulting from the pressure of a confined fluid. Generally high pressures or stress relates to increasing the tendency of a gasket to creep. Stress relaxation data to the designer provides a guide to developing a suitable design compromise, without overdesigning. These data show that the thinner the gasket, the less stress relaxation occurs. A gasket is redesign to be stronger and more expensive in construction. In material evaluations, stress relaxation can be related with geometric variables by means of a shape factor such as:

Shape factor = Annular area/total lateral area = $(OD - ID)/4t$ (4.23)

where OD = outside diameter, ID = inside diameter, and t = thickness.

The trend of this factor is generally consistent with plastics' behaviors as reviewed in Chapters 1 & 2. The relaxation-test data has to be determined for the plastic to be used. Individual behavior of one plastic is usually different when compared to another plastic.

Shape

Overview

In addition to what has been reviewed in meeting structural requirements, analyses of product shapes also includes factors such as the size available equipment can handle including thicknesses and product complexity and capability to package and ship to the customer

(Chapters 1 and 2). The ability to achieve specific shapes and design details is dependent on the way the process operates and plastics to be processed. Generally the lower the process pressure, the larger the product that can be produced. With most labor-intensive fabricating methods, such as RP hand lay-up with TS plastic there is virtually no limit on size.

An important requirement concerns meeting dimensional tolerances of shaped products. Reported are different shrinkages for different plastics per standard tests that may have a relation to the designed product. The probability is that experience with prototyping will only provide the true shrinkage conditions of the shaped products. Minimum shrink values are included in the design of mold cavities and die openings so that if the processed plastic does not meet required dimensions all that is required is to cut the metal in the tools.

If the reverse occurs, expensive tool modifications may be required, if not replacing the complete tool. It is vital to set up a complete checklist of product requirements, to preclude the possibility that a critical requirement may be overlooked initially. Fortunately there are occasions where changes in process control during fabrication can be used to produce the required dimensional product.

Filament Wound Shape
Filament winding (FW) shapes are principally circular (cylinders, pipes, tubing, etc.) or enclosed vessel (storage tanks, oxygen tanks, etc.). They produce spherical, conical, and geodesic shapes. The fabricating process permits tightly controlled fiber netting orientation and exceptional quality control in different fiber-resin matrix ratios required by design. Structures can be fabricated into shapes such as rectangular or square beams or boxes, longitudinal leaf or coil springs, etc. Filaments can be set up in a part to meet different design stresses.

There are two basic patterns used by industry to produce FW structures, namely, circumferential winding and helical winding. Each winding pattern can be used alone or in various combinations in order to meet different structural requirements. The circumferential winding pattern involves the circumferential winding at about a $90°$ angle with the axis of rotation interspersed with longitudinal reinforcements. Maximum strength is obtainable in the hoop direction. This type of pattern generally does not permit winding of slopes over $20°$ when using a wet winding reinforcement or $30°$ when using a dry winding process. It also does not result in the most efficient structure when end closures are required. With end closures and/or steep slopes, a combination of helical and circumferential winding is used.

With helical winding, the reinforcements are applied at any angle from 25° to 85° to the axis of rotation. No longitudinal filament need be applied because low-winding angles provide the desired longitudinal strength as well as the hoop strength. By varying the angle of winding, many different ratios of hoop to longitudinal strengths can be obtained.

Two different techniques of applying the reinforcements in helical windings are used by industry. One technique is the application of only one complete revolution around the mandrel from end to end. The other technique involves a multi-circuit winding procedure that permits a greater degree of flexibility of wrapping and length of cylinder.

Netting Analysis
Continuous reinforced filaments should be used to develop an efficient high-strength to low-weight FW structure. Structural properties are derived primarily from the arrangement of continuous reinforcements in a netting analysis system in which the forces, owing to internal pressure, are resisted only by pure tension in the filaments (applicable to internal-pressure systems).

There is the closed-end cylinder structure that provides for balanced netting of reinforcements. Although the cylinder and the ends require two distinctly different netting systems, they may be integrally fabricated. The structure consists of a system of low helix angle windings carrying the longitudinal forces in the cylinder shell and forming integral end closures which retain their own polar fittings. Circular windings are also applied to this cylindrical portion of the vessel, yielding a balanced netting system. Such a netting arrangement is said to be balanced when the membrane generated contains the appropriate combination of filament orientations to balance exactly the combination of loadings imposed.

The girth load of the cylindrical shell is generally two times the axial load. The helical system is so designed that its longitudinal strength is exactly equal to the pressure requirement. Such a low-angle helical system has a limited girth strength. The circular windings are required in order to carry the balance of the girth load.

The end dome design contains no circular windings since the profile is designed to accommodate the netting system generated by the terminal windings of the helical pattern. It is termed an ovaloid: that is, it is the surface of revolution whose geometry is such that fiber stress is uniform throughout and there is no secondary bending when the entire internal pressure is resisted by the netting system.

There is the ovaloid netting system that is the natural result of the reversal of helical windings over the end of the vessel. The windings become thicker as they converge near the polar fittings. In order to resist internal pressure by constant filament tension only, the radius of curvature must increase in this region. It can also be equal to one half the cylinder radius when the helix angle $\alpha = 0°$, and equal to the cylinder radius when $\alpha = 45°$. The profile will also be affected by the presence of an external axial force.

In the application of bidirectional patterns, the end domes can be formed by fibers that are laid down in polar winding patterns. The best geometrical shape of the dome is an oblated hemispheroid. Theoretically, the allowable stress level in the two perpendicular directions should be identical. However, the efficiency of the longitudinal fibers is less than that of the circumferential fibers. It is possible to estimate an optimum or length-to-diameter ratio of a cylindrical case for a given volume.

The filament-wound sphere design structure provides another example of a balanced netting analysis system. It is simpler in some respects than the closed-end cylinder. The sphere must be constructed by winding large circles omni-directionally and by uniform distribution over the surface of the sphere. In practice, distribution is limited so that a small polar zone is left open to accommodate a connecting fitting.

The netting pattern required generates a membrane in which the strength is uniform in all directions. The simplest form of such a membrane would have its structural fibers running in one direction and the other half at right angles to this pattern. This layup results in the strength of the spherical membrane being one-half of the strength of a consolidated parallel fiber system.

The oblated spheroid design structure relates to special spherical shapes. Practical design parameters have shown that the sphere is the best geometric shape when compared to a cylinder for obtaining the most efficient strength-to-weight pressure vessel. The fiber RPs is the best basic constituents. Certain modifications of the spherical shape can improve the efficiency of the vessel. One modification involves designing the winding pattern of the fibers so that unidirectional loading can be maintained. In this type of structure, it is generally assumed that the fibers are under equal tension. This type of structure is identified as an isotensoid. The geometry of this modified sphere is called oblated spheroid, ovaloid, or ellipsoid.

The term isotensoid identifies a pressure vessel consisting entirely of filaments that are loaded to identical stress levels. The head shape of an isotensoid is given by an elliptical integral, which can most readily be solved by a computer. Its only parameter is the ratio of central opening to vessel diameter. This ratio determines the variation of the angle of winding for the pressure vessel. During pressurization the vessel is under uniform strain; consequently, no bending stresses or discontinuity stresses are induced.

A short polar axis and a larger perpendicular equatorial diameter characterize the vessel. The fibers are oriented in the general direction of a polar axis. Their angle with this axis depends on the size of the pole openings (end closures). For glass fiber-TS polyester RP vessels levels of 200,000 psi (1.4 GPa) can be obtained.

The toroidal design structure is a pressure vessel made with two sets of filaments symmetrically arranged with respect to the meridians. They meet two basic requirements: (1) static equilibrium at each point, which determines the angle between the two filaments, and (2) stability of the filaments on the surface, which requires the filaments to follow geodesic paths on the surface. When the equation of the surface is given, these two requirements are generally incompatible. One way to reconcile the correct angularity of the filaments (equilibrium) with the correct paths of the filaments (stability) is to take some freedom in determining the geometry of the surface.

Cylinder

Standard engineering analysis can be used. Consider a cylinder of inside radius r, outside radius R, and length L containing a fluid under pressure p. The circumferential or hoopwise load in the wall (t = thickness) is proportional to the pressure times radius = pr, and the hoop stress:

$$f_h = \text{hoopwise load/cross sectional area} = pr/t \text{ or } = pd/2t \qquad (4\text{-}24)$$

similarly, the longitudinal stress:

$$f_l = pd/4t \qquad (4\text{-}25)$$

assuming $\pi(R^2 - r^2) = 27\pi rt$ for a thin-walled tube.

Thus this condition of the hoop stress being twice the longitudinal stress is normal for a cylinder under internal pressure forces only. The load in pounds acts on the tube at a distance from one end and a bending moment M is introduced. This produces a bending stress in the wall of the cylinder of :

$$f_b = My/I \qquad (4\text{-}26)$$

where $y = R$ and I= moment of inertia. For a cylinder:

$$I = \pi(D^4 - d^4)/64 \text{ in.}^4 \qquad (4\text{-}27)$$

Figure 4.16 Cylinder comparison of thickness for a flat end and a hemispherical end

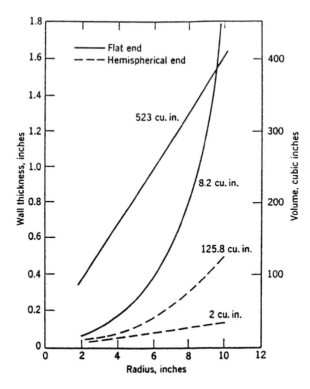

This stress must then be considered in addition to the longitudinal stress already presented because of internal pressure. If the end closures are in the form of flat plates, bending stresses due to the internal pressure are introduced as:

$$F_b = 1.25 \left(pr^2/t_l \right) \tag{4-28}$$

where t_l = thickness of end. This necessitates the wall of a flat disc-type end being extremely thick compared with a hemispherical end which is found to be the most efficient shape where the stress in the wall is:

$$pd/4t \tag{4-29}$$

Fig. 4.16 compares the thicknesses and corresponding volumes of the two types of ends for varying values of r (assuming p = 2,000 psi and ultimate stress in the wall material of 100,000 psi).

The volume of the flat end is found to be approximately four times the volume of the hemispherical end for any given radius of tube, resulting in increased weight and material cost. Other end shapes such as ellipse will have a volume of weight somewhere between the two, depending on the actual shape chosen.

Sphere

Circumferential load in the wall of all the spheres under internal pressure is equal to the pressure times the internal cross-sectional area, and the hoop stress, using the previous engineering assumption, is found to be:

$$f_h = pd/4t \qquad (4\text{-}30)$$

It will readily be seen that no matter which section is chosen, provided the plane of the section passes through the center of the sphere, the condition will be the same, and it can be said that the hoop stress will be the same in all directions.

When it is assumed that:

$$\pi (R^2 - r^2) = 27 \pi r t \qquad (4\text{-}31)$$

it is determined that for wall thicknesses up to approximately 3 inches, the error is negligible. It can also be determined that the percent of error decreases as the ratio r/t increases. Table 4.6 provides size vs. weight of RP spheres.

Table 4.6 Data for 3000 psi glass fiber-epoxy spheres

Capacity, cubic inches	Outside diameter, inches	Nominal weight, pounds	Maximum weight, pounds	Overall length, inches
50	$5^1/_4$	1.50	1.60	$5^5/_8$
100	$6^1/_2$	2.50	2.63	$6^{15}/_{16}$
200	$8^1/_8$	4.44	4.62	$8^1/_2$
300	$9^1/_4$	6.25	6.56	$9^5/_8$
400	$10^1/_8$	8.06	8.48	$10^9/_{16}$
500	$10^{15}/_{16}$	9.87	10.35	$11^5/_{16}$
650	$11^{15}/_{16}$	12.56	13.18	$12^5/_{16}$
880	$13^1/_{16}$	16.00	16.80	$13^3/_8$
1,070	14	20.06	21.07	$14^3/_8$
1,325	$15^1/_8$	24.50	25.73	$15^7/_{16}$
1,575	$15^7/_8$	28.75	30.15	$16^1/_4$
1,800	$16^5/_8$	32.75	34.42	$16^{15}/_{16}$
2,500	$18^1/_2$	44.81	47.06	$18^7/_8$
3,200	20	56.50	59.32	$20^7/_{16}$

For a sphere with the stresses uniform in all directions, it follows that the fibers require equal orientation in all directions. The problem of orientation resolves itself purely into one of practical application of the fibers. In the cylinder, the fibers are specifically oriented to meet any condition of stressing. The simplest method of doing this is to employ a single helical pattern.

Theory shows that this is highly sensitive to variations in the longitudinal hoop-stress ratio and also the accuracy of the angle wound. The addition of pure hoop windings to the helix gives a theoretical gain in stability with no loss of strength or efficiency. In order to develop the most satisfactory orientation, the winding is performed so those two different helix angles are used.

Tank

Fabricating RP Tank

Classical engineering stress analysis shows that hoop stress (stress trying to push out the ends of the tank) is twice that of longitudinal stress. To build a tank of conventional materials (steel, aluminum, etc.) requires the designer to use sufficient materials to resist the hoop stresses that results in unused strength in the longitudinal direction. In RP, however, the designer specifies a laminate that has twice as many fibers in the hoop direction as in the longitudinal direction.

An example is a tank 0.9 m (3 ft) in diameter and 1.8 m (6 ft) long with semi-spherical ends. Such a tank's stress calculations (excluding the weight of both the product contained in it, and the support for the tank) are represented by the formulas:

$$s = pd/2t \quad \text{for the hoop stress} \tag{4-32}$$

and:

$$s = pd/4t \quad \text{for the end and longitudinal stresses} \tag{4-33}$$

where s = stress, p = pressure, d = diameter, and t = thickness.

Tensile stresses are critical in tank design. The designer can assume the pressure in this application will not exceed 100 psi (700 Pa) and selects a safety factor of 5. The stress must be known so that the thickness can be determined. The stress or the strength of the final laminate is derived from the makeup and proportions of the resin, mat, and continuous fibers in the RP composite material.

Representative panels must be made and tested, with the developed tensile stress values then used in the formula. Thus, the calculated tank thickness and method of lay-up or construction can be determined based on:

$$t_h = pd/2s_h \text{ or } t_h = pd/4s_h \tag{4-34}$$

where:

$$t_h = \frac{100 \times 3 \times 12}{2 \times \dfrac{20 \times 10^3}{5}} = 0.450 \text{ in.} \tag{4-35}$$

t_l = 1/2 t_h = 0.225 in. (or the same thickness with half the load or stress)
t_h = hoop thickness
t_l = longitudinal thickness
s_h = hoop stress
s_l = longitudinal stress
s_h = 20×10^3 psi (140 MPa)
safety factor = 5
p = 100 psi (700 Pa)
d = 3 ft (0.9m)

If the stress values had been developed from a laminate of alternating plies of woven roving and mat, the lay-up plan would include sufficient plies to make 1 cm (0.40 in.) or about four plies of woven roving and three plies of 460 g/m² (1½ oz) mat. However, the laminate would be too strong axially. To achieve a laminate with 2 to 1 hoop to axial strength, one would have to carefully specify the fibers in those two right angle directions, or filament wind the tank so that the vector sum of the helical wraps would give a value of 2 (hoop) and 1 (axial), or wrap of approximately 54° from the axial.

Another alternative would be to select a special fabric whose weave is 2 to 1, wrap to fill, and circumferentially wrap the cylindrical sections to the proper thickness, thus getting the required hoop and axial strengths with no extra, unnecessary strength in the axial (longitudinal) direction, as would inevitably be the case with a homogeneous metal tank.

As can be seen from the above, the design of RP products, while essentially similar to conventional design, does differ in that the materials are combined when the product is manufactured. The RP designer must consider how the load-bearing fibers are placed and ensure that they stay in the proper position during fabrication.

Underground Storage Tank

GFRP (glass fiber-TS polyester RP) underground tanks for storing gasoline and other materials have been in use worldwide since at least the 1950s. Experience with them initiated many tank standards for different materials. RPs provided much longer life than their steel counterparts. In fact, steel tanks previously had no "real life" or no requirement standards until the RPs entered the market. It has been estimated that more than 200,000 GFRP tanks were installed in the USA from 1960–1990. A previous study by the Steel Tank Institute (Lake Zurich, IL) reported 61% were of steel and 39% of GFRP. At present, at least 50% of all tanks are GFRP. This RP vs. steel debate escalated when the EPA gave service stations and fleet refueling areas 10 years to remove steel tanks that leaked.

Historically a Chicago service station documented the long life of RPs. A May 1963 installation remained leak tight and structurally sound when unearthed in May 1988. After testing the vessel, engineers buried it at another gas station. This tank was one of sixty developed by Amoco Chemical Co. It was fabricated in two semi-cylinder sections of glass fiber woven roving and chopped strand mat impregnated by an unsaturated isophthalic TS-polyester resin selected for its superior resistance to acids, alkalis, aromatics, solvents, and hydrocarbons. Two sections were bonded to each other and to end caps with RP lap joints. Today, the tanks are fabricated by using chopped glass fiber mixed with the isophthalic resin. This mixture is dropped from above onto a rotating steel mandrel. The glass-resin mix is sprayed to make the end caps.

Demand for this type of petroleum storage tank has grown rapidly as environmental regulations have become more stringent. Marina installation have taken advantage of these RP tanks. They permit for boat owners to purchase gasoline at the pier. Before they were installed, gasoline either had to be carried to the marina or purchased elsewhere, because of corrosive conditions underground for metal or other tanks, particularly ones next to salt water.

Standards require that today's underground tanks must last thirty or more years without undue maintenance. To meet these criteria, they must be able to maintain structural integrity and resist the corrosive effects of soil and gasoline, including gasoline that has been contaminated by moisture and soil. The tank just mentioned that was removed in 1991 met these requirements, but two steel tanks unearthed from the same site at that time failed to meet them. One was dusted with white metal oxide and the other showed signs of corrosion at the weld line. Rust had weakened this joint so much that it could be scraped away with a pocketknife. Tests and evaluations were conducted on the RP tank that had been in the ground for 25 years; tests were also conducted on similarly constructed tanks unearthed at 51 and 71 years that showed the RP tanks could more than meet the service requirements.

Prior to the development of the GFRP tanks, no standards were required for buried tanks such as loads or loading conditions, minimum depths of earth cover, or structural safety factors were available. At that time, sizes were 22,704 to 45,408 liter (6,000 to 12,000 gal), with a nominal width and height of 2.44 m (8 ft) for truck shipments to local gasoline stations. Standards have developed listing requirements for stored fluid type, environmental resistance, minimum earth cover, ground water submerged limits, and surface wheel load over tank. Increasing acceptance of buried GFRP tanks has widened the size range from 2,081 to 181,632 liter (550 to at least 48,000 gal) and the range in typical diameters from about 1.22 to 3.35 m (4 ft to at least 11 ft).

The tank configuration is basically cylindrical, in order to provide the required design volumes within the established envelope of heights and widths. Length ranges are from 5.5 to 11 m (18 to 36 ft); they are well within practical truck shipment limits. A circular shape is required to support the substantial internal and external fluid and earth pressures with good structural efficiency. Other considerations in selection of an efficient configuration are used.

With a vertical axis, tank underlay requires that much less land area than with axis horizontal, but very deep excavation is required where expensive ledge or ground water conditions will frequently be encountered. Both internal and external pressures are large, requiring a substantial increase in wall thickness and rib stiffness, compared with a horizontally-placed tank. With axis horizontal, maximum external and internal pressures do not vary with size (length), and can be resisted with economically feasible wall thickness and rib proportions. Tanks underlay a larger ground area. Uniform bedding is more difficult to attain.

Hemispherical shells, low rise dished-shaped heads and flat plate closures were all considered. Hemispherical shells were found structurally very efficient because of good buckling resistances under external fluid and earth pressure, good strength under internal pressure, no requirement for edge ring, and low discontinuity stresses at the junction with the cylinder. Flat end closures result in excessive deflections and large edge bending moments on the cylinder. Sandwich construction could be used to improve structural efficiency of flat ends. Sandwich wall construction was also investigated for attaining necessary buckling resistance of spherical shell end closures and found to be feasible but less cost effective.

Use of rib stiffening was required. It was found necessary to stiffen the cylindrical shell against buckling under external pressure from ground water and dimpling from local earth pressure due to surface wheel loads. Sandwich wall construction was investigated as an alternative to use of stiffening ribs and found to be feasible but less cost effective.

Shape-wise, a hollow trapezoid provides efficient bending strength and stiffness, a wide base for proper spacing of cylinder shell support against

Figure 4.17 Example of a design for 10,000 gallon gasoline RP storage tank

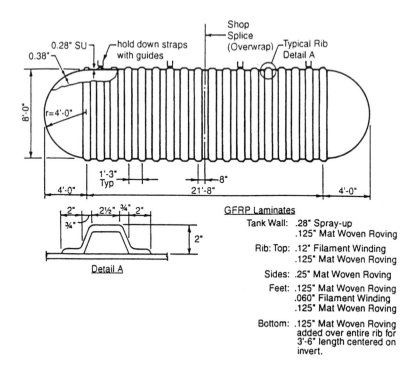

GFRP Laminates

Tank Wall: .28" Spray-up
.125" Mat Woven Roving

Rib: Top: .12" Filament Winding
.125" Mat Woven Roving

Sides: .25" Mat Woven Roving

Feet: .125" Mat Woven Roving
.060" Filament Winding
.125" Mat Woven Roving

Bottom: .125" Mat Woven Roving
added over entire rib for
3'-6" length centered on
invert.

local buckling, and a narrower top to resist local buckling with high circumferential flange forces. Fig. 4.17 provides a design concept for a tank with hollow trapezoidal ribbing.

Base width and clear spacing between ribs are established to minimize the number of ribs while providing adequate local buckling resistance for a cylinder wall of approximately the minimum practical shell thickness. Clear spacing must also be sufficient to permit installation of sleeves or nozzles for fill pipes and vents between ribs.

The RPs selected for detailed consideration are designed to provide both structural and liquid-sealing qualities. For example, a smooth liquid tight inner surface is obtained with a resin rich surface layer reinforced with glass filament surfacing veil. It is backed up by a 3.2 mm (1/8 in.) thick liquid seal layer of chopped glass/polyester spray-up. Discontinuous fibers are provided to avoid a continuous path for liquid migration into and along the reinforcement. Additional thickness for structural purposes is provided, either by adding more chopped fiber reinforced spray-up, or by filament winding. An outer resin rich surface layer is reinforced with a surfacing veil. A silica sand filler may be used of bulk, improved compressive stiffness and economy.

Minimum practical total RP thickness is established as 4.8 mm (³⁄₁₆ in.) for the combined spray-up liquid seal and filament wound structural layers and 6.4 mm (¼ in.) for an all-chopped fiber spray-up laminate with sand filler. The choice for any construction is made on the basis of comparative design thickness, weight, and fabrication costs. The all-chopped fiber reinforced construction using somewhat greater wall thickness than the composite filament wound-chopped fiber wall is determined to provide the lowest tank cost; filament winding provides lower weight.

Hopper Rail Car Tank
In the past (1973) a severe shortage of railroad covered hopper cars for the transportation of grain developed. Cargill, Inc. provided a contract to Structural Composites, Inc. for determine feasibility studies on the potential of using RP in the design and fabrication of these cars. Test results showed structural deficient existed. By 1978 an acceptable design resulted fabricating the Glasshopper (registered name). It was used in rail service March 1981. Cargill Inc., Southern Pacific Transportation Co., and ACF Industries, Inc. (Fig. 4.18). It was larger and lighter in weight than the conventional steel covered hopper car resulting in being able to deliver more commodity per fuel dollar. Other advantages included corrosion resistance, and lower maintenance costs.

The first to be built was Glasshopper I. It successfully passed all of the

Figure 4.18 RP railroad covered hopper car

required American Association of Railroads (AAR) tests including the 454,000 kg (1,000,000 lb) static end compression test and the 568,000 kg (1,250,000 lb) coupler force impact test in the laboratory, and then successfully completed a round trip between St. Louis, MO and Oakland, CA [9700 km (6000 mile)].

From outward appearance, the RP designs were very similar to the standard ACF steel-covered hopper car. The first RP prototype, Glasshopper I that was in grain service, had four compartments. The car had a total capacity of 142 m³ (5000 ft³) and an overall length of about 16 m (53 ft). Its basic specifications are shown in Table 4.7.

Table 4.7 Glasshopper I basic specifications

Length inside	50 ft 3½ in
Length over end sills	51 ft 5⅝ in
Length over strikers	52 ft 11 in
Length over coupler pulling face	55 ft 6½ in
Length over running boards	53 ft ⅞ in
Length between truck centres	42 ft 3 in
Extreme width	10 ft 8 in
Height, rail to top of running boards	15 ft 1 $^{27}/_{32}$ in
Height, rail to bottom of outlet	12 in
Extreme height, rail to top of hatch bumper	15 ft 6 in
Number of discharge outlets	4
Roof hatch opening, continuous	20 in x 44 ft 7¾ in
Curve negotiability, uncoupled	150 ft
Cubic capacity	500 ft³
Tare weight	59,000 lb
Gross rail load	263,000 lb
AAR clearance diagram	Plate 'C'

The second prototype car Glasshopper II that was later put into service had three compartments. The tare weight of the second car was 24,600 kg (54,200 lb), which was 4000 kg (8800 lb) lighter than a standard steel car weight of 28,600 kg (63,000 lb).

Construction details for Glasshopper I consist of a filament wound (FW) RP car body, RP/balsawood core sandwich panel bulkheads and slope sheets, steel side sills and shear plates, steel bolster webs, and RP hatch covers. Standard running gear and safety appliances were utilized, as were standard gravity outlets. Several changes in construction details such as the use of single laminate slope sheets were made in the design of Glasshopper II to reduce weight and manufacturing costs.

Table 4.8, shows the weight percentages of steel or RP materials. A significant amount (30 wt%) of the RP car structure is fabricated using RP materials. By subtracting the trucks steel weight, the remaining structure is RP. This construction allows the significant weight reduction to be possible. Finite element analysis (FEA) modeling was used throughout the design stages of the program to aid the structural analysis effort. The structural response in both static and dynamic loading conditions was characterized prior to initiation of the car construction.

Table 4.8 Glasshopper I component weight summary

Component	Material	Weight lbs	$\dfrac{Component\ weight}{Total\ weight} \times 100$
Car body	RP	6800	11.5
Sandwich panels	RP	4410	7.5
Wide flange beams	RP	640	1.1
Stiffeners	RP	2910	4.9
Top sill	RP	460	0.8
Roof/side angles	RP	510	0.9
Adhesive/bonding strip	RP	1070	1.8
Hatches	RP	860	1.5
Outlets	Steel	1800	3.0
End arrangement	Steel	9640	16.3
Side sills	Steel	4420	7.5
Running boards/safety appln.	Steel	1650	2.8
Brake system	Steel	1570	2.7
Misc. hardware	Steel	1060	1.8
Trucks	Steel	21200	35.9
	Total weight	59000	100.0

Σ RP component percentages = 30
Σ Steel component percentages = 70

FW process was used to fabricate both Glasshopper car bodies. It was determined that this process afforded the best mechanical properties for the lowest cost. Fabricating processes exist that can be highly automated which would help towards having RP-covered hopper cars compete economically with conventional steel-covered hopper cars in the marketplace.

Resin matrix material system chosen for fabrication of the car bodies was a proprietary isophthalic polyester resin system developed by Cargill specifically for the Glasshopper project. PPG, Certain Teed, and

OCF supplied the reinforcement of E-glass rovings. Both OCF 450 and 675 yield glass was used successfully in conjunction with the Cargill resin during FW operation. To provide adequate mechanical properties in the directions required to withstand the externally applied service loads, the FW apparatus was programmed to provide the multi-axial filament directional orientation capability.

Secondary bonding operations involving the attachment of stiffeners, etc., for the first RP car, used Hysol's epoxy adhesive (EA 919). This same adhesive was used in joints where both bonding and bolting with mechanical fasteners were employed. Lord Corp.'s acrylic adhesive system (TS 3929-70) was used successfully for Glasshopper II. Hat section stiffeners and wide flange beams were fabricated using the hand lay-up and pultrusion processes, respectively. The material used in the construction of the hat stiffeners included 1½ oz mat, 24 oz woven rovings, 22½ oz unidirectional fabric, and the isophthalic polyester resin. Pultrusions were purchased finished, and were fabricated using standard pultrusion processes.

In order to demonstrate structural adequacy, Glasshopper I was tested in the ACF test laboratory located in St. Charles, MO. The test program was designed to show that the car meets and exceeds all requirements as specified by the AAR. Both static and dynamic tests were included in the testing. To determine the car's structural response under various applied loading conditions, Glasshopper I was instrumented with a total of 224 strain gauges, located at various areas determined through structural analysis to be of greatest importance and to provide maximum information. Glasshopper II, instrumented with 310 strain gauges, successfully completed the test program in 1983.

A series of six different static tests were successfully passed by Glasshopper I, including end compression, draft, vertical coupler-up, vertical coupler-down, coupler shank, and torsional jacking. The end compression test consisted of "squeezing" the car, while empty, with a hydraulic ram until a coupler force of 1,000,000 lb was measured. The draft test was conducted on the loaded car (105.9 tons) and consisted of pulling on the coupler until a force of 630,000 lb was experimentally observed. The remaining static tests were all conducted on the loaded car and involved using calibrated hydraulic rams to:

1. Jack the car upward with a vertical force of 22,700 kg (50,000 lb) applied at the coupler pulling face.

2. Jack the car downward with a vertical force of 22,700 kg (50,000 lb) applied at the coupler pulling face.

3. Lift the car free of the truck bolster by jacking at the coupler shank, a vertical force of 50,400 kg (111,000lb) was required.

4. Lift the car free of the truck bolster by jacking at the lifting lug/jacking pad assembly to verify torsional rigidity and stability, a vertical force of 31,780 kg (70,000 lb) was required.

An analysis of test results show the experimentally observed strains to be very close to those predicted using FEA techniques and "hand" calculations. This fact made it possible to use these techniques to further optimize the Glasshopper II design.

After successfully passing all required static tests, Glasshopper I was subjected to a series of impact tests. For these tests, the car that was fully loaded was pulled by cable up an inclined ramp and released to impact another fully loaded standing car that had its brakes released. Velocity of the car at impact was controlled by its height on the ramp at the time of release. Car's velocity was incrementally increased until an experimentally measured coupler force of 113,500 kg (250,000 lb) was developed during the impact.

Velocity of 14.9 km/h (9.24 m/h) was required to obtain the AAR specified load. It is noted that this velocity is significantly higher than the velocity required to reach the specified force with conventional steel covered hopper cars, which is about 12.1 km/h (7.5 m/h). Glasshopper I, with modified bulkhead joints, successfully passed the AAR impact test required and was subsequently prepared for the extended road test.

Following completion of laboratory testing, Glasshopper I was tested over a 9700 km (6000 m) route on the Southern Pacific system. Fully loaded car with 9,600 kg (211,000 lb) made the trip from St. Charles, MO to Oakland, CA and back to Houston, TX. The car was unloaded at the Cargill export grain terminal in the Houston area and then returned empty to the facility in St. Charles, MO. The car was accompanied on the trip by the fully instrumented ACF test car used for data acquisition that monitored key strain gauges and load cells throughout the trip. All test results and visual observations showed the car performed well, and as predicted. During certain segments of the testing, speeds of 113 km/h (70 m/h) were reached with no dynamic problems (flutter, hunting, etc.) being observed.

It was determined that two major advantages of the RP covered hopper rail car are its tare weight and its corrosion resistance. As a result of its significantly lower weight and large size, the car is capable of carrying more payload per fuel dollar. This fact is extremely important in today's conditions of escalating and high fuelled prices. Glasshopper is able to carry many highly corrosive commodities (salts, potash, fertilizer, ore, etc.) without the need for expensive linings and with significantly reduced car maintenance costs. Also, the car's service lifetime would be greatly extended in these severe service environments. Other advantages include the potential to eliminate painting requirements, reduced labor costs in manufacturing, lower center of gravity in the unloaded condition, ability to easily adapt to internal pressure designs, and rapid production changeover to alternate capacity cars.

Highway Tank
RP tanks on firm ground have been holding corrosive materials safely since the 1940s. Later the same technology, with some enhancements material-wise and design-wise, was applied to over-the-road tankers. Some tankers fabricated by Comptank Corp., Bothwell, Ontario, Canada have been on the road in the USA and Canada carrying a wide range of corrosive and hazardous liquids. These RP tank trailers are coded 312 for hauling corrosive and hazardous materials; special designed models haul acids or other corrosive chemicals; they unload by pressure, vacuum, or gravity.

The tankers are filament wound using E-glass rovings with polyester resin (Reichhold Atlac 4010 AC) and surfacing veil. RP moldings are integrated parts of the shell that is usually 15.88 mm (0.625 in.) thick. These parts include external rings/ribs, covers for steel rollover guards, spill dam, catwalk, hose trays, etc.

Very Large Tank
Large filament wound 150,000 gal (568 m^3) tank has been fabricated by the Rucker Co. for Aerojet-General (1966). An example is shown schematically in Fig. 4.19 of the so-called racetrack-fabricating machine. A fabricated 56 m (22 ft) high, 152 m (60 ft) wide, 318 m (125 ft) long that weighed 32 ton, of all RP, large tank is shown in Fig. 4.20.

Just the mandrel for this FW machine weighed 100 ton all of metal. Total weight of the steel-constructed machine was 200 ton. The tank contained about 251 million km (158 million miles) of glass fiber, used 8 ton of textile creel containing 60 spools of glass fiber moving up to 7.24 km/h (4^1/2 mph), and took three weeks to manufacture the epoxy-glass fiber RP tank in the Todd shipyard in Los Angeles, California.

Corrosive Resistant Tank
Chemical and corrosive resistant property of many plastics make them useful to contain different liquids ranging from water to acids. They are used extensively in water treatment plants and piping to handle

Figure 4.19 Tank's fabricating process

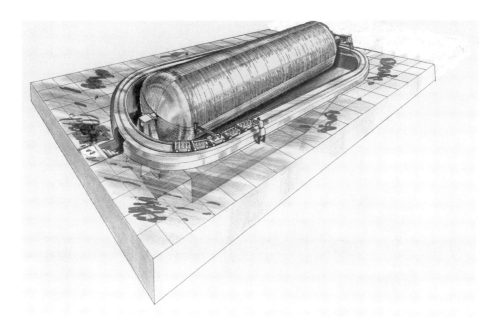

Figure 4.20 RP 150,000 gal tank

Figure 4.21 Large water filtration tank with 6 ft opening

drainage, sewage, and water supply. Glass fiber TS polyester RP water filtration tank is shown in Fig. 4.21. It is a 20 ft diameter, 32 ft high structure made in sections by a low pressure RP fabricating method. This bonded, assembled tank was shipped on a water barge to its destination. Structural shapes such as this tank for use under corrosive conditions often takes advantage of the properties of RPs and other plastics.

Pipe

Thermoplastic Pipe

Extensive amount of plastic pipe is used worldwide to move different types of liquids, gases, and solids. With the different properties of plastics (such as corrosion resistance, toughness, and strength), pipes can be fabricated to handle practically any type of material. A major and important market for plastics is in producing pipe (tube) for use such as on the ground, underground, in water, and electrical conduits. The largest use is in transporting water, gas, waste matter, industrial mining,

Figure 4.22 Pipe wall thickness determination based on internal pressure

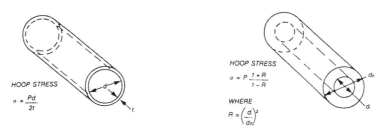

etc. Use of thermoplastic, such as HDPE, PVC, and PP, provide most of the world using extruders. Plastic pipe represents about 30% of the dollar share compared to other materials (iron/steel at 45%, copper at 12%, concrete at 8%, aluminum at 4%, etc.).

Fig. 4.22 provides a method to determine pipe thickness subjected to uniform internal pressure P using the standard engineered thin-wall-tube hoop-stress equation. Top view provides an equation that is approximately accurate for $t < d/10$. However when the wall thickness increases the error becomes large. It is useful in determining an approximate wall thickness, even when condition $(t<d/10)$ is not met. Bottom view provides an equation for the maximum hoop stress that occurs on the surface of the inside wall of the pipe. After the thin-wall stress equation is applied, the thick-wall stress equation can be used to verify the design.

RP Pipe

An important product even though it represents a small portion of the market is reinforced TS (RTS) plastic; also called reinforced thermoset resin (RTR) according to ASTM standards. Its major material construction is glass fiber with TS polyester plastic that uses fabricating methods ranging from bag molding to filament winding (Chapters 1 and 2). These RTR pipes provide high load performance both internally and externally.

There are large diameter filament-wound pipes (RTRs) used and accepted in underground burial because they provide conditions such as corrosion resistance and installation-cost savings. Pipe design equations have been used that specifically provide useful information to meet internal and/or external pressure loads. More recently finite element analysis (FEA) has been used to design RP pipes and other structural products. These design approaches utilize performance standards based upon internal pressures and pipes' stiffness. Other requirements must be met such as longitudinal effects of internal pressure, temperature

Figure 4.23 Loading (26 psi) rigid concrete pipe (left) and flexible plastic pipe

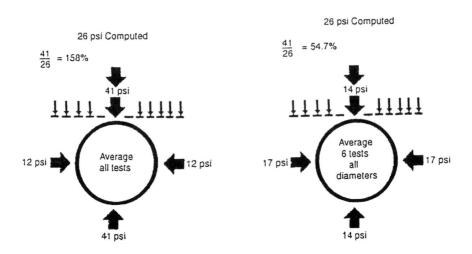

gradients, and pipe bridging.

When compared to steel pipes there are similarities and dissimilarities. They both differ from concrete (includes asbestos filled type) pipe which is a rigid pipe that cannot tolerate bending or deflection to the same extent as RTR and steel pipe. The following review provides information on the design approach and results of tests conducted on these type pipes (rigid RTR, rigid steel, and flexible concrete). They were buried in trenches under 25 ft of the same dirt and subjected to actual load testing. Specific pressures varied from installation to installation, but the relationship in the way these pipes react to the same burial condition generally remains constants.

As shown in Fig. 4.23 the load on the surface of the (a) rigid pipe (concrete) is higher at the crown and is transmitted through the pipe directly to the bed of the trench in which the pipe rests and using some side support. The RTR or steel flexible conduit (b) deflects under covering load of earth, this deflection transfers portions of the load to the surrounding envelope of soil that increases the strength of the flexible conduit. Analyzing the type and consolidation of backfill materials is to be considered an integral part of the design process. Because less of a load on the trench bed occurs the trench requires less bedding bearing strength reducing the installed cost.

Steel pipe is considered a homogeneous isotropic material (equally strong in both hoop and longitudinal directions) where RTR is an anisotropic material (different strength in both the hoop and longitudinal directions). These directional behavior results in the

Figure 4.24 Cross-section of filament wound layup of the RTR pipe

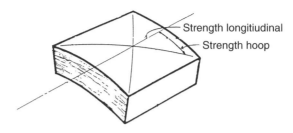

Strength longitiudinal
Strength hoop

modulus of elasticity (E) to be equal in all directions for steel and not equal for RTR.

The RTR pipe structure is shown in Fig. 4.24 where the glass fibers are filament wounded at a helix angle that is at 55 to 65° to the horizontal to maximize hoop and longitudinal stress efficiency. Glass fiber content is at a minimum of 45wt%. An internal corrosion resistant leak proof barrier liner is usually included that is not included in the stress analysis.

Required is to design a pipe wall structure of sufficient stiffness and strength to meet the combined loads that the pipe will experience in long time service. One design is a straight wall pipe in which the wall thickness controls the stiffness of the pipe. Another way is to design a rib-wall pipe on which reinforcement ribs of a specific shape and dimension are wound around the circumference of the pipe at precisely designed intervals. As previously reviewed (RIB section) the advantage of a rib wall pipe is that the wall thickness of the pipe can be reduced (also reducing costs) while maintaining or even increasing its overall strength-to-weight ratio. Where burial conditions are extreme or difficult underwater installations exist a rib-wall pipe design should be considered.

Maximum allowable pipe deflection should be no more than 5% using the Fig. 4.25 equation where ΔX = deflection. This value is the standard

Figure 4.25 Pipe deflection equation

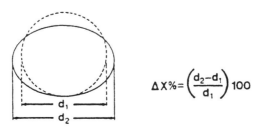

$$\Delta X\% = \left(\frac{d_2 - d_1}{d_1}\right) 100$$

of the pipe industry for steel conduit and pipe (AWWA M-11, ASTM, and ASME). Deflection relates to pipe stiffness (El), pipe radius, external loads that will be imposed on the pipe, both the dead load of the dirt overburden as well as the live loads such as wheel and rail traffic, modulus of soil reaction, differential soil stress, bedding shape, and type of backfill.

To meet the designed deflection of no more than 5% the pipe wall structure could be either a straight wall pipe with a thickness of about 1.3 cm (0.50 in.) or a rib wall pipe that provides the same stiffness. It has to be determined if the wall structure selected is of sufficient stiffness to resist the buckling pressures of burial or superimposed longitudinal loads. The ASME Standard of a four-to-one safety factor on critical buckling is used based on many years of field experience. To calculate the stiffness or wall thickness capable of meeting that design criterion one must know what anticipated external loads will occur (Fig. 4.26).

As reviewed the strength of RTR pipe in its longitudinal and hoop directions are not equal. Before a final wall structure can be selected, it is necessary to conduct a combined strain analysis in both the longitudinal and hoop directions of the RTR pipe. This analysis will consider longitudinal direction and the hoop direction, material's allowable strain, thermal contraction strains, internal pressure, and pipe's ability to bridge soft spots in the trench's bedding. These values are determinable through standard ASTM tests such as hydrostatic testing, parallel plate loading, coupon test, and accelerated aging tests.

Stress-strain (S-S) analysis of the materials provides important information. The tensile S-S curve for steel-pipe material identifies its yield point that is used as the basis in their design. Beyond this static loaded yield point (Chapter 2) the steel will enter into the range of plastic deformation that would lead to a total collapse of the pipe. The allowable design strain used is about two thirds of the yield point.

Figure 4.26 Buckling analysis based on conditions such as dead loads, effects of possible flooding, and the vacuum load it is expected to carry

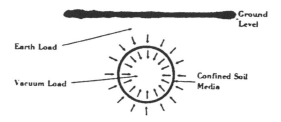

RTR pipe designers also use a S-S curve but instead of a yield point, they use the point of first crack (empirical weep point). Either the ASTM hydrostatic or coupon test determines it. The weep point is the point at which the RTR matrix (plastic) becomes excessively strained so that minute fractures begin to appear in the structural wall. At this point it is probable that in time even a more elastic liner on the inner wall will be damaged and allow water or other liquid to weep through the wall. Even with this situation, as is the case with the yield point of steel pipe, reaching the weep point is not catastrophic. It will continue to withstand additional load before it reaches the point of ultimate strain and failure. By using a more substantial, stronger liner the weep point will be extended on the S-S curve.

The filament-wound pipe weep point is less than 0.009 in./in. The design is at a strain of 0.0018 in./in. providing a 5 to 1 safety factor. For transient design conditions a strain of 0.0030 in./in. is used providing a 3 to 1 safety factor.

Stress or strain analysis in the longitudinal and hoop directions is conducted with strain usually used, since it is easily and accurately measured using strain gauges, whereas stresses have to be calculated. From a practical standpoint both the longitudinal and the hoop analysis determine the minimum structural wall thickness of the pipe. However, since the longitudinal strength of RTR pipe is less than it is in the hoop direction, the longitudinal analysis is first conducted that considers the effects of internal pressure, expected temperature gradients, and ability of the pipe to bridge voids in the bedding. Analyzing these factors requires that several equations be superimposed, one on another. All these longitudinal design conditions can be solved simultaneously, the usual approach is to examine each individually.

Poisson's ratio (Chapter 3) can have an influence since a longitudinal load could exist. The Poisson's effect must be considered when designing long or short length of pipe. This effect occurs when an open-ended cylinder is subjected to internal pressure. As the diameter of the cylinder expands, it also shortens longitudinally. Since in a buried pipe movement is resisted by the surrounding soil, a tensile load is produced within the pipe. The internal longitudinal pressure load in the pipe is independent of the length of the pipe.

Several equations can be used to calculate the result of Poisson's effect on the pipe in the longitudinal direction in terms of stress or strain. Equation provides a solution for a straight run of pipe in terms of strain. However, where there is a change in direction by pipe bends and thrust blocks are eliminated through the use of harness-welded joints, a different analysis is necessary. Longitudinal load imposed on either side of an elbow is high. This increased load is the result of internal pressure, temperature gradient, and/or change in momentum of the fluid. Because of this increased load, the pipe joint and elbow thickness may have to be increased to avoid overstraining.

The extent of the tensile forces imposed on the pipe because of cooling is to be determined. Temperature gradient produces the longitudinal tensile load. With an open-ended cylinder cooling, it attempts to shorten longitudinally. The resistance of the surrounding soil then imposes a tensile load. Any temperature change in the surrounding soil or medium that the pipe may be carrying also can produce a tensile load. Engineering-wise the effects of temperature gradient on a pipe can be determined in terms of strain.

Longitudinal analysis includes examining bridging if it occurs where the bedding grade's elevation or the trench bed's bearing strength varies, when a pipe projects from a headwall, or in all subaqueous installations. Design of the pipe includes making it strong enough to support the weight of its contents, itself, and its overburden while spanning a void of two pipe diameters.

When a pipe provides a support the normal practice is to solve all equations simultaneously, then determine the minimum wall thickness that has strains equal to or less than the allowable design strain. The result is obtaining the minimum structural wall thickness. This approach provides the designer with a minimum wall thickness on which to base the ultimate choice of pipe configuration. As an example, there is the situation of the combined longitudinal analysis requiring a minimum of $5/8$ in. (1.59 cm) wall thickness when the deflection analysis requires a $1/2$ in. (1.27 cm) wall, and the buckling analysis requires a $3/4$ in. (1.9 cm) wall. As reviewed the thickness was the $3/4$ in. wall. However with the longitudinal analysis a $5/8$ in. wall is enough to handle the longitudinal strains likely to be encountered.

In deciding which wall thickness, or what pipe configuration (straight wall or ribbed wall) is to be used, economic considerations are involved. The designer would most likely choose the $3/4$ in. straight wall pipe if the design analysis was complete, but it is not since there still remains strain analysis in the hoop direction. Required is to determine if the combined loads of internal pressure and diametrical bending deflection will exceed the allowable design strain.

There was a tendency in the past to overlook designing of joints. The performance of the whole piping system is directly related to the performance of the joints rather than just as an internal pressure-seal pipe. Examples of joints are bell-and-spigot joints with an elastomeric seal or weld overlay joints designed with the required stiffness and longitudinal strength. The bell type permits rapid assembly of a piping system offering an installation cost advantage. It should be able to rotate at least two degrees without a loss of flexibility. The weld type is used to eliminate the need for costly thrust blocks.

Spring

There is a difference when comparing the plastic to metal spring shape designs. With metals shape options are the usual torsion bar, helical coil, and flat-shaped leaf spring. The TPs and TSs can be fabricated into a variety of shapes to meet different product requirements. An example is TP spring actions with a dual action shape (Fig. 4.27) that is injection molded. This stapler illustrates a spring design with the body and curved spring section molded in a single part. When the stapler is depressed, the outer curved shape is in tension and the ribbed center section is put into compression. When the pressure is released, the tension and compression forces are in turn released and the stapler returns to its original position.

Other thermoplastics are used to fabricate springs. Acetal plastic has been used as a direct replacement for conventional metal springs as well providing the capability to use different spring designs such as in zigzag springs, un-coil springs, cord locks with molded-in springs, snap fits, etc. A special application is where TP replaced a metal pump in a PVC plastic bag containing blood. The plastic spring hand-operating pump (as well as other plastic spring designs) did not contaminate the blood.

RP leaf springs have the potential in the replacements for steel springs. These unidirectional fiber RPs have been used in trucks and automotive suspension applications. Their use in aircraft landing systems dates back to the early 1940s taking advantage of weight savings and

Figure 4.27 TP Delrin acetal plastic molded stapler (Courtesy of DuPont)

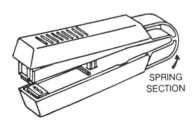

SPRING
SECTION

performances. Because of the material's high specific strain energy storage capability as compared to steel, a direct replacement of multileaf steel springs by monoleaf composite springs can be justified on a weight-saving basis.

The design advantages of these springs is to fabricate spring leaves having continuously variable widths and thicknesses along their length. These leaf springs serve multiple functions, thereby providing a consolidation of parts and simplification of suspension systems. One distinction between steel and plastic is that complete knowledge of shear stresses is not important in a steel part undergoing flexure, whereas with RP design shear stresses, rather than normal stress components, usually control the design.

Design of spring has been documented in various SAE and ASTM-STP design manuals. They provide the equations for evaluating design parameters that are derived from geometric and material considerations. However, none of this currently available literature is directly relevant to the problem of design and design evaluation regarding RP structures. The design of any RP product is unique because the stress conditions within a given structure depend on its manufacturing methods, not just its shape. Programs have therefore been developed on the basis of the strain balance within the spring to enable suitable design criteria to be met. Stress levels were then calculated, after which the design and manufacture of RP springs became feasible.

Leaf Spring

RP/composite leaf springs constructed of unidirectional glass fibers in a matrix, such as epoxy resin, have been recognized as a viable replacement for steel springs in truck and automotive suspension applications. Because of the material's high specific strain energy storage capability compared with steel, direct replacement of multi-leaf steel springs by mono-leaf composite springs is justifiable on a weight saving basis. Other advantages of RP springs accrue from the ability to design and fabricate a spring leaf having continuously variable width and/or thickness along its length. Such design features can lead to new suspension arrangements in which the composite leaf spring will serve multiple functions thereby providing part consolidation and simplification of the suspension system.

The spring configuration and material of construction should be selected so as to maximize the strain energy storage capacity per unit mass without exceeding stress levels consistent with reliable, long life operation. Elastic strain energy must be computed relative to a particular stress state. For simplicity, two materials are compared, steel and unidirectional glass fibers in an epoxy matrix having a volume fraction of 0.5 for the stress state of uniaxial tension. If a long bar of either material is loaded axially the strain energy stored per unit volume of material is given by

$$U = (\delta_A^2/2E) \text{ (in-lb/in}^3) \tag{4-36}$$

where δ_A is the allowable tensile stress and E is Young's modulus for the material.

In Table 4.9 the appropriate E for each material has been used and a conservative value selected for δ_A. On a volume basis the RP is about twice as efficient as steel in storing energy; on a weight basis it is about eight times as efficient.

Table 4.9 Glass fiber-epoxy RP leaf spring design

Material	σ_A(ksi)	U(lb/in²)	U/w* (in)
Steel	90	135	470
Glass/epoxy	60	325	4880

* w = specific weight

The RP has an advantage because it is an anisotropic material that is correctly designed for this application whereas steel is isotropic. Under a different loading condition (such as torsion) the results would be reversed unless the RP were redesigned for that condition. The above results are applicable to the leaf spring being reviewed because the principal stress component in the spring will be a normal stress along the length of the spring that is the natural direction for fiber orientation.

In addition to the influence of material type on elastic energy storage, it is also important to consider spring configuration. The most efficient configuration (although not very practical as a spring) is the uniform bar in uniaxial tension because the stresses are completely homogeneous. If the elastic energy storage efficiency is defined as the energy stored per unit volume, then the tensile bar has an efficiency of 100%. On that basis a helical spring made of uniform round wire would have an efficiency of 32% (the highest of any practical spring configuration) while a leaf spring of uniform rectangular cross section would be only 11% efficient.

The low efficiency of this latter configuration is due to stress gradients through the thickness (zero at the mid-surface and maximum at the upper and lower surfaces) as well as along the length (maximum at mid-span and zero at the tips). Recognition of this latter contribution to inefficiency led to development of so-called constant strength beams which for a cantilever of constant thickness dictates a geometry of triangular plan-form. Such a spring would have an energy storage efficiency of 33%. A practical embodiment of this principle is the multi-leaf spring of constant thickness, but decreasing length plates, which for a typical five leaf configuration would have an efficiency of about 22%.

More sophisticated steel springs involving variable leaf thickness bring improvements of energy storage efficiency, but are expensive since the leaves must be forged rather than cut from constant thickness plate. However, a spring leaf molded of the RP can have both thickness and width variations along its length. For instance, a practical RP spring configuration having a constant cross-sectional area and appropriately changing thickness and width will have an energy storage efficiency of 22%. This approaches the efficiency of a tapered multi-leaf configuration and is accomplished with a material whose inherent energy storage efficiency is eight times better than steel.

In this design, the dimensions of the spring are chosen in such a way that the maximum bending stresses (due to vertical loads) are uniform along the central portion of the spring. This method of selection of the spring dimensions allows the unidirectional long fiber reinforced plastic material to be used most effectively. Consequently, the amount of material needed for the construction of the spring is reduced and the maximum bending stresses are evenly distributed along the length of the spring. Thus, the maximum design stress in the spring can be reduced without paying a penalty for an increase in the weight of the spring. Two design equations are given in the following using the concepts described above.

To develop design formulas for RP springs, we model a spring as a

Figure 4.28 RP spring model

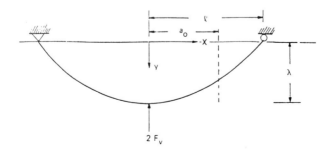

circular arc or as a parabolic arc carrying a concentrated load $2F_v$ at mid-length (Fig. 4.28).

The governing equation for bending of the spring

$$\frac{1}{\rho} - \frac{1}{R} = \frac{M}{EI} \qquad (4\text{-}37)$$

where R = radius of curvature of unloaded spring; ρ = radius of curvature of deformed spring; M = bending moment, E = Young's modulus; and I = moment of inertia of spring cross section.

Using the coordinate system shown in Fig. 4.28, equation 4-37 is rewritten as

$$\frac{d^2y}{dx^2} = \frac{F_v(l-x)}{EI} - \frac{1}{R}, \quad 0<x<l \qquad (4\text{-}38)$$

where the coordinate y is used to denote the deformed configuration of the spring. Once the maximum allowable design stress in the spring is chosen, equation 4-38 will be used to determine the load carrying capability of the spring. Due to the symmetry of the spring at $x = 0$, only half of the spring needs to be analyzed. It should be noted that equation 4-38 is only an approximate representation of the deformation of the spring. However, for small values of λ/l, it is expected to give reasonably good prediction of the spring rate. Here λ is the arc height and $2i/l$ is the chord length of the spring. Although a nonlinear relation can be used in place of equation 4-38, it would be difficult to derive simple equations for design purposes.

For this particular design, the thicknesses of the spring decreases front the center to the two ends of the spring. Hence, the cross-sectional area of the spring varies along its length. The maximum bending stresses at every cross section of the spring from $x = 0$ to $x = \alpha_o$ are assumed to be identical (Fig. 4.28). The value of α_o is a design parameter that is used to control the thickness and the load carrying capability of the spring. If α_o is the maximum allowable design stress, then the thickness of the spring for $0 \leq x \leq \alpha_o$ is determined by equating the maximum bending stress in the spring to α_o, thus:

$$h = \left[\frac{6(1-v^2)F_v(l-x)}{b\sigma_0} \right]^{1/2}, \quad x \leq \alpha_0 \qquad (4\text{-}39)$$

where v is Poisson's ratio, and b and h are the width and thickness of the spring, respectively.

The factor $(1 - v_2)$ is introduced to account for the fact that b could be several times larger than h. If b and h are of the same order of magnitude, a zero value of v is suggested to be used with equation 4-

39. This equation shows that the thickness of the spring should be a function of F_v, σ_o, l, and b. Once F_v, σ_o, and l are fixed, then the value of h is inversely proportional to the square root of the width of the spring.

For $x > \sigma_o$, the thickness of the spring is assumed to remain constant. The minimum value of h is governed by the ability of the unidirectional composite to carry shear stresses. Using equation 4-38 and the appropriate boundary and continuity conditions, the following equation for the determination of the spring rate is obtained,

$$k_b = \frac{16R^2\sigma_0^3}{27(1 - v^2)\lambda l^4 E^2} \, [2^{\beta/2} - (l - \alpha_0)^{3/2}]^2 \tag{4-40}$$

where k_b is the spring rate per unit width of the spring in lb per in. of vertical deflection. In deriving equation 4-40, the maximum bending stress σ_o is assumed to develop when $y = 0$ at which the center of the spring rate has undergone a deflection equal to λ. If the actual design value of $2F_y$ is less than or greater than bk_b, the appropriate value of σ_o to be used in equation 4-74 can be determined easily from the maximum design stress by treating σ_o as a linear function of $2F_y$.

A constraint on the current fabricating method of the RP leaf spring is that the cross-sectional area of the spring has to remain constant along the length of the spring. This imposes a restriction on the use of variable cross-sectional area design since additional work is required to trim a constant cross-sectional area spring to fit a variable cross-sectional area design. Unless the design stresses in the spring are excessively high, it is preferable to use the less labor-intensive constant cross-sectional area spring. This section describes the design formulae for this type of spring design.

Using the same coordinate system and symbols as shown in Fig. 4.28, equations 4-37 and 4-38 remain valid for the constant cross-sectional area spring. The mid-section thickness of the spring h_o is related to the maximum bending stress σ_o by:

$$h_0^2 = \frac{6(1 - v^2)F_v l}{b_0 \, \sigma_0} \tag{4-41}$$

where b_o is the corresponding mid-section width of the spring. Imposing the constant cross-sectional area constraint,

$$b_0 h_0 = bh \tag{4-42}$$

the thickness of the spring at any other section is given by:

$$h = \frac{6(1 - v^2)F_v(l - x)}{b_0 h_0 \sigma_0} \tag{4-43}$$

The corresponding width of the spring is then obtained from equation

4-42. Based on equations 4-42 and 4-43 that the width of the spring will continue to increase as it moves away from the mid-section. In general there is a limitation on the maximum allowable spring width. Using b_a, to denote the maximum width, the value of x beyond which tapering of the spring is not allowed can be determined by imposing the constant cross-sectional area constraint. One can use α_0 to denote this value of x, then:

$$\alpha_0 = l \left[1 - \frac{b_0}{b_a}\right] \tag{4-44}$$

Thus, equation 4-43 holds only for $x \leq$. Beyond $x = a_0$, the thickness of the spring remains constant and is given by:

$$h = \frac{6(1 - v_2)F_v(l \ \alpha_0)}{b_0 h_0 \sigma_0}, \quad x \geq \alpha_0 \tag{4-45}$$

An implication of equations 4-43 and 4-44 is that the maximum bending stresses will remain constant along the length of the spring for $[x] \geq \alpha_0$. Equation 4-38 with the appropriate boundary and continuity conditions, the spring rate, k, can be shown to be:

$$k = \frac{3EI_0}{R\lambda l(1 - v_2)[1 + 2(1 - b_0/b_a)]} \tag{4-46}$$

where I_0 is the moment of inertia of the cross-section of the spring at mid-section. In the design of a spring, the values of b_a, l, R, λ, and k are usually given and it is required to determine the values of h_0 for a desirable value of σ_0. The following equation has been obtained for the determination of h_0:

$$h_0^3 - \frac{4R\sigma_0 h_0^2}{E} + \frac{8Rk\lambda l(1 - v^2)}{b_a E} = 0 \tag{4-47}$$

Once ho is determined, the corresponding value of b_0 is then obtained from equation 4-37. In equation 4-47, the value of σ_0 corresponds to a center deflection equal to λ. If the actual design value of $2F_v$, is less than or greater than $k\lambda$, the appropriate value of σ_0 to be used in equation 4-47 can be determined easily from the maximum design stress by treating σ_0 as a linear function of $2F_v$.

Consider, as an example, the design of a pair of longitudinal rear leaf springs for a light truck suspension. The geometry of the middle surface of the springs is given as:

$$l = 23 \text{ in}$$
$$\lambda = 6 \text{ in}$$
$$R = 44.08 \text{ in}$$
$$b_a = 4.5 \text{ in}$$

The design load per spring is 2200 lb and a spring rate of 367 lb/in. is required. If σ_o is set equal to 53 ksi in equation 4-47, two possible design values of h_o are obtained. Using equation 4-41, the corresponding values of b_o are determined. Thus, there are two possible constant cross-sectional area designs for this particular spring: (S) h_o = 1.074 in, b_o = 2.484 in. and (SS) h_o = 1.190 in., b_o = 2.023 in. A value of Young's modulus of 5.5 x 10^6 psi is used in the design of these springs. This corresponds to the modulus of a unidirectional RP with 50vol% of glass fibers. If a value of σ_o less than 53 ksi is used in the design, negative and complex values of h_o are obtained from equation 4-47.

This indicates that it is impossible to design a constant cross-sectional area spring to fit the given design parameters with a maximum bending stress of less than 53 ksi. If a constant width design is required, it can be shown from equation 4-40 that a spring with a constant width of 2.484 in. and a maximum thickness of 1.074 in. will satisfy all the design specifications. The corresponding value of a_o is 18 in. If a constant width of greater than 2.484 in. is allowed, then a maximum design stress of less than 53-ksi can be obtained.

The above example shows that two plausible constant cross-sectional area designs are obtained to satisfy all the design requirements. If the spring were subjected only to vertical loading, the second design would be selected since it involves less material. However, if the spring is expected to experience other loadings in addition to the vertical load, then it is necessary to investigate the response of the spring to these loadings before a decision can be made.

The effects of these loadings can be determined easily using Castigliano's Theorem, together with numerical integration. For illustration, a comparison summation of the responses of the two constant cross-sectional area spring designs are reviewed:

1. *Rotation due to axle torque, M_T* : The rate of rotation of the center portion of the spring due to the axle torque, M_T, is: design (S) = 1.901×10^5 in-lb/radians and design (SS) = 1.895×10^5 in-lb/radians.

   If an axle torque of 15,000 in-lb is used for M_T, the rotation and the maximum bending stresses for these two springs are in table form:

	Rotation, degree	Maximum stress
design (S)	4.5	15.7
design (SS)	4.5	15.7

The responses of these two designs to the axle torque are, for all practical purposes, identical. As in the case of transverse loading, the maximum bending stresses are uniform along the springs for $[x] \leq \alpha_o$.

2. *Effect of longitudinal force, F_L:* The longitudinal force F_L, will produce a longitudinal and vertical displacement of the spring. Using k_L and k_v, to denote the corresponding spring rate associated with F_L, results in:

	k_L	k_v
design (S)	2663 lb/in.	1033 lb/in.
design (SS)	2516 lb/in.	1008 lb/in.

Assuming that a maximum value of F_L equal to the design load is expected to be carried by the spring, the deflection and the maximum stress experienced by the spring are:

	Longitudinal disp., in.	Vertical disp., in.	Maximum stress, ksi
design (S)	0.83	2.13	13.8
design (SS)	0.87	2.18	13.8

The responses of the two designs to the longitudinal force are essentially identical, The maximum bending stresses are uniform along the springs for $[x] \leq \alpha_o$.

3. *Effect of twisting torque, M_L:* In the usual suspension applications, leaf springs may be subjected to twisting, for example, by an obstacle under one wheel of an axle. For the two springs studied here, the rate of twist is: design (S) $= 1.47 \times 10^4$ in-lb/radians and design (SS) $= 1.23 \times 10^4$ in-lb/radians.

In addition, due to the geometry of the spring, the twisting torque M_L will cause the spring to deflect in the transverse direction. The rate of transverse deflection is: design (S) $= 3319$ in-lb/in.and design (SS) $= 2676$ in-lb/in.

If a maximum total angle of twist of 10 degrees is allowed, the response of the spring will be:

	Twisting torque in-lb	Lateral deflection in.	Maximum shear stress, ksi
design (S)	2559	0.77	2.55
design (SS)	2150	0.80	2.58

In calculating the effect of the twisting torque, the transverse shear modulus of the unidirectional RP has been used. For an RP with

50vol% of glass fibers, the modulus has a value of 4.6×10^5-psi. The maximum shearing stress occurs at $[x] = \alpha_o$. For designs (S) and (SS), the values of α_o are 10.3 in. and 12.66 in., respectively. The values of the bending stresses associated with the twisting torque are negligibly small.

4. *Effect of transverse force, F_T*: As in the case of the twisting torque, the transverse force, F_T, will cause the spring to twist as well as to deflect transversely. The spring rates associated with the transverse force are:

	Twist (in-lb/radian)	Deflection (lb/in)
design (S)	3319	600
design (SS)	2676	458

Assuming that a maximum value of F_T is equal to 0.5 times the design load expected to be carried by the spring, the deflection and the maximum stress experienced by the spring will be:

	Angle of twist (degree)	Transverse deflection (in.)	Max. bending stress (ksi)	Max. shear stress (ksi)
design (S)	19	1.83	11.5	6.05
design (SS)	23.6	2.40	15.6	7.38

The angle of twist and the maximum shear stresses associated with this lateral force are rather high. In practice, the spring will have to be properly constrained to reduce the angle of twist and the maximum shear stress to lower values. Assuming that a maximum angle of twist of no more than 10 degrees is allowed, the deflection and the maximum stresses experienced by the spring are:

	Angle of twist (degree)	Transverse deflection (in.)	Max. bending stress (ksi)	Max. shear stress (ksi)
design (S)	10	1.14	11.5	3.93
design (SS)	10	1.31	15.6	4.09

The maximum bending stresses occur at the center of the spring while the maximum shear stresses occur at the ends of the spring. Based on the above numerical simulations, it appears that both designs respond approximately the same to all different types of loadings. However, design (S) will be preferred since it provides a better response to the lateral and twisting movement of the vehicle.

The maximum bending stress that will be experienced by the spring is obtained by assuming a simultaneous application of the vertical and the longitudinal forces together with the axle torque. A maximum bending stress of 82.5 ksi is obtained. This bending stress is uniform along the spring for $[x] \le \alpha_o$. In view of the infrequent occurrence of this maximum bending stress, it is expect that the service life of the spring is guaranteed to be long in service. However, a maximum shearing stress close to 6.3 ksi can be reached when the spring is subjected to both twisting torque and transverse force at the same time. The value of this shearing stress may be too high for long life application. However, a more complete assessment of the suitability of the design can only be obtained through interaction with the vehicle chassis designers.

Special Spring

As RP leaf springs find more applications, innovations in design and fabrication will follow. As an example, certain processes are limited to producing springs having the same cross-sectional area from end to end. This leads to an efficient utilization of material in the energy storage sense. However, satisfying the requirement that the spring become increasingly thinner towards the tips can present a difficulty in that the spring width at the tip may exceed space limitations in some applications. In that case, it will be necessary to cut the spring to an allowable width after fabrication. There are special processes such as basic filament winding that can fabricate these type structures.

A similar post-molding machining operation is required to produce variable thickness/constant width springs. In both instances end to end continuity of the fibers is lost by trimming the width. This is of particular significance near the upper and lower faces of the spring that are subject to the highest levels of tensile and compressive normal stresses. A practical compromise solution is illustrated in Fig. 4.29. Here excess material is forced out of a mid-thickness region during molding that maintains continuity of fibers in the highly stressed upper and

Figure 4.29 Spring with a practical loading solution

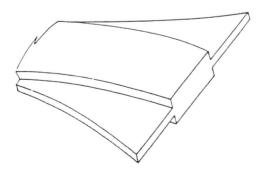

Figure 4.30 Spring has a bonded bracket

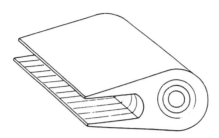

lower face regions. A further advantage is that a natural cutoff edge is produced. The design of such a feature into the mold must be done carefully so that the molding pressure (desirable for void-free parts) can be maintained.

An area of importance is that of attaching the spring to the vehicle. Since the RP spring is a highly anisotropic part especially designed as a flexural element, attachments involving holes or poorly distributed clamping loads may be detrimental. For example, central clamping of the spring with U bolts to an axle saddle will produce local strains transverse to the fibers that in combination with transverse strains due to normal bending may result in local failure in the plastic matrix. The use of a hole for a locating bolt in the highly stressed central clamped region should also be avoided.

Load transfer from the tips of the RP spring to the vehicle is particularly difficult if it is via transverse bushings to a hanger bracket or shackles since the bushing axis is perpendicular to all the reinforced fibers. One favorable design is shown in Fig. 4.30. This design utilizes a molded random fiber RP (SMC; Chapter 1) bracket that is bonded to the spring. Load transfer into this part from the spring occurs gradually along the bonded region and results in shear stresses that are conservative for the adhesive as well as both composite parts.

Cantilever Spring

The cantilever spring (unreinforced or reinforced plastics) can be employed to provide a simple format from a design standpoint. Cantilever springs, which absorb energy by bending, may be treated as a series of beams. Their deflections and stresses are calculated as short-term individual beam-bending stresses under load.

The calculations arrived at for multiple-cantilever springs (two or more beams joined in a zigzag configuration, as in Fig. 4.31) are similar to,

Figure 4.31 Multiple-cantilever zigzag beam spring (Courtesy of Plastics FALLO)

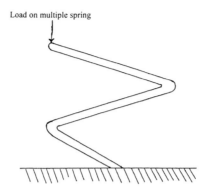

but may not be as accurate as those for a single-beam spring. The top beam is loaded (F) either along its entire length or at a fixed point. This load gives rise to deflection y at its free end and moment M at the fixed end. The second beam load develops a moment M (upward) and load F (the effective portion of load F, as determined by the various angles) at its free end. This moment results in deflection y_2 at the free end and moment M_2 at the fixed end (that is, the free end of the next beam). The third beam is loaded by M_2 (downward) and force F_2 (the effective portion of F_1). This type action continues.

Total deflection, y, becomes the sum of the deflections of the individual beams. The bending stress, deflection, and moment at each point can be calculated by using standard engineering equations. To reduce stress concentration, all corners should be fully radiused. The relative lengths, angles, and cross-sectional areas can be varied to give the desired spring rate F/y in the available space. Thus, the total energy stored in a cantilever spring is equal to:

$$E_c = {}^1\!/_2\, F_y \qquad\qquad\qquad (4\text{-}48)$$

where F = total load in lb, y = deflection in., and E, = energy absorbed by the cantilever spring, in-lbs.

Torsional Beam Spring

Torsional beam spring design absorbs the load energy by its twisting action through an angle zero. Fig. 4.32 is an example of its behavior is that of a shaft in torsion so that it is considered to have failed when the strength of the material in shear is exceeded.

For a torsional load the shear strength used in design should be the value obtained from the industry literature (material suppliers, etc.) or one half the ultimate tensile strength, whichever is less. Maximum shear

Figure 4.32 Example of a shaft under torque

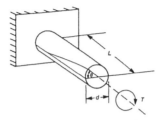

stress on a shaft in torsion is given by the following equation using the designations from Fig. 4.32:

$$\tau = T_c/J \tag{4-49}$$

where T = applied torque in in-lb, c = the distance from the center of the shaft to the location on the outer surface of shaft where the maximum shear stress occurs, in. and J = the polar moment of inertia, in.[4].

Using mechanical engineering handbook information the angular rotation of the shaft is caused by torque that is developed by:

$$\theta = TL/GJ \tag{4-50}$$

where L = length of shaft, in., G = shear modulus, psi = $E/2 (1 + v)$, E = tensile modulus of elasticity, psi, and v = Poisson's ratio.

The energy absorbed by a torsional spring deflected through angle θ equals:

$$E_t = {}^1\!/2 \, M_\theta \times \theta \tag{4-51}$$

where M_θ = the torque required for deflection θ at the free end of the spring, in-lb.

Hinge

Since many different plastics are flexible (Chapter 1) they are used to manufacture hinges. They can operate in different environments. Based on the plastic used they can meet a variety of load performance requirements. Land length to thickness ratio is usually at least 3 to 1.

Hinges can be fabricated by using different processes such as injection molding, blow molding, compression molding, and cold worked. So called "living hinges" use the TP's molecular orientation to provide the bending action in the plastic hinge. With proper mold design (proper melt flow direction, eliminate weld line, etc.) and process control fabricating procedures these integral hinge moldings operate efficiently.

Otherwise problems in service immediately or shortly after initial use delamination occurs. Immediate post-mold flexing while it is still hot is usually required to ensure its proper operation.

Hinges depend not only on processing technique but using the proper dimensions based on the type plastic used. Dimensions can differ if the hinge is to move 45° to 180°. If the web land length is too short for the 180° it will self-destruct due to excessive loads on the plastic's land.

Press fit

Press fits that depend on having a mechanical interface provide a fast, clean, economical assembly. Common usage is to have a plastic hub or boss that accepts either a plastic or metal shaft or pin. Press-fit procedure tends to expand the hub, creating tensile or hoop stress. If the interference is too great, a high strain and stress will develop. Thus it may fail immediately, by developing a crack parallel to the axis of the hub to relieve the stress, which is a typical hoop-stress failure. It could survive the assembly process, but fail prematurely in use for a variety of reasons related to its high induced-stress levels. Or it might undergo stress relaxation sufficient to reduce the stress to a lower level that can be maintained (Chapter 2).

Hoop-stress equations for press-fit situations are used. Allowable design stress or strain will depend upon the particular plastic, the temperature, and other environmental considerations. Hoop stress can be obtained by multiplying the appropriate modulus. For high strains, the secant modulus will give the initial stress; the apparent or creep modulus should be used for longer-term stresses. The maximum strain or stress must be below the value that will produce creep rupture in the material. There could be a weld line in the hub that can significantly affect the creep-rupture strength of most plastics.

Complications could develop during processing with press fits in that a round hub or boss may not be the correct shape. Strict processing controls are used to eliminate these type potential problems. There is a tendency for a round hub to be slightly elliptical in cross-section, increasing the stresses on the part. For critical product performance and in view of what could occur, life-type prototyping testing should be conducted under actual service conditions in critical applications.

The consequences of stress occurring will depend upon many factors, such as temperature during and after assembly of the press fit, modulus of the mating material, type of stress, usage environment and probably the most important is the type of material being used. Some substances will creep or stress relaxes, but others will fracture or craze if the strain is too high. Except for light press fits, this type of assembly design can be risky enough for the novice, because a weld line might already weaken the boss.

Associated with press fit assembly methods are others such as molded-in inserts usually used to develop good holding power between the insert and the molded plastic.

Snap fit

Snap fits are used in all kinds of products ranging from toys to highly loaded mechanical tools. There are both temporary and permanent assemblies, principally in injection and blow molded products. The following guidelines are recommended regarding the position of the snap joint to the injection molded gate and the choice of the wall thicknesses in the area of flow to the place of joining: (1) there should be no binding seams at critical points; (2) avoid binding seams created by stagnation of the melt during filling; (3) the plastic molecules and the filler should be oriented in the direction of stress; and (4) any uneven distribution of the filler should not occur at high-stress points.

Use of snap fits provides an economical approach where structural and nonstructural members can be molded simultaneously with the finished product and provide rapid assembly when compared with such other joining processes as screws. As in other product design approaches (nothing is perfect), snap fits have limitations such as those described in Table 4.10.

Snap fits can be rectangular or of a geometrically more complex cross-section. The design approach for the snap fit beam is that either its thickness or width tapers from the root to the hook. Thus, the load-bearing cross-section at any location relates more to the local load. Result is that the maximum strain on the plastic can be reduced and less material will be used. With this design approach, the vulnerable cross-section is always at the root.

Geometry for snap joints should be chosen in such a manner that excessive increases in stress do not occur. The arrangement of the undercut should be chosen in such a manner that deformations of the molded product from shrinkage, distortion, unilateral heating, and loading do not disturb its functioning

Snap fits can be applied to any combination of materials, such as plastic

Table 4.10 Snap fit behaviors

Advantages

Compact, space-saving form

Takes over other functions like bearing, spring cushioning, fixing

Higher forces can also be transmitted with proper designing

Small number of individual parts

Assembly of a construction system with little expenditure of production facilities and time

Can be easily integrated into the structural member

Disadvantages

Influence of environmental effects (such as distortion due to temperature differences) on the functioning

Effects of processing on the properties of the snap joints (orientation of the molecules and of the filler, distribution of the filler, binding seams, shrinkage, surface, roughness and structure)

The fixing of the joined parts is weaker than in welding, bonding, and screw joining

The conduct of force at the joining place is lesser than in areal joining (bonding, welding)

---

and plastic, metal and plastics, glass and plastics, and others. All types of plastics can be used. Their strength comes from its mechanical interlocking, as well as from friction. Pullout strength in a snap fit can be made hundreds of times larger than its snap in force. In the assembly process, a snap fit undergoes an energy exchange, with a clicking sound. Once assembled, the components in a snap fit are not under load, unlike the press fit, where the component is constantly under the stress resulting from the assembly process. Therefore, stress relaxation and creep over a long period may cause a press fit to fail, but the strength of a snap fit will not decrease with time. They compete with screw joints when used as demountable assemblies. The loss of friction under vibration can loosen bolts and screws where as a snap fit is vibration proof.

The interference in a snap fit is the total deflection in the two mating members during the assembly process. Note that too much interference will create difficulty in assembly, but too little will cause low pullout strength. A snap fit can also fail from permanent deformation or the breakage of its spring action components. A drastic change in the amount of friction., created by abrasion or oil contamination, may ruin the snap. These conditions influence the successful designing of snap fits that basically depend on observing their shape, dimensions, materials, and interaction of the mating parts.

Most common snaps are the cantilever type, the hollow-cylinder type (as in the lids of pill bottles) and the distortion type. Cantilever category includes any leaf-spring components, and the cylinder type is used also to include noncircular section tubes. These snaps include those in any shape that is deformed or deflected to pass over interference. The shapes of the mating parts in a hollow cylinder snap is the same, but the shapes of the mating parts in a distortion snap are different, by definition.

Snap fits flex like a spring and quickly return, or at least nearly return to its unflexed position. Target is to provide sufficient holding power without exceeding the elastic limits of the plastics. Using the engineering beam equations one can calculate the maximum stress during assembly. If it stays below the yield point of the plastic, the flexing beam will return to its original position. However, for certain designs there will not be enough holding power, because of the low forces or small deflections. It has been found that with many plastics the calculated flexing stress can far exceed the yield point stress if the assembly occurs too rapidly. The flexing finger will just momentarily pass through its condition of maximum deflection or strain, and the material will not respond as if the yield stress had been greatly exceeded.

Another popular approach to evaluate the design of snap fits is to calculate their strain rather than their stress. Then compare this value with the allowable dynamic strain limits for the particular plastics. In designing the beams it is important to avoid having sharp comers or structural discontinuities that can cause stress risers. Tapered finger provides a more uniform stress distribution, which makes it advisable to use where possible.

Tape

Plastic tapes are used to meet many different requirements that range from being flexible to strong, water to chemical and other environmental resistance, soft to wear resistance, and so on. This review is on the overall performance of tapes. Even though tape is a market in its own, there are other markets such as for belts that have some similar features that they both meet rigid and versatile requirements.

One of many different performing types is a low-profile long conveyor belts with prolonged high-speed operating life and minimal maintenance for use in plants, in underground mines, etc. The conveyors can have different belts for different applications including part accumulation, hot or cold processes, or chemical resistance. Closed-top and open-mesh versions are available. There are accumulation belts that use a blend of acetal and Teflon and meets FDA standards. It can withstand temperatures up to 180°C. For higher temperatures, a flat or cleated line of nylon belts operates at temperatures as high as 375°F. Chemical-resistant applications can use a flat, side walled, or cleated belt that resists bleaches and acids while functioning. For electronic applications, the flat or cleated static-conductive belt made from polypropylene meets FDA standards for Class 11 type charges.

This review on tapes highlights the historically Du Pont's research leading to Dymetrol® elastomeric tape that began in 1974. The General Motors Corp. in the USA had developed a new lightweight window regulator, to replace the heavy metal segment window regulators, but cold not make it work adequately with the metal and plastic tapes used at that time. Using its plastics processing know-how, Du Pont developed what is now known as highly engineered oriented elastomeric tape, or as the Dymetrol mechanical drive tape, and General Motors have been using it since 1979 in manual and electric window car regulators. Today this tape with its applications has evolved into a multi-tape/multi-application proposition that include safety passive restrainers, windshield wiper linkages, sliding car roofs, garage door openers, vending machines, etc.

This high modulus material composition provides tape with steep stress/strain characteristics. In other words higher dimensional stability under applied loads or higher tensile loading capability at the same elongation (vs. the standard material composition tapes). High modulus material composition tapes also have higher stiffness, resulting in a much-improved push vs. pull load transfer efficiency. In practice this means for instance that window regulator mechanisms can be constructed with tape lifting the window as well in the compressive as in the tension mode which provides the automotive design engineer with more possibilities and flexibility to conceive car doors with optimum cost, performance and design characteristics.

Another novelty is the abrasion resistant material composition option that confers much improve abrasion resistance and somewhat lower coefficient of friction. The mechanical drive tape will also transfer tension and compressive forces when used in non-linear directions. Contributing factors are not only the tape's axial stiffness, providing the push and pull, but also its torsional and edge-bend flexibility.

Fig. 4.33 shows the flexibility of high modulus vs. standard tape

Figure 4.33 Dymetrol mechanical drive tape: (a) flexural modules and (b) beam flexure versus tape thickness (unpunched)

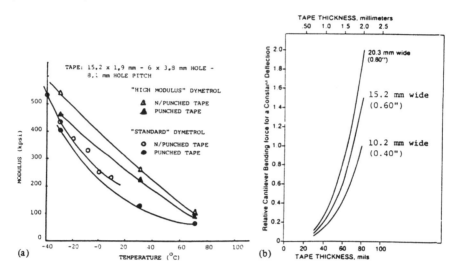

materials and also the effect of punching and of temperature on tape stiffness. Naturally tape thickness and width also affect a tape's stiffness. The obvious user benefit of this tape flexibility is greater versatility in the design of energy transfer mechanisms, since allowing for push/pull in non-linear modes, thereby advantageously replacing more complex movement transmission devices such as lever arm systems, pulleys, or gear systems. By using this feature, a cigarette vending machine offers a 50% increase in brand choice without increase in machine dimensions. Metal wire cable may rapidly fail in energy transfer mechanisms due to its frequently inadequate alternate flex cycle life, and spiral metal cable due to its spiral collapse, these mechanical drive tapes have proved to be extremely tough.

From these data, taken from various points of the tape's stress/strain curve, it can be concluded that the strength of the mechanical drive tapes. and in particular of the high modulus materials composition, is appreciable and adequate indeed for the low to medium load transfer service applications to which it addressed itself, and more than what is needed for window regulators for instance. The user benefit here naturally is long-term performance dependability or tape driven energy transfer mechanisms and proved for instance by the low GM car after-sale replacement rates of tape driven window regulators. Another example is the. use of tape to drive outdoor venetian blinds in which the previous drive device failed frequently causing expensive repair.

Contributing significantly to this tape toughness is its property to recover from strain caused by permanent or intermittent operating stress, even after exposure to temperatures as high as 80°C.

There is virtually no creep and very little permanent deformation. Similar tests have furthermore shown that there is not much more deformation even after 8000 hours exposure to 4000 psi. The added benefit of this low creep and strain recovery characteristic of tape is that it confers operating shock absorbency and smoothness to energy transfer mechanisms, not or less provided by other energy transmission devices since featuring steeper stress/strain characteristics or no stress elasticity at all.

Packaging

Plastics are used to package many different forms and shapes of products. Their performance requirements are very diversified ranging from relatively no strength to extremely high strength, flexible to rigid, non-permeable to permeable (in many different environments), and so on. They require design performance requirements that include many reviewed in this book. There is extensive literature on the subject of packaging and all their ramifications.

Packaging industry and its technology is the major outlet for plastics where it consumes about 30wt% of all plastics (yearly sales above $40 billion) with building and construction in second place consuming about 20wt%. If plastic packaging were not used, the amount of packaging contents (food, soaps, etc.) discarded from USA households would more than double. Plastics are the most efficient packaging materials due to their higher product-to-package ratio as compared to other materials. One ounce of plastic packaging can hold about 34 ounces of product. A comparison of product delivered per ounce of packaging material shows 34.0 plastics, 21.7 aluminum, 6.9 paper, 5.6 steel, and 1.8 glass.

When packaging problems are tough, plastics often are the answer and sometimes the only answer. They can perform tasks no other materials can and provide consumers with products and services no other materials can provide. As an example plastics have extended the life of vegetables after they are packaged.

If plastic packaging were not used, the amount of packaging contents (food, etc.) discarded from USA households would more than double. Plastics are the most efficient packaging materials due to their higher product-to-package ratio as compared to other materials. One ounce of plastic packaging can hold about 34 ounces of product. A comparison of product delivered per ounce of packaging material shows 34.0 plastics, 21.7 aluminum, 6.9 paper, 5.6 steel, and 1.8 glass.

Different designs and processing techniques are used to produce many different packaging products. These different products show how innovative designs have created different products based on plastic behaviors and they're processing capabilities. Most of these products are extruded film and sheet that are usually thermoformed. Other processes are used with injection molding and blow molding being the other principal types used.

The largest market at 35% of the total is for carded blister packs. The second major product is window packaging at 24%. The others are clamshell packaging at 20%, skin packaging at 18%, and others at 3%. The following information provides examples of packaging products that meet different performance requirements that relates to the capability of the plastic used: *aseptic* in food processing; *bag-in-box* refers to a sealed, sprouted plastic film bag inside a molded rigid container; *beverage can* with aluminum cans dominating the USA market for soft drink containers with about 70% of the market, PET plastic and glass compete for second place-note that most aluminum cans have an inside coating, usually epoxy, to protect its contents from the aluminum; *beverage container* with carbonated soft drinks being the largest market at about 50% followed by beer at about 25%, fruit juices and drinks, and milk; *beer bottle* potential in bioriented stretched plastic bottles in USA is on the horizon using coinjection or coextruded plastics such as PET plastic and/or PEN plastic using various barrier plastics or systems; *biological substance* that are classified as hazardous requiring specialty packaging where plastics play an important role to meet strict requirements; *blister* also called blister carded packaging; *bubble pack* is plastic cushioning material used in packaging; *contour packaging* is also called skin packaging; *dual-ovenable tray* are used for frozen foods; *electronic packaging* with plastic ease of processing and low cost has given them a wide application in solving problems in electronic packaging; *film breathable* identified as controlled-atmospheric packaging (cap); *food packaging* with plastics provides the most efficient packaging materials due to their higher product-to-package ratio as compared to other materials; *food oxygen scavenger* impregnated plastics with chemically reactive additives that absorb oxygen, ethyl, and other agents of spoilage inside the package once it has been sealed; *grocery bag*; *hot fill package* injection & blow molded bottles, thermoformed containers, etc.; *modified atmosphere packaging* (mat) is a packaging method that uses special mixtures of gases (carbon dioxide, nitrogen, oxygen, or their combinations; hermetically sealed mat extends the shelf life of red meat, skinless turkey breast, chicken, half-baked bread, pizza's crust, bagels, etc.; *peelable film* for case-ready ground beef package add color and shelf life; *pouch* heat-sealed, wrap, and reusable container help keep food fresh and free of contamination; and *retortable pouch* has superior flavor retention and long shelf life.

Permeability

There are different approaches for providing permeability resistance in plastic packaging. The more extensively used is barrier plastics with nonbarrier types to meet cost-to-performance requirements. Most plastics can be considered barrier types to some degree, but as barrier properties are maximized in one area (as gases such as O_2, N_2, or CO_2), other properties as permeability and moisture resistance diminish. This is achieved through coextrusion, coinjection, corotation, and other such processes. The other chemically modifies the plastics' surfaces.

There are plastics that have different behaviors to protect and preserve products in storage and distribution. They provide different diffusion (transport) action of gases, vapors, and other low molecular weight species through the materials. Important selection in packaging (food, etc.) is based on the permeability of the materials to factors such as oxygen and water vapor. There are special applications such as packaging bananas where an ethylene gas remains in a package is used to artificially ripen the bananas. There is other industrial gas separation systems that use the selective permeability of plastics to separate their constituents.

This diffusion rate is related to the resistance, within the plastic wall, to the movement of gases and vapors. Important aspects of the diffusion process are permeability and migration of additives. Possible migrants from plastics can include the many different additives and fillers used (Chapter 1).

The conditions inside and outside packages relate to the gas and vapor pressure forces penetrating or diffusing through permeable packages. As the engineering handbooks report diffusing substance's transmission rate is expressed by mathematical equations commonly called Fick's First and Second Laws of Diffusion very popular in the industry:

$$F = D\,(dC/dX) \tag{4-52}$$

$$dC/dt = D(d^2C/dX^2) \tag{4-53}$$

where: F = flux (the rate of transfer of a diffusing substance per unit area),
D = diffusion coefficient, C = concentration of diffusing substance,
t = time, and X = space coordinate measured normal to the section.

Different test methods are used to conduct permeability behaviors of plastics to measure gas, water, and other material vapor permeability. Permeability test procedures used to measure the permeability of plastic films include those identified as the absolute pressure method, the isostatic method, and the quasi-isostatic method. Basic approach mounts a plastic film sample between two cell chambers of a permeability cell (two piece closed container). One chamber holds the gas or vapor to be used as the permeant. The permeant then diffuses through the film into a second chamber, where a detection method such as optical devices, infrared spectroscopy, a manometric, gravimetric, or coulometric method; isotopic counting; or gas-liquid chromatography provides a quantitative measurement.

The absolute pressure method (ASTM D1434, Gas Transmission Rate of Plastic Film and Sheeting) is used when no gas other than the permeant in question is present. In this test the change in pressure on the volume of the low-pressure chamber measures the permeation rate. Between the two chambers a pressure differential provides the driving force for developing permeation.

Isostatic testing equipment has been used for measuring the oxygen and carbon dioxide permeability of both plastic films and complete plastic packages. Pressure in each test chamber are held constant by keeping both chambers at atmospheric pressure. With gas permeability measurement, there must be a difference in permeant partial pressure or a concentration gradient between the two cell chambers. Gas that permeates through the film into the lower-concentration chamber is then conveyed to a gas sensor or detector by a carrier gas for quantification.

A variation of the isostatic method is the quasi-isostatic method where at least one chamber is completely closed and no exposure to atmospheric pressure. However, there is a difference in penetrant partial pressure or a concentration gradient between the two cell chambers. The concentration of permeant gas or vapor that has permeated through into the lower-concentration chamber can be quantified by a technique such as gas chromatography.

There is the chemically modifying approach. It changes the plastic's surface during or after fabrication permiting the control of the permeation behavior of such products as film, sheets, diaphragms, and containers. These techniques are becoming increasingly important. There is an endless search for better barrier materials for packaging applications. As an example in blow-molded gasoline containers/tanks, the amount of gasoline permeation through HDPE even though it is very low, is still excessive per the standard requirements, thus has required some type of barrier. Including a barrier in a multilayer construction has created a satisfactory solution. The approach to functionalized PE formed tanks on the inside wall is by a chemical reaction, mostly by exposing the surface to sulfonation or fluorination.

Oxifluorination has the fluorine gas combined with nitrogen to which several percent of oxygen by volume have been added. Subjecting PE to fluorine and oxygen at the same time leads to functionalization of the PE making it impermeable. This method permits substantially reducing the required amount of fluorine, resulting in a cost-to-performance improvement. Barrier plastics using oxifluorination are used for foods. They provide barriers that are needed to protect them against spoilage from oxidation, moisture loss or gain, and changes or losses in favor, aroma, or color.

Cushioning

Different plastic foams are used in many different products meeting different performance requirements (Chapter 1). In the packaging industry they provide different performances that includes cushioning. They are used in different weights to meet different product requirements. Regarding cost and performance it is sometimes believed that the lower density closed-cell foams that are usually priced lower provide superior cushioning performance. This assumption is usually incorrect as shown in Fig. 4.34. Even though it contains less plastic, the fabricating rates, the amount and cost of the blowing agent, and the amount and cost of the base plastic all influence the final cost. As a

Figure 4.34 Cushioning effect of polyethylene foam density is influenced by loading

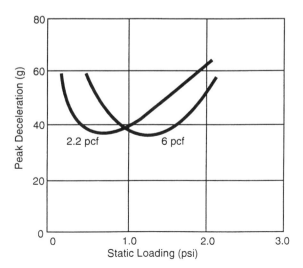

result, very low-density foams can actually be more costly to make than others. Cost of a foam is usually not proportional to its density.

The same type and thickness closed-cell polyethylene foams with differing densities have been subjected to dynamic static load cushioning conditions during impact. Curves relate mechanical shock experienced during an impact where the lower the curve goes, the greater the cushioning efficiency. Curves for foam densities above 6 pcf (lb/ft³) the maximum cushioning efficiency of each material is not significantly different, however a dramatic change occurs with a change in the applied load.

If a 40 g package were to be designed according to Fig. 4.34 using a 6.0 pcf foam, the foam would measure 3 in. thick resulting in 40 g shock. If a similar package were then produced using a 2.2 pcf foam, its shock performance would not go as low as 40 g but would instead produce about 60 g's, or 50% more shock or a 50% loss in shock efficiency.

To meet 40 g's, the 2.2 pcf package would need to be redesigned. Greatly increasing the thickness of the pads constructed from the lower density foam can be used to provide adequate protection. However the result would of increasing the package size, impair handling and shipping efficiency, and possibly result in higher costs. The 6.0 pcf foam could, however, be reliably used.

To be at 40g level and keep the 2.2 pcf foam thickness the same, reduce loading from 1.35 to 0.87 psi can be used. Although this approach

Figure 4.35 Comparison of different foam densities

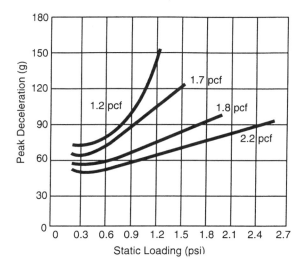

Static Loading (psi)

keeps the package size the same, nearly twice as much foam must be used to meet the lower loading. Lowering the density produces a considerably higher deceleration and reduces cushioning performance. Fig. 4.35 compares the effect of density. The lower density foam cost less than half on a cost-per-unit volume basis resulting in a cost savings. Below a density of about 2.2 pcf the cushioning efficiency can begin to change with the density.

Significant in this figure is the narrower range of usable static loadings at the bottoms of the curves that resulted when the density is reduced. Important consideration in comparing foams of different densities is their compressive creep resistance, and their ability to resist undergoing a permanent thickness loss during their time under load. As the density decreases, so does the creep resistance.

Building

There are many applications of structural and nonstructural plastics being used in the building and construction industry worldwide. Plastic use continues to expand in present and new products in homes, bridge structures, etc. They continue to be based on a combination of factors

Figure 4.36 All-plastic GE house in Pittsfield, MA (Courtesy of GE)

that include understanding building requirements, reliable static and dynamic testing of products, material reliability, public acceptability, feasibility, keeping up-to-date on customer problems, product identification, quality control establishment of engineering standards, approval of regulatory agencies, supervise installations, accurate cost and time estimation, organizational responsibility defined, meeting delivery schedule, development of proper marketing and sales approach, resolution of profit potential based on careful selection of application, economics, and acknowledge competition exists. The trend is to resist change in appearance so that plastic originally had to look like something else. With time passing the beauty and aesthetic of plastic was accepted and at the same time usually at lower costs and more benefits developed to the consumer (Fig. 4.36).

This industry consumes about 20wt% of all plastics produced. It is the second-largest consumer of plastics following packaging. This amount of plastics only represents about 5% of all materials consumed. Plastics will eventually significantly expand in this market. Its real growth will occur when plastic performance is understood by the building industry (meeting their specifications, etc.) and more important when the price is right in order to compete with other materials. Interesting that due to the destruction caused by wars, particularly WW II, use of plastics in most of those countries significantly escalated due to their desire to change specifications and standards allowing the use of plastics where it was technically beneficial.

The present and growing large market for plastics in building and construction is principally due to its suitability in different internal and external environments. The versatility of different plastics to exist in different environments permits the ability for plastics to be maintenance-free when compared to the more conventional and older materials such as wood. So it is said (factual) if wood did not have its excellent record of performances and costs for many centuries, based on present laws and regulations they could not be used. They burn, rot, etc. Regardless it would be ridiculous not to use wood. The different plastics inherently have superior properties that make them useful in other markets such as high strength and stiffness, durability, performances, insulation, cosmetics, etc. so eventually their use in building and construction will expand.

The plastic industry understands that the subjects important to the architect and builder includes information such as code requirements, setting up building standards and logical identification of over 25,000 plastics, static and dynamic load capabilities in structural parts, use of multifunctional parts, products provide new or better solutions to the myriad problems that plaguing the construction industry, performance data rather than sales type data, realistic and understandable technical data, limits on service life, fire safety, educating labor on benefits in using plastics, consumer acceptance, legal risk, competitive products, and costs.

Regarding codes unfortunately, particularly in the past, USA codes and other standards would contain requirements that certain plastics could meet with flying colors except those that would specify a specific material such as steel could only be used. So plastics could not be used until other materials such as plastics would be included or no specific material was specified. Outside USA the changes in most cases were immediate.

House of the Future

Engineering ideas using unconventional approaches to building houses have evolved since at least the 1940s. Designers have applied spray- and solid-foamed homes that were originally introduced by the US Army during the 1950s. At the October 1965 annual National Decorative and Design Show in New York City products were on display of entire rooms and furnishings molded-in-place were found appealing and could survive the environment. Latter all structures (wall, roof, etc.) buildings were successfully built from interlocking extruded PVC

Figure 4.37 Monsanto house in Disneyland, California designed by MIT (Courtesy of Monsanto)

hollowed sandwich-ribbed panels that were filled with insulation material (PUR foam, etc.), concrete, or other materials. Also building structures were made from extruded PS foam logs that were heat-bonded wrapped in dome shaped structures, RP filament wound room structures and support pillars, and so on Fig 4.37 and 4.38.

Monsanto's house of the future (Fig. 4.37) located in Disney Land, FL was built during 1957 of practically all glass fiber-TS polyester RP; other plastics also used and reinforced concrete (foundation). MIT professors designed it. View (a) shows the four U-shaped monocoque structural cantilever beams each 16 ft from the square center foundation. View (b) shows load bearing requirements of the cantilever beams. View (c) is a sectional view of the house.

This main attraction was subjected to winds, tornadoes, earthquakes, and subjected to the movement of people that simulated a century of use. After two decades it was replaced to locate another scene. At that time it was determine the structure remained in tact as it was produced initially. It basically did not deflect. It was almost impossible to destroy by the conventional wrecking procedures. They had to actually cut it up

Figure 4.38 Dow house of the future

into small sections and then use wrecking balls, etc.

The Dow's house of the future (Fig. 4.38) located in Lafayette, IN was built during 1966 of practically all polystyrene foam board patented spiral generation technique producing high structurally loaded self-supporting domes. They provided their own insulation. View (a) is a model of what the finished building complex resembled. View (b) shows the lay up procedure where the PS boards are heated on site using a heating device. The operation starts on a support circular steel foundation room. These layers bonded to each other with the applied heat formed the required contours with a continuous board structure.

These self-supporting domes required no internal or external support during or after manufacture. View (c) shows the ease in cutting windows, doors, connecting hallways from dome to dome, etc. The exterior of each dome was covered with steel wire mesh and concrete followed with waterproofing. Unfortunately its limitation in building structures was its potential fire hazard. However this basic concept with appropriate plastic foam material protection provides another approach to the low cost plastic house, etc.

House Top

House top and top of other buildings is designed to take static loads and endure the outside environment such as sunlight, rain, wind, falling objects, temperature changes, hurricanes, and/or snow. It must also support loads imposed by people walking on it. This house top structure represents a simple type of a design project in static loading in that the loads are clearly long term and well defined. Creep effects can be easily predicted and the structure can be designed with a sufficiently large safety factor such as 3 to avoid the probability of failure. These type data (static, creep, etc.) are available of the material suppliers (Chapter 7).

A roof design could use a translucent corrugated RP panel structure. This construction material has been used for roofs to admit daylight on a porch, shanty, canopy, and patio, shelter tops such as those used at a bus stop, industrial building, etc. Also used on transportation vehicles such as a boat, truck, airplane, bus, and train. Corrugated materials are available in sheets of different sizes such as from 4 ft × 8 ft to as large as 10 ft × 20 ft. A typical material is 0.1 in. thick with 2 to 4 in. corrugations and a corrugation depth of 1 to 4 in. The usual material of construction is glass fiber-TS polyester RP. Standardized panels are used that have specific physical and mechanical properties. Usual attachment is by using nails or screws to wooden or pultruded plastic supports.

Design approach takes into account the weight of the material that is the major static load applied to the structure itself. Other weight such as snow is available from experience obtained in the area where the structure is to be used. Similarly, weights due to wind load and people load can be determined from experience factors that are generally known.

Main stress caused by the wind is at the anchorage points of the roof to the rest of the structure. They should be designed to take lifting forces as well as bearing forces. By using a lower angle of top structure, the less wind lifting stress. Proper anchorage of the support structure to the ground is also essential.

Localized loads represented by people walking on the roof can be solved by assuming concentrated loads at various locations such as between supports and by doing a short time solution to the bending problem and the extreme fiber stress condition. The local bearing loads and the localized shear should also be examined since it may cause possible local damage to the structure.

Requirements for the design of the structure included factors such as building and fire codes, and those that are reported by the owner of the property. As an example the material selected is to take the required loads without severe sagging (amount to be specified) for a 20 year period with no danger that the structure will collapse due to excessive stress on the material. One typical way in which excessive loading for a single section is handled is to bond two layers of the corrugated panel together with the corrugations crossed. This approach result in a very stiff section capable of substantially greater weight bearing than a single sheet and it will meet the necessary requirements. It also provides thermal insulation because of the trapped air space between the two layers particularly if they are edged sealed.

When using a large surface area of the RP panels expansion with a temperature rise occurs. Recognize that very few of the traditional building materials, such as wood, have significant expansion under normal temperature shifts. The RP materials, when compared to wood, does not have a problem because of its low thermal coefficients of expansion. What also keeps the dimensional changes low is that the corrugated shape can flex and accommodate the temperature changes. When compared to plastic materials such as polyvinyl chloride (PVC) siding, the expansion factor becomes significant and is an important consideration in the PVC method of fastening.

Data on the material (from the material supplier and/or data determined) will have been determined that it meets the static and dynamic loads required based on engineering analysis. Exposure to the environment (includes water and sunlight) can have a significant effect on the properties of the materials. In this example it is assumed that a 50% or more drop in the mechanical properties is expected to occur in 10 years. Based on knowledge of the materials behavior in service this loss of property levels off to a low rate of deterioration so that any potential failure will not occur in 20 years. If necessary increasing the panel strength requirements can compensate loss of properties by a suitable factor of safety that could be 3 in this example. However one knows that with a protective coating on the panel will minimize the effects of weathering so that only 20% loss in properties occurs after 10 years so that initial requirements for the panel performance will be reduced accordingly. The preferred type of coating could be a fluoropolymer that has the best resistance to sunlight and other weathering factors of all of the plastics and will last 20 years. If a coating is applied there usually is a requirement that the panel be recoat after a specified period in service.

Transportation

Plastics and RPs offer a wide variety of benefits to the different transportation vehicles (automobiles, trucks, motorcycles, boats, airplanes, and so on). They range from assembling separate parts to providing safety (impact, etc.), aesthetics, lightweight, durability, corrosion resistance, recyclability, and fuel savings. Growth in the use of plastics tends to be a continuous process.

Structural analysis methods coupled with composite mechanics play a unique role during design in that they are used to formally evaluate alternative design concepts prior to committing some of these design concepts to costly testing or fabrication. Important material properties are not known during design in many cases. Frequently, some of these properties are assumed and structural analyses are performed to assess their effects on the design as well as their effects on the structural performance of the designed vehicle component. Also structural analysis methods are used to establish acceptable ranges of key factors (parameters) in a given design.

With RPs there can be the intraply hybrid composites have two kinds of fibers embedded in the matrix in general with the same ply providing different orientation performances. They have evolved as a logical sequel to conventional composites and to interply hybrids. Intraply hybrid composites have unique features that can be used to meet diverse and competing design requirements in a more cost effective way than either advanced or conventional composites. Some of the specific advantages of intraply hybrids over other composites are balanced strength and stiffness, balanced bending and membrane mechanical properties, balanced thermal distortion stability, reduced weight and/or cost, improved fatigue resistance, reduced notch sensitivity, improved fracture toughness and/or crack-arresting properties, and improved impact resistance. With intraply hybrids, it is possible to obtain a viable compromise between mechanical properties and cost to meet specified design requirements.

Structural mechanics analyses are used to determined design variables such as displacements, forces, vibrations, buckling loads, and dynamic responses, including application of corresponding special areas of structural mechanics for simple structural elements. General purpose finite element programs such as NASTRAN are used for the structural analysis of complex structural shapes, large structures made from simple structural elements, And structural parts made from combinations of simple elements such as bars, rods and plates.

Composite mechanics in conjunction with structural mechanics can be used to derive explicit equations for the structural response of simple structural elements. These explicit expressions can then be used to perform parametric studies (sensitivity analyses) to assess the influence of the hybridization ratio on structural response. For example the

Figure 4.39 Loads related to flexural modulus

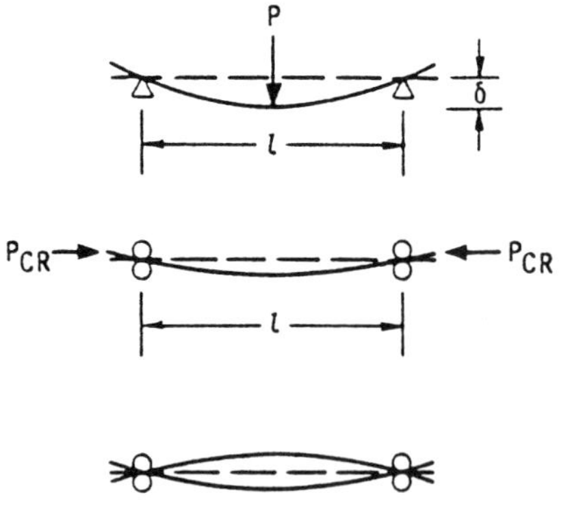

$$\delta = \frac{Pl^3}{48E_F I} = \frac{Pl^3}{48E_{PC} I\left[1 + V_{SC}\left(\frac{E_{SC}}{E_{PC}} - 1\right)\right]}$$

$$P_{CR} = \frac{\pi^2 E_F I}{l^2} = \frac{\pi^2 E_{PC} I\left[1 + V_{SC}\left(\frac{E_{SC}}{E_{PC}} - 1\right)\right]}{l^2}$$

$$\omega_n = (n\pi)^2 \sqrt{\frac{E_F I}{\rho l^2}}$$

$$= (n\pi)^2 \left(\frac{P_{PC} l^2}{I E_{PC}}\right)^{1/2} \left[\frac{1 + V_{SC}\left(\frac{E_{SC}}{E_{PC}} - 1\right)}{1 + V_{SC}\left(\frac{P_{SC}}{P_{PC}} - 1\right)}\right]^{1/2}$$

structural response (behavior variables) equations for maximum deflection, buckling load and frequency of a simply supported beam made from intraply hybrids are summarized in Fig. 4.39 Flexural modulus is used to determine the maximum deflection, buckling load, and frequency of a simple supported beam made from intraply hybrids.

The notation in these equations and others that follow are as follows:

a_1, a_2	Correlation coefficients for longitudinal compressive strength.
$B_{c1, k}$	Buckling limit (buckling behavior constraint) due to loading condition k.
$B'_{c2, k}$	Strength limit (strength behavior constraint) due to loading condition k.
$B'c3, k$	Interply delamintion limit (delamination constraint) due to loading condition k.
E_F	Flexural modulus.
E_{f11}, E_{f22}	Longitudinal and transverse fibre moduli.
E_{m11}, E_{m22}	In situ matrix longitudinal and transverse moduli.
E_{PC}	Modulus primary composite.
E_{SC}	Modulus secondary composite.
G_{f12}	Fibre shear modulus.
G_{m12}	Matrix shear modulus.
H_j	Matrix interply layer effect.
I	Moment of inertia.
$K_{l12\alpha\beta}$	Correlation coefficients in the combined-stress failure criterion; α, β = T or C denoting stress direction.
k_f	Fiber volume ratio.
k_v	Void volume ratio.
k	K = 1, 2, 3, denotes loading condition index.
l	Length.
N	Number of plies.
N_{xk}, N_{yk}	Inplane loads – x and y directions corresponding to k.
P	Load.
P_{cr}	Buckling load.
P_{cro}	Buckling load of reference composite.
S_{ft}	Fiber tensile strength.
Spc	Strength primary composite.
S_{sc}	Strength secondary composite.
V_{sc}	Volume fraction secondary composite..
W	Panel cost units per unit area.
β, β'	Correlation coefficients in composite micromechanics to predict ply elastic behavior.
β_{del}	Interply delamination factor.
δ	Displacement.
δ_o	Displacement of reference composite
ε_{mpc}, S, T	In situ allowable matrix strain for compression, shear and tension.
θ	Ply angle measured from x-axis.

Figure 4.40 Effects of hybridizing ratio and constituent composites on center deflection of intraply hybrid composite beams

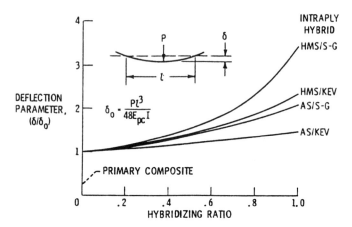

Figure 4.41 Effects of hybridizing ratio and constituent composites on buckling load of intraply hybrid composite beams

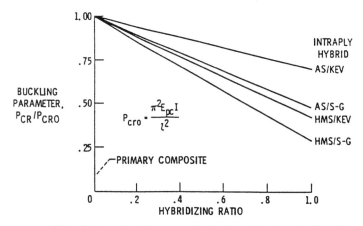

v_f	Fiber Poisson's ratio, numerical subscripts denote direction.
v_m	Matrix Poisson's ratio, numerical subscripts denote direction.
ω_n	Frequency of the nth vibration mode.
ω_{no}	Frequency of the nth vibration mode of the reference composite.

The equations are first expressed in terms of E_f, the equivalent flexural modulus, and then in terms of the moduli of the constituent composites (E_{pc} and E_{sc}) and the secondary composite volume ratio (V_{sc}). These equations were used to generate the parametric nondimensional plots shown in Figs 4.40 to 4.42.

The nondimensionalized structural response is plotted versus the

Figure 4.42 Effects of hybridizing ratio and constituent composites on frequences of intraply hybrid composite beams

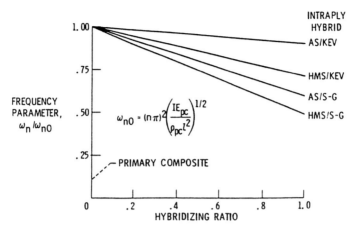

Figure 4.43 Effects of hybridizing ratio and constituent composites on Izod-type energy density of intraply hybrids

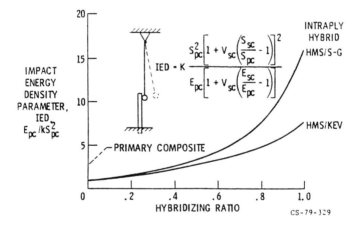

CS-79-329

hybridizing ratio V_{sc} for four different intraply hybrid systems. These figures show that small amounts of secondary composite (V_{sc} 0.2) have negligible effect on the structural response. However, small amounts of primary composite (V_{sc} 0.8) have a substantial effect on the structural response. A parametric plot of Izod-type, impact energy density is illustrated in Fig. 4.43. This parametric plot shows also negligible effects for small hybridizing ratios (V_{sc} 0.2) and substantial effects for hybridizing ratios (V_{sc} 0.2).

The parametric curves in Figs 4.40 to 4.43 can be used individually to select hybridization ratios to satisfy a particular design requirement or

Figure 4.44 Schematic of strip hybrids

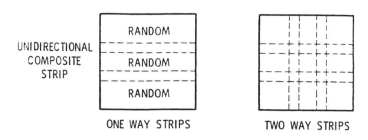

they may be used jointly to satisfy two or more design requirements simultaneously, for example, frequency and impact resistance. Comparable plots can be generated for other structural components. such as plates or shells. Also plots can be developed for other behavior variables (local deformation, stress concentration, and stress intensity factors) and/or other design variables, (different composite systems). This procedure can be formalized and embedded within a structural synthesis capability to permit optimum designs of intraply hybrid composites based on constituent fibers and matrices.

Low-cost, stiff, lightweight structural panels can be made by embedding strips of advanced unidirectional composite (UDC) in selected locations in inexpensive random composites. For example, advanced composite strips from high modulus graphite/resin, inter-mediate graphite modulus/resin, and Keviar-49 resin can be embedded in planar random E-glass/resin composite. Schematics showing two possible locations of advanced UDC strips in a random composite are shown in Fig. 4.44 to illustrate the concept. It is important to note that the embedded strips do not increase either the thickness or the weight of the composite. However, the strips increase the cost.

It is important that the amount, type and location of the strip reinforcement be used judiciously. The determination of all of these is part of the design and analysis procedures. These procedures would require composite mechanics and advanced analyses methods such as finite element. The reason is that these components are designed to meet several adverse design requirements simultaneously. Henceforth, planar random composites reinforced with advanced composite strips will be called strip hybrids. Chamis and Sinclair give a detailed description of strip hybrids. Here, the discussion is limited to some design guidelines inferred from several structural responses obtained by using finite element structural analysis. Structural responses of panels structural components can be used to provide design guidelines for sizing and designing strip hybrids for aircraft engine nacelle, windmill

Figure 4.45 Structural responses of strip hybrid plates with fixed edges

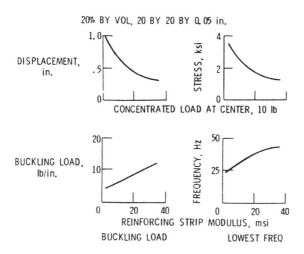

blades and auto body applications. Several examples are described below to illustrate the procedure.

The displacement and base material stress of the strip hybrids for the concentrated load, the buckling load, and the lowest natural frequency are plotted versus reinforcing strip modulus in Fig. 4.45. As can be seen the displacement and stress and the lowest natural frequency vary nonlinearly with reinforcing strip modulus while the buckling load varies linearly. These figures can be used to select reinforcing strip moduli for sizing strip hybrids to meet several specific design requirements. These figures are restricted to square fixed-end panels with 20% strip reinforcement by volume. For designing more general panels. suitable graphical data has to be generated.

The maximum vibratory stress in the base material of the strip hybrids due to periodic excitations with three different frequencies is plotted versus reinforcing strip modulus in Fig. 4.46. The maximum vibratory stress in the base material varies nonlinearly and decreases rapidly with reinforcing strip modulus to about 103 GPa (15×10^6 psi). It decreases mildly beyond this modulus. The significant point here is that the modulus of the reinforcing strips should be about 103 GPa (15×10^6 psi) to minimize vibratory stresses (since they may cause fatigue failures) for the strip hybrids considered. For more general strip hybrids, graphical data with different percentage reinforcement and different boundary conditions are required.

The maximum dynamic stress in the base material of the strip hybrids

Figure 4.46 Maximum stress in base material die to periodic vibrations

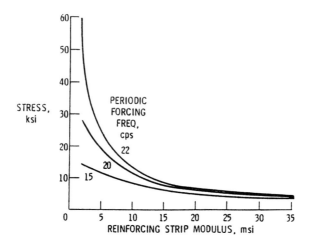

Figure 4.47 Maximum impulse stress at center

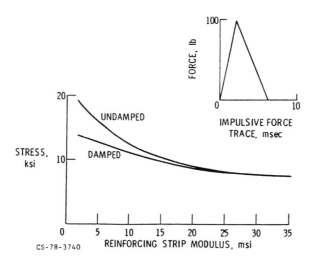

due to an impulsive load is plotted in Fig. 4.47 versus reinforcing strip modulus for two cases: (1) undamped and (2) with 0.009% of critical damping. The points to be noted from this figure are: (a) the dynamic displacement varies nonlinearly with reinforcing strip modulus and (b) the damping is much more effective in strip hybrids with reinforcing strip moduli less than 103 GPa (15×10^6 psi). Corresponding displacements are shown in Fig. 4.48. The behavior of the dynamic displacements is similar to that of the stress as would be expected. Curves comparable to those in Figs 4.46 and 4.47 are needed to size

Figure 4.48 Maximum impulse displacement

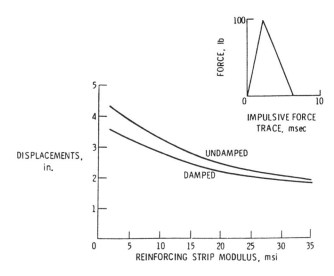

and design strip hybrid panels so that impulsive loads will not induce displacements or stresses in the base material greater than those specified in the design requirements or are incompatible with the material operational capabilities.

The previous discussion and the conclusions derived therefrom were based on panels of equal thickness. Structural responses for panels with different thicknesses can be obtained from the corresponding responses in Fig. 4.47 as follows (let t = panel thickness):

1　The displacement due to a concentrated static load varies inversely with t^3 and the stress varies inversely with t^2.

2.　The buckling load varies directly with t^3.

3.　The natural vibration frequencies vary directly with t.

No simple relationships exist for scaling the displacement and stress due to periodic excitation or impulsive loading. Also, all of the above responses vary inversely with the square of the panel edge dimension. Responses for square panels with different edge dimensions but with all edges fixed can be scaled from the corresponding curve in Fig. 4.45. The significance of the scaling discussed above is that the curves in Fig. 4.45 can be used directly to size square strip hybrids for preliminary design purposes.

The effects of a multitude of parameters, inherent in composites, on the structural response and/or performance of composite structures, and/or structural components are difficult to assess in general. These parameters include several fiber properties (transverse and shear moduli), in situ matrix properties, empirical or correlation factors used in the micromechanical. equations, stress allowables (strengths), processing variables, and perturbations of applied loading conditions. The difficulty in assessing the effects of these parameters on composite structural response arises from the fact that each parameter cannot be isolated and its effects measured independently of the others. Of course, the effects of single loading conditions can be measured independently. However. small perturbations of several sets of combined design loading conditions are not easily assessed by measurement.

An alternate approach to assess the effects of this multitude of parameters is the use of optimum design (structural synthesis) concepts and procedures. In this approach the design of a composite structure is cast as a mathematical programming problem. The weight or cost of the structure is the objective (merit) function that is minimized subject to a given set of conditions. These conditions may include loading conditions, design variables that are allowed to vary during the design (such as fiber type, ply angle and number of plies), constraints on response (behavior) variables (such as allowable stress, displacements, buckling loads, frequencies, etc.) and variables that are assumed to remain constant (preassigned parameters) during the design.

The preassigned parameters may include fiber volume ratio, void ratio, transverse and shear fiber properties, in situ matrix properties, empirical or correlation factors, structure size and design loads. Once the optimum design for a given structural component has been obtained, the effects of the various preassigned design parameters on the optimum design are determined using sensitivity analyses. Each parameter is perturbed about its preassigned value and the structural component is re-optimized. Any changes in the optimum design are a direct measure of the effects of the parameter being perturbed. This provides a formal approach to quantitatively assess the effects of the numerous parameters mentioned previously on the optimum design of a structural component and to identify which of the parameters studied have significant effects on the optimum design of the structural component of interest. The sensitivity analysis results to be described subsequently were obtained using the angle plied composite panel and loading conditions as shown in Fig. 4.49.

Sensitivity analyses are carried out to answer, for example, the following questions:

1. What is the influence of the preassigned filament elastic properties on the composite optimum design?

Figure 4.49 Schematic of composite panel used in structural synthesis

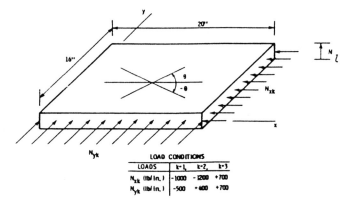

LOADS	k-1	k-2	k-3
N_{xk} (lb/in.)	-1000	-1200	+700
N_{yk} (lb/in.)	-500	+400	+700

LOAD CONDITIONS

2. What is the influence of the various empirical factors/correlation coefficients on the composite optimum design?

3. Which of the preassigned parameters should be treated with care or as design variables for the multilayered-filamentary composite?

4. What is the influence of applied load perturbations on the composite optimum design?

The load system for the standard case consisted of three distinct load conditions as specified in Fig. 4.49. The panel used is 20 in. × 16 in. made from an $[(+\theta)_n]_s$. angle plied laminate. The influence of the various preassigned parameters and the applied loads on optimum designs is assessed by sensitivity analyses. The sensitivity analyses consist of perturbing the preassigned parameters individually by some fixed percentage of that value which was used in a reference (standard) case. The results obtained were compared to the standard case for comparison and assessment of their effects.

Introductory approaches have been described to formally evaluate design concepts for select structural components made from composites including intraply hybrid composites and strip hybrids. These approaches consist of structural analysis methods coupled with composite micromechanics, finite element analysis in conjunction with composite mechanics, and sensitivity analyses using structural optimization. Specific cases described include:

1. Hybridizing ratio effects on the structural response (displacement, buckling, periodic excitation and impact) of a simply supported beam made from intraply hybrid composite.

2. Strip modulus effects on the structural response of a panel made

Figure 4.50 Graphite fiber RP automobile (Courtesy of Ford Co.)

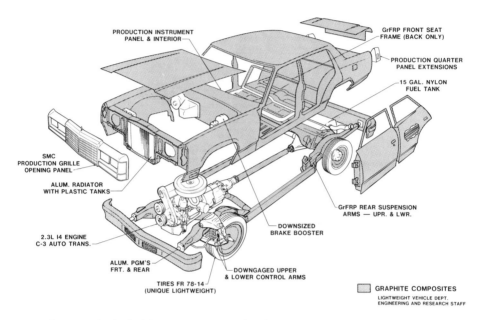

from strip hybrid composite and subjected to static and dynamic loading conditions.

3. Various constituent material properties, fabrication processes and loading conditions effects on the optimum design of a panel subject to three different sets of biaxial in-plane loading conditions.

Automobile

Plastics play a very important role in vital areas of transportation technology by providing special design considerations, process freedom, novel opportunities, economy, aesthetics, durability, corrosion resistance, lightweight, fuel savings, recyclability, safety, and so on. Designs include lightweight and low cost principally injection molded thermoplastic car body to totally eliminate metal structure to support the body panels such as the concept in Fig. 4.50. Other processes include thermoforming and stamping. With more fuel-efficiency regulation new developments in lightweight vehicles is occurring with plastics. Plastics used include ABS, TPO, PC, PC/ABS, PVC, PVC/ABS, PUR, and RPs.

Different cars, worldwide have been designed and fabricated such as those that follow. (1) Chrysler's light-weight (50wt% reduction)

Composite Concept Vehicle (CCV) uses large injection molded glass fiber-TP structural body panels with only a limited amount of metal underneath/assembled by adhesive bonding or fusion welding. (2) Ford has plastic parts in its 2001 Explorer Sport Trac sport utility vehicle replaces the steel open cargo area with RP (SMC), and other cars. (3) Daimler-Benz's (Stuttgart, Germany) light-weight 2-seat coupe, called the Smart car, has injection molded outer body panels/unitizes TP body ties together the front fender, outer door panels, front panels, rear valence panels, and wheel arch in one wrap-around package. (4) GM focusing with plastics in their electric vehicle. (5) Asha/Taisun of Singapore producing taxi cabs for China with thermoformed body panels mounted on a tubular stainless steel space frame. NA Bus Industries of Phoenix is delivering buses in USA and Europe with all RP bodies. Brunswick Tech. Inc. of Brunswick, ME produces-weight30 ft RP buses except for the metallic engine. Sichuan Huatong Motors Group's (Chengdu, China) 4-door/5-passenger midsize vehicle all-plastic car, called Paradigm, has glass fiber-TS polyester RP sandwich chassis and thermoformed coextruded ABS body panels/chassis features single thermoformed lower tub and an upper skeleton X-brace roof/monocoque structure where body panels are stitched-bonded to the chassis, forming a unitized structure.

Truck

Since mid 1040s plastics and RPs have been used in trucks and trailers. In use are long plastic floors, side panels, translucent roofs, aeronautical over-the-cabin structures, insulated refrigerated trucks, etc. (that were initially installed on Strick Trailers by DVR during the late 1940s). The lighter weight plastic products permitted trailers to carry heavier loads, conserve fuel, refrigerated trucks traveled longer distance (due to improved heat insulation), etc. Different plastics continued to be used in the different truck applications to meet static and dynamic loads that includes high vibration loads. Pickup trucks make use of 100 lb box containers using TPs and for the tougher requirements RPs are used.

Aircraft

Plastics continue to expand their use in primary and secondary aeronautical structures that include aircraft, helicopters, and balloons, to missiles space structures. Lightweight durable plastics and high performance reinforced plastics (RPs) save on fuel while resisting all kinds of static and dynamic loads (creep, fatigue, impact, etc.) in different and extreme environments. Certain military planes contain up to 60wt%

Figure 4.51 McDonald-Douglas AV-8B Harrier plastic parts (Courtesy of McDonald-Douglas)

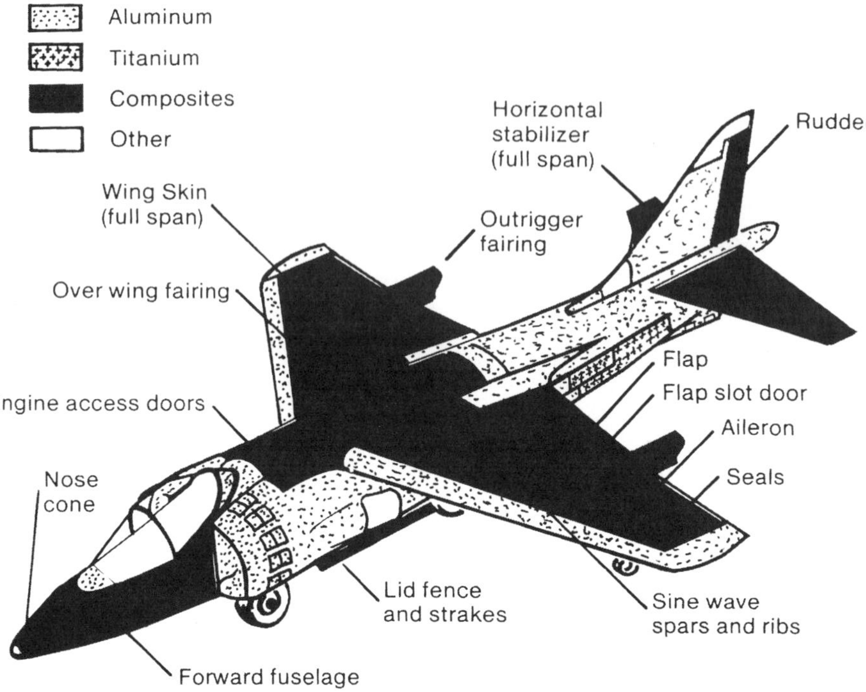

plastics. Other airplanes take advantage of plastics performances such as the McDonald-Douglas AV-8B Harrier with over 26 % of this aircraft's weight using carbon fiber-epoxy reinforced plastics; other plastics also used (Fig. 4.51). Aircraft developments at the present time are extensively using cost-effective reinforced plastics and hybrid composites.

A historical event occurred during 1944 at U. S. Air Force, Wright-Patterson AF Base, Dayton, OH with a successful all-plastic airplane (primary and secondary structures) during its first flight. This BT-19 aircraft was designed, fabricated, and flight-tested in the laboratories of WPEFB using RPs (glass fiber-TS polyester hand lay-up that included the use of the lost-wax process sandwich constructions for the fabrication of monocoque fuselage, wings, vertical stabilizer, etc. Sandwich (cellular acetate foamed core) construction provides meeting the static and dynamic loads that the aircraft encountered in flight and on the ground. This project was conducted in case the aluminum that was used to build airplanes became unavailable. The wooden airplane, the Spruce Goose built by Howard Hughes was also a contender for replacing aircraft aluminum.

Extensive material testing was conducted to obtain new engineering data applicable to the loads the sandwich structures would encounter;

Figure 4.52 Example of orientation of fibers (fabrics) in the all-plastic airplane wing construction

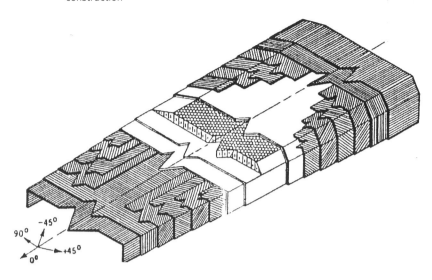

data was extrapolated for long time periods. Short term creep and fatigue tests conducted proved to be exceptionally satisfactory. Later 50 of this type of aircraft were built by Grumman Aircraft that also resulted in more than satisfactory technical performance going through different maneuvers.

In order to develop and maximize load performances required in the aircraft structures, glass fabric reinforcement laminated construction (with varying thickness) was oriented in the required patterns (Chapter 2). Fig. 4.52 shows an example of the fabric lay-out pattern for the wing structure. It is a view of a section of the wing after fabrication and ready for attachments, etc.

Developments of aircraft turbine intake engine blades that started during the early 1940s may now reach fulfillment. Major problem in the past has been to control the expansion of the blades that become heated during engine operation. The next generation of turbine fan blades should significantly improve safety and reliability, reduce noise, and lower maintenance and fuel costs for commercial and military planes because engineers will probably craft them from carbon fiber RP composites. Initial feasibility tests by University of California at San Diego (UCSD) structural engineers, NASA, and the U.S. Air Force show these carbon composite fan blades are superior to the metallic, titanium blades currently used.

Turbine fan blades play a critical role in overall functionality of an airplane. They connect to the turbine engine located in the nacelle, a large chamber that contains wind flow to generate more power. These usually 6 ft long blades create high wind velocity and 80% of the plane's thrust.

It is reported that the leading cause of engine failure is damaged fan blades. Failure may occur from the ingestion of external objects, such as birds, or it may be related to material defects. If it's a metallic blade and it breaks, it can tear through the nacelle as well as the fuselage and damage fuel lines and control systems. When this happens, the safety of the aircraft and its passengers is threatened, and the likelihood of a plane crash increases.

In contrast, if an RP blade breaks, it simply crumbles to bits and does not pose a threat to the structure of the plane. However, breakage is less likely because composite materials are tougher and lighter than metallic blades and exhibit better fatigue characteristics. A multiengine plane can shut down an engine and continue to fly if a blade is lost and no other damage has occurred. A composite blade disintegrates into many small pieces because it is really just brittle graphite fibers held together in a plastic. A titanium blade, however, will fail at the blade root, causing large, 4- to 6-foot blades to fly through the air.

As designed, the RP blades are essentially hollow with an internal rib structure. These rib like vents direct, mix, and control airflow more effectively which reduces the amount of energy needed to turn the blades and cuts back on noise. Most engine noise actually comes from wind turbulence that collides with the nacelle. By directing air out the back of the fan blades, the noise can be reduced by a factor of two. And by drawing more air into the blades, engine efficiency is improved by 20%.

There also exists embedded elastic dampening materials in the blades, which minimizes vibrations to improve resiliency. Because the blade is lighter and experiences lower centrifugal forces, further enhanced the blade's durability occurs. Small-scale wind tunnel tests show they last 10 to 15 times longer than any existing blade. The No. 1 maintenance task is the constant process of taking engines apart to check the blades.

These new blades should lend themselves to more efficient production techniques. If you use titanium, you need to buy a big block of it and machine it down to size, wasting a lot of material. As reported, this is very time consuming, and one has to worry about thermal warping. The RP allows for mass production. It is fabricated into a mold, making the process more precise and ensuring the blades are identical. NASA will test the new blades in large-scale wind tunnels at the NASA Glenn Research Center in Cleveland. If successful, they could see installation by 2004.

Over the years innovations in aircraft designs have given rise to more new plastic developments and have kept the plastics industry profits at a higher level than any other major market principally since they can meet different load and environmental conditions. Virtually all plastics have received the benefit of the aircraft industry's uplifting influence.

Marine Application

From ships to submarines to mining the sea floor worldwide, certain plastics and RPs can survive the sea environment. This environment can be considered more hostile than that on earth or in space. For water surface vehicles, many different plastics have been used in designs in successful products in both fresh and the more hostile seawater. Boats have been designed and fabricated since at least the 1940s. Anyone can now observe that practically all boats, at least up to 9 m (30 ft) are made from RPs that are usually hand lay-up moldings from glass rovings, chopper glass pray-ups, and/or glass fiber mats with TS polyester resin matrices. Because of the excellent performance of many plastics in fresh and sea water, they have been used in practically all structural and nonstructural applications from ropes to tanks to all kinds of instrument containers.

Boat
In addition to their use in boat hull construction, plastics and RPs have been used in a variety of shipboard structures (internal and external). They are used generally to save weight and to eliminate corrosion problems inherent in the use of aluminum and steel or other metallic constructions.

Plastic use in boat construction is in both civilian and military boats [28 to 188 ft. (8.5 to 55 m)]. Hulls with non-traditional structural shapes do not have longitudinal or transverse framing inside the hull. Growth continues where it has been dominating in the small boats and continues with the longer boat boats. The present big boats that are at least up to 188 ft long have been designed and built in different countries (USA, UK, Russia, Italy, etc.). In practically all of these boats low pressure RP molding fabrication techniques were used.

Examples of a large boat are the U.S. Navy's upgraded minehunter fleet, the "Osprey" class minehunter that withstands underwater explosions. Design used longitudinal or transverse framing inside the piece hull. It has a one piece RP super structure. Material of construction used was glass fiber-TS polyester plastic. The designer and fabricator was Interimarine S.P.A., Sarzana, Italy. The unconventional,

Figure 4.53 Examples of materials for deep submergence vehicles

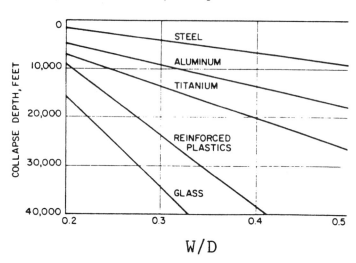

unstiffened hull with its strength and resiliency was engineered to deform elastically as it absorbs the shock waves of a detonated mine. Its design requirements included to simplify inspection and maintenance from within the structure.

Underwater Hull

On going R&D programs continue to be conducted for deep submergence hulls. Materials of construction are usually limited to certain steels, aluminum, titanium, glass, fiber RPs, and other composites (Fig. 4.53). There is a factor relating material's strength-to-weight characteristics to a geometric configuration for a specified design depth. Ratio showing the weight of the pressure hull to the weight of the seawater displaced by the submerged hull is the factor referred to as the weight displacement (W/D) ratio. Submergence materials show the variation of the collapse depth of spherical hulls with the weight displacement of these materials. All these materials, initially, would permit building the hull of a rescue vehicle operating at 1800 m (6000 ft) with a collapse depth of 2700 m (9000 ft).

When analyzing materials for an underwater search vehicle operating at 6000 m (20,000 ft) with collapse depth of 9000 m (30,000 ft), metals are not applicable. Materials considered are glass and RP. The strength-to-weight values for metals potentially are not satisfactory. One of the advantages of glass is its high compressive strength; however, one of its major drawbacks is its lack of toughness and destructive effect if any twist, etc. occurs other than the compression load. It also has difficulty if the design requires penetrations and hatches in the glass hull. A solution could be filament winding RP around the glass or using a tough plastic skin.

These glass problems show that the RP hull is very attractive on weight-displacement ratio, strength-weight ratio, and for its fabrication capability. By using the higher modulus and lower weight advanced designed fibers (high strength glass, aramid, carbon, graphite, etc.) additional gains will occur.

Depth limitations of various hull materials in near-perfect spheres superimposed the familiar distribution curve of ocean depths. To place materials in their proper perspective, as reviewed, the common factor relating their strength-to-weight characteristics to a geometric configuration for a specified design depth is the ratio showing the weight of the pressure hull to the weight of the seawater displaced by the submerged hull. This factor is referred to as the weight displacement (W/D) ratio. The portions the vehicles above the depth distribution curve correspond to hulls having a 0.5 W/D ratio; portion beneath showing the depth attainable by heavier hulls with a 0.7 W/D.

Based on test programs the ratio of 0.5 and 0.7 is not arbitrary. For small vehicles they can be designed with W/D ratios of 0.5 or less, and vehicle displacements can become large as their W/D approach 0.7. By using this approach these values permits making meaningful comparisons of the depth potential for various hull materials. With the best examination data reveals that for the metallic pressure-hull materials, best results would permit operation to a depth of about 18,288 m (20,000 ft) only at the expense of increased displacement. RPs (those with just glass fiber-TS polyester plastic) and glass would permit operation to 20,000 ft or more with minimum displacement vehicles.

The design of a hull is a very complex problem. Under varying submergence depths there can be significant working of the hull structure, resulting in movement of the attached piping and foundation. These deflections, however slight, set up high stresses in the attached members. Hence, the extent of such strain loads must be considered in designing attached components.

Missile and Rocket

Different plastics, particularly high performance plastics and RPs are required in missiles (Fig. 4.54) and rockets as well as outer space vehicles. Parts in a missile are very diverse ranging from structural and nonstructural members, piping systems, electrical devices, exhaust insulators, ablative devices, personnel support equipment, etc.

Figure 4.54 Missile in flight includes the use of plastics

Electrical/Electronic

With the diverse electrical properties of plastics, extensive use of plastics has been made since the first plastics was produced. Plastics permits the operation of many electrical and electronic devices worldwide. As it has been said many times most of the electrical/ electronic equipment and devices used and enjoyed today would not be practical, economical, and/or some even possibly exist without plastics. Plastics offer the designer a great degree of freedom in the design and particularly the fabrication of products requiring specific electrical properties and usually requiring special and accurately fabricated products. Their combination of mechanical and electrical properties makes them an ideal choice for everything from micro electronic components and fiber optics to large electrical equipment enclosures.

Development of many different polymers and plastic compounds (via additives, fillers, and reinforcements) continues to expand the use of plastics in electrical applications. By including fillers/additives, such as glass in plastics, electrical properties can considerably extend performances of many plastics (Fig. 4.55).

The electrical properties of plastics vary from being excellent insulators to being quite conductive in different environments. Depending on the application, plastics may be formulated and processed to exhibit a single property or a designed combination of electrical, mechanical, chemical,

Figure 4.55 Dielectric constant

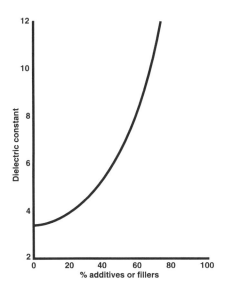

thermal, optical, aging properties, and others. The chemical structure of polymers and the various additives they may incorporate provide compounds to meet many different performance requirements.

Plastic provides ideas for advancing electrical and electronic systems from conducting electricity to the telephone to electronic communication devices. Thousands of outstanding applications use plastics in electrical products. The users' and designers' imaginations have excelled in developing new plastic products.

Shielding Electrical Device

With the extensive use of plastics in devices such as computers, medical devices, and communication equipment the issue of electromagnetic compatibility (EMC) exists that in turn relate to electromagnetic interference (EMI) and radio-frequency interference (RFI).

EMC identifies types of electrical device's capability to function normally without interference by any electrical device. These devices are designed to minimize risks associated with reasonably foreseeable environmental conditions. They include magnetic fields, external electrical influences, electrostatic discharge, pressure, temperature, or variations in pressure and acceleration, and reciprocal interference with other devices normally used in investigations or treatment.

EMI or RFI as well as static charge is the interference related to accumulated electrostatic charge in a nonconductor. As electronic products become smaller and more powerful, there is a growing need for higher shielding levels to assure their performance and guard against failure. From the past 40 dB shielding, the 60 dB is becoming the normal higher value. There is EMI shielding-effectiveness (SE) that defines the ratio of the incident electrical field strength to the transmitted electrical field strength. Frequency range is from 30 MHz to 1.5 GHz (ASTM D 4935-89).

Many plastics are electrical insulators because they are nonconductive. They do not shield electronic signals generated by outside sources or prevent electromagnetic energy from being emitted from equipment housed in a plastic enclosure. Government regulations have been set up requiring shielding when the operating frequencies are greater than 10 kHz.

The plastic shielding material used may include the use of additives. Designs may include board-level shielding of circuit, bondable gaskets, and locating all electrical circuits in one location so only that section requires appropriate shielding. Designers of enclosures for electronic devices should be aware of changes in EMC that tend to continually develop worldwide.

Conductive plastics provide EMI/RFI shielding by absorbing electromagnetic energy (EME) and converting it into electrical or thermal energy. They also function by reflecting EME. This action ensures operational integrity and EMC with existing standards. Conductive plastics are generally designed to meet specific performance requirements (physical, mechanical, etc.) in addition to EMI/RFI or static control. Often these plastics have to perform structural functions, meet flammability or temperature standards, and provide wear or corrosive resistant surfaces, etc.

The usual plastics alone lack sufficient conductivity to shield EMI and RFI interference. Designers can reduce or eliminate sufficiently electro-magnetic emissions from plastic housings like those of medical devices and computers just by shielding the inner emission sources with metal shrouds in the so-called tin can method. The same effect can be obtained by designing electronics to keep emissions below standard limits or by incorporating shielding into the plastic housing itself. Designers will often employ all these strategies in a single design. What is most important is to attempt to locate all the shielding in a relatively small volume within the larger housing and then tin can it to provide a simplified solution rather than spreading it out.

Every electronic system has some level of electromagnetic radiation associated with it. If this level is strong enough to cause other equipment to malfunction, the radiating device will be considered a noise source and usually subjected to shielding regulations. This is especially true when EMI occurs within the normal frequencies of communication. When the electronic noise is sufficient to cause malfunctioning in equipment such as data processing systems, medical devices, flight instrumentation, traffic control, etc. the results could prove damaging and even life threatening. Reducing the emission of and susceptibility to EMI or radio frequency interference (RFI) to safe levels is thus the prime reason to shield medical devices (and other devices) in whatever type of housing exist, including plastic.

In addition to compounding additives for shielding, there is the technology of applying conductive coatings, such as vacuum systems or paint systems (sprays, etc.). Other methods include the use of conductive foils or molded conductive plastics, silver reduction, vacuum metalization, and cathode sputtering. Although zinc-arc spraying once accounted for about half the market, others have surpassed it. Other conductive coatings are also used. Unlike other shielding methods, conductive coatings are usually applied to the interiors of housings and do not require additional design efforts to achieve external aesthetic goals. All systems offer trade-offs in shielding performance, the physical properties of the plastics, ease in production, and cost.

Designers have to confirm the suitability of a material's shielding performance for each system through such conventional means as screen-room or open-field testing. Each approach to shielding should also be subjected to simulated environmental conditions, to determine the shield's behavior during storage, shipment, and exposure to humidity. Some times comparison of shielding materials becomes difficult. ASTM has a standard that defines the methods for stabilizing materials measurement, thus allowing relative measurements to be repeated in any laboratory. These procedures permit relative performance ranking, so that comparisons of materials can also be made.

Organizations involved in conducting and/or preparing specifications/ standards on the electrical properties on plastics include the Underwriters Laboratories (UL), American Society for Testing and Materials (ASTM), Canadian Standards Association (CSA), International Electrotechnical Commission (IEC), International Organization for Standardization (ISO), and American National Standards Institute (ANSI).

UL has a combination of methods for environmental conditioning and adhesion testing to evaluate various approaches to shielding and to determine the plastic types that are suitable. The primary concern is safety. Should a metalized plastic delaminate or chip off, an electrical short is formed that could cause a fire.

Radome

Radome (radiation dome) is used to cover a microwave electronic communication antenna. It protects the antenna from the environment such as the ground, underwater, and in the air vehicles. To eliminate any transmission interference, it would be desirable not to use a radome since transmission loss of up to 5% occurs with the protective radome cover material. The radome is made to be as possibly transparent to electromagnetic radiation and structurally strong. Different materials can be used such as plastics, wood, rubber-coated air-supported fabric, etc. To meet structural load requirements such as an aircraft radome to ground radomes subjected to wind loads, use is made of RPs that are molded to very tight thickness tolerances. Fig. 4.56 shows a schematic of a typical ground radome that protects an antenna from the

Figure 4.56 Antenna (150ft) protected by a plastic radome

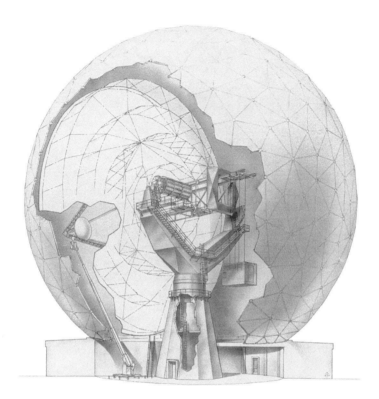

environment (withstand over 150 mph winds and temperatures from arctic to tropical conditions, sand/dirt, etc.) using RP-honeycomb sandwich curved panels This schematic represents protecting in service 150 ft (46 m) antennas. Since that time the most popular is the use of glass fiber-TS polyester RPs. The shape of the dome, that is usually spherical, is designed not to interfere with the radiation transmission.

The use of the secondary load structure RP aircraft radomes have been used since the early 1940s. At that time the problem of rain erosion developed on their front of the radome. It first appeared on the RP "eagle wing" radome located below the B-29 bomber aircraft. It had an airfoil-shaped radome that was 6 m (20 ft) long located about 0.5 m (11 ft) below its wing. On its initial flight over the Pacific Ocean upon encountering rain, the RP radome (and its radar capacity) was completely destroyed. This introduced the era of rain erosion damage to plastics in using a rain erosion elastomeric plastic coating (Chapter 2).

Medical

Plastic devices of all types have become vital in the medical industry. Products range from disposables (medical supplies, drug delivery devices, ointments, etc.) to non-disposables and packages to containers to body parts. Packaged drugs include premeasured single-dose disposable units. The diversified properties and behaviors of plastics have developed into an important market for plastics.

Plastic applications are very diversified ranging from band-aids to parts of the heart. Many examples exist in addition to those being reviewed.

1. Abiomed (Danvers, MA) is the manufacturer of the self-contained artificial heart. Of the six patients who received the grapefruit-sized plastic and titanium heart, three remain alive (2003).

2. Developments have found certain plastics existing in the environment of living tissues.

3. The heart valve that is often used in surgery to correct heart deficiencies was a contribution to medicine. In order for it to be successful it required ingenuity in design that would function as a replacement for the mitral valve and to perform as well as the one replaced long enough to justify the risk involved in the operation. It also included using a bioplastic material that would function in the highly complex environment of the human circulatory system without being degraded and without causing harm to the circulatory system.

4. Many developments occur in the area of implants that include the use of plastics. Examples include pacemaker, surgical prosthesis devices to replace limbs, use of plastic tubing to support damaged blood vessels, and work with the portable artificial kidney.

5. There are applications based on the membrane qualities of plastics. They can control such things as the chemical constituents that pass from one part of a system to another, the electrical surface potential in a system, the surface catalytic effect on a system, and in some cases the reaction to specific influences such as toxins or strong radiation.

6. Polyelectrolytes plastics are chemically active. They have been used to make artificial mechanical power muscle materials. They create motion by the lengthening and shortening of fibers made from the chemically active plastic by changing the composition of the surrounding liquid medium, either directly or by the use of electrolytic chemical action. It is no a competitor to thermal energy sources, but it is potentially valuable in detector equipment that

would be sensitive to the changing composition of a water stream or other environmental flow situation.

7. There is the application of extruded high- and low-pressure plastic balloons used in angioplasty catheters. Use of these balloons has extended into many applications such as other catheters (dilatation, heat transfer, laser, cryogenic, etc.), photodynamic therapy devices, drug delivery devices, etc.

8. Plastics in bioscience have potential to be used for mechanical implants in living systems (includes animals and plants) where they can serve as repair parts or as modifications of the system.

9. Kidney applications involve more than the mechanical characteristics of potential plastic use. The kidney machine consists of large areas of a semi-permeable membrane, a cellulosic material in some machines, where the kidney toxins are removed from the body fluids by dialysis based on the semi-permeable characteristics of the plastic membrane. Different plastics are being study for use in this area, but the basic unit is a device to circulate the body fluid through the dialysis device to separate toxic substances from the blood. The mechanical aspects of the problem are minor but do involve supports for the large amount of membrane required.

10. Surgical implants are essentially plastic repair parts for worn out parts of the body. It is possible to conceive of major replacements of an entire organ such as a kidney or a heart by combining the plastic skills with tissue regeneration efforts that may extend life. This is used to time the heart action. Extensively used are plastic corrugated, fiber (silicone or TP polyester) braided aortas.

11. Different customized developments exist and are being used. An example is a porous (foam type) ultrahigh-molecular-weight polyethylene (UHMWPE) that is an FDA-compliant material. Its porosity and pore size can be adjusted per the end-users' requirements. The porosity is uniform in all three (X, Y, Z) axes, which is vital to constant liquid flow in filtration and separation. It is already used in a wide range of medical and laboratory filtration and separation applications by providing customizing processing where chemical purity of the material is maintained (no additives are used). UHMWPE is a chemically resistant, long-chain polymer of ethylene with an extremely high molecular weight of 3.1 million amu or above. Because of its high molecular weight, the polymer maintains abrasion resistance and strength even when it is made porous for filtration or separation applications. The porous form, which can be pleated, is easily handled in manufacturing. It can be precision skived

into films as thin as 0.002 in. (skiving consists of shaving off a thin film layer from a large block of solid plastic, usually a round billet).

Doctors with long, intensive training as basic scientists make them uniquely suitable as product designers. They become involved in designing new products that in turn could require the plastic industry in developing new/modified plastics. With all this action in developing medical products and devices the FDA usually requires approval that takes time.

Surgical Product

The wide range of forms (film, tube, or fiber) and mechanical properties available in plastics continues to make them attractive candidates for such uses (Fig. 4.57). These plastics are required to possess desired physical/mechanical properties and the assurance that they may be successfully utilized in the body. To be successful plastic implants involve the cross-fertilization of different disciplines (chemists, designers, physician, fabricator, etc.).

Developing of surgical implants confronts major problems because human bodies have extremely complex environments. They could be identified as having the most horrible environmental situation on earth. In addition different human bodies have different environmental requirements. Thus what can survive in one body usually does not survive in other bodies. This type of reaction requires extensive R&D to ensure that a medical product can survive and meet its requirements in all human or specific bodies.

Dental Product

Dentures continue to be an important and major market for plastics. Use is made of acrylics (PMMAs) that includes full dentures, partial dentures, teeth, denture reliners, fillings and miscellaneous uses. PMMAs provide strength, exceptional optical properties, low water absorption and solubility, and excellent dimensional stability. In the past plastics used included nitrocellulose, phenol-formaldehyde, and vinyls as denture base materials. Results, however, were not satisfactory because these plastics did not have the proper requisites of dental plastics. Since then, PMMAs have kept their lead as the most useful dental plastics, although many new plastics have appeared and others are being tested.

Todate plastics are not very useful as filling materials with only about 2wt% of all fillings using plastics. The low mechanical properties of

Figure 4.57 Examples of surgical implants (Courtesy of Plastics FALLO)

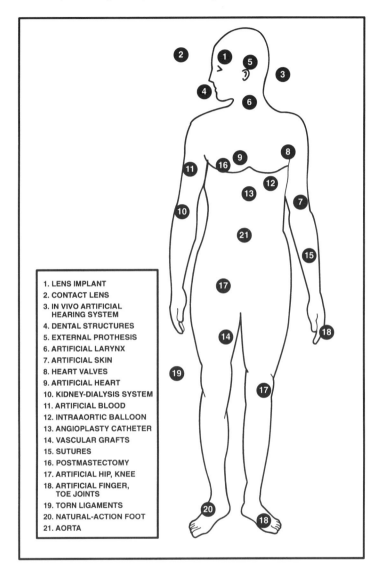

1. LENS IMPLANT
2. CONTACT LENS
3. IN VIVO ARTIFICIAL HEARING SYSTEM
4. DENTAL STRUCTURES
5. EXTERNAL PROTHESIS
6. ARTIFICIAL LARYNX
7. ARTIFICIAL SKIN
8. HEART VALVES
9. ARTIFICIAL HEART
10. KIDNEY-DIALYSIS SYSTEM
11. ARTIFICIAL BLOOD
12. INTRAAORTIC BALLOON
13. ANGIOPLASTY CATHETER
14. VASCULAR GRAFTS
15. SUTURES
16. POSTMASTECTOMY
17. ARTIFICIAL HIP, KNEE
18. ARTIFICIAL FINGER, TOE JOINTS
19. TORN LIGAMENTS
20. NATURAL-ACTION FOOT
21. AORTA

plastics in comparison with metals limit their application where stresses are great. It is interesting to note development efforts has taken place in the use of whiskers for reinforcing dental plastic, metal, and ceramic fillings. Some preliminary test results on the addition of randomly distributed chopped, short whiskers to a coating plastic have reversed the previous proportional loss of strength with powder additions. Although this is far from theoretical, it is already quite significant in that it allows the addition of pigment for coloring purposes and a restoration of the loss of strength with whisker additions.

Health Care

Here they have made many major contributions to the contemporary scene. Health-care professionals depend on plastics for everything from intravenous bags to wheelchairs, disposable labware to silicone body parts, and so on. The diversity of plastics allows them to serve in many ways, improving and prolonging lives [such as a braided, corrugated (Du Pont's Dacron polyester) TP aorta tube].

Thousands of biodegradable plastics have been analyzed. An example is at the Massachusetts Institute of Technology (MIT) test program that may save lives in the form of a medical implant. Tests include its effectiveness as a drug-releasing implant in brain cancer patients. These implants, roughly the size of a quarter, are being placed in patients' brains to release the chemotherapy drug BCNU (Carmustine). Todate these biocompatible implants have been found to be safer than injections, which can cause the BCNU to enter the bone marrow or lungs, where the drug is toxic.

Use is being made of the polyanhydride plastic that was designed shapewise so that water would trigger its degradation but would not allow a drug to be released all at once. The implant degrades from the outside, like a bar of soap, releasing the drug at a controlled rate even as it becomes smaller. The rate at which the drug is released is determined by the surface shaped area of the implant and the rate of plastic degradation that is customized to release drugs at rates varying from one day to many years. This design approach also holds promise for use with different drugs for various other-medical systems. Interesting that in the past these plastics were used with explosives for use in a war. When the plastic degraded by water or sunlight, the device would explode.

Recreation

Plastics provide structurally sound, durability, and safe equipment for sports and recreation facilities. The broad range of properties available from plastics has made them part of all types of sports and recreational equipment worldwide for land, water, and airborne activities. Roller-skate wheels have become abrasion- and wear-resistant polyurethane, tennis rackets are molded from specially reinforced plastics (using glass, aramid, graphite, and/or hybrid fibers), skis are laminated with RPs, RP pole-vaults extensively used to go "higher", and many more.

Appliance

By far most appliances use plastics to take advantage of the properties they inherently provide ranging from aesthetics to structural integrated (corrosion resistance, tough, etc.) to insulation. When examining the plastics in an appliance in addition too different plastics being used, there are the many different fabricating methods used to produce simple to complex parts.

Many success stories exit that include the benefits gained in the household and other places. In the past washing, drying, and ironing clothes was a rigorous, two-day affair involving the filling of metal tubs, scrubbing by hand, hanging clothes to dry, and heating cast-iron flat irons on a stove. With new technology and "plastics" laundry rooms and kitchens worldwide are operating in relatively a few minutes and looking better than ever before.

As one of many thousands of success examples, Milwaukee Electric Tool Corp. in the past found itself on the short end of the age-old supply-and-demand equation. That is, it was unable to keep up with demand for its heavy-duty electric power tools. Problem was that their machining operations could not turn out enough aluminum die-cast motor housings to keep up with market demand. The firm briefly considered what would have been a long-term solution; a state-of-the-art machining center. But a feasibility study showed that capital costs for such a facility would run into hundreds of thousands of dollars, while resulting savings would amount to a few cents per part.

Fortunately, there was the other option of using plastic motor housings. Du Pont agreed to produce plastic prototypes of the housing in Zytel nylon 82G plastic. Prototypes were quickly assembled; then they passed demanding drop tests and other field tests that are standard for Milwaukee Electric tools. When the housings of impact-resistant Zytel passed the tests with no problem, the firm had a new, lower-cost solution to its machining problem: a plastic housing, produced from a production mold that required no machining.

Redesign presented several additional opportunities. Initial target was to replace aluminum die-casting, and thereby eliminate machining as well as deflashing, trimming, and spadoning (a surface treatment that imparts a matte finish). But they also wanted to eliminate as many parts as possible, simplify the assembly, and use a product that worked as well or better than aluminum. Achieving these goals produced some spectacular benefits; parts costs dropped by two-thirds while manufacturing throughput rates increased. Savings in labor, machining, and assembly operations were augmented by lower capital and maintenance costs. As many as one million plastic housings were injection molded without major tool repair or replacement, vs. 100,000 parts for the die-casting operation.

Six parts in the housing were eliminated. Because plastic used is not conductive, designers were able to do away with insulating parts, such as a coil shield that separated the electric brush holder area from the aluminum, and the cardboard insulating sleeve that went between the copper wiring of the field core and the housing. Removing the sleeve had the added benefit of creating better airflow inside the housing, so the motor ran cooler under load. Press-fitting a rear ball bearing into the housing and keeping the bearing securely in place proved to be a major obstacle. The solution was to use eight small ribs inside the rear-bearing pocket. The ribs increase the amount of interference that can be overcome when press-fitting the ball bearing, and keep the bearing in its pocket with a strong, uniform force.

Another concern was achieving overall perpendicularity of the housing face where it fitted with a mating gear case. A molder solved this problem by repeatedly adjusting molded housing dimensions by a few thousands of an inch. Key to this fine-tuning was to establish three adjustment spots; one at each screw hole location. Thus it was much easier to design mating parts so that they sat on the lands, specific points, rather than trying to align a complete surface. Accurately repeating such minute dimensions required batch-to-batch plastic consistency and process control.

This example has been repeated many times over in many electrical devices. Many different electric appliance devices have been design and put in long production runs worldwide.

Furniture

Plastic office furniture, principally just for chairs, panels, and laminates for desks, represents about a $10 billion market or just 3½ wt% of all plastics. Unreinforced and glass reinforced PPs are principally used. Lawn and garden furniture consumes about 2½ wt% plastics also principally using PPs. All these products as well as others have a good growth rate.

An exercise follows in designing and fabricated furniture racks. Their function is to serve as a support that can hold several objects in a desired location for storage or for display. Unless a more specific function is defined, the rack can take on a wide range of possible shapes, structures, and materials subjected to very little damaging loads; racks can fall, be twisted during relocations, and so on. One type of rack is a bookshelf compatible with the environment found in a library, suitable for holding five or more books per foot of shelf in a certain location, and for constant reference use rather than for storage.

In setting requirements for this type rack factors to consider includes defining the environment, set load levels and type of loading situation, and with the library environment provide some idea of the possible aesthetics of the rack. A wide range of design choices remain such as to the material to be used, structure, and shape but they would be limited to those normally used in a library environment. The more accurately and completely the function is defined, the more restricted are the design possibilities and the more detailed the specification requirements become.

This product has to fit its function within the confines of the space in which it is used that could involve how many books it will hold or by stating the size of the supporting rack that will be used. The size can then be decided either by burden or by space restrictions. There will be those where one or the other of these considerations will apply. Also both may apply.

The width of the shelf may be determined by the width of a book. In this example it ranges from 6 to 11 in. (15 to 28 cm). The typical bookshelf is supported at 3 ft (0.9 m) intervals so that the shelf would be sufficiently wide to fit a typical book rack in the library. The shelf will hold about 5 books per foot with each having an average weight of 2 lb (0.91 kg). Maximum load on the shelf becomes 30 lb (13.6 kg). If the shelf were completely filled, it could be considered a distributed load, or it can be considered as a set of discreet loads.

The type of shelf design is another consideration. The shelf can be a solid plate of plastic material, an inverted pan-like structure with reinforcing ribs, a sandwich-type structure with two skins and an expanded core, or even a lattice type sheet that has a series of openings. The choice is usually dictated by a number of factors. An important one is appearance or aesthetics.

The design includes developing its appearance that is to match its performance. There are lattice-type shelf, hanging solid structures from the ceiling, etc. They may all be functionally but may not look appropriate for a bookshelf in a library. A simple plastic beam that will function adequately in terms of strength and stiffness may be rather thin. A shelf of this type can look weak even if it is more than functional. However the solid plate approach results in an un-economical use of material.

One of the requirements included that it look like a wood shelf to match the atmosphere of the library. To produce the desired thickness appearance using plastics either a lipped edge with internal reinforcement can be used or a sandwich-type structure (Fig. 4.10). In either case the displacement of the material from the plane of bending will improve the stiffness efficiency of the product.

In the materials selection properties to be considered or analyzed are static tensile and compressive strength, stiffness (modulus), impact properties, temperature resistance, expansion during temperature changes, and even dynamic properties. In this study the only major concern is the long-term stiffness since this is a statically loaded product with minimum heat and environmental exposure. While some degree of impact strength is desirable to take occasional abuse, it is not really subjected to any significant impact. Using several materials such as PP, glass-filled PS, and PS molded structural foam sandwich type panel material, the design procedure follows to determine the deflection and stress limitations of the material in the several designs.

From a load carrying requirement the rack is to meet a maximum deflection and relate it to the cost of fabrication (material, labor, amortization costs of equipment including tools, and the usual plant costs of overhead, etc.). If the designer does not have this type experience it is best done with the assistance of a reliable fabricator who potentially will make the product.

The various designs and costs can be tabulated and the ones that are the most economical can be determined. It may become evident that the design life can be long and the cost of increasing the design life small, or, alternatively, it may be that the cost of small increments in the design life are quite costly. In the latter case, the design life should be limited to the acceptable minimum.

A value judgment can be made as to the product quality requirements and the final design made to meet requirements. It may included reinforced areas, coating or plating, inserts, and so on. With the product dimension a performance evaluation can be conducted. The colored plastic and fabricating process is selected.

In this evaluation what is being presented appears unwanted complications that are not needed for such a simple design. However when going into a complicated design one can apply this basic approach to the more complicated designs. Thus the next move is to produce prototypes to ensure that the product will meet performance requirements (Chapters 2 and 3). Prototypes made by machining or other simplified model making techniques do not have the same properties as the product made by molding or extrusion or whatever process is to be used. Simplified prototype procedures may reduce trial mold cost and produce adequate test data in some cases. Main value in this study is appearance and feel to determine whether the aesthetics are correct. Any testing has to be done with considerable reservation and caution.

Following testing of the prototypes for appearance and structural integrity field is made to ensure all is well. Locations and individuals selected require closely monitored or controlled environment. Perhaps a requirement was overlooked such as learning that the library shelves are often cleaned with a solution that disintegrates the plastic being used requiring a new analysis. Unless there has been a serious misjudgment by the product development designers, the field tests do not lead to redesign. The results of the field-testing can help define the basis for labeling and the instructions to be used with the product so that it can properly operate in service.

Water filter

Worldwide everybody desires pure drinking water. Pure water is obtainable and literally all around us if we want to convert the ocean waters with their dissolved salts. Areas such as the Middle East, where fresh water is very scarce, ocean salt water has been filtered by different techniques for the past many decades. PVC and/or RP piping is used to direct the flow of the salt and filter water.

Todate conventional converting systems are wasteful energywise and costwise. Plastic membranes are being used in systems that could go into large-scale water recovery systems from the sea. There is the process called reverse osmosis (RO) where water is separated from a concentrated solution of a salt by a semi-permeable membrane. As an example the DuPont Permasep permeators (Figs. 4.58 and 4.59) has pressure that drives the pure water into the solution for dilution. The driving force is the concentration gradient and it is in the form of a pressure that is related to the difference in the vapor pressure of the water and the vapor pressure of the solution at the temperature at which the process takes place. By applying a pressure greater than the osmotic pressure to a solution, the direction of flow of the water is reversed and pure water is removed from the solution.

These permeators come in different product types that are used

Figure 4.58 Water filter through system (Courtesy of DuPont)

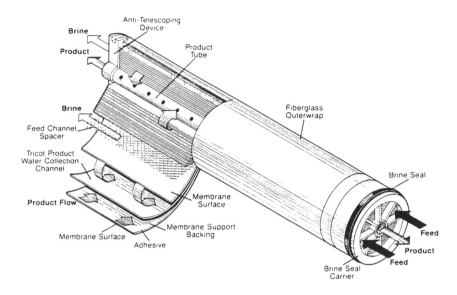

according to the type of water to be treated. There is seawater and brackish water. Brackish water is lower in salinity than normal salt water and higher in salinity than freshwater, ranging from 0.5 to 30 parts salt per 1000 parts water located between sea water and fresh water.

Lumber

Different recyclable scrap plastics (PE, PP, etc.) can be made in what is usually called plastic lumber that is take off from wood lumber. To increase their mechanical properties short glass fibers (as low as 3wt%) are added to the recycled plastic resulting in doubling strength. Most are extruded into boards and other shapes. Another process used is injection molding. Fabricated products compete with wood lumber on land and particularly in the water (fresh, brackish water, and saltwater). Uses includes bridge decks, boat docks, pallets, outdoor furniture, and so on. Compared to wood advantages exist such as being relatively maintenance-free for at least half a century, as opposed to 15 years for treated wood and 5 years for untreated wood. Todate the major disadvantage has been cost. Extensive use is made in applying plastics in wood to improve their structural and decorative properties (compreg, etc.).

Plastic lumber are now available in stores such as Home Depot Inc., USA Plastic Lumber Co. (Boca Raton, Fl), Loewes, and Menard who

Figure 4.59 Water filter deflector block system (Courtesy of DuPont)

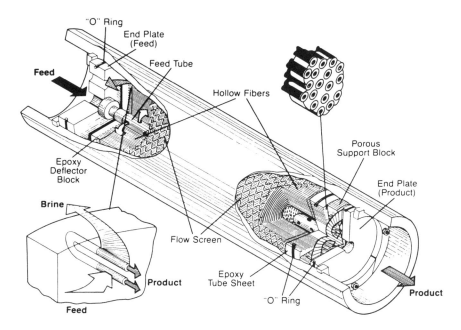

are large home-improvement retailers

Throughout the years there have been used different forms and/or terms of plastic lumber that usually provide significant increases in physical, mechanical, chemical resistant, long life, and other properties. Compressed wood (also called densified wood or laminated wood) is wood that has been subjected to high compression pressure to increase its density with or without plastics. Composition boards are small particles of wood bonded together that usually use a plastic to bond them. Laminated wood is a high pressure bonded wood product composed of layers of wood with plastic such as phenolic as the laminating agent. Modified wood is any wood where its properties have been improved by plastics. Compreg (also called impregnated wood-plastic) is a contraction of compressed plastic impregnated wood.

Metal

The mechanical behavior of metals in service can often be assumed to be that of a linear, isotropic, and elastic solid. Thus, design analysis can be based on classical strength of materials theory extensively reviewed in textbooks and literature. Such uses are most appropriate to components of simple geometric shapes for which standard solutions exist, or for more complex shapes where they can possibly be used for initial approximate design calculations.

For the more complex, and shapes that do not exist, the solution of the applicable elasticity equations may require some form of numerical procedure, such as finite element analysis (FEA) or finite difference analysis (FDA). If design analysis involves frequent consideration of similar problems, then the burden on the designer can be reduced by generating a set of solutions presented as a set of design charts.

An alternative is to provide a method in the form of a computer program for which die design analyst requires minimal familiarity with the design method. However, in critical situations, there may be no alternative to a detailed FEA with evaluation of the manufactured product to ensure meeting performance requirements.

Under appropriate conditions, metal design involves plasticity, creep, and geometric nonlinearity. These topics are treated in standard texts and have been put into computer software. However, such complexities are necessarily modeled in a simple technical format (Chapter 5).

Metal Replacement with Plastic

This review shows how to design a plastic to replace a metal in such a way as to optimize weight and cost reductions. The plastic used is a stampable reinforced thermoplastic sheet (STX) material (STX is a registered tradename). It is composed of a glass fiber-nylon RTP. Replacing metal with plastic while maintaining material efficiency and achieving cost reduction requires a degree of ingenuity and practicality in design. Direct replacement of 21 gauge (0.032 in.) sheet steel having a modulus of 30×10^6 psi would require that an STX material with a modulus of 1×10^6 psi (1/30th that of steel) have a thickness of 0.099 in. (3 times that of steel). These figures are derived as follows:

$$E_1 I_1 = E_2 I_2 \tag{4-54}$$

where E = flexural modulus and I = moment of inertia

$E_1 = 30,000,000 \qquad\qquad E_2 = 1,000,000$

$$I_1 = \frac{bh^3}{12} = 2.7 \times 10^{-6}\, b \qquad I_2 = \frac{h3}{12}\, b$$

$(30,000,000)(2.7 \times 10^{-6})b = 1,000,000(h^3/12)b$

$81.92 = (h^3/12) \times 10^6$

$h^3 = 9.83 \times 10^{-4}$

$h = 0.099''$

While this increase in thickness may have some disadvantages, (space limitations and cost among them), plastics can still be weight and energy effective. The beam problem in Fig. 4.104 illustrates the weight effectiveness achievable by using plastic instead of metal. To achieve comparable deflection, a plastic beam would require three times the height of a steel beam, but importantly, the weight of the plastic would be only 62% of the metal.

To use material effectively and, at the same time, overcome the space utilization of packaging problem that probably would be encountered with a large section beam, the designers can resort to the extremely simple addition of ribbing to a plastic part. What would be a very complex operation in sheet metal is an attractive option in plastic because the fabrication method permits incorporation of ribs in the molding process. Ribbing increases part section modulus with minimum weight increase. In most cases, ribbing can be very simply and easily incorporated into an STX or other RP constructions part with minimal weight addition and without a penalty in molding cost.

This principle of ribbing can be applied to other mechanical strengthening and stiffening requirements of an STX component. For example, to replace a normal sheet metal flange, a plastic flange would be designed, incorporating several modifications that are easy to make in plastic. The flange is thickened and an outer, down or upturned lip is added, increasing the moment of inertia and thus limiting torque type racking; in addition, gussets are added to distribute the bolt loading. Corrugations in the wall are valuable to provide stiffness to reduce vibration noises or tin canning, and in flange design, to increase the effectiveness of bolt loading by moving the wall to the bolt centerline, thereby utilizing the wall section's stiffness to distribute the bolt loading between bolt holes.

In contrast to sheet metal, where the thickness is uniform, STX can be molded with sections of varying thickness. While it is always recommended in design of plastic parts that section thicknesses be uniform, it is always the first rule to be violated. It is this, very design flexibility that becomes advantageous, and designers are usually unable to resist the ability to strategically modify section thickness to meet the load or stress requirements. If stress, temperature, creep, impact or any other environmental conditions require a heavier or specific section thickness, it can be done within prevalent design parameters.

Performance Behavior

All structures under load will deflect in some manner and the resultant moment will create bending and shear stresses. Therefore, it is important that the load analysis be executed correctly. There are several major considerations that must be addressed before any mathematical analysis can be performed. The following conditions directly affect the design stresses that will be obtained by computations.

Moisture Effect
Nylon, the resin base for the first series of STX composites, is a hygroscopic material. Therefore, STX materials exhibit the normal property and dimensional modifications characteristic of nylon as they absorb moisture. Dimensional changes in improperly designed rigidly affixed parts can develop excessive stresses. (Note that the effects of temperature are similar).

Long Term Vs. Short Term Loading
This is a very important consideration since all materials will exhibit creep or relaxation deformation: with plastic materials, it occurs at lower temperatures and stress levels. Hookean stresses, which are instantaneous stresses, are shown in Fig. 4.60 for reference levels only, or for very short-term loadings. However, it is the flexural creep characteristics that govern the usable design stresses, especially when dealing with continuous loading such as is encountered in beams and spring members. The specific curves representing percentage strain vs. time at a given stress of 7,000 psi, is shown in Fig. 4.61.

Figure 4.60 Stress-strain relations

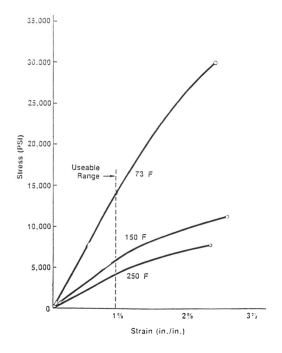

Figure 4.61 Flexural creep resistance at 7,000 psi initial stress

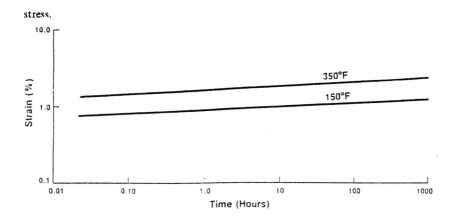

Once these stresses are allowed by the design, the proper geometry (moment of inertia) and deflection can be obtained from the known equations of mechanics of solids. Since stress and deflections are related to both flexural modulus and moment of inertia, the design can now be optimized. It is recommended to use a value for E (modulus) of about 700,000 psi at room temperature and 50% R.H. For other conditions, review Figs. 4.62 and 4.63. The moment of inertia (I) can be changed substantially by the addition of ribs or gussets or a combination of both.

Figure 4.62 Effect of temperature on the flexural modulus

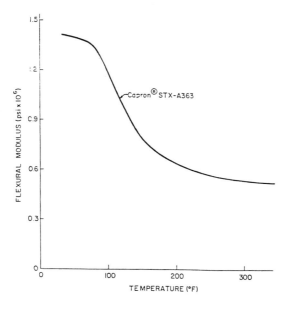

Figure 4.63 Example of temperature effect on flexural properties

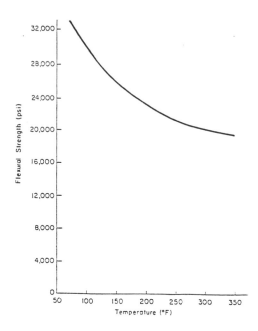

The effectiveness of gussets is very easily analyzed considering that the deflection of any beam is a function of the unsupported length (L) at the third power. By incorporating a gusset at the midpoint of the span, the unsupported length is reduced by half, therefore, the deflection is now decreased by a factor of 8 $[(L/2)^2]$. These simple examples show how both stress and deflection can easily be minimized by taking advantage of the material manufacturing characteristics that allow the forming of ribs and gussets during the molding cycle. Although beam problems were discussed (they are the most commonly used structural members), the same reasoning can be applied to other structural members such as plates or shell sections. Flanges of covers for example, can be handled in that way.

Stress Concentration
The addition of ribs and gussets while reducing deflection can also increase the bending stress if care is not taken in the design stage. Since the ribs or gussets are always in a plane perpendicular to the original surface, the intersection of the planes must comprise a radius to avoid stress concentration. Consequently, it is suggested that a minimum radius of ⅛ in. be incorporated in all designs. With specific reference to ribbing, the shape and size of the ribs must be carefully analyzed to prevent an increase in the bending stress (Machine Design books review this subject).

Coefficient of Expansion

Since the coefficient of expansion of metals and plastics is different, stresses and deformation can result. A given plastic beam clamped to a steel support at both ends by fasteners, will elongate about three times more than the steel support: α steel = 6.5×10^6 in/in/°F, α STX = 16.2×10^6 in/in/°F. If the plastic beam is not allowed to slide under the fasteners then end stresses will occur. These stresses can be computed simply as follows:

$$\Delta L = L \times (\alpha_p - \alpha_s) \times \Delta T \tag{4-5}$$

$$\varepsilon = \frac{\Delta L}{L}$$

$$S = \varepsilon \times E_p$$

$$P_{cr} = \frac{4 \times \pi^2 \times E_p \times I}{L^2} \quad \text{(Euler equation)}$$

where L = distance between bolts = strain, E = flexural modulus, p = plastic, s = steel, and S = expansion stress

Since this stress is directed in the plane parallel to the beam axis, the beam becomes a column fixed at both ends. Applying the standard Euler equation, the critical buckling load (P_{cr}) and corresponding stress can now be computed ($S_{cr} = P_{cr} /A$). The Euler equation comprises several parameters: E, L, I. Of these three, E is fixed, and L (distance between bolts) is usually fixed, (in most cases more bolts means additional costs). I is therefore the only possible variable. Adding one or more ribs will change I and substantially reduce the induced expansion load so that the beam will remain flat and free of distortion since the expansion stress is now much smaller than the critical stress.

Bolt Torque Effect

Since there be STX and other plastic components that are bolted to other components, it is important to consider the effects of the bolt torque, namely the compressive and shear stresses under the head of the bolt. To retain the maximum torque, the following parameters must be considered: (1) STX material, when loaded will tend to relax (long term effects) and (2) based on the fastener requirements, there is a minimum torque suggested to stretch the bolt (bolt preload).

When the torque drops below this minimum value, the fastener tends to loosen. If the STX part is a vessel for containment of fluids, leakage could result at the flanges even with a gasket. The load due to the torque is computed by

$$F = T / (0.2 \times \text{dia. of bolt}) \tag{4-56}$$
$$\text{where T = torque}$$

This load (F) is then used to calculate the compressive and shear stress as shown by the following equations respectively:

$$S_c = \frac{F}{net\ A\ washer} \qquad (4\text{-}57)$$

$$S_s = \frac{F}{\pi \times t \times O.D.\ washer} \qquad (4\text{-}48)$$

where t = thickness of material; and
A = 0.785 (O.D.^{2} – (I.D.^{2}))

These equations show that the use of a flat washer to distribute the load and reduce the respective stresses is recommended (permitted S_c design = 7 000 psi, and permitted S_s design = 6 000 psi). To illustrate the torque and relaxation relationship, see Fig. 4.64. The ordinate represents the ratio of original stress to the actual or time dependent stress and the abscissa is the time cycle. The relaxation curve shows that there is no difference in relaxation between stress levels of 5 000 psi and 1 5 000 psi and that most of the relaxation takes place during the first twenty hours.

Incorporating a gasket between the two surfaces changes the shape of these curves depending upon the thickness and hardness of the gasket material. The analysis is now further complicated and must be handled by the theory of beams on elastic foundations to obtain moment distribution and deflection. Good results can be obtained by using an elastomer such as RTV as gasketing material. The effects of gasket behavior are shown in Fig. 4.65.

Figure 4.64 Compressive relaxation at 73°F (0.2655 in2 compressed area)

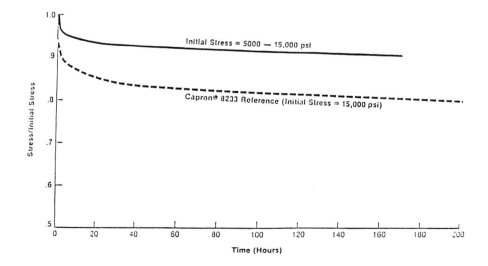

Figure 4.65 Compressive relaxation at 73F between STX without and with gasket

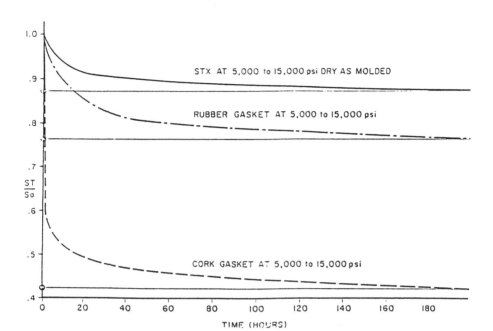

Impact Barrier
The glass mat at the core of the STX sheet was shown to impart some very unusual impact properties to the material. The testing performed in the laboratory (Gardner drop test) and the field results from production components have shown that, in contrast to glass reinforced plastics, the failure mode is elastic rather than brittle. The resultant damage is a hole with little or no crack propagation around the impact area whereas the other reinforced materials will shatter under similar conditions. The STX failure mode applies not only at room temperature, but is equally valid at –40°F. STX impact resistance increases with the thickness of the material, i.e., ¼ in. thick plate will resist much higher impact loads than ⅛ in. thick plate.

However, there is a better solution that result both in lower material costs and lower weight. By keeping the impact area to a minimum thickness and adding ribs, strength is dramatically improved. As an example of this approach, a 0.10 in. thick plate was pierced when impacted by a ½ in. steel ball (1.85 × 10^{-2} lb) travelling at 164 mph (198 in-lb of K.E.). By adding small ribs, 0.10 in. wide × 0.10 in. high spaced 0.5 in. apart, (the plate thickness remained at 0.10 in.), the steel ball had to travel at 272 mph (552 in-lb of K.E.) to duplicate the break resulting in a substantial improvement.

Vehicle Oil Pan

The oil pan is a critical component in auto and truck applications because any failure would be catastrophic. It must withstand high temperature and impact, keep the bolt torque intact for good sealing, resist vibration and abuse, especially during assembly, and in many instances, functions as a structural member. The pan is divided into basic elements and analyzed as follows: (1) sides are considered to be flat plates uniformly loaded with rigid supports. (2) flanges are analyzed as beams on elastic foundations because of the gasket/metal inter-facing, and (3) bottom section is considered as a plate but needs to withstand high impact loads (dynamic loading).

After the bolt torque is established, the washer size must be determined by computation based upon compressive and shear stresses and flange deflection. The flange thickness will also be determined from calculations. After it has been established that these values are safe, the remaining sections such as the sides and bottom can now be designed for thickness and shape based on the values of the static and/or dynamic loads (impact) supplied as part of the input data. If some elements show either high stress or excessive deflection (1% elongation in the elastic range) ribs or gussets can be added and/or the wall thickness increased in selective areas only. History has shown that theoretical analysis yields a design very close to the final manufactured product and meets intended performance requirements dictated by laboratory as well as field testing.

Attachment

There will be cases where attachments become necessary. One of the most frequent types is threaded parts. In most instances, these fasteners are used repeatedly to attach or remove components. To ensure that the reliability and the life of the frequently used joint are maximized, a threaded insert is used rather than threading the plastic material. In this manner, the wall thickness of the insert and plastic part is reduced, or, if a boss is used, it is smaller in diameter and shorter in height compared to its plastic counterpart. Since E_{metal} is much larger than $E_{plastic}$ a higher torque can be applied (shear stress is also much larger).

Extremely close tolerances such as might be necessary for a seal or a bearing may not be within the capabilities of the STX molding process. Machining is not recommended because it would break the nylon surface and expose the glass fibers that then act as wicks for fluid. An aluminum insert that has been finish machined is used as a substitute. Since the ratios of the moduli are quite large (about 20:1 for steel and 10:1 for aluminum), there will be no deformation of the metal insert and only the plastic will be highly stressed. It is most important that these stresses are calculated so that the boss does not split during assembly or that the metallic ring does not become loose because of long term relaxation effects.

Design limitation and constraint

As reviewed throughout this book, designing acceptable products requires knowledge of the different plastics and their processing limitations such as individual advantages and disadvantages. Although there is no limit theoretically to the shapes that can be created, practical considerations must be met such as available and size of processing equipment and cost. These relate not only to the product design, but also the mold or die design, since they must be considered as one entity in the total creation of a usable, economically feasible product.

One of the earliest steps in product design is to establish the configuration that will form the basis on which strength calculations will be made and a suitable material selected to meet the anticipated requirements. During the sketching and drawing phase of working with shapes and cross-sections there are certain design features with plastics that have to be kept in mind to obtain the best cost-performances and avoid degradation of the properties. Such features may be called property detractors or constraints. Most of them are responsible for the unwanted internal stresses that can reduce the available stress level for load-bearing products. Other features may be classified as precautionary measures that may influence the favorable performance of a product if they are properly incorporated.

As an example a weld line(s) can exist in a product that could have met design requirements if the weld line(s) did not develop. The designer did not contemplate the potential for weld line (s). However the person designing the mold took a logical approach to simplify its construction and reduce cycle time to mold the product based on the requirements specified for the product. Result was in reducing the cost of the mold and fabrication time. To meet this objective the design of the mold caused one or more weld lines to develop in the product. With conventional injection molding, molded products can be designed that create unwanted weld line(s). The so-called line forms when two melt flow fronts meet during the filling of an injection mold cavity. This action can also occur during extrusion through a die, etc. Depending on how the weld line forms it could have very little strength and under the most ideal molding conditions it may obtain up to possibly 85% strength retention. To eliminate any problem the product requirements have to account for loss in properties if weld line(s) occur or specify that no weld lines are to occur.

Weld lines are also called knit lines. During processing, such as by injection molding and extrusion, weld lines can occur. They can form during molding when hot melts meet in a cavity because of flow patterns caused by the cavity configuration or when there are two or more gates. With extrusion dies, such as those with "spiders" that hold a center metal core, as in certain pipe dies, the hot melt that is separated momentarily produces a weld line in the direction of the extrudate and machine direction. The results of these weld lines could be a poor bond at the weld lines, dimensional changes, aesthetic damages, a reduction of mechanical properties, and other such conditions.

The top set has a single gate for each specimen, the center set has double gates that are opposite each other for each specimen, and the bottom set has fan gates on the side of each specimen. The highest mechanical properties come with the top set of specimens, because of its melt orientation being in the most beneficial direction. The bottom set of specimens, with its flow direction being limited insofar as the test method is concerned, results in lower test data performance. With the double-gated specimens (the center set) weld lines develop in the critical testing area that usually results in this set's having the potential lowest performance of any of the specimens in this diagram.

Fabricating techniques can be used to reduce this problem in a product. However, the approach used in designing the product, particularly its mold (relocate gates), is most important to eliminate unwanted orientation or weld lines. This approach is no different from that of designing with other materials like steel, aluminum, or glass.

With moldings that include openings (holes), problems can develop. In the process of filling a cavity the flowing melt is obstructed by the core, splits its stream, and surrounds the core. The split stream then reunites and continues flowing until the cavity is filled. The rejoining of the split streams forms a weld line. It lacks the strength properties that exist in an area without a weld line because the flowing material tends to wipe air, moisture, and/or lubricant into the area where the joining of the stream takes place and introduces foreign substances into the welding surface. Furthermore, since the plastic material has lost some of its heat, the temperature for self-welding is not conducive to the most favorable results.

A surface that is to be subjected to load bearing should be targeted not to contain weld lines. If this is not possible, the allowable working stress should be reduced by at least 15%. Under the ideal molding conditions up to about 85% of available strength in the solidified plastic can be developed. At the other extreme where poor process controls exist the weld line could approach zero strength. In fact the two melt fronts could just meet and not blend so that there is relatively a microscopic space. Other problems occur such as influencing aesthetics.

Prior to designing a product, the designer should understand such basic factors as those reviewed in this book. Recognize that success with plastics, or any other material for that matter, is directly related to observing design details.

The important factors to consider in designing can be categorized as follows: shape, part thickness, tolerances, ribs, bosses and studs, radii and fillets, drafts or tapers, holes, threads, colors, surface finishes and gloss levels, decorating operations, parting lines, shrinkages, assembly techniques, production volumes, mold or die designs, tooling and other equipment amortization periods, as well as the plastic and process selections. The order that these factors follow can vary, depending on the product to be designed and the designer's familiarity with particular materials and processes.

5 COMPUTER–AIDED DESIGN

Technology overview

Computer technology requires a completely different methodology of engineering design. It has revolutionized the speed and efficiency of the plastic design functions. The more the entire design function is studied, the more repetitive tasks are uncovered in that function. The computer's ability to perform these tasks untiringly and with blazing speed is the basis for these productivity gains.

The computer continues to provide the engineer with the means to simplify and more accurately develop a design timewise and costwise. It provides a better understanding of the operating requirements for a product design, resulting in maximizing the design efficiency in meeting product requirements. The computer is able to convert a design into a fabricated product providing a faster manufacturing startup. Other benefits resulting from the computer technology include (1) ease of developing and applying new innovative design ideas, (2) fewer errors in drawings; (3) good communications with the fabricator, (4) improved manufacturing accuracy; and (5) a faster response to market demand.

Many of the individual tasks within the overall design process can be performed using a computer. As each of these tasks is made more efficient, the efficiency of the overall process increases as well. The computer is suited to aid the designer by incorporating customer inputs, problem definitions, evaluations, and final product designs.

Computer-aided design (CAD) uses the mathematical and graphic-processing power of the computer to assist the mechanical engineer in the creation, modification, analysis, and display of designs. Many factors have contributed to CAD technology becoming a necessary tool in the engineering world, such as the computer's speed at processing complex equations and managing technical databases. CAD combines the characteristics of designer and computer that are best applicable to the design process.

There is also the combination of human creativity with computer technology that provides the design efficiency that has made CAD such a popular design tool. CAD is often thought of simply as computer-aided drafting, and its use as an electronic drawing board is a powerful tool in itself. The functions of a CAD system extend far beyond its ability to represent and manipulate graphics. Geometric modeling, engineering analysis, simulation, and communication of the design information can also be performed using CAD.

In every branch of engineering, prior to the implementation of CAD, design has traditionally been accomplished manually on the drawing board. The resulting drawing, complete with significant details, was then subjected to analysis using complex mathematical formulae and then sent back to the drawing board with suggestions for improving the design. The same procedure was followed and, because of the manual nature of the drawing and the subsequent analysis, the whole procedure was time-consuming and labor-intensive.

For many decades CAD has allowed the designer to bypass much of the manual drafting and analysis that was previously required, making the design process flow more smoothly and much more efficiently. It is helpful to understand the general product development process as a step-wise process. However, in today's engineering environment, the steps outlined have become consolidated into a more streamlined approach called concurrent engineering. This approach enables teams to work concurrently by providing common ground for interrelated product development tasks.

Product information can be easily communicated among all development processes: design, manufacturing, marketing, management, and supplier networks. Concurrent engineering recognizes that fewer alterations result in less time and money spent in moving from design concept to manufacture and from manufacturing to market. The related processes of computer-aided engineering (CAE), computer-aided manufacturing (CAM), computer-aided assembly (CAA), computer-aided testing (CAT), and other computer-aided systems have become integral parts of the concurrent engineering design approach. Design for manufacturing and assembly methods use cross-disciplinary input from a variety of sources (design engineers, manufacturing engineers, materials & equipment suppliers, and shop floor personnel) to facilitate the efficient design of a product that can be manufactured, assembled, and marketed in the shortest possible period of time.

CAD, CAE, CAM, CAA, and CAT are the directions all types of plastics product design, mold or die making, and the fabricating line. The number and complexity of plastic products being produced are greater every year, but the number of experienced product designers, mold/die designers, and fabricators generally have not kept pace. The answer to this "people power" shortage has been to increase "design to productivity" through the use of CAD/CAE/CAM/CAA/CAT.

Computers and people

Computers have their place but most important is the person involved with proper knowledge in using and understanding its hardware and software in order to operate them efficiently. The computer basically supports rather routine tasks of embodiment and detailed operation rather than the human creative activities of conceptual human operation. Recognize that if the computer can do the job of a designer, fabricator, and others there is no need for these people. The computer is another tool for the designer, fabricator, and others to use. It makes it easier if one is knowledgeable on the computer's software capability in specific areas of interest such as designing simple to complex shapes, product design of combining parts, material data evaluation, mold design, die design, finite element analysis, etc. By using the computer tools properly, the results are a much higher level of product designing and processing that will result in no myths.

Successful products require the combination of various factors that includes sound judgment and knowledge of processing. Until the designer becomes familiar with processing, a fabricator must be taken into the designer's confidence early in development and consulted frequently. It is particularly important during the early design phase when working with conditions such as shapes and sizes. There are certain features that have to be kept in mind to avoid degradation of plastic properties. Most of these detractors or constrains are responsible for the unwanted internal stresses that can reduce the available stress for load bearing purposes.

The industrial production process as practiced in today's business is based on a smooth interaction between regulation technology, industrial handling applications, and computer science. Particularly important is computer science because of the integrating functions it performs that includes the tool manufacture, primary processing equipment, auxiliary equipment, material handling, and so forth up to business management itself. This means that CIM (computer-integrated manufacturing) is very realistic to maximize reproducibility that results in producing successful products.

The use of computers in design and related fields is widespread and will continue to expand. It is increasingly important for designers to keep up to date continually with the nature and prospects of new computer hardware and software technologies. For example, plastic databases, accessible through computers, provide product designers with up-dated property data and information on materials and processes. To keep material selection accessible via computer terminal and a modem, there are design database that maintain graphic data on thermal expansion, specific heat, tensile stress and strain, creep, fatigue, programs for doing fast approximations of the stiffening effects of rib geometry, educational information and design assistance, and more.

Today's software developers are laced with a serious challenge concerning how to produce a safe and reliable product in the shortest possible time frame. This is not a new problem; it has simply been exaggerated in recent years by pressures from the marketplace, and the manufacturing industry certainly is not immune to those pressures. Manufacturers including throwing large budgets into software development tools and manpower have sought many solutions.

Geometric modeling

Geometric modeling is one of the major uses of the CAD systems. It uses mathematical descriptions of geometric elements to facilitate the representation and manipulation of graphical images on the computer's screen. While the central processing unit (CPU) provides the ability to quickly make the calculations specific to the element, the software provides the instructions necessary for efficient transfer of information between user and the CPU.

There are three types of commands used by the designer in CAD geometric modeling. Its first allows the user to input the variables needed by the computer to represent basic geometric elements such as points, lines, arcs, circles, splines, and ellipses. The second is used to transform these elements that include scaling, rotation, and translation. The third allows the various elements previously created by the first two commands to be joined into a desired shape. During the whole geometric modeling process, mathematical operations are at work that can be easily stored as computerized data and retrieved as needed for review, analysis, and modification.

There are different ways of displaying the same data on the CRT (cathode ray tube) screen, depending on the needs or preferences of the designer. One method is to display the design as a 2-D representation of a flat object formed by interconnecting lines. Another method displays the design as a 3-D view of the product. In 3-D representations, there are the four types of modeling of wireframe modeling, surface modeling, solid modeling, and hybrid solid modeling.

The wireframe model is a skeletal description of a 3-D part. It consists only of points, lines, and curves that describe the geometric boundaries of the object. There are no surfaces in a wireframe model. The 3-D wireframe representations can be confusing because all of the lines defining the object appear on the 2-D display screen. This makes it difficult for the viewer to tell whether the model is being viewed from above or below, inside or outside. It is the simplest of the CAD/CAM modeling methods. The simplicity of this modeling method also implies simplicity in the database.

With the surface modeling one defines not only the edge of the 3-D part, but also its surface. One of its major benefits is that it allows mass-related properties to be computed for the product model (volume, surface area, moment of inertia, etc.) and allows section views to be automatically generated. The surface modeling is more sophisticated than wireframe modeling. In surface modeling, there are the two different types of surfaces that can be generated: faceted surfaces using a polygon mesh and true curve surfaces. NLTRBS (Non-Uniform Rational B-Spline) is a B-spline curve or surface defined by a series of weighted control points and one or more knot vectors. It can exactly represent a wide range of curves such as arcs and cones. The greater flexibility for controlling continuity is one advantage of NURBS. It can precisely model nearly all kinds of surfaces more robustly than the polynomial-based curves that were used in earlier surface models.

The computer still defines the object in terms of a wireframe but can generate a surface to cover the frame, thus giving the illusion of a real product. However, because the computer has the image stored in its data as a wireframe representation having no mass, physical properties cannot be calculated directly from the image data. Surface models are very advantageous due to point-to-point data collections usually required for numerical control (NC) programs in CAM applications. Most surface modeling systems also produce the stereolithographic data required for rapid prototyping systems.

An important technique is the solid modeling that defines the surfaces of a product with the added advantages of volume and mass. It takes the surface model one step further in that it assures that the product being modeled is valid and realizable. This allows image data to be used in calculating the physical properties of the final product. Solid modeling software uses one of two methods: constructive solid geometry (CSG) or boundary representation (B-rep). CSG method uses engineering Boolean operations (union, subtraction, and intersection) on two sets of objects to define composite models. B-rep is a representation of a solid model that defines a product in terms of its surface boundaries that are faces, edges, and vertices.

Hybrid solid modeling allows the user to represent a product with a mixture of wireframe, surface modeling, and solid geometry.

By using CAD software, its hidden-line command can remove the background lines of the part in a model. Certain features have been developed to minimize the ambiguity of wireframe representations. These features include using dashed lines to represent the background of a view, or removing those background lines altogether. This hidden-line removal feature makes it easier to visualize the model because the back faces are not displayed. Shading removes hidden lines and assigns flat colors to visible surfaces. Rendering adds and adjusts lights and materials to surfaces to produce realistic effects. Shading and rendering can greatly enhance the realism of the 3-D image.

Design accuracy and efficiency

CAD permits reviewing a design quickly and permits ease in accomplishing the design evaluation. Design accuracy can be checked using automated tolerancing and dimensioning routines to reduce the possibility of error. Layering is a technique that allows the designer to superimpose images upon one another. This can be quite useful during the evaluative stage of the design process by allowing the designer to check the dimensions of a final design visually against the dimensions of stages of the design's proposed fabricator, ensuring that sufficient material is present in preliminary stages for the correct fabrication.

CAD permits checking on interference potential problems. This procedure involves making sure that no two parts of a design occupy the same space at the same time. Automated drafting capabilities in CAD systems facilitate the design presentation, which is the final stage of the design process. CAD data, stored in computer memory, can be sent to a pen plotter or other hard-copy device to produce a detailed drawing quickly and easily. In the early days of CAD, this feature was the primary rationale for investing in a CAD system.

Drafting conventions, including but not limited to dimensioning, crosshatching, scaling of the design, and enlarged views of parts or other design areas, can be included automatically in nearly all CAD systems. Detail and assembly drawings, bills of materials, and cross-sectioned views of design products are also automated and simplified through CAD. In addition, most systems are capable of presenting as many as six views of the design automatically. Drafting standards defined by a company can be programmed into the system such that all final drafts will comply with the standard.

Documentation of the design is also simplified using CAD. Product data management (PDM) has become an important application associated with CAD. PDM allows companies to make CAD data available inter-departmentally on a computer network. This approach holds significant advantages over conventional data management. PDM is not simply a database holding CAD data as a library for interested users. PDM systems offer increased data management efficiency through a client-server relationship among individual computers and a networked server.

Benefits exist when implementing a PDM system. It provides faster retrieval of CAD files through keyword searches and other search features; automated distribution of designs to management, manufacturing engineers, and shop-floor workers for design review; record keeping functions that provide a history of design changes; and data security functions limiting access levels to design files. PDM facilitates the exchange of information characteristic of the emerging workplace. As companies face increased pressure to provide clients with customized solutions to their individual needs, PDM systems allow an increased level of teamwork among personnel at all levels of product design and manufacturing, cutting the costs often associated with information lag and rework.

Although CAD has made the design process less tedious and more efficient than traditional methods, the fundamental design process in general remains unchanged. As reviewed it still requires human input and ingenuity to initiate and proceed through the many iterations of the process. Nevertheless, CAD is such a powerful, timesaving design tool that it is now difficult to function in a competitive engineering world without such a system in place.

Input/output device

The computer systems all share a dependence on components that allow the actual interaction between computer and users. These electronic components are categorized under two general headings: input devices and output devices. Input devices transfer information from the designer into the computer's central processing unit (CPU) so that the data, encoded in binary sequencing, may be manipulated and analyzed efficiently. Output devices do exactly the opposite. They transfer binary data from the CPU back to the user in a usable (usually visual) format. Both types of devices are required in a CAD system. Without an input device, no information can be transferred to the CPU for processing, and without an output device, any information in the CPU is of little use to the designer because binary code is lengthy and tedious.

Central Process Unit

The computer's central processing unit (CPU) is the portion of a computer that retrieves and executes instructions. The CPU is essentially the brain of a CAD system. It consists of an arithmetic and logic unit (ALU), a control unit, and various registers. The CPU is often simply referred to as the processor. The ALU performs arithmetic operations, logic operations, and related operations, according to the program instructions.

The control unit controls all CPU operations, including ALU operations, the movement of data within the CPU, and the exchange of data and control signals across external interfaces (system bus). Registers are high-speed internal memory-storage units within the CPU. Some registers are user-visible; that is, available to the programmer via the machine instruction set. Other registers are dedicated strictly to the CPU for control purposes. An internal clock synchronizes all CPU components. The clock speed (number of clock pulses per second) is measured in megahertz (MHz) or millions of clock pulses per second. The clock speed essentially measures how fast an instruction the CPU processes.

Software

Today's CAD software is often sold in packages that feature all of the programs needed for CAD applications. These fall into two categories: graphics software and analysis software. Graphics software makes use of the CPU and its peripheral input/output devices to generate a design and represent it on-screen. Computer graphics software, including that used in CAD systems, enables designs to be represented pictorially on the screen, so as the human mind may create perspective, thus giving the illusion of 3-D pictorial on a 2-D screen. Analysis software makes use of the stored data relating to the design and applies them to dimensional modeling and various analytical methods using the computational speed of the CPU.

The electronic drawing board's feature is one of the advantages of CAD. The drawing board available through CAD systems is largely a result of the supporting graphics software. That software facilitates graphical representation of a design on-screen by converting graphical input into Cartesian coordinates along x-, y-, and sometimes z-axes.

Design elements such as geometric shapes are often programmed directly into the software for simplified geometric representation. The coordinates of the lines and shapes created by the user can then be organized into a matrix and manipulated through matrix multiplication, and the resulting points, lines, and shapes are relayed back to the graphics software and, finally, the display screen for simplified editing of designs. Because the whole process can take as little as a few nanoseconds, the user sees the results almost instantaneously.

Some basic graphical techniques that can be used in CAD systems include rotation and translation. All are accomplished through an application of matrix manipulation to the image coordinates. While matrix mathematics provides the basis for the movement and manipulation of a drawing, much of CAD software is dedicated to simplifying the process of drafting itself because creating the drawing line by line, shape by shape is a lengthy and tedious process in itself. CAD systems offer users various techniques that can shorten the initial drafting time. The user must usually specify the variables specific to the desired element. For example, the CAD software might have, stored in the program, the mathematical definition of a circle, square, etc.

There are many off-the-shelf software programs, with many more always arriving, in addition to some companies developing their own. They include product design, processing techniques, mold and die design, management control, storage control, testing, quality control, cost analysis, and so on. The software tasks vary so that if you need a particular program, one should be available. You may not be successful in your selection since you probably did not set up the complete requirements. Remember we do not need humans if the software does all the jobs of product design, mold or die deign, material selection, processing setup, and so on. Software programs are useful tools and can perform certain functions. The key to success is the peoples capability in using what is available, that includes software programs. This section on computers refers to a few programs.

Mathematical models are particularly useful because of the large body of mathematical and computational theory that exists for the study and solution of equations. Based on these theories, a wide range of techniques has been developed. In recent years, computer software programs have been written that implements virtually all of these techniques. Computer software packages are now widely available for both simulation and computational assistance in the analysis and design of control systems.

Programs

Literally thousands of off-the-shelf software programs are available (and more always on the horizon) to meet different requirements such as product/mold/die designing, engineering, processing operations, testing, quality control, cost analysis, and management. They are guides that provide a logical approach that range from training to conducting research. Design software programs allow the fabrication of different designs using different types of plastics and processes. All kinds of solutions to engineering equations and mathematical models applicable to static and dynamic loading conditions are available. There are simulated fabricating process controls that permit processing operators to make changes and see the effects that occur on a fabricated product, such as thickness or tolerance.

The software tasks vary so that if you need a particular program, one should be available or can approximate it. Consider software that can easily accommodate change. The probability is that if you are not successful in your selection, you probably did not set up the complete requirements. Examples of a few software programs follow:

ABAQUS A world leader in advanced finite element analysis program. It is used routinely to solve large, complex engineering problems that typically include nonlinear effects by Hibbitt, Karlsson & Sorensen, Inc., Pawtucket, RI 02860 www.albaqs.com

ABAQUS/MOLFLOW It is an interface between ABAQUS and MOLDFLOW. MOLDFLOW to be reviewed.

Alibre Design 3-D Algor reports that it has added support for the Alibre Design 3-D parametric modeling package from Alibre, Inc. The new software provides capabilities for opening Alibre design assembly and part geometry in Algor, a midplane mesh engine for converting thin solid features in a model to plate or shell elements, and a joint creation utility for quickly adding pin and ball joints to models. Algor reports that its intuitive finite element analysis (FEA) and mechanical event simulation (MES) solutions for Alibre Design geometry support analyses including static stress with linear and non-linear material models, linear dynamics, steady-state and transient heat transfer, steady and unsteady fluid flow, electrostatic, MEMS (Micro Electro Mechanical Systems) simulation, and full multiphysics. Alibre design is described as an affordable, easy-to-use application for mechanical design and collaboration. Algor, Inc. Pittsburgh, PA (tel. 800-48-ALGOR) www.algor.com

CADalog A parts library for AutoCAD users, which includes news and reviews, classified ads, a software store, CAD shareware/freeware, a book store, employment center and CAD sharing center with CAD/CAM/CAE links by Cadalog, Bellingham, WA www.cadalog.com

CAD^*Plus* *SOLID EDGE* Advanced mechanical simulation via finite element analysis by Algor, Inc. Pittsburgh, PA (tel. 800-48-ALGOR) www.algor.com

ContentCentral, 3-D It can quickly find and download solid models of parts in their designs to check compatibility and ensure accuracy. SolidWorks Corp., Concord, MA, 01742 USA (tel. 800-693-9000) www.solidworks.com.

CoCREATE Provides the tools to automatically and securely distribute selected product design and specification data by PlanetCAD Inc., Boulder, CO www.planetcad.com.

COSMIC NASA's software catalog, via the University of Georgia, Computer Software Management and Information Center has over 1300 programs. They include programs on training, management procedures, thermodynamics, structural mechanics, heat transfer/fluid flow, etc.

ISA TechNetwork System sources for many different types of technical information. Instrumentation, Systems, and Automation Society, (tel. 919-549-8411) www.isa.org/techcommunities.

Linux Operating system features stable, multi-tasking, virtual memory, fast networking, and multi-user capability. It can ease networking and software development. jrose@modplas.com.

Prospector Examines and provides tabular, single-point (for preliminary material evaluation) and multi-point data (predict structural performance of a material under actual load conditions) for its 35,000 plastics by IDES Inc., Laramie, WY.

Search Engines (www.altavista.com) (www.dejanews.com) (www.excite.com)

TMconcept This molding and cost optimization (MCO) software from Plastics & Computer, Inc. is designed as a practical working tool for application by any engineer who bears responsibility for a molding project. It provides a rather complete molding simulation with over 300 variables.

WebDirect This business module software is designed to enable manufacturers to collaborate with supply chain partners (suppliers, vendors, customers and employees) and to quickly respond to market demands and provide an environment for real-time interactions over the Internet. The secure employee portal allows employees to view a listing of current employee benefits; access current employee deductions, and make requests or changes. It also identifies what percent of income is going into what accounts and request necessary changes; see current year-to-date totals of gross and net income, as well as year-to-date figures for withholdings and additional user-defined HR information. This IQMS software has enhanced their original WebDirect business module of its Enterprise IQ enterprise resource planning (ERP) and supply chain software. Enterprise IQ software was formerly known as IQWin32. IQMS, Pasco Robles, CA 93446 (tel. 805-227-1122).

Database/General Information

In addition to the databases and general information provided throughout this book (Chapter 6, etc.) other examples follow:

CAMPUS This internationally known database software Computer-Aided Material Selection uses uniform standards of testing methods comparing different plastics available from different material suppliers. It was developed by close cooperation with leading plastics producing companies. Special CAMPUS pages are on their websites, updated each time they finish further testing of present and new materials. Its data can be directly merged into CAE programs. CAMPUS provides comparable property database on a uniform set of testing standards on materials along with processing information. The database contains single-point data for mechanical, thermal, rheological, electrical, flammability, and other properties. Multipoint data is also provided such as secant modulus vs. strain, tensile stress-strain over a wide range of temperatures, and viscosity vs. shear rate at multiple temperatures. Software initially developed by BASF, Bayer, Hoechst, and Hulls; followed with Dow, GE, Ciba, etc. The CAMPUS Plastics Database is a registered trademark of CWFG GmbH, Frankfurt/Main, Germany, tel: +49 241 963 1450, Fax: +49 241 963 1469, *http://www.CAMPUSplastics.com*

CENBASE The database is available on CD-ROM, and contains the equivalent of over 150,000 pages of data. http://www.centor.com/cbmat/

DART A diagnostic software expert system, developed by IBM, which is used to diagnose equipment failure problems. It is unique in that it does not hold information about why equipment fails. Instead, it contrasts the expected behavior with the actual behavior of the equipment in order to diagnose the problem.

DATAPOINT Extensive laboratory equipment and research support services by Datapoint Testing Services, Ithaca, NY 14850 www.datapointlabs.com.

DFMA Design for Manufacture & Assembly provides determinants of costs associated with processes by Boothroyd Dewhurst Inc., Wakefield, RI www.dfma.com

EnPlot This is ASM's analytical engineering graphics software used to transform raw data into meaningful, presentation-ready plots and curves. It offers users a wide array of mathematical functions used to fit data to known curves; includes quadratic Bezier spline, straight-line polynomial, Legendre polynomial, Nth order, and exponential splines.

GAIM This is the Gas-Assisted Injection Molding software from Advanced CAE Technology Inc., Ithaca, NY. GAIM helps overcome the lack of experience with the gas-assisted IM process, helping user evaluate alternative designs and determine the best processing conditions.

Globalability: The Key to International Compliance Is the world your marketplace? If so, consider learning how to identify and comply with regulatory requirements in numerous markets around the globe. Develop and implement a global compliance strategy that will reduce your costs and speed time-to-market. You will get the facts about China's 3C Mark, Japan's DENAN Program, and Europe's CE Marketing along with key information on markets including Australia, Argentina, Mexico, Russian Federation, and South Korea. Get the latest information on product legislation, certification schemes, regional trade arrangements, and international standards development. Underwriters Laboratory, Northbrook, IL, USA (tel. 847-272-8800) www.ul.com.seminars.

Injection Molding Operator IBM's molders training programs.

IBM Patents The IBM Intellectual Property Network (IPN) has evolved into a premier Website for searching, viewing, and analyzing patent documents. The IPN provides you with free access to a wide variety of data collections and patent information. http://www.patents.IBM.com

Maintenance Professional-Main Spirex (Youngstown, OH) provides maintenance and inventory control programs for molders http://www.msdssearch.com/

Mastercam Software updates toolpaths to reflect changes in a model change. It can easily select an entire set of operations from another similar part to apply to the CAD model. CNC Software, Inc., Tolland, CT 06084 USA (tel. 800-228-2877) www. mastercam.com.

MoldCAE Provides CAE solutions for the moldmaking industry and offers information on software, services, ordering capabilities and downloads by MoldCAE, Brampton, Ontario, Canada www. moldcar.com

Moldflow This is a series of software modules to analyze melt flow, cooling, shrinkage, warpage. MoldMaking provides a global information center for moldmaking tips, trends and technologies including events, news and new products and offers subscription capabilities, an online buyers guide, an outsourcing directory, and its new MoldMaker's Forum by MoldMaking Technology magazine, Doylestown, PA www.moldmakingtechnology.com/nbm

MOLDEST Provides product design, mold design, and injection molding process control by Fujitsu Ltd., Tokyo, Japan.

MPI LiTE A maintenance scheduling program from Spirex for injection molders.

Nypro Online Nypro (Clinton, MA) molders training programs that provide basics to technological advances.

PDLCOM Published by the Plastics Design Library, PDLCOM is an exhaustive reference source of how exposure environments influence the physical characteristics of plastics. http://www.nace.org/naceframes/Store/pdlindex.htm

PDM It is for product development management and training as opposed to product data or document management. It extends CAD data to a manufacturing organization's non-design department such as analysis, tooling development, manufacturing/assembly, quality control, maintenance, and sales/marketing.

PennStateCool Program involves corner cooling to warpage analysis.

PICAT Molders training programs from A. Routsis Associates

PLA-Ace Software package from Daido Steel Co., Tokyo, Japan. It provides the basic information that encompasses selections that include a mold base, cavity, and core pin(s).

PLASCAMS Computer-aided materials selector. Access is regulated by user ID and password). RAPRA Technology Ltd. Shawbury, Shrewsbury, Shropshire SY4 4NR, U.K., tel: +44-1939-250-383,

Fax: +44-1939-251-118, http://www.rapra.net

PLASPEC It is a Materials Selection Database tel: 212-592-6570, http://www.plaspec.com

Plastics Design Library The PDL Electronic Databooks (also available in hardcopy) provide properties of thermoplastics, elastomers, and rubbers. The world's largest collection of phenomenological data, information is provided as concise textual discussions, tables, graphs and images on chemical resistance, creep, stress strain, fatigue, tribology, the effects of UV light and weather, sterilization methods, permeability, film properties, thermal aging, effects of temperature. The Databooks are available on a single CD-ROM as a complete set or as individual topics. They are updated annually. William Andrew Inc., NY, http://www.williamandrew.com

Plastics Materials Resources This website can be accessed by members and nonmembers alike; however, there are several areas that have restricted access, i.e., for SPE members only. Society of Plastics Engineers, 14 Fairfield Dr., Brookfield, CT 06804-0403 USA, tel. 1 203 775 8490, Fax 203 775 8490, http://www.4spe.org.

PMP The McGill University, Montreal, Canada PMP Software packages (initially known as CBT) addresses a wide variety of topics associated with plastic materials. They include their introduction, classes/types, processing, technical photographs, and properties (mechanical, physical, electrical, etc.)

POLYMAT Fiz Chemie Berlin, Postfach 12 03 37, D-10593 Berlin, tel: +49 (0)30 / 3 99 77-0, Fax: +49 (0)30 / 3 99 77-134, E-mail: Info@FIZ-CHEMIE.DE, http://www.fiz-chemie.de/en/katalog/

Polymer Search on the Internet This is the RAPRA free internet search engine. The number of plastic-related websites is increasing exponentially, yet searching for relevant information is often laborious and costly. During 1999 RAPRA Technology Ltd., the UK-based plastics and rubber consultancy, launched what is believed to be the first free Internet search engine focused exclusively in the plastics industry. It is called (PSI). It is accessible at www.polymersearch.com. Companies involved in any plastic-related activity are invited to submit their web-site address for free inclusion on PSI. The USA office is RAPRA Technology's USA office is in Charlotte, NC (tel. 704-571-4005).

Polymer Software PC-based polymer research tools. DTW Associates, Inc. P.O. Box 916, Ardmore, PA 19003 USA, tel: 1 610 642 0380, Fax: +1 610 642 2599, http://www.dtwassociates.com

ProHelp EPM Features a powerful shop-floor algorithms scheduler that monitors machines in real time with Windows-based software, updating its drag-and-drop bar charts during production by Mattec Corp., Loveland, OH 45140 (tel. 800-966-1301).

Prospector Web and *Prospector Desktop* The Prospector Web is an interactive database used to find and compare over 35,000 plastic materials. The Prospector Desktop is a disk-based version of the popular Prospector Web. Prospector Desktop also contains multi point data graphs. Available on CD-ROM or diskette for Windows and Macintosh. IDES INC., tel: 800-788-4668/307-742-9227, Fax: 307-745-9339, http://www.idesinc.com/Products_1.htm

RMA Resinate Material Advisor is resin evaluation and selection tool. This version 4.0 operates in conjunction with its 3-D mechanical design software called Autodesk Inventor. RMA streamlines the plastic-material selection process by integrating a material database with leading computer-aided design software. The modeled parts and assemblies can be assigned the correct plastic material properties directly, using a database of more than 13,000 materials. The information then can be used downstream easily without manufacturing inputting the data. Version 4.0 users can access the material database over the Internet. Resinate Corp., Andover, MA, USA.

RUBSCAMS Computer-aided materials selector for elastomers. RAPRA Technology Ltd. Shawbury, Shrewsbury, Shropshire SY4 4NR, UK, tel: +44-1939 250 383, Fax: +44-1939 251 118, http://www.rapra.net

SimTech This molding simulator from Paulson Training Programs (Chester, CT) links injection molding with production floor experience. It is designed to provide realistic setup and problem solving training for setup personnel, technicians, and process engineers.

SpirexLink An inventory control software package from Spirex for your plant's plasticating components.

SpirexMoldFill A comprehensive, time-saving assistance tool for molders from Spirex with the added advantage of a mold filling analysis program built-in.

Tech Connect Interactive troubleshooting software that allows supervisors, operators, and set-up personnel to identify and correct common injection molded part defects. Syscon-PlantStar, South Bend, IN, USA (tel. 574-232-3900), www.plantstar.org.

Topaq This mold pressure analysis system is used in conjunction with the companies Pressurex stres- indicating films. It provides a perspective of the distribution and actual magnitude (psi) of pressure between any contacting or impacting surfaces. Films change colors in proportion to the amount of pressure applied. A

Window-based system scans and interprets the exposed film, rendering high definition, digital enhanced images, and statistical reports. Sensor Products, Inc., East Hanover, NJ 07936 (tel. 973-884-1755)

Troubleshooting IM Problems Molders training programs from SME.

Supply Chain Software

IQMS are developers of ERP (enterprise resource planning) and supply chain management software for plastic processors and other repetitive manufacturers. The ERP 11 growth continues with the introduction of the IQ Human Resources Suite, or IQ HR. Traditional ERP systems include some human resources functionality, but are typically limited by integration issues (getting two different companies' software to talk to each other). Enterprise IQ, however, allows companies to completely manage HR with the same integrated system they use to manage their manufacturing, accounting, production monitoring and customer/vendor relations. This is a first in the ERP software industry.

The expanded IQ HR Suite of modules includes all the functionalities repetitive manufacturers require to effectively manage human resources. Included in the suite are two existing modules: IQ Payroll and IQ Time & Attendance, and one new module, IQ Workforce. IQ Payroll manages employee compensation and planning, including job performance, salary history, benefit management and administration reporting. IQ Time & Attendance quickly and easily allows management to allocate, track and apply time to a work order for a manufactured item, production process, tooling project or preventative maintenance routine.

IQ HR helps complete the supply chain loop for repetitive manufacturers. "For years, the supply chain has been defined solely as vendors and customers," says Terry Cline, IQMS vice-president of operations. "With IQ HR, we give manufacturers' the power to efficiently and effectively manage the third and often-overlooked link in their supply chain: the workforce."

Finite element analysis

The name FEA refers to an object or structure to be modeled with a finite number of elements. FEA can be defined as a numerical technique, involving breaking a complex problem down into small subproblems, via computer models that can be solved by a computer.

The key to effective FEA modeling is to concentrate element details at areas of highest stress. This approach produces maximum accuracy at the lowest cost.

FEA is one of the major advancements in engineering analysis. When first introduced, the cost of computer equipment and FEA software limited its use to high budget projects such as military hardware, space-craft, and aircraft design. With the cost of both computer time and software significantly reduced, FEA began to be used for high volume product designs in such markets as automobiles, large buildings, and civil engineering structures.

During this time period of about two decades, there occurred a dramatic increase in the power of desktop personal computers. Simultaneously, advances in relatively low priced software has made it possible for the designer to use advanced engineering techniques that at one time were restricted to these high budget projects and designs. All this action has permitted more use of desktop computers running FEA software that can be run with relative ease. At present helping this expansion is due to an increase in users and much more competition both in hardware/ software and design projects to meet fast manufactured plastic products.

In its most fundamental form, FEA is limited to static, linear elastic analyses. However, there are advanced finite element computer software programs that can treat highly nonlinear (plastics viscoelastic behavior) dynamic problems efficiently. Important features of these programs include their ability to handle sliding interfaces between contacting bodies and the ability to model elastic-plastic material properties. These program features have made possible the analysis of impact problems that only a few years ago had to be handled with very approximate techniques. FEA have made these analyses much more precise, resulting in better and more optimum designs.

Application

FEA with a computer analysis provides a means to theoretically predict the structural integrity of a product using mathematical geometry and load simulation. A stress analysis can be taken of finite sections for analysis of the forces and loads the part will experience in service. It generates an analysis that shows the force concentrations in the section and determines if the material and design shape selected will meet product performance requirements.

In reviewing mechanical engineering analysis, one can perform using one or two approaches, namely analytical or experimental. Using the analytical method, the design is subjected to simulated conditions, using any number of analytical formulae. By contrast, the experimental approach to analysis, requires that a prototype be constructed and subsequently subjected to various experiments, to yield data that might not be available through purely analytical methods.

There are various analytical methods available to the designer using a CAD system. FEA and static and dynamic analysis are all commonly performed analytical methods available in CAD. FEA is a computer numerical analysis program used to solve the complex problems in many engineering and scientific fields, such as structural analysis (stress, deflection, vibration), thermal analysis (steady state and transient), and fluid dynamics analysis (laminar and turbulent flow).

The FEA method divides a given physical or mathematical model into smaller and simpler elements, performs analysis on each individual element, using the required mathematics. It then assembles the individual solutions of the elements to reach a solution for the model. FEA software programs usually consist of three parts: the preprocessor, the solver, and the postprocessor.

The program inputs are prepared in the preprocessor. Model geometry can be defined or imported from CAD software. Meshes are generated on a surface or solid model to form the elements, Element properties and material descriptions can be assigned to the model. Finally, the boundary conditions and loads are applied to the elements and their nodes. Certain checks must be completed before the analysis calculation. These include checking for duplication of nodes and elements and verifying the element connectivity of the surface elements so that the surface normals are all in the same direction.

To optimize disk space and running time, the nodes and elements should usually be renumbered and sequenced. Many analysis options are available in the analysis solver to execute the model. The FEA stiffness matrices can be formulated and solved to form a stiffness value for the model solution. The postprocessor then interprets the results of the analysis data in an orderly manner. The postprocessor in most FEA applications offers graphical output and animation displays. Vendors of CAD software developing pre- and post processors that allow the user to visualize their input and output graphically.

Designing

There are the practical and engineering approaches used to design products. Both have their important place in the world of design. With experience most products usually use the practical approach since they are not subjected to extreme loading conditions and require no computer analysis. Experience is also used in producing new and complex shaped products usually with the required analytical evaluation that involves minor evaluation of stress-strain characteristics of the plastic materials.

When required the engineering approach is used. It involves the use of applicable to stress-strain static and dynamic load equations and formulas such as those in this book and from engineering handbooks. FEA can help a designer to take full advantage of the unique properties of plastics by making products lighter, yet stronger while at the same time also saving money and time to market. The use of FEA has expanded rapidly over the past decades. Unlike metals, plastics are nonlinear [viscoelastic (Chapters 1 and 2)], so they requite different software for analysis, The early software programs were difficult and complex, but gradually the software for plastics has become easier to use. Graphic displays are better organized and are easier to understand.

FEA consumes less time resulting in shortening the lead-time to less than half. Other advantages include increased accuracy, improving reliability, reducing material costs while reducing the expense of building prototypes and remachining tools. By eliminating excess material, it can save weight. It can simulate what will happen, allowing immediate redesign to prevent premature failure. This capability exists because the computer solves simultaneously hundreds of equations that would take literally years to solve without the computer.

With FEA one constructs a model that reduces a product into simple standardized shapes that are called elements. They are located in common coordinate grid system. The coordinate points of the element corners, or nodes, are the locations in the model where output data are provided. In some cases, special elements can also be used that provide additional nodes along their length or sides. The node stiffness properties are identified. They are arranged into matrices and are loaded into a computer. To calculate displacements and strains imposed by the loads on the nodes the computer processes the applied loads. This modeling technique establishes the structural locations where stresses will be evaluated. A cost-effective model concentrates on the smallest elements at areas of highest stress. This configuration provides greater detail in areas of major stress and distortion, and minimizes computer time in analyzing regions of the component where stresses and local distortions are smaller.

Modeling can set up problems because the process of separating a component into elements is not essentially straightforward. Some degree of personal insight, along with an understanding of how materials behave under strain, is required to determine the best way to model a component for FEA. The procedure can be made easier by setting up a few ground rules before attempting to construct the model. An inadequate model could be quite expensive in terms of computer time. As an example if a component is modeled inadequately for a given problem, the resulting computer analysis could be quite misleading in its prediction of areas of maximum strain and maximum deflection.

For a plane stress analysis, if possible quadrilateral elements should be used. These elements provide better accuracy than the more popular corresponding triangular elements without adding significantly to calculation time. Element size should be in inverse proportion to the anticipated strain gradient with the smallest elements in regions of highest strain. The 2-D and 3-D elements should have corners that are approximately right-angled. They should resemble squares and cubes as much as possible in regions of high strain gradient.

Models used for reinforced plastic (RP) tanks normally range from 1000 to 10,000 elements. Each element is defined by nodes ranging from 3 to 8 per element for a typical shell element. Depending on the type of element, each node will have a given number of degrees of freedom. A 3-D shell element can have 6 degrees of freedom at each node, 3 degrees of translation, and 3 degrees of rotation. Each degree of freedom is described by an equation. Thus, the solution of a finite element evaluation requires the simultaneous solution of a set of equations equal to the number of degrees of freedom.

The RP tank shell element model with 10,000 elements will have more than 50,000 degrees of freedom. Solving such a huge set of equations in a reasonable amount of time, even with the most refined of matrix techniques, requires a great deal of computational power.

Graphics

Viewing many imaginative variations would blunt the opportunity for creative design by viewing many imaginative variations if each variation introduced a new set of doubts as to its ability to withstand whatever stress might be applied. From this point of view the development of computer graphics has to be accompanied by an analysis technique capable of determining stress levels, regardless of the shape of the part. This need is met by FEA.

Structural Analysis

The FEA computer-based technique determines the stresses and deflections in a structure. Essentially, this method divides a structure into small elements with defined stress and deflection characteristics. The method is based on manipulating arrays of large matrix equations that can be realistically solved only by computer. Most often, FEA is performed with commercial programs. In many cases these programs require that the user know only how to properly prepare the program input.

FEA is applicable in several types of analyses. The most common one is static analysis to solve for deflections, strains, and stresses in a structure that is under a constant set of applied loads. In FEA material is generally assumed to be linear elastic, but nonlinear behavior such as plastic deformation, creep, and large deflections also are capable of being analyzed. The designer must be aware that as the degree of anisotropy increases the number of constants or moduli required describing the material increases.

Uncertainty about a material's properties, along with a questionable applicability of the simple analysis techniques generally used, provides justification for extensive end use testing of plastic products before approving them in a particular application. As the use of more FEA methods becomes common in plastic design, the ability of FEAs will be simplified in understanding the behavior and the nature of plastics.

FEA does not replace prototype testing; rather, the two are complementary in nature. Testing supplies only one basic answer about a design that either passed or failed. It does not quantify results, because it is not possible to know from testing alone how close to the point of passing or failing a design actually exists. FEA does, however, provide information with which to quantify performance.

Software Analysis

An important part of the design process is the simulation of the performance of a designed device. As an example a fastener or snap-fit is designed to work under certain static or dynamic loads (Chapter 4). The temperature distribution in an electronic chip may need to be calculated to determine the heat transfer behavior and possible thermal stress. Turbulent flow over a turbine blade controls cooling but may induce vibration. Whatever the device being designed, there are many possible influences on the device's performance.

These types of loads can be calculated using FEA. The analysis divides a given domain into smaller, discrete fundamental parts called elements. An analysis of each element is then conducted using the required mathematics. Finally, the solution to the problem as a whole is determined through an aggregation of the individual solutions of the elements. In this manner, dividing the problem into smaller and simpler problems upon which approximate solutions can be obtained can solve complex problems. General-purpose FEA software programs have been generalized such that users do not need to have detailed knowledge of FEA. A FEA model can be thought of as a system of solid blocks (elements) assembled together. Several types of elements that are available in the finite-element library are FEA packages such as NASTRAN and ANSYS.

When a structure is modeled, individual sets of matrix equations are automatically generated for each element. The elements in the model share common nodes so that individual sets of matrix equations can be combined into a set of matrix equations. This set relates all of the nodal deflections to the nodal forces. Nodal deflections are solved simultaneously from the matrix. When displacements for all nodes are known, the state of deformation of each element is known and stress can be determined through stress-strain relations.

With a 2-D structure problem, each node displacement has three degrees of freedom, one translational in each of x and y directions and a rotational in the (x-y) plane. In a 3-D structure problem, the displacement vector can have up to six degrees of freedom for each nodal point. Each degree of freedom at a nodal point may be unconstrained (unknown) or constrained. The nodal constraint can be given as a fixed value or a defined relation with its adjacent nodes. One or more constraints must be given prior to solving a structure problem.

FEA obtains stresses, temperatures, velocity potentials, and other desired unknown variables in the analyzed model by minimizing an energy function. The law of conservation of energy is a well-known principle of physics. It states that, unless atomic energy is involved, the total energy of a system must be zero. Thus, the finite element energy functional must equal zero. The finite element method obtains the correct solution for any analyzed model by minimizing the energy functional. Thus, the obtained solution satisfies the law of conservation of energy.

Synthesizing design

CAD with FEA is a powerful tool in effectively synthesizing a mechanical system design into an optimized product. A few of these are reviewed. With the dynamic analysis method one combines motion with forces in a mechanical system to calculate positions, velocities, accelerations, and reaction forces on parts in the system. The analysis is performed stepwise within a given interval of time. Each degree of freedom is associated with a specific coordinate for which initial position and velocity must be supplied.

The computer model from which the design is analyzed creates by defining the system in various ways. Generally, data relating to individual parts, the user must supply joints, forces, and overall system coordination, either directly or through a manipulation of data within the software. The results of all of these methods of analyses are typically available in many forms, depending on the needs of the designer. As an example the kinematic analysis and synthesis method studies the motion or position of a set of rigid bodies in a system without reference to the forces causing that motion or the mass of the bodies. It allows engineers to see how the mechanisms they design will function in motion.

This approach gives the designer the ability to avoid faulty designs and also to apply the design to a variety of approaches without constructing a physical prototype. Synthesis of the data extracted from kinematic analysis in numerous approaches of the design process leads to optimization of the design. The increased number of trials that kinematic analysis allows the mechanical engineer to perform will provide results in optimizing the behavior of the resulting product before actual production.

There is the static analysis method that determines reaction forces at the attachment positions of resting mechanisms when a constant load is applied. As long as zero velocity is assumed, static analysis can be performed on mechanisms at different points of their range of motion. Static analysis allows the designer to determine the reaction forces on whole mechanical systems as well as interconnection forces transmitted to their individual joints. The data extracted from static analysis can be useful in determining compatibility with the various criteria set out in the problem definition. These criteria may include reliability, fatigue, and performance considerations to be analyzed through stress analysis methods.

With the experimental analysis one involves fabricating a prototype and subjecting it to various experimental test methods. Although this usually takes place in the later stages of design, CAD systems enable the designer to make more effective use of experimental data, especially where analytical methods are thought to be unreliable for the given model. CAD also provides a useful platform for incorporating experimental results into the design process when experimental analysis is performed in earlier approaches of the process.

CAD special use

The use of CAD data in conjunction with specialized applications falls outside the usual realm of CAD software. However, they provide opportunities for the designer to use the data generated through CAD in innovative techniques that can influence design efficiency. Some of these applications are presented.

Optimization

FEA is a major tool used to identify and solve design problems. Increased design efficiency provided by CAD has been augmented by the application of finite element methods to analysis, but engineers still often use a trial-and-error method for correcting the problems identified through FEA. This method inevitably increases the time and effort associated with design because it increases the time needed for interaction with the computer. Also, solution possibilities are often limited by the designer's personal experiences. As reported by many design optimizations seeks to eliminate much of this extra time by applying a logical mathematical method to facilitate modification of complex designs.

The optimization approaches were difficult to implement in the engineering environment because the process tends to be academic in nature and not viewed as easily applicable to design practices. However, if viewed as a part of the process itself, optimization techniques can be readily understood and implemented in the design process. The objectives and constraints upon the optimization must first be defined. The program then evaluates the design with respect to the objectives and constraints and makes automated adjustments in the design. Because the process is automatic, engineers should have the ability to monitor the progress of the design during optimization, stop the program if necessary, and begin again.

Target for optimization is to minimize or maximize a quality, such as weight or physical size, that is subjected to constraints on one or more parameters. Either the size, shape, or both determines the approach used to optimize a design. Optimizing the size is usually easier than optimizing the shape of a design. Optimizing the thickness of a plate does not significantly change its geometry. However, optimizing a design parameter, such as the radius of a hole, does change the geometry during shape optimization.

Preliminary design data are used to meet the desired design goals through evaluation and revision. Acceptable tolerances must then be entered along with imposed constraints on the optimization. The engineer should be able to choose from a large selection of design objectives and behavior constraints and use these with ease. Also, constraints from a variety of analytical procedures should be supported so that optimization routines can use the data from previously performed analyses.

Although designers tend to find shape optimization more difficult to perform than of size, the use of parametric modeling capabilities in some CAD software minimizes this difficulty. Shape optimization is an important tool in many industries, including shipbuilding, aerospace, and automotive manufacturing. The shape of a model can be designed using any number of parameters, but as few as possible should be used, for the sake of simplicity. If the designer cannot define the parameters, neither design nor optimization can take place. The designer's input is crucial during an optimization program.

CAD Prototyping

Creating physical models can be time-consuming and provide limited evaluation. By employing kinematic (branch of dynamics that deals with aspects of motion apart from considerations of mass and force) and dynamic analyses on a design within the computer, time is saved and often the result of the analysis is more useful than experimental results from physical prototypes. Physical prototyping often requires a great deal of manual work, not only to create the parts of the model, but to assemble them and apply the instrumentation needed as well.

CAD prototyping uses kinematic and dynamic analytical methods to perform many of the same tests on a design model. The inherent advantage of CAD prototyping is that it allows the engineer to fine-tune the design before a physical prototype is created. When the prototype is eventually fabricated, the designer is likely to have better information with which to create and test the model.

Engineers increasingly perform kinematic and dynamic analyses on a CAD prototype because a well-designed simulation leads to information that can be used to modify design parameters and characteristics that might not have otherwise been considered. Kinematic and dynamic analysis methods apply the laws of physics to a computerized model in order to analyze the motion of pans within the system and evaluate the overall interaction and performance of the system as a whole.

One advantage of kinematic/dynamic analysis software is that it allows the engineer to overload forces on the model as well as change location of the forces. Because the model can be reconstructed in an instant, the engineer can take advantage of the destructive testing data. Physical prototypes would have to be fabricated and reconstructed every time the test was repeated. There are many situations in which physical prototypes must be constructed, but those situations can often be made more efficient and informative by the application of CAD prototyping analyses.

Physical models can provide the engineer with valuable design data, but the time required to create a physical prototype is long and must be repeated often through iterations of the process. A second disadvantage is that through repeated iterations, the design is usually changed, so that time is lost in the process when parts are reconstructed as a working model. Too often, the time invested in prototype construction and testing reveals less useful data than expected.

CAD prototyping employs computer-based testing so that progressive design changes can be incorporated quickly and efficiently into the prototype model. Also, with virtual prototyping, tests can be performed on the system or its parts in a way that might not be possible in a laboratory setting. For example, the instrumentation required to test the performance of a small part in a system might disrupt the system itself, thus denying the engineer the accurate information needed to optimize the design. It can also apply forces to the design that would be impossible to apply in the laboratory.

Prototyping and testing capabilities have been enhanced by rapid prototyping systems with the ability to convert CAD data quickly into solid, full-scale models that can be examined and tested. The major advantage of rapid prototyping is in the ability of the design to be seen and felt by the designer and less technically adept personnel, especially when esthetic considerations must come into play. There exists limitations in testing operations for rapid prototyping technology. For example, in systems with moving parts, joining rapid prototype models can be difficult and time-consuming. With a CAD prototyping system, connections between parts can be made with one or two simple inputs. Since the goal is to provide as much data in as little time as possible, use of virtual prototyping before a prototype is fabricated can significantly benefit the design program.

Rapid Prototyping

An important application of CAD technology has been in the area of rapid prototyping. Physical models traditionally have the characteristic of being one of the best evaluative tools for influencing the design process. Unfortunately, they have also represented the most time-consuming and costly stage of the design process. Rapid prototyping, such as stereolithography, addresses this problem, combining CAD data with sintering, layering, or deposition techniques to create a solid physical model of the design or part. This industry is able to produce small-scale production of real products, as well as molds and dies that can then be used in subsequent traditional fabricating methods. The rapid prototyping industry is specializing in the production of highly accurate, structurally sound products to be used in the manufacturing process.

CAD standard and translator

In order for CAD applications to run across systems from various vendors, four main formats facilitate this data exchange: IGES (Initial Graphics Exchange Specification), STEP (Standard for the Exchange of Product Model Data), DXF (Drawing Exchange Format), and ACIS (American Committee for Inter-operable Systems)

IGES is an ANSI standard for the digital representation and exchange of information between CAD/CAM systems. A 2-D geometry and 3-D constructive solid geometry (CSG) can be translated into IGES format. New versions of IGES also support boundary representation (B-rep) solid modeling capabilities. Common translators (IGES-in and IGES-out) function available in the IGES library.

STEP is an international standard. It provides one natural format that can apply to CAD data throughout the life cycle of a product. STEP offers features and benefits that are absent from IGES. STEP is a collection of standards. The user can pull out an IGES specification and get all the data required in one document. STEP can also transfer B-rep solids between CAD systems. STEP differs from IGES in how it defines data. In IGES, the user pulls out the specification, reads it, and implements what it says. In STEP, the implementor takes the definition and moves it through a special compiler that then delivers the code. This process assures that there is no ambiguous understanding of data among implementors.

DXF, developed by Autodesk, Inc. for AutoCAD software, is the de facto standard for exchanging CAD/CAM data on a PC-based system. Only 2-D drawing information can be converted into DXF, either in ASCII or in binary format.

The ACTS modeling function N a set of software algorithms used for creating solid-modeling packages. Software developers license ACTS routines from the developer Spatial Technology Corp. to simplify the task of writing new solid modelers. The key benefit of this approach is that models created using software based on ACTS should run unchanged with other brands of ACTS-based modelers. This eliminates the need to use IGES translators for transferring model data back and forth among applications. ACIS-based packages are commercially available for CAD/CAM and FEA software packages.

Data sharing

In the World of Computers an important subject is the collaboration required in storing and subsequent sharing of information. The technical barriers to this are enormous in the general situation when the users are on different types of computer platforms on different networks, using applications that produce proprietary data types, and so on. The challenge is to make data-sharing possible even when there is heterogeneity of one kind or another among those who need to share data.

There is a consensus that such heterogeneity is the rule, rather than the exception, even within one company or one department and today's collaboration worldwide and cross the boundaries of companies and networks. A variety of techniques and software to share information are now available. They include:

1. *Acrobat*: Software to render the output of any software package into a neutral form, devised by the Adobe Corp., so that it can be viewed without the software that created the output.

2. *Notes*: Forms-based e-mail and database software useful to create shared databases over a wide area network and create workflow applications.

3. *EDI*: Electronic data interchange, a host of formats standardized by various interest groups under ANSI to exchange commercial information of different types among companies who wish to do business with each other electronically.

4. *Web*: The Internet de facto standard, consisting of a transport protocol called HM, a document format description standard called HTML, and a variety of graphic and video standards, so that users can access linked multimedia documents placed on Web servers from anywhere in the world. Currently, there are many application packages, such as databases and spreadsheets that are readily accessible using the Web interface, without extra work.

5. *PDM*: Product Data Management Systems that store information about products, their decomposition structure, and revision history, in order to support a company's manufacturing, inventory control, maintenance, and other technical activities.

6. *DMS*: Document-management systems that perform much the same as PDMS for the more general class of documents occurring in a company. Also enables fast retrieval by indexes, keywords, document types, and even text search."

7. *ISS*: An information-sharing system that constructs a model of the information lying in different databases, builds gateways to each of them, and provides a single system image to external users who wish to access data from any of the constituent databases.

Engineered personal computer

CAD projects often range from simple 2-D drawings to graphics-intensive engineering applications. Computationally intensive number crunching in 3-D surface and solid modeling, photo-realistic rendering, and finite element analysis applications demand a great deal from a personal computer (PC). Careful selection of a PC for these applications requires an examination of the capabilities of the CPU, RAM capacity, disk space, operating system, network features, and graphics capabilities. The industry advances quickly, especially in microprocessor capabilities.

There are minimum requirements of PC configurations for various CAD applications. This area of development has more capability than what is reviewed here and because of its rapid advances whatever listing exists is continually outdated.

For 2-D drafting applications, a low-end PC is sufficient. This denotes a PC equipped with a 486 processor running on a 16-bit DOS operating system with 16 MB of RAM, or Microsoft Windows with 32 MB of RAM. The operating system needs 4 MB, most drafting applications require at least 8 MB of RAM, and Windows holds as much data as possible in RAM. Eight K of on-board cache memory and 256 K to 512 K of external cache for faster response is recommended as a minimum. A 500-MB, fast SCSI (smaller computer system interface) hard drive is the minimum. SCSI is a type of bus used to support local disk drives and other peripherals. Five hundred-MB is considered minimum because CAD files require approximately 150 MB, and the operating system itself usually requires about 100 MB. A 16 inch high-resolution (1024 × 768), 256-color monitor should also be considered a minimum requirement. CD-ROM drives and fast modems with transfer rates of 28,800 band are essential for non-networked tasks.

For 3-D modeling and FEA applications, a Pentium 100 MHz processor running on a 32-bit Windows NT operating system works significantly better. The minimum memory requirement for Windows NT is 32 MB and the operating system requires 160 MB of hard-drive space. The system should have a 16-K internal cache with 256 K to 512 K of external cache for increased performance. The monitor for these applications should be 21 inch with 0.25 ultrafine dot pitch, high resolution (1600 × 1280), and 65,536 colors Peripheral component interconnect (PCI) local bus (a high-bandwidth, processor-independent bus) and a minimum I-GB SCSI-2 hard drive are also required.

CAD editing

CAD systems offer the engineer powerful editing features that reduce the design time by avoiding all the manual redrawing that was traditionally required. Common editing features are performed on cells of single or conglomerate geometric shape elements. Most of the editing features offered in CAD is transformations performed using algebraic matrix manipulations. Transformation in general refers to the movement or other manipulation of graphical data.

Most CAD systems offer all of the following editing functions, as well as others that might be specific to a program being used: (1) Movement: Allows a cell to be moved to another location on the display screen. (2) Duplication: Allows a cell to appear at a second location without deleting the original location. Rotation: Rotates a cell a given angle about an axis. (4) Mirroring: Displays a mirror image of the cell about a plane. (5) Deletion: Removes the cell from the display and the design data file. (6) Removal: Erases the cell from the display, but maintains it in the design data file. (7) Trim: Removes any part of the cell extending beyond a defined point, line, or plane. (8) Scaling: Enlarges or reduces the cell by a specified factor along x-, y-, and z-axes. (9) Offsetting: Creates a new object that is similar to a selected object at a specified distance. (10) Chamfering: Connects two nonparallel objects by extending or trimming them to intersect or join with a beveled line. (11) Filleting: Connects two objects with a smoothly fitted arc of a specified radius. (12) Hatching: User can edit both hatch boundaries and hatch patterns

CIM changing

The computer-integrated Manufacturing pyramid of the 1980s has been crumbled to make way for a variety of better models for manufacturing information technology in the 2000s. The Supply Chain Operations Reference (SCOR) model; the Manufacturing Execution Systems Association (MESA) model; and the AMR Research's Ready, Execute, Process, Analyze, & Coordinate (REPAC) model all define manufacturing applications from a functional point of view. Meanwhile, you can define manufacturing applications from the point of view of vertical markets, specific implementation models, and a broad range of functional category.

There have been many acronyms and models in the past two decades that describe the topic of manufacturing application software. However, regardless of naming and modeling, manufacturers fundamental needs have not changed significantly. What has changed is the availability of commercial software, experience in applying software applications to manufacturing, and the emergence of standards for applying software and computer technology to manufacturing.

Today, many well-developed tools are available that can be successfully applied to meet the functional needs of manufacturing processes. Experience gained applying software and computers to manufacturing has been well documented, and international standards communicate generally accepted best practices in manufacturing systems integration. Manufacturers today can take advantage of experience gained from early adopter's efforts and apply current technology with a high degree of confidence that the application will successfully meet requirements.

Computer-based training

While a variety of approaches are available to train people such as a classroom with an instructor, books, etc. Each have their place but available and affordable CBT video technology makes possible training a large number of people with ease and efficiently. They provide opportunities to create plants own in-house training programs with ease of updating. However, video-based training can have limitations such as being sequential; teaching units logically follow and build on one another. User attempt to circumvent the predetermined sequence can be both time consuming and frustrating. Also, it is a one-way communication tool with the user maintaining a relatively passive role in the learning process.

A major advantage of CBT is that learning can take place at the conveyance of the consumer. These training methodologies can take a back seat to the multimedia-based training. MMBA (computers and media working together) uses audio, video, text, and graphics to take full advantage of a PC's ability to capture, reconfigure, and display data. They are efficient and effective particularly in the area of technical training such as running complex or dangerous equipment in a safe environment. Real life situations are depicted and the learner asked to respond, etc.

IBM advances computer

IBM researchers have created what they say is the world's first logic-performing computer circuit within a single molecule. This development can lead into a new class of smaller and faster computers that consume less power than today's machines. They made a voltage inverter. It is one of three fundamental logic circuits that are the basis for today's computers; from a carbon nanotube, a tube-shaped molecule of carbon atoms 100,000 times thinner than a human hair.

IBM scientists presented this achievement at the 2001 National Meeting of the American Chemical Society in Chicago. It was the second major research development last year by IBM scientists using carbon nanotubes to make tiny electronic devices. Carbon nanotubes are the top candidate to replace silicon when current chip features just cannot be made any smaller; a physical barrier expected to occur in about 10 to 15 years. Beyond this silicon nanotube electronics may then lead to unimagined progress in computing miniaturization and power.

Artificial intelligence

AI is an interdisciplinary approach to understanding human intelligence that has as its common connection the computer as an experimental vehicle. This definition emphasizes the fact that many disciplines contribute to the field of AI. They include computer science, engineering, business, psychology, mathematics, physics, and philosophy. It uses symbolic pattern-matching methods to describe objects, events, or processes, and to make inferences. There is the aspect of computer science that is concerned with building computer systems that emulate what is commonly associated with human intelligence. A hypothetical machine processing superhuman intelligence for which supposedly all artificial intelligence researchers are striving. It is argued that we ought to be thinking about the desirability of such machines and how we would cope with them.

Plastic Toys–Smart computer

The big market of toys includes the use of "smart" microprocessor-based technology. Foremost player in this innovative technology is the MIT Media Laboratory's Toys of Tomorrow (TOT) consortium that was organized April 1998. Members include Acer, Bandai America, Deutsche Telekom, Energizer, Intel, Disney, LEGO, Mattel, Motorola, Polar Electro Oy, TOMY, and the International Olympic Committee.

This consortium envisions toys of the future that carry new forms of networking into the home. They report that toys will lead the way to bring a home networking technology infrastructure faster than other means. This action will be different than what the consumers have become accustomed to tolerating home computers that have their share of problems.

Computer devices via DNA

DNA moves from biological to material to meet the speed of miniaturization. Ever had the creepy feeling that your computerized machine is taking on a life of its own, your coffeemaker is developing a personality, or your digital stereo is exhibiting its own taste in music? That day may be much closer to reality when DNA is used to build electronic process control circuits. Yes, that's right, deoxyribonucleic acid, known as DNA and long held in awe as the wonder molecule of life, the software of choice for most living organisms, now shows promise for use in creating tiny electronic devices. Building on earlier work: with carbon nanotubes, Swiss physicists Hans-Wemer Fink and Christian Schonenberger, working at the University of Basel in Switzerland, modified a low-energy electron point source (LEEPS) microscope to measure electrical conductivity across a few strands of DNA. In a "Eureka!" moment, they discovered that DNA conducts electricity as well as being a good semiconductor.

The ability to measure DNA's conductance, coupled with existing technology able to make DNA strands of specified length, moves DNA from the biological world into the materials world. The conductance discovery also represents a big step forward in the current push to find ways to build ever-smaller electronic devices. Self-assembly, a manufacturing process whereby wires, switches, and memory elements are chemically synthesized and connected to form working computers or other electronic circuits, is a promising alternative to standard silicon-based semiconductor manufacturing methods, which are reaching a practical limit of miniaturization. And DNA promises to help make self-assembly a reality.

DNA's power lies in directing the assembly, not necessarily in serving as a semiconductor. As reviewed by Dutch physicist Leo Kouwenhoven of the University of Delft, the Netherlands, much better semiconductors already exist. As in the cellular machinery of living things, DNA in nanodevices will be the brain of the outfit. The high fidelity with which the base pairing occurs is the main reason DNA is so promising in electronic self-assembly. By attaching single-stranded short pieces of DNA, called oligomers, to nanocrystals and then allowing the single strand to seek another single strand with its complementary sequence and perhaps with a different type of nanocrystal attached, it is possible to assemble circuits in solution.

It is believed that the immediate next step will be to find appropriate switching functions and then build an integrated circuit structure using DNA molecules with anchors at their ends to assemble a microstructured chip automatically.

Stan Williams, head of basic research at Hewlett, Packard (HP) Laboratories in Palo Alto, CA, reports DNA will be used to create structure in self-assembly, probably more as a tool than a main player. Williams and his colleagues at HP tested principles of self assembly in building their experimental Teramac computer, which operates 100 times faster that a high end single processor workstation. While DNA was not used to create this computer, he believes it will take a minimum of a decade before DNA is incorporated. Williams sees the use of DNA in nanodevices, although in need of more experimental work, as a step forward in finding ways to create many devices within a device that needs to talk to each other. The favored software of life could soon become the favored software of machines, further blurring the disappearing distinctions between the living and the nonliving worlds.

Design via internet

As Tom Rodak (Commerx, Inc.) reported in today's time-constrained workplace, you can spend a great deal of valuable time trying to find the information you need to make product design decisions. Unfortunately, not many have the luxury of time. Unforgiving deadlines and customer demands make the ability to find information quickly a necessity. Over the past few years, the Internet has rapidly evolved as an ideal tool for locating this needed data. However, with the incredible vastness of the Internet, knowing where to go is key to success.

Currently, a growing number of sites cater to the needs of product designers and engineers. From material selection and design software to educational programs and article archives, the Internet can provide a great deal of information at the click of a mouse.

The Plastics Network from Commerx, Inc. (www. plasticsnet.com) features a Sourcing Center that allows users to search for specific plastics related products and services. The site, which provides secure online ordering, enables users to compare vendors of similar products and services to get the best value.

Regarding materials selection, there are a number of company-specific sites on the Internet that allow you to search product lines by brand name, intended application, and properties. These include: GE Plastics (www.plasticsnet.com/ge), Bayer Corporation (www.polymers-usa. bayer.com), BASF (www.basf.com), Polymerland (www.polymerland. com), and M.A. Hanna (www.plasticsnet.com/mahanna). When searching for materials from a multiple number of vendors, there are several online material databases to visit. Some offer free access to information, while others require a fee for their information. Some of these include: IDES (idesinc.com), and PLASPEC (www2.plaspec.com).

In addition, the Material Engineering Center at Dow Plastics offers its PAMS (Processes and Materials Selection) system on the Internet to help designers match material and fabrication requirements with product and economic requirements (www. plasticsnot.com/moc).

Regarding supplier and product selection in addition to materials selection, there are several sites that allow users to locate and interact with suppliers of products and services. Some of these sites support online ordering as well. Developages (www.developages.com) allows users to locate companies that can assist with all areas of product development from design and prototyping through sales and logistics.

Regarding articles, educational information, and Networking in addition to sourcing vendors and selecting materials, the Internet makes it easy to locate article archives, register for educational programs, and network with other professionals.

Many industry trade associations have Web sites that provide a number of resources for designers. For example, the Web site of the PD3 (Product Design and Development Division) of the Society of Plastics Engineers (www-pd3.org) contains a Design Forum or chat area where users can discuss design challenges and exchange advice. They also provide a schedule of educational programs and links to helpful design articles.

The IDSA (industrial Designers Society of America) (wwwidsa.org) provides similar links, as well as opportunities to locate reference materials, job openings, and suppliers.

6 PLASTIC PERFORMANCE

OVERVIEW

Throughout this book many different properties are reviewed. What follows provides additional information on the properties for different plastics. As a construction material, plastics provide practically unlimited benefits to the design of products, but unfortunately, as with other materials, no one specific plastic exhibits all these positive characteristics. The successful application of their strengths and an understanding of their weaknesses (limitations) will allow designers to produce useful and cost efficient products. With any material (plastic, steel, etc.) products fail not because of the material's disadvantage(s). They fail because someone did not perform their design approach in the proper manner to meet product performance requirements. The design approach includes meeting required performance of material and its fabricating process that operate within material and process controllable variables (Chapter 1, Variables).

There is a wide variation in properties among the over 35,0000 commercially worldwide available materials classified as plastics. They now represent an important, highly versatile group of commodity and engineering plastics. Like steel, wood, and other materials, specific groups of plastics can be characterized as having certain properties.

Many plastics (that are extensively used worldwide) are typically not as strong or as stiff as metals and they are prone to dimensional changes especially under load or heat. They are used instead of metals, glass, etc. (in millions of products) because their performances meet requirements. However there are plastics that have very high properties (Fig. 6.1), meet dimensional tight requirements, dimensional stability, and are stronger or stiffer, based on product shape, than other materials.

Figure 6.1 Mechanical and physical properties of materials (Courtesy of Plastics FALLO)

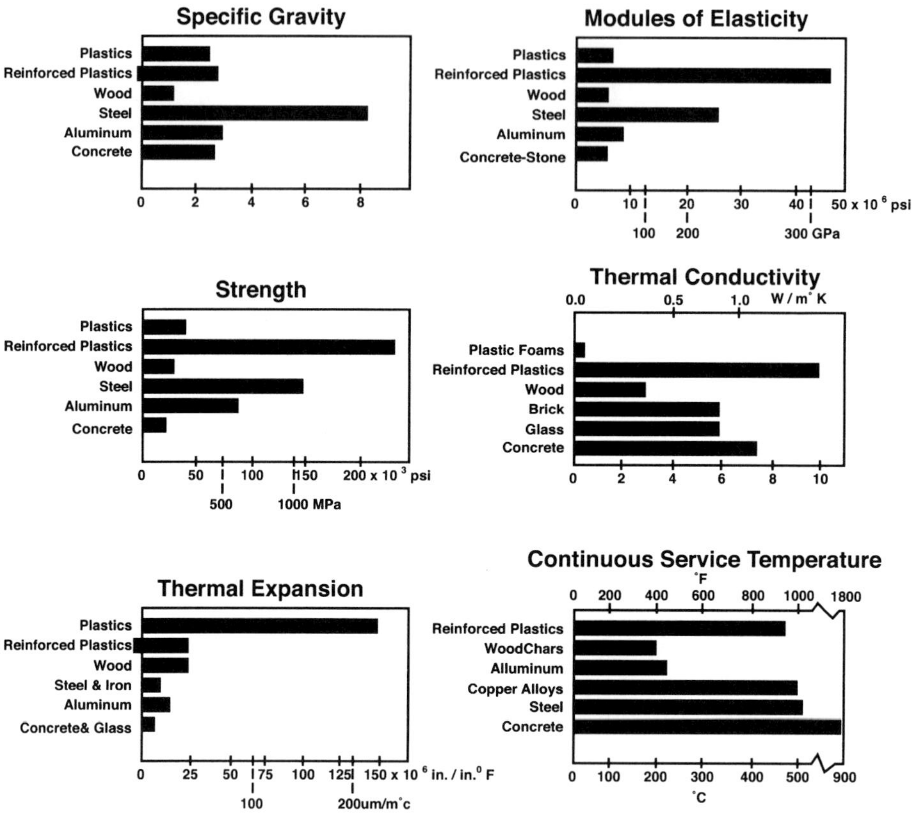

Highly favorable conditions such as less density, strength through shape, good thermal insulation, high degree of mechanical dampening, high resistance to corrosion and chemical attack, and exceptional electric resistance exist for many plastics. There are also those that will deteriorate when exposed to sunlight, weather, or ultraviolet light, but then there are those that resist such deterioration.

For room-temperature applications most metals can be considered to be truly elastic. When stresses beyond the yield point are permitted in the design permanent deformation is considered to be a function only of applied load and can be determined directly from the usual tensile stress-strain diagram. The behavior of most plastics is much more dependent on the time of application of the load, the past history of loading, the current and past temperature cycles, and the environmental conditions. Ignorance of these conditions has resulted in the appearance on the market of plastic products that were improperly designed.

The plastics material properties information and data presented are provided as comparative guides; readers can obtain the latest and more detailed information from suppliers and/or software programs (Chapter 5). Since new developments in plastic materials are always on the horizon it is important to keep up to date. It is important to ensure that the fabricating process to be used to produce a product provides the properties desired. Much of the market success or failure of a plastic product can be attributed to the initial choices of material, process, and cost.

For many materials (plastics, metals, etc.) it can be a highly complex process if not properly approached particularly when using recycled plastics. As an example, its methodology ranges from a high degree of subjective intuition in some areas to a high degree of sophistication in other areas. It runs the gamut from highly systematic value engineering or failure analysis such as in aerospace to a telephone call for advice from a material supplier in the decorative houseware business. As reviewed at the end of this chapter there are available different publications, seminars, and software programs that can be helpful.

Plastics are families of materials each with their own special advantages and drastically different properties. An example is polyethylene (PE) with its many types that include low density PE (LDPE), high density PE (HDPE), High molecular weight PE (HMWPE), etc. The major consideration for a designer and/or fabricator is to analyze what is required as regards to performances and develop a logical selection procedure from what is available.

Recognize that most of the plastic products produced only have to meet the usual requirements we humans have to endure such as the environment (temperature, etc.). Thus there is no need for someone to identify that most plastics cannot take heat like steels. Also recognize that most plastics in use also do not have a high modulus of elasticity or long creep and fatigue behaviors because they are not required in their many respective designs. However there are plastics with extremely high modulus and very long creep and exceptional high performance fatigue behaviors. These type products have performed in service for long periods of time with some performing well over a half-century. For certain plastic products there are definite properties (modulus of elasticity, temperature, chemical resistance, load, etc.) that have far better performance than steels and other materials.

The designer can use plastics that are available in sheet form, in I-beams, or other forms as is common with many other materials. Although this approach with plastics has its place, the real advantage with plastic lies in the ability to process them to fit the design shape, particularly when it comes to complex shapes. Examples include two or more products with mechanical and electrical connections, living hinges, colors, snap fits that can be combined into one product, and so on.

Designing is the process of devising a product that fulfills as completely as possible the total requirements of the user, and at the same time satisfies the needs of the fabricator in terms of cost-effectiveness (return on investment). The efficient use of the best available material and production process should be the goal of every design effort. Product design is as much an art as a science. Guidelines exist regarding meeting and complying with art and science.

Influencing Factor

Design guidelines for plastics have existed for over a century producing many thousands of parts meeting service requirements, including those subjected to static and dynamic loads requiring long life. Basically design is the mechanism whereby a requirement is converted to a meaningful plan. The basic information involved in designing with plastics concerns the load, temperature, time, and environment. As reviewed throughout this book there are other important performance requirements that may exist such as aesthetics, non-permeability, and cost.

In evaluating and comparing specific plastics to meet these requirements, past experience and/or the material suppliers are sources of information. It is important to ensure that when making comparisons the data is available where the tests were performed using similar procedures. Where information or data may not be available some type of testing can be performed by the designer's organization, outside laboratory (many around), and/or possible the material supplier if it warrants their participation (technicalwise and/or potential costwise). If little is known about the product or cannot be related to similar products prototype testing is usually required.

When required, plastics permit a greater amount of structural design freedom than any other material (Chapter 4). Products can be small or large, simple or complex, rigid or flexible, solid or hollow, tough or brittle, transparent or opaque, black or virtually any color, chemical resist or biodegradable, etc. Materials can be blended to achieve different desired properties. The final product performance is affected by interrelating the plastic with its design and processing method. The designer's knowledge of all these variables can profoundly affect the ultimate success or failure of a consumer or industrial product.

For these reasons design is spoken of as having to be appropriate to the materials of its construction, its methods of manufacture, and the loads (stresses/strains) involved in the product's environment. Where all these aspects can be closely interwoven, plastics are able to solve design problems efficiently in ways that are economically advantageous. It is important to recognize that these characteristics of plastics exist. This book starting with Chapter 1 provides their characteristics and behavior.

Selecting plastic

It is unfortunate that plastics do not have all the advantages and none of the disadvantages of other materials but often overlooked is the fact that there are no materials that do not suffer from some disadvantages or limitations. The faults of materials known and utilized for hundreds of years are often overlooked; the faults of the new materials (plastics) are often over-emphasized.

As examples, steel is attacked by the elements of fire [1500 to 2500°F (815 to 1370°C)]. They lose all their strength, modulus of elasticity, etc. Common protective practice includes the use of protective coatings (plastic, cement, etc.) and then forgetting their susceptibility to attack is all too prevalent. Wood and concrete are useful materials yet who has not seen a rotted board (wood on fire, etc.) and cracked concrete. Regardless this lack of perfection does not mean that no steel, wood, or concrete should be used. The same reasoning should apply to plastics. In many respects, the gains made with plastics in a short span of time far outdistance the advances made in these other technologies.

To significantly extend the life of structural beams, hardwood (thicker than steel, etc.) can be used; thus people can escape even though the wood slowly burns. The more useful and reliable structural beams would be using reinforced plastics (RPs) that meet structural performance requirements with even a more extended supporting life than wood. To date these RPs are not used in this type of fire environment primarily because of their high cost.

Even though the range of plastics continues to be large and the levels of their properties so varied that in any proposed application only a few of the many plastics will be suitable. A compromise among properties, cost, and manufacturing process generally determines the material of construction. Selecting a plastic is very similar to selecting a metal. Even within one class, plastics differ because of varying formulations (Chapter 1), just as steel compositions vary (tool steel, stainless steel, etc.).

For many applications plastics have superseded metal, wood, glass, natural fibers, etc. Many developments in the electronics and transportation industries and in packaging and domestic goods have been made possible by the availability of suitable plastics. Thus comes the question of whether to use a plastic and if so, which one.

As an initial step, the product designer must know and/or anticipate the conditions of use and the performance requirements of the product, considering such factors as life expectancy, size, condition of use, shape, color, strength, and stiffness. These end use requirements can be ascertained through market analysis, surveys, examinations of similar products, testing, and/or experience. A clear definition of product requirements will often lead directly to choice of the material of construction. At times incomplete or improper product requirement analysis is the cause for a product to fail.

As a general rule, until experience is developed, it is considered desirable to examine the properties of three or more materials before making a final choice. Material suppliers should be asked to participate in type and grade selection so that their experience is part of the input. The technology of manufacturing plastic materials, as with other materials (steel, wood, etc.) results in that the same plastic compounds supplied from various sources will generally not deliver the same results in a product. As a matter of record, even each individual supplier furnishes their product under a batch number, so that any variation can be tied down to the exact condition of the raw-material production. Taking into account manufacturing tolerances of the plastics, plus variables of equipment and procedure, it becomes apparent that checking several types of materials from the same and/or from different sources is an important part of material selection.

Experience has proven that the so-called interchangeable grades of materials have to be evaluated carefully as to their affect on the quality of a product. Another important consideration as far as equivalent grade of material is concerned is its processing characteristic. There can be large differences in properties of a product and test data if the processability features vary from grade to grade. It must always be remembered that test data has been obtained from simple and easy to process shapes and does not necessarily reflect results in complex product configurations. This situation is similar to those encountered with other materials (steel, wood, glass, etc.).

Most plastics are used to produce products because they have desirable mechanical properties at an economical cost. For this reason their mechanical properties may be considered the most important of all the physical, chemical, electrical, and other considerations for most applications. Thus, everyone designing with such materials needs at least some elementary knowledge of their mechanical behavior and how they can be modified by the numerous structural geometric shape factors that can be in plastics.

Comparison

The following information provides examples of guidelines on performance comparisons of different plastics. As an example, if the product requires flexibility, examples of the choices include polyethylene, vinyl, polypropylene, EVA, ionomer, urethane-polyester, fluorocarbon, silicone, polyurethane, plastisols, acetal, nylon, or some of the rigid plastics that have limited flexibility in thin sections.

The subject of strength can be complex since so many different types exist: short or long term, static or dynamic, etc. Some strength aspects are interrelated with those of toughness. The crystallinity of TPs is important for their short-term yield strength. Unless the crystallinity is impeded, increased molecular weight generally also increases the yield strength. However, the crosslinking of TSs increases their yield strength substantially but has an adverse effect upon toughness (Chapter 1).

Increasing the secondary bonds' strength and crystallinity than by increasing the primary bond strength increases long-term rupture strengths in TPs much more readily. Fatigue strength is similarly influenced, and all factors that influence thermal dimensional stability also affect fatigue strength. This is a result of the substantial heating that is often encountered with fatigue, particularly in TPs.

Polystyrene, styrene-acrylonitrile, polyethylene, acrylic, ABS, polysulfone, EVA, polyphenylene oxide, and many other TPs are satisfactorily odor-free. FDA approvals are available for many of these plastics. There are food packaging and refrigerating conditions that will eliminate certain plastics. Melamine and urea compounds are examples of suitable plastics for this service.

Thermal considerations will eliminate many materials. Examples for products operating above 450°F (232°C) include the silicones, fluoroplastics, polyirnides, hydrocarbon resins, methylpentene cold mold, or glass-bonded mica plastics may be required. A few of the organic plastic-bonded inorganic fibers such as bonded ceramic wool, perform well in this field. Epoxy, diallyl phthalate, and phenolic-bonded glass fibers may be satisfactory in the 450 to 550°F (232 to 288°C) ranges. A limited group of ablation material is made for outer space reentry use.

Between 250 and 450°F (121 and 232°C) glass or mineral-filled phenolics, melamine, alkyd, silicone, nylon, polyphenylene oxide, polysulfone, polycarbonate, methylpentene, fluorocarbon, polypropylene, and diallyl phthalate can be considered. The addition of glass fillers to the thermoplastics can raise the useful temperature range as much as 100°F (212°C) and at the same time shorten the fabricating cycle.

In the 0 to 212°F range, a broad selection of materials is available. Low temperature considerations may eliminate many of the thermoplastics. Polyphenylene oxide can be used at temperatures as low as –275°F. Thermosetting materials exhibit minimum embrittlement at low temperature.

Underwriters' Laboratory (UL) ruling on the use of self-extinguishing plastics for contact-carrying members and many other components introduces critical material selection problems. All thermosets are self-extinguishing. Nylon, polyphenylene oxide, polysulfone, polycarbonate, vinyl, chlorinated polyether, chlorotrifluoroethylene, vinylidene fluoride, and fluorocarbon are thermoplastics that may be suitable for applications requiring self-extinguishing properties. Cellulose acetate and ABS are also available with these properties. Glass reinforcement improves these materials considerably.

Many TPs will craze or crack under certain environmental conditions, and products that are highly stressed mechanically must be checked very carefully. Polypropylene, ionomer, chlorinated polyether, phenoxy, EVA, and linear polyethylene offer greater freedom from stress crazing than some other TPs. Solvents may crack products held under stress.

Toughness behaviors and evaluation can be rather complex. A definition of toughness is simply the energy required to break the plastic. This energy is equal to the area under the tensile stress-strain curve. The toughest plastics should be those with very great elongations to break, accompanied by high tensile strengths; these materials nearly always have yield points. One major exception to this rule is RPs that use reinforcing fibers such as glass and graphite that have low elongation. For high toughness a plastic needs both the ability to withstand load and the ability to elongate substantially without failing except in the case of RPs (Fig. 6.2).

It may appear that factors contributing to high stiffness are required. This is not true because there is an inverse relationship between flaw sensitivity and toughness; the higher the stiffness and the yield strength of a TP, the more flaw sensitive it becomes. However, because some load-bearing capacity is required for toughness, high toughness can be achieved by a high trade-off of certain factors.

Figure 6.2 Toughness behaviors (courtesy of Plastics FALLO)

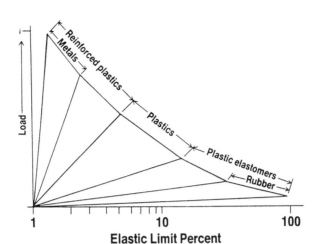

Crystallinity increases both stiffness and yield strength, resulting usually in decreased toughness. This is true below its glass transition temperature (T_g) in most noncrystalline (amorphous) plastics, and below or above the Tg in a substantially crystalline plastic (Chapter 1). However, above the T_g in a plastic having only moderate crystallinity, increased crystallinity improves its toughness. Furthermore, an increase in molecular weight from low values increases toughness, but with continued increases, the toughness begins to drop.

Deformation is an important attribute in most plastics, so much so that it is the very factor that has led them to be called plastic. For designs requiring such traits as toughness or elasticity this characteristic has its advantages, but for other designs it is a disadvantage. However, there are plastics, in particular the RPs, that have relatively no deformation or elasticity and yet are extremely tough where (a) toughness is related to heat deflection or rigidity and (b) toughness or impact is related to temperature for polystyrene (PS) and high impact polystyrene (HIPS).

This type of behavior characterizes the many different plastics available. Some are tough at room temperature and brittle at low temperatures. Others are tough and flexible at temperatures far below freezing but become soft and limp at moderately high temperatures. Still others are hard and rigid at normal temperatures but may be made flexible by copolymerization or adding plasticizers.

By toughness is meant resistance to fracture. However, there are those materials that are nominally tough but may become embrittled due to processing conditions, chemical attack, prolonged exposure to constant stress, and so on. A high modulus and high strength with ductility is the desired combination of attributes. However, the inherent nature of plastics is such that their having a high modulus tends to associate them with low ductility, and the steps taken to improve the one will cause the other to deteriorate.

Soft, weak materials have a low modulus, low tensile strength, and only moderate elongation to break. According to ASTM standards, the elastic modulus or the modulus or elasticity is the slope of the initial straight-line portion of the curve. Hard, brittle materials have high moduli and quite high tensile strengths, but they break at small elongations and have no yield point. Hard, strong plastics have high moduli, high tensile strengths, and elongations of about 5% before breaking. Their curves often look as though the material broke about where a yield point might have been expected.

Soft, tough plastics are characterized by low moduli, yield values or plateaus, high elongations of 20 to 1,000%, and moderately high breaking strengths. The hard, tough plastics have high moduli, yield points, high tensile strengths, and large elongations. Most plastics in this category show cold drawing or necking during the stretching operation. The RPs will have at least one modulus but some materials can have two or three.

Although impact strength of plastics is widely reported, the properties have no particular design values and can be used only to compare relative response of materials (toughness, etc.). Even this comparison is not completely valid because it does not solely reflect the capacity of the material to withstand shock loading, but can pick up discriminatory response to notch sensitivity.

A better value is impact tensile, but unfortunately this property is not generally reported. The impact value can broadly separate those that can withstand shock loading vs. those that are poor in this response. Therefore, only broad generalizations can be obtained on these values. Comparative tests on sections of similar size which are fabricated in accordance with the proposed product must be tested to determine the impact performance of a plastics material. The laminated plastics, glass-filled epoxy, melamine, and phenolic are outstanding in impact strength. Polycarbonate and ultrahigh molecular weight PE are also outstanding in impact strength.

In general, rigid plastics are superior to elastomers in radiation resistance but are inferior to metals and ceramics. The materials that will respond satisfactorily in the range of 10^{10} and 10^{11} erg per gram are glass and asbestos-filled phenolics, certain epoxies, polyurethane, poly-styrene, mineral-filled polyesters, silicone, and furane. The next group of plastics in order of radiation resistance includes polyethylene, melamine, urea formaldehyde, unfilled phenolic, and silicone plastics. Those materials that have poor radiation resistance include methyl methacrylate, unfilled polyesters, cellulosics, polyamides, and fluoro-carbons.

Maximum transparency is available in acrylic, polycarbonate, polyethylene, ionomer, and styrene compounds. Many other thermoplastics may have adequate transparency.

Urea, melamine, polycarbonate, polyphenylene oxide, polysulfone, polypropylene, diallyl phthalate, and the phenolics are needed in the temperature range above 200°F (93°C) for good color stability. Most TPs will be suitable below this range.

Deteriorating effects of moisture are well known. For high moisture applications, polyphenylene oxide, polysulfone, acrylic, butyrate, diallyl phthalate, glass-bonded mica, mineral-filled phenolic, chlorotrifluoro-ethylene, vinylidene, chlorinated polyether chloride, vinylidene fluoride, and the fluorocarbons should be satisfactory. Diallyl phthalate, polysulfone, and polyphenylene oxide have performed well with moisture/steam on one side and air on the other (a troublesome combination), and they also will withstand repeated steam autoclaving. Long-term studies of the effect of water have disclosed that chlorinated polyether gives outstanding performance. Impact styrene plus 25wt% graphite and high density polyethylene with 15% graphite give long-term performance in water.

Depending on what is required, the different plastics can provide different rates of permeability properties. As an example certain polyethylenes will pass wintergreen, hydrocarbons, and many other chemicals. It is used in certain cases for the separation of gases since it will pass one and block another. Chlorotrifluoroethylene and vinylidene fluoride, vinylidene chloride, polypropylene, EVA, and phenoxy merit evaluation.

There are materials with low or no permeability to different environ-ments or products. Barrier plastics are used with their technology not becoming more complex but more precise. Different factors influence performance such as being pinhole-free; chemical composition, crosslinking, modification, molecular orientation; density, and thickness. The coextrusion and coinjection processes are used to reduce permeability while retaining other desirable properties. Total protection against vapor transmission by a single barrier material increases linearly with increasing thickness, but it usually is not economical. Thus extensive use is made of multiple layer constructions. This composite would include low cost as well as recycled plastics that provide mechanical support, etc. while an expensive barrier material thickness is significantly reduced.

With crystalline plastics, the crystallites can be considered impermeable. Thus, the higher the degree of crystallinity, the lower the permeability to gases and vapors. The permeability in an amorphous plastic below or not too far above its glass transition temperature (T_g) is dependent on the degree of molecular orientation. It is normally reduced when compared to higher temperatures, although small strains sometimes increases the permeability of certain plastics. The orientation of elastomers well above their T_g has relatively less effect on the overall transport property. Crosslinking thermoplastics will decrease permeability due to the decrease in their diffusion coefficient. The effect of crosslinking is more pronounced for large molecule size vapors. The addition of a plasticizer usually increases the rates of vapor diffusion and permeation (Chapter 1).

The permeation of vapors includes two basic processes: the sorption and diffusion of vapors in the plastic. As an example in the packaging industry, the resistance of moisture is essential for the preservation of many products. The loss of moisture, flavor, etc. through packaging materials may damage foodstuff. The prevention of the ingress of moisture by a barrier is essential for the storage of dry foods and other products. In other applications, the degree of resistance to water and oxygen is important for the development of corrosion resistance coatings, electrical and electronic parts, etc.

Fluorination is the process of chemically reacting a material with a fluorine-containing compound to produce a desired product. As an example it can improve the gasoline barrier of PE to nonpolar solvents. A barrier is created by the chemical reaction of the fluorine and the PE, which form a thin (20 to 40 mm) fluorocarbon layer on the surface. Two systems can be used to apply the treatment depending on the results desired. With the "in-process" system, such as that used during blow molding PE gasoline tanks, fluorine is used as a part of the parison expanding gas in the blowing operation (the result is no gasoline leakage). The barrier layer is created only on the inside. In a post-treatment system, bottles and other products are placed in an enclosed chamber filled with fluorine gas. This method forms barrier layers on both the inside and outside surfaces.

Worksheet

The first step in selecting a plastic for a product to be fabricated is to determine its complete requirements. Since there could be a tendency to overlook certain properties because they may appear to be insignificant or overlooked, it is vital to ensure that the product will perform during packaging, shipment, and/or in service. Selecting an optimal material for a given product must obviously be based on analysis of the requirements to be met. A simplified approach involves comparing the specific service requirements to the potential properties of a plastic. What follows is a simplified but practical material-selection approach. This "longhand" system has been used for almost a half century during which time it became a basis in many fast computerized software material selection databases.

A simplified approach is where one starts by selecting the design criteria as well as potential plastics of interest and incorporating them into a table format checking off only the major criteria across the worksheet. Follow by setting up a comparison of the performance requirements for the potential plastics being considered and transfer the bold-faced numerical rating in each selected criteria column to the worktable. Add these numbers across the worktable to determine the plastic group with the lowest-point subtotal that will be the best plastic for a given application on a performance basis. Next add in the cost factor and total it to find the plastic group with the lowest number that results the best choice based on a cost-performance evaluation.

Follow by determining the specific plastic within the plastic group selected. The plastic with the lowest final total will be the best for the application on a cost-performance basis.

Temperature

Temperature is the thermal state of matter as measured by a specific scale. Basically it is a measure of the intensity of the molecular energy in a substance. The higher temperatures have more molecular movement. The temperature at which molecular movement ceases completely is absolute zero; it has been reached theoretically but not yet in actuality. Ambient temperature, usually synonymous with room temperature, denotes the surrounding environmental conditions such as pressure and temperature.

Plastics behave differently when exposed to temperatures; most plastic can take greater heat than humans. There are some plastics that cannot take boiling water and others operate at 150°C (300°F) with a few up to 540°C (1000°F). Most are not effected by low temperature (below freezing). The flexible (elastomer) plastics at room temperature become less flexible as they are cooled, finally becoming brittle at a certain low temperature. Then there are plastics that reach 1370°C (2500°F) with exposures in fractions of a second. Performance is influenced by short to long time static and dynamic mechanical requirements. An excellent test if a plastic can take heat is put in your automobile trunk or a railroad boxcar where temperatures can reach 55°C (130°F).

Important to understand that there is a temperature transition in plastics; also called ductile-to-brittle transition temperature. It is temperature at which the properties of a material change. Depending on the material, the transition change may or may not be reversible. A few other characteristics are presented. The plastic softening range temperature is the temperature at which a plastic is sufficiently soft to be distorted easily. A number of tests exist and the temperatures arrived at may vary according to the particular test method. Softening range is sometimes erroneously referred to as the softening point. Temperature stability identifies the percent change usually in tensile strength or in percent elongation as measured at a specified temperature and compared to values obtained at the standard conditions of testing.

Data obtained by testing different impact properties at various temperatures produces information that is similar to an elongation vs. temperature curve. As temperatures drop significantly below the ambient temperatures, most TPs lose much of their room-temperature impact strength. A few, however, are on the lower, almost horizontal portion of the curve at room temperature and thus show only a gradual decrease in impact properties with decreases in temperature. One major exception is provided by the glass fiber RPs, which have relatively high Izod impact values, down to at least –40°C (–40°F). The S-N (fatigue) curves for TPs at various temperatures show a decrease in strength values with increases in temperature. However the TSs, specifically the TS RPs, in comparison can have very low losses in strength.

Plastics can be affected in different ways by temperature. It can influence short- and long-time static and dynamic mechanical properties, aesthetics, dimensions, electronic properties, and other characteristics. Fig. 6.3 provides a guide relating time at temperature vs. 50% retention mechanical and physical properties. Testing temperature was at the exposure temperature of test specimens.

As the temperature rises thermoplastics (TPs) are effected. In comparison thermosets (TSs) are not affected. The maximum temperatures

Figure 6.3 Guide to temperature versus plastic properties (Courtesy of Plastics FALLO)

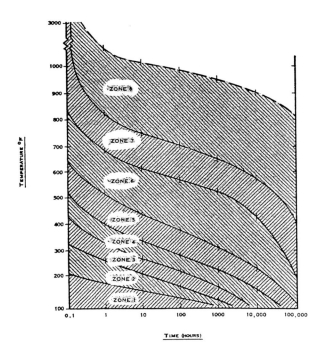

under which plastics can be employed are generally higher than the temperatures found in buildings, including walls and roofs, but there are those such as LDPE that are marginal and cannot carry appreciable stresses at these moderately elevated temperatures without undergoing noticeable creep. Many plastics can take shipping conditions that are more severe than their service conditions. With a closed automobile trunk or railroad boxcar temperatures reach at least 52°C (126°F); a temperature endurance test could be run in these closed containers or other containers.

Plastic strength and modulus will decrease and its elongation increase with increasing temperature at constant strain. Curves for creep isochronous stress and isometric stress are usually produced from measurements at a fixed temperature (Chapter 3). Complete sets of these curves are sometimes available at temperatures other than the ambient. It is common to obtain creep rupture or apparent modulus curves plotted against log time, with temperature as a parameter (Fig. 6.4).

A set of creep-rupture curves at various temperatures (Fig. 6.5) can be extended to provide data to obtain longer time data. With these data projecting the lowest-temperature curves to longer times as a straight line would produce a dangerously high prediction of rupture strength.

Figure 6.4 Effect of temperature on creep modulus

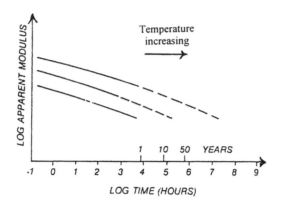

An advantage of conducting complete creep-rupture testing at elevated temperatures is that although such testing for endurance requires long times, the strength levels of the plastic at different temperatures can be developed in a relatively short time of usually just 1,000 to 2,000 h. The Underwriters Laboratories and other such organizations have employed such a system for many decades.

Thermal Property

Different plastics provide a wide range of temperature capabilities with a wide difference between TPs and TSs. The more common TP follows different phases as it is subjected to heat. Fig. 6.6 shows a plot for a TP

Figure 6.5 Temperature effect on creep-rupture

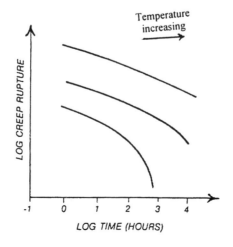

where the modulus of elasticity (relating to viscosity) is effected by temperature changes. As the temperature is increased the plastic changes through different stages from a rigid solid to a liquid through the stages of being glassy, in transition, rubbery, and flow. TSs when compared to TPs remains solid retaining properties as the heat increases up to the temperature where it disintegrates.

TPs' properties (and processes) are influenced by their thermal characteristics such as melt temperature (T_m), glass-transition temperature (T_g), dimensional stability, thermal conductivity, thermal diffusivity, heat capacity, coefficient of thermal expansion, and decomposition (T_d). Table 6.1 provides some of this data on different plastics (also applicable data for aluminum and steel). All these thermal properties relate to how to determine the best useful processing conditions to meet product performance requirements. There is a maximum temperature or, to be more precise, a maximum time-to-temperature relationship for all materials preceding loss of performance or decomposition.

Heat history or residence time develops when TPs repeatedly exposed to heating and cooling cycles such as when recycled. Certain TPs can be indefinitely granulating (scrap, defective products, and so on). During the heating and cooling cycles the performance and properties of certain plastics will not change or be insignificantly affected. However there are TPs that have minor to completely destructive results. TPs that are heat sensitive or those with certain additives and/or fillers are subject to destruction on their first recycling. If incorrect methods were used in granulating recycled material, more degradation will occur.

TPs are subjected to various degrees of dimensional stability. Dimensional stability is the temperature above which plastics lose their dimensional stability. For most plastics the main determinant of dimensional stability is their T_g. With highly crystalline plastics is T_g not a major problem (Chapter 1). The crystalline plastics in the range between T_g and T_m are referred to as leathery, because they are made up of a combination of rubbery noncrystalline regions and stiff crystalline regions. The result is that such plastics as PE and PP are still useful at lower temperatures and nylon is useful to moderately elevated temperatures even though those temperatures may be above their respective T_g.

Plastic memory is another behavior of TPs. When subjected to heat they can be bent, twisted, stretched, compressed, or squeezed into various useful shapes, but eventually, especially if you add heat, they return to their original form. This behavior, can be annoying. When

Table 6.1 Thermal properties of materials (Courtesy of Plastics FALLO).

Plastics (morphology)		Density g/cm³ (lb./ft.³)		Melt temperature T_m °C (°F)		Glass transition temperature T_g °C (°F)		Thermal conductivity (10⁻⁴ cal/s cm°C) (BTU/lb.°F)		Heat capacity cal/g°C (BTU/lb.°F)		Thermal diffusivity 10⁻⁴ cm²/s 10⁻³ ft.²/hr.		Thermal expansion 10⁻⁶cm/cm°C (10⁻⁶in./in. °F)	
PP	(C)	0.9	(56)	168	(334)	5	(41)	2.8	(0.068)	0.9	(0.004)	3.5	(1.36)	81	(45)
HDPE	(C)	0.96	(60)	134	(273)	−110	(−166)	12	(0.290)	0.9	(0.004)	13.9	(5.4)	59	(33)
PTFE	(C)	2.2	(137)	330	(626)	−115	(−175)	6	(0.145)	0.3	(0.001)	9.1	(3.53)	70	(39)
PA	(C)	1.13	(71)	260	(500)	50	(122)	5.8	(0.140)	0.075	(0.003)	6.8	(2.64)	80	(44)
PET	(C)	1.35	(84)	250	(490)	70	(158)	3.6	(0.087)	0.45	(0.002)	5.9	(2.29)	65	(36)
ABS	(A)	1.05	(66)	105	(221)	102	(215)	3	(0.073)	0.5	(0.002)	3.8	(1.47)	60	(33)
PS	(A)	1.05	(66)	100	(212)	90	(194)	3	(0.073)	0.5	(0.002)	5.7	(2.2)	50	(28)
PMMA	(A)	1.20	(75)	95	(203)	100	(212)	6	(0.145)	0.56	(0.002)	8.9	(3.45)	50	(28)
PC	(A)	1.20	(75)	266	(510)	150	(300)	4.7	(0.114)	0.5	(0.002)	7.8	(3.0)	68	(38)
PVC	(A)	1.35	(84)	199	(390)	90	(194)	5	(0.121)	0.6	(0.002)	6.2	(2.4)	5.0	(128)
Aluminum		2.68	(167)	1,000				3000	(72.50)	0.23		4900	(1900)	19	(10.6)

* = Crystalline resin. A = Amorphous resin.

Figure 6.6 Example of TP modulus of elasticity versus temperature

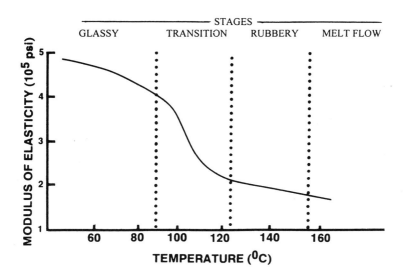

property applied plastic memory offers interesting design possibilities for plastic products. Their time/temperature-dependent change in mechanical properties results from stress relaxation and other viscoelastic phenomena typical of plastics. When the change is an unwanted limitation it is called creep. When the change is skillfully adapted to the overall design, it is called plastic memory.

During this shaping they do not alter their molecular structure or grain orientation to accommodate the deformation permanently. Plastics temporarily assume the deformed shape but always maintain internal stresses that want to force the material back to its original shape.

From a design approach plastic memory can be built into the product during fabrication. The tendency for the product to move into a new shape is included as an integral part of the design. After the product is assembled in place, a small amount of heat can coax that part to change shape. TP products can be deformed during assembly then allowed returning to their original shape. In this example products can be stretched around obstacles or made to conform to unavoidable irregularities without permanent damage.

Most TPs naturally have this memory capability. Polyolefins, neoprene, silicone, and other crosslinkable polymers can be given a memory either by radiation or by chemically curing. Fluorocarbons, however, need no curing. With fluorocarbons such as TFE, FEP, ETFE, ECTFE, CTFE, and PVF useful high temperature or wear resistant applications are possible.

Thermal Conductivity

When using plastics in a design requiring heat insulation or heat dissipation thermal conductivity (TC) is useful. It is the rate at which a material will conduct heat energy along its length or through its thickness. The heat capacity or specific heat of a unit mass of material is the amount of energy required to raise its temperature 1°C. It can be measured either at constant pressure or constant volume. At constant pressure it can be larger than at constant volume, because additional energy is required to bring about a volume change against external pressure.

Where as heat capacity is a measure of energy, thermal diffusivity is a measure of the rate at which energy is transmitted through a given plastic. In contrast, metals have values hundreds of times larger than those of plastics. Thermal diffusivity determines plastics' rate of change with time. Although this function depends on thermal conductivity, specific heat at constant pressure, and density, all of which vary with temperature, the thermal diffusivity is relatively constant.

The specific heat of amorphous plastics increases with temperature in an approximately linear fashion below and above T_g, but a steplike change occurs near the T_g. No such stepping occurs with crystalline types. The high degree of the molecular order for crystalline TPs makes their values tend to be twice those of the amorphous types. The TSs has the highest values. To increase TC the usual approach is to add metallic fillers, glass fibers, foamed structure, or electrically insulating fillers such as alumina.

In general, TC is low for plastics and the plastic's structure does not alter its value significantly. TC of plastics depends on several variables and cannot be reported as a single factor. But it is possible to ascertain the two principal dependencies of temperature and molecular orientation (MO). In fact, MO may vary within a product producing a variation in TC. It is important for the product designer and processor to recognize such a situation. Certain products require personal skill to estimate a part's performance under steady-state heat flow.

Thermal Expansion/Contraction

These values are based on the coefficient of linear thermal-expansion (CLTE). It is the ratio between the change of a linear dimension to the original dimension of the material per unit change in temperature (per ASTM D 696 standard). It is generally given as cm/cm/C or in./in./F. Plastics CLTE behaviors vary because different plastics have different CLTEs. Overall the TPs tend to expand while the TSs have little if any expansion.

There are plastics that have CLTE equal or less than those of other materials of construction (metals, glass, or wood). In fact with certain additives such as graphite powder contraction can occur rather than the expected expansion when heat is applied. Many plastics typically have CLTE that are considerably higher than those of other materials of construction such as metals, glass, or wood. This difference may amount to a factor of 10 to 30. However the design of thousands of TP products do not have a requirement that they have low or zero change so, guess what, it does not matter if they have considerable CLTE. However even though many TPs have a greater change than metals they can be made to behave like metals by adding fillers and reinforcements.

The designer takes thermal expansion and contraction into account if critical dimensions and clearances are to be maintained during use where material is in a restricted design. They recognize (as with metals, etc.) the fact that products may develop high stresses when they are constrained from freely expanding or contracting in response to temperature changes. These temperature-induced stresses can cause material failure if proper design is not used.

Plastic products are often constrained from freely expanding or contracting by rigidly attaching them to another structure made of a material with a lower CLTE. When such composite structures are heated, the plastic component is placed in a state of compression and may buckle. When such composite structures are cooled, the plastic component is placed in a state of tension, which may cause the material to yield or crack. The precise level of stress in the plastic depends on the relative compliance of the component to which it is attached, and on assembly stress.

CLTE is an important consideration if dissimilar materials like one plastic to another or a plastic to metal and so forth are to be assembled where material expansion or contraction is restricted. The type of plastic and RP, particularly the glass fibers content and its orientation influences the CLTE. It is especially important if the temperature range includes a thermal transition such as T_g. Normally, all this activity with dimensional changes is available from material suppliers readily enough to let the designer apply a logical approach and understand what could happen.

The goal is to eliminate or significantly reduce all sources of thermal stress. This can be achieved by keeping the following factors in mind:

(1) when adding material for local reinforcement, select a material with the same or a similar CLTE, (2) where plastic is to be attached to a more-rigid material, use mechanical fasteners with slotted or oversized holes to permit expansion and contraction to occur, (3) do not fasten dissimilar materials tightly, and (4) adhesives that remain ductile, such as urethane and silicone, through the product's expected end-use temperature range can be used without causing stress cracking or other problems.

In addition to dimensional changes from changes in temperature, other types of dimensional instability are possible in plastics as in other materials. Water-absorbing plastics, such as certain nylons, may expand and shrink as they gain or lose water, or even as the relative humidity changes. The migration or leaching of plasticizers, as in certain PVCs, can result in slight dimensional change.

Hyperenvironment

During the past decades progress in aeronautics and astronautics has been remarkable because people have learned to master the difficult feat of hypervelocity flight. A variety of manned and unmanned aircraft have been developed for faster transportation from one point on earth to another. Similarly, aerospace vehicles have been constructed for further exploration of the vast depths of space and the neighboring planets in the solar system. Plastics have found numerous uses in specialty areas such as hypersonic atmospheric flight and chemical propulsion exhaust systems. The particular plastic employed in these applications is based on the inherent properties of the plastics or the ability to combine it with another component material to obtain a balance of properties uncommon to either component.

Plastics have been developed for uses in very high temperature environments. It has been demonstrated that plastic materials are suitable for thermally protecting structures during intense rocket and missile propulsion heating. This discovery became one of the greatest achievements of modern times, because it essentially initially eliminated the thermal barrier to hypersonic atmospheric flight as well as many of the internal heating problems associated with chemical propulsion systems.

Modern supersonic aircraft experience appreciable heating. This incident flux is accommodated by the use of an insulated metallic structure, which provides a near balance between the incident thermal pulse and the heat dissipated by surface radiation. The result is that only a small amount of heat has to be absorbed by mechanisms other than radiation. With speeds increasing (8,000 fps) heating increases to a point where some added form of thermal protection is necessary to prevent thermostructural failure. Hypervelocity vehicles transcending through a planetary atmosphere also encounter gas-dynamic heating. The magnitude of heating is very large, however, and the heating period is much shorter.

This latter type of thermal problem is frequently referred to as the re-entry heating problem, and it posed one of the most difficult engineering problems of the twentieth century. A vehicle entering the earth's atmosphere at 25,000 fps has a kinetic energy equivalent to 12,500 Btu/lb of vehicle mass. Assuming the vehicle weighs a ton, it possesses a thermal energy equivalent to 25,000,000 Btu. This magnitude of energy greatly exceeds that required to completely vaporize the entire vehicle. Fortunately, only a very small fraction of the kinetic energy converted to heat reaches the body while the remainder is dissipated in the gas surrounding the vehicle. Materials performance during hypersonic atmospheric flight depends upon certain environmental parameters. These thermal, mechanical, and chemical variables differ greatly in magnitude and with body position.

In general, they are concerned with temperatures from about 2,000 to over 20,000F (1,100 to 11,000°C), gas enthalpies up to 40,000 Btu/lb, convective/radiative heating from 10 to over 10,000 Btu/ft^2/see. The stagnation pressures is less than 1 to over 100 atm., surface shear stresses up to about 900 psf, heating times from a few to several thousand seconds, and gaseous vapor compositions involving molecular, dissociated, and ionized species. To operate in these extreme conditions ablative materials can be used.

During ablation its surface material is physically removed. The injected vapors alter the chemical composition, transport properties, and temperature profile of the boundary layer, thus reducing the heat transfer to the material surface. At high ablation rates, the heat transfer to the surface may be only 15% of the thermal flux to a non-ablating surface. Up to tens of thousands of Btu's of heat can be absorbed, dissipated and blocked per pound of ablative material through the sensible heat capacity, chemical reactions, phase changes, surface radiation and boundary layer cooling of the ablator.

The heating rate or environmental temperature does not limit ablative systems, but rather by the total heat load. In spite of this limitation, however, the versatility of ablation has permitted it to be used on various hypervelocity atmospheric vehicles. No single, universally acceptable ablative material has been developed. Nevertheless, the interdisciplinary efforts of materials scientists and engineers have resulted in obtaining a wide variety of ablative compositions and constructions. These thermally protective materials have been arbitrarily categorized by their matrix composition, and typical materials are given in Table 6.2.

A popularly used plastic ablative heat protective material is plastic-base composites that include a TS plastic organic matrix. Their response to the heat abrasion occurs in a variety of ways. There are the depolymerization-vaporization (polytetrafluoroethylene), pyrolysis-vaporization (phenolic, epoxy) and decomposition melting vaporization (nylon fiber reinforced plastic). The principal advantages of plastic-base ablators are their high heat shielding capability and low thermal conductivity, however they are limited in accommodating very high heat loads due to the high erosion rates. Most TS plastics and highly crosslinked plastics (especially those with aromatic ring structures) form a hard surface residue of porous carbon. The amount of char formed depends upon various factors:

1. Carbon-to-hydrogen ratio present in the original plastic structure.

2. Degree of crosslinking and tendency to further crosslink during ablative heating.

3. Presence of foreign elements like the halogens, asymmetry and aromaticity of the base plastic structure.

4. Degree of vapor pyrolysis of the ablative hydrocarbon species percolating through the char layer.

5. Type of elemental bonding.

The behavior of the char during flight is pertinent to its success as an ablative material. Once the carbonaceous layer forms, the primary region of pyrolysis gradually shifts from the surface to a substrate zone beneath the char layer. The newly formed char structure is attached to the virgin substrate material and remains thereon for at least a short period of time. Meanwhile, its refractory nature serves to protect the temperature-sensitive substrate from the environment.

Gaseous products formed in the substrate pass through the porous char plastic layer, undergo partial vapor phase cracking, and deposit pyrolytic carbon (or graphite) onto the walls of the pores. As the organic plastic or its residual char are removed by the ablative aspects of the hyper-environment, the reinforcing fibers or particle fillers are left exposed and unsupported. Being vitreous in composition, they undergo melting. The resultant molten material covers the surface as liquid

Table 6.2 Plastics and other high temperature performance materials (Courtesy of Plastics FALLO).

Ablative Plastics	Elastomer	Ceramic	Metal
Polytetrafluoroethylene	Silicone rubber filled with microspheres and reinforced with a plastic honeycomb	Porous oxide (silica) matrix infiltrated with phenolic resin	Porous refractory (tungsten infiltrated with a low melting point metal (silver)
Epoxy-polyamide resin with a powdered oxide filler	Polybutadiene-acrylo-nitrile elastomer modified phenolic resin with a subliming powder	Porous filament wound composite of oxide fibers and an inorganic adhesive, impregnated with an organic resin	Hot-pressed refractory metal containing an oxide filler
Phenolic resin with an organic (nylon), inorganic (silica), or refractory (carbon) reinforcement		Hot pressed oxide, carbide, or nitride in a metal honeycomb	
Precharred epoxy impregnated with a noncharring resin			

Major property Of interest	Type of plastics	Propulsion system application
Ablative	Phenol-formaldehyde	Charring resin for rocket nozzle
Chemical resistance	Fluorosilicone	Seals, gaskets, hose linings for liquid fuels
Cryogenic	Polyurethane	Insulative foam for cryogenic tankage
Adhesion	Epoxy	Bonding reinforcements on external surface of combustion chamber
Dielectric	Silicone	Wire and cable electrical insulation
Elastomeric	Polybutadiene-acrylonitrile	Soli propellant binder
Power transmission	Diesters	Hydraulic fluid
Specific strength	epoxy-novolac	Resin matrix for filament wound motor case
Thermally nonconductive	Polyamide	Resin modifier for plastic thrust chamber
Absorptivity : emissivity ratio	Alkyd silicone	Thermal control coating
Gelling agent	Polyvinyl chloride	Thixotrophic liquid propellant

droplets, irregular globules, and/or a thin film. Continued addition of heat to the surface causes the melt to be vaporized. A fraction of the melt may be splattered by internal pressure forces, or sloughed away when acted upon by external pressure and shear forces of the dynamic environment.

Thermoplastic and elastomeric plastics tend to thermally degrade into simple monomeric units with the formation of considerable liquid and a lesser amount of gaseous species. Little or no solid desired residue generally remain on the ablating surface. Elastomeric-base materials represent a second major class of ablators. They thermally decompose by such processes as depolymerization, pyrolysis, and vaporization. Most of the interest to date has been focused on the silicone plastics because of their low thermal conductivity, high thermal efficiency at low to moderate heat fluxes, low temperature properties, elongation of several hundred percent at failure, oxidative resistance, low density, and compatibility with other structural materials. They are generally limited by the amount of structural quality of char formed during ablation, that restricts their use in hyperthermal environments of relatively low mechanical forces.

Flammability

The fire or flammability properties of plastics vary similarly to the way their other properties vary. There are those that burn very easily to those that do not burn. There are also plastics that cause no smoke and those that release large amount of smoke. Like other materials, hot enough fires can destroy all plastics. Some burn readily, others slowly, others only with difficulty; still others do not support combustion after the removal of the flame. As reviewed there are certain plastics used to withstand the re-entry temperature of 2,500°F (1,370°C) that occurs when a spacecraft returns into the earth's atmosphere; the time exposure is part of a millisecond. Different industry standards can be used to rate plastics at various degrees of combustibility.

Steel and Plastic

Plastics' behavior in fire depends upon the nature and scale of the fire as well as the surrounding conditions and how the products are designed. For example, the virtually all-plastic 35 mm slide projectors use a very hot electric bulb. When designed with a metal heat reflector with an air-circulating fan next to its very high heat dissipating light, the all-plastic projector operates with no fire hazard.

Steel structural beams cannot take the heat of a fire operating at and above 1500°F (816°C); they just loose all their strength, modulus of elasticity, etc.(Fig. 6.7). The unfortunate disaster of the Twin Towers in New York City on September 11, 2001; fires after being hit by planes is an example. As reviewed to protect steel from this environment they can obtain a temporary short time protection by being covered with products such as concrete and certain plastics. To significantly extend the life of structural beams hard wood (thicker, etc.) can be used; thus people can escape even though the wood slowly burns. The more useful and reliable structural beams would be using reinforced plastics (RPs) that meet structural performance requirements with a much more extended supporting life than wood. To date these RPs are not used in this type of fire environment because their costs are very high.

Test

When plastics are used, their behavior in a fire situation must be understood. Ease of ignition, the rate of flame spread and of heat release, smoke release, toxicity of products of combustion, and other factors must be taken into account. A plastic's behavior in fire depends upon the nature and scale of the fire as well as the surrounding conditions. Fire is a highly complex, variable phenomenon, and the behavior of plastics in a fire is equally complex and variable. Fire tests of plastics, like fire tests generally, are frequently highly specific, with the results being specific to the tests. The results of one type of test do not

Figure 6.7 Strength vs. temperature of steel and plastics (Courtesy of Plastics FALLO)

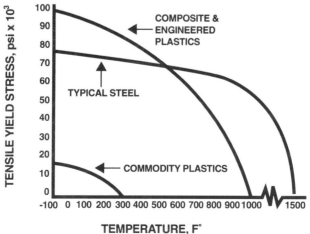

in fact often correlate directly with those of another and may bear little relationship to actual fires. There are tests that are intended for screening purposes during R&D. Tests such as large-scale product tests, are designed to nearly approximate actual fires.

Terms used that relate to fires include self-extinguishing, nonburning, flame spread, and toxicity. They are to be understood in the context of the specific tests with which they are used. Some materials may burn quite slowly but may propagate a flame rapidly over their surfaces. Thin wood paneling will burn readily, yet a heavy timber post will sustain a fire on its surface until it is charred, then smolder at a remarkably slow rate of burning. Bituminous materials may spread a fire by softening and running down a wall. Steel of course does not burn, but as reviewed, is catastrophically weakened by the elevated temperatures of a fire. PVC, silicone and fluorine does not burn, but it softens at relatively low temperatures. Other plastics may not burn readily but still emit copious amounts of smoke. And some flammable plastics, such as polyurethane, may be made flame retardant (FR) by incorporating in them additives such as antimony oxide.

In applying fire safety in a design requires information on where and how the product is to be used. According to those requirements the principles of good design for fire safety can be applied as they relate to plastics as to other materials. It is often helpful to select plastic materials for specific applications by first evaluating the flammability of the plastics in laboratory tests if the data is not available. These tests, often used for specifying materials, fall into the category either of small- scale or large-scale tests. Of course, as in evaluating any properties, having prior knowledge or obtaining reliable data applicable to fire or other requirements is the ideal situation.

There are different products that have specific fire tests. As an example for appliance safety the Underwriters Laboratory (UL) have published more than four hundred safety standards to assess the hazards associated with manufacturing appliances. These standards represent basic design requirements for various categories of products covered by the organization. For example, under UL's Component Plastics Program a material is tested under standardized, uniform conditions to provide preliminary information as to a material's strong and potentially weak characteristics.

The UL's plastics program is divided into two phases. The first develops information on a material's long- and short-term properties. The second phase uses this data to screen out and indicate a material's strong and weak characteristics. For example, manufacturers and safety engineers can analyze the possible hazardous effects of potentially weak characteristics, using UL standard 746°C.

It is the general consensus within the worldwide "fire community" that the only proper way to evaluate the fire safety of products is to conduct full-scale tests or complete fire-risk assessments. Most of these tests were extracted from procedures developed by the American Society for Testing and Materials (ASTM) and the International Electrotechnical Commission (IEC). Because they are time tested, they provide generally accepted methods to evaluate a given property. Where there were no universally accepted methods the UL developed its own.

Smoke

There are plastics that have different behaviors to smoke, going from no smoke to large amounts of smoke. Whether a plastic gives off light or heavy smoke and toxic or noxious gases depends on the plastic used, its composition of additives and fillers, and the conditions under which its burning occurs. Some plastics burn with a relatively clean flame, but some may give off dense smoke while smoldering.

Smoke is recognized by firefighters as being in many ways more dangerous than actual flames. It obscures vision, making it impossible to find safe means of egress, thus often leading to panic and not being able to rescue victims. Smoke from plastics, wood, and other materials usually contains toxic gases such as carbon monoxide (CO), which has no odor, often accompanied by noxious gases that may lead to nausea and other debilitating effects as well as panic, warning the fire victim of danger. With only CO the victim would die whereas the start of a fire with noxious gases could alert a person that a fire has started and to leave the area.

One of the most stringent and most widely accepted tests is UL 94 that concerns electrical devices. This test, which involves burning a specimen, is the one used for most flame-retardant plastics. In this test the best rating is UL 94 V-0, which identifies a flame with a duration of 0 to 5 s, an afterglow of 0 to 25 s, and the presence of no flaming drips to ignite a sample of dry, absorbent cotton located below the specimen. The ratings go from V-0, V-1, V-2, and V-5 to HB, based on specific specimen thicknesses. Details on fire testing and evaluating plastics are provided by UL who have extensive history on the effect and evaluation of fire as it effects plastics and other materials.

Electrical/Electronic

The electrical and electronic industry continues to be not only one of the major areas for plastic applications, but a necessity in many applications worldwide with their many diversified electrical performance capabilities. They principally provide dielectric or insulation capabilities. The field can be a steady direct current (DC) field or an alternating current (AC) field and the frequency range may vary such as from 0 to 10^{10} Hz.

The usual plastics are good insulators, however there are plastics that conduct electricity using certain plastics but more so by the addition of fillers such as carbon black and metallic flake. The type and degree of interaction depends on the polarity of the basic plastic material and the ability of an electrical field to produce ions that will cause current flows. In most applications for plastics, the intrinsic properties of the plastics are related to the performance under specific test conditions.

The properties of interest are the dielectric strength, the dielectric constant at a range of frequencies, the dielectric loss factor at a range of frequencies, the volume resistivity, the surface resistivity, and the arc resistance. The last three are particularly sensitive to moisture content in many materials. These properties are determined by the use of standardized tests such as those described by ASTM or UL. The properties of the plastics are temperature and/or moisture dependent as are many of their other properties. Temperature and/or moisture dependence must be recognized to avoid problems in electrical products made of plastics.

Electric currents can vary from fractions of a volt such as in communications signals to millions of volts in power systems. The currents carried by the conductor range from microamperes to millions of amperes. With this wide range of electrical conditions the types of plastic that can be used are different; no one plastic meets the different operating conditions. The selection of the materials and the configuration of the dielectric to perform under the different voltage, current, and frequency stresses are the primary design problem in electrical applications for plastics.

The dielectric materials interact with the electrical fields and alter the characteristics of the electric field. In some cases this is desirable and in others it is deleterious to the operation of the system and must be minimized. Both the selection of the plastic and the configuration of the dielectric can meet required performances.

An important area for the use of plastics in electrical applications is at the terminations of the conductors. The connectors that are used to tie the wires into the equipment using the power, or used to connect the wires to the power source, are rigid members with spaced contacts. These are designed to connect with a mating unit and to the extension wires. The other type of wire termination is terminal boards where there are means to secure the ends of the wire leading to the equipment and the internal wiring in the equipment. These termination units require adequate dielectric strength to resist the electric field between the conductors, good surface resistivity to prevent leakage of current across the surface of the material of the connector, good arc resistance to prevent permanent damage to the surface of the unit in case of an accidental arc over, and good mechanical properties to permit accurate alignment of the connector elements so that the connectors can be mated properly.

Electrical devices often require arc resistance, as a high current, high-temperature will ruin many plastics. Some special arc resisting plastics are available. The most serious cases may require cold mold, glass-bonded mica, or mineral-filled fluorocarbon products. Lesser arcing problems may be solved by the use of polysulfone, polyester glass, DAP-glass, alkyd, melamine, urea, or phenolics. With low-current arcs, general-purpose phenolic and glass-filled nylon or polycarbonate, acetal, and urea may be used very satisfactorily. A coating of fluorocarbon film will improve arc resistance in some cases. All circuit breaker problems must be scrutinized with respect to product performance under short-circuit conditions and mechanical shock.

Electromagnetic interference (EMI) or radio frequency interference (RFI) as well as static charge is the interference related to accumulated electrostatic charge in a nonconductor. As electronic products become smaller and more powerful, there is a growing need for higher shielding levels to assure their performance and guard against failure. Conductive plastics provide EMI/RFI shielding by absorbing electromagnetic energy (EME) and converting it into electrical or thermal energy.

Corrosion resistance

Plastics are basically noncorrosive. However, there are compounds that can be affected when exposed to corrosive environments. The corrosion resistance of most plastics have provided outstanding performance in all types of products worldwide.

Since plastics (not containing metallic additives) are not subjected to electrolytic corrosion, they are widely used where this property is required alone as a product or as coatings and linings for material subjected to corrosion such as in chemical and water filtration plants, mold/die, etc. Plastics are used as protective coatings on products such as steel rod, concrete steel reinforcement, mold cavity coating, plasticator screw coating, etc.

Complex corrosive environments results in at least 30% of total yearly plastics production being required in buildings, chemical plants, transportation, packaging, and communications. Plastics find many ways to save some of the billion dollars lost each year by industry due to the many forms of corrosion.

Chemical resistance

Many plastics have the ability to withstand attack of acids, alkalis, solvents, and other chemicals. Generally plastics have good chemical resistance. Part of the wide acceptance of plastics is from their relative compatibility to chemicals, particularly to moisture, as compared to that of other materials. Because plastics are largely immune to the electrochemical corrosion to which metals are susceptible, they can frequently be used profitably to contain water and corrosive chemicals that would attack metals. Plastics are often used in applications such as chemical tanks, water treatment plants, and piping to handle drainage, sewage, and water supply.

Structural shapes for use under chemical and corrosive conditions often take advantage of the properties of glass fiber-thermoset RPs. Today's RP (not steel tanks) gasoline underground tanks must last thirty or more years without undue maintenance. To meet these criteria they must be able to maintain their structural integrity and resist the corrosive effects of soil and gasoline including gasoline that has been contaminated with moisture and soil.

Some plastics like HDPE are immune to almost all commonly found solvents. PTFE (polytetrafluoroethylene) in particular is noted principally for its resistance to practically all-chemical substances. It includes what has been generally identified as the most inert material known worldwide.

Friction

Although plastics may not be as hard as metal products, there are those that have excellent resistance to wear and abrasion. Plastic hardware products such as cams, gears, slides, rollers, and pinions frequently provide outstanding wear resistance and quiet operation. Smooth plastic surfaces result in reduced friction, as they do in pipes and valves.

The frictional properties of TPs, specifically the reinforced and filled types, vary in a way that is unique from metals. In contrast to metals, even the highly reinforced plastics have low modulus values and thus do not behave according to the classic laws of friction. Metal-to-thermoplastic friction is characterized by adhesion and deformation resulting in frictional forces that are not proportional to load, because friction decreases as load increases, but are proportional to speed, The wear rate is generally defined as the volumetric loss of material over a given unit of time. Several mechanisms operate simultaneously to remove material from the wear interface. However, the primary mechanism is adhesive wear, which is characterized by having fine particles of plastic removed from the surface.

Presence of this powder is a good indication that rubbing surfaces are wearing properly. Conversely, the presence of melted plastic or large gouges or grooves at the interface normally indicates that the materials are abrading, not wearing, or the pressure velocity (PV) limits of the materials may be exceeded.

The ease and economy of manufacturing gears, cams, bearings, slides, ratchets, and so on with injection-moldable TPs have led to a widespread displacement of metals in these types of applications. In addition to their inherent processing advantages, the products made from these materials are able to dampen shock and vibration, reduce product weight, run with less power, provide corrosion protection, run quietly, and operate with little or no maintenance, while still giving the design engineer tremendous freedom.

These characteristics can be further enhanced and their applications widened by fillers, additives, and reinforcements. Compounding properly will yield an almost limitless combination of an increased load-carrying capacity, a reduced coefficient of friction, improved wear resistance, higher mechanical strengths, improved thermal properties, greater fatigue endurance and creep resistance, excellent dimensional stability and reproducibility, and the like.

Tolerance

The specific dimensions that can be obtained on a finished, processed plastic product basically depend on the performance and control of the plastic material, the fabrication process and, in many cases, upon properly integrating the materials with the process. In turn, a number of variable characteristics exist with the material itself. Unfortunately, some designers tend to consider dimensional tolerances on plastic products to be complex, unpredictable, and not susceptible to control.

If steel, aluminum, and ceramics were to be made into complex shapes but no prior history on their behavior during processing existed, a period of trial and error would be required to ensure their meeting the required measurements. If relevant processing information or experience did exist, it would be possible for these metallic products to meet the requirements with the first part produced. This same situation exists with plastics. To be successful with this material requires experience with their melt behavior, melt-flow behavior during processing, and the process controls needed to ensure meeting the dimensions that can be achieved in a complete processing operation. Based on the plastic to be used and the equipment available for processing, certain combinations will make it possible to meet extremely tight tolerances, but others will perform with no tight tolerances or any degree of repeatability.

Fortunately, there are many different types of plastics that can provide all kinds of properties, including specific dimensional tolerances. It can thus be said that the real problem is not with the different plastics or processes but rather with the designer, who requires knowledge and experience to create products to meet the desired requirements. The designer with no knowledge or experience has to become familiar with the plastic-design concepts expressed throughout this book and work with capable people such as the suppliers of plastic materials.

Some plastics, such as the TSs and in particular the TS-RP composites, can produce parts with exceptionally tight tolerances. In the compression molding of relatively thin to thick and complex shapes, tolerances can be held to less than 0.001 in. or to even zero, as can also be done using hand layup fabricating techniques. At the other extreme are the unfilled, unreinforced extruded TPs. Generally, unless a very thin uniform wall is to be extruded, it is impossible to hold to such tight tolerances as just given. The thicker and more complex an extruded shape is, the more difficult it becomes to meet tight tolerances without experience or trial and error. What is important is to determine the tolerances that can be met and then design around them.

Limit

To maximize control in setting tolerances there is usually a minimum and a maximum limit on thickness, based on the process to be used. Available from the literature and material suppliers is extensive information on tolerances based on plastic related to fabricating process. Examples are provided in Tables 6.3 to 6.5. Each specific plastic has its own range that depends on its chemical structure and melt-processing characteristics. Outside these ranges, melts are usually uncontrollable. Any dimensions and tolerances are theoretically possible, but they could result in requiring special processing equipment, which usually becomes expensive. There are of course products that require and use special equipment.

One factor in tolerances is shrinkage. Generally, shrinkage is the difference between the dimensions of a fabricated part at room temperature and the cooled part, checked usually twelve to twenty-four hours after fabrication. Having an elapsed time is necessary for many plastics, particularly the commodity TPs, to allow parts to complete their inherent shrinkage behavior. The extent of this postshrinkage can be near zero for certain plastics or may vary considerably. Shrinkage can also be dependent on such climatic conditions as temperature and humidity, under which the part will exist in service, as well as its conditions of storage.

Plastic suppliers can provide the initial information on shrinkage that has to be added to the design shape and will influence its processing. The shrinkage and postshrinkage will depend on the types of plastics and fillers and/or reinforcements. The amount of filler and reinforcement can significantly reduce shrinkage and tolerances.

Another influence on dimensions and tolerances involves the coefficient of linear thermal expansion or contraction. This CLTE value usually has to be determined at the part's operating temperature. So it is important to include in the design specifications the operating temperature conditions, to specify a plastic that will do the job. Plastics can provide all the extremes in CLTEs, including graphite-filled compounds that could work in reverse. Upon heating, they contract rather than expand, and vice-versa.

To assist the designer a Society of the Plastics Industry (SPI) bulletin is available that specifies the limits for certain dimensions. Each material supplier converts this data to suit their specific plastics.

Table 6.3 Wall thickness tolerance guide for thermoplastic moldings

Dimensions, in.	ABS		Acetal		Nylon		Polycarbonate	
	Commercial	Fine	Commercial	Fine	Commercial	Fine	Commercial	Fine
To 1.000	0.005	0.003	0.006	0.004	0.004	0.002	0.004	0.0025
1.000–2.000	0.006	0.004	0.008	0.005	0.006	0.003	0.005	0.003
2.000–3.000	0.008	0.005	0.009	0.006	0.007	0.005	0.006	0.004
3.000–4.000	0.009	0.006	0.011	0.007	0.009	0.006	0.007	0.005
4.000–5.000	0.011	0.007	0.013	0.008	0.010	0.007	-0.008	0.005
5.000–6.000	0.012	0.008	0.014	0.009	0.012	0.008	0.009	0.006
6.000–12.000, for each additional inch add	0.003	0.002	0.004	0.002	0.003	0.002	0.003	0.015
0.000–0.125	0.004	0.002	0.004	0.002	0.004	0.003	0.003	0.002
0.125–0.250	0.003	0.002	0.004	0.002	0.005	0.003	0.003	0.002
0.250–0.500	0.002	0.001	0.002	0.001	0.002	0.001	0.002	0.001
0.500 and over	0.002	0.002	0.003	0.002	0.003	0.002	0.003	0.002
0.000–0.250	0.003	0.002	0.004	0.002	0.003	0.002	0.003	0.002
0.250–0.500	0.004	0.002	0.005	0.003	0.004	0.003	0.002	0.002
0.500–1.000	0.005	0.003	0.006	0.004	0.005	0.004	0.004	0.003
0.000–3.000	0.015	0.010	0.011	0.006	0.010	0.004	0.005	0.003
3.000–6.000	0.030	0.020	0.020	0.010	0.015	0.007	0.007	0.004
Total Indicator Reading	0.009	0.005	0.010	0.006	0.010	0.006	0.005	0.003

Table 6.3 continued

Dimensions, in.	Polyethylene, high-density		Polyethylene, low-density		Polystyrene		Vinyl, flexible		Vinyl, rigid	
	Commercial	Fine	Commercial	Fine	Commercial	Fine	Commercial	Fine	Commercial	Fine
To 1.000	0.008	0.006	0.007	0.004	0.004	0.0025	0.011	0.007	0.008	0.0045
1.000–2.000	0.010	0.008	0.010	0.006	0.005	0.003	0.012	0.008	0.009	0.005
2.000–3.000	0.013	0.011	0.012	0.008	0.007	0.004	0.014	0.009	0.010	0.006
3.000–4.000	0.015	0.013	0.015	0.010	0.008	0.005	0.015	0.011	0.012	0.007
4.000–5.000	0.018	0.016	0.017	0.011	0.010	0.006	0.017	0.012	0.013	0.008
5.000–6.000	0.020	0.018	0.020	0.013	0.011	0.007	0.018	0.013	0.014	0.009
6.000–12.000, for each additional inch add	0.006	0.003	0.005	0.004	0.004	0.002	0.005	0.003	0.005	0.003
	0.006	0.004	0.005	0.004	0.0055	0.003	0.007	0.003	0.007	0.003
	0.006	0.004	0.005	0.004	0.007	0.0035	0.007	0.003	0.007	0.003
0.000–0.125	0.003	0.002	0.003	0.002	0.002	0.001	0.004	0.003	0.004	0.003
0.125–0.250	0.005	0.003	0.004	0.003	0.002	0.001	0.005	0.004	0.004	0.003
0.250–0.500	0.006	0.004	0.005	0.004	0.0035	0.0015	0.006	0.005	0.005	0.004
0.500 and over	0.008	0.005	0.006	0.005	0.0035	0.002	0.008	0.006	0.006	0.005
0.000–0.250	0.005	0.003	0.003	0.003	0.004	0.002	0.004	0.003	0.004	0.003
0.250–0.500	0.007	0.004	0.004	0.004	0.005	0.002	0.005	0.004	0.005	0.004
0.500–1.000	0.009	0.006	0.006	0.005	0.007	0.003	0.006	0.005	0.006	0.005
0.000–3.000	0.023	0.015	0.020	0.015	0.013	0.004	0.010	0.007	0.015	0.010
3.000–6.000	0.037	0.022	0.030	0.020		0.005	0.020	0.015	0.020	0.015
Total Indicator Reading	0.027	0.010	0.010	0.008	0.010	0.008	0.015	0.010	0.010	0.005

Table 6.4 Wall thickness tolerance guide for thermoset plastic moldings

	Minimum Thickness in. (mm)	Average thickness in. (mm)	Maximum thickness in. (mm)
Alkyd–glass filled	0.040 (1.0)	0.125 (3.2)	0.500 (13)
Alkyd–mineral filled	0.040 (1.0)	0.187 (4.7)	0.375 (9.5)
Diallyl phthalate	0.040 (1.0)	0.187 (4.7)	0.375 (9.5)
Epoxy-glass filled	0.030 (0.76)	0.125 (3.2)	1.000 (25.4)
Melamine-cellulose filled	0.035 (0.89)	0.100 (2.5)	0.187 (4.7)
Urea–cellulose filled	0.035 (0.89)	0.100 (2.5)	0.187 (4.7)
Phenolic–general purpose	0.050 (1.3)	0.125 (3.2)	1.000 (25.4)
Phenolic–glass filled	0.030 (0.76)	0.093 (2.4)	0.750 (19)
Phenolic–fabric filled	0.062 (1.6)	0.187 (4.7)	0.375 (9.5)
Silicone glass	0.050 (1.3)	0.125 (3.2)	0.250 (6.4)
Polyester premix	0.040 (1.0)	0.070 (1.8)	1.000 (25.4)

Table 6.5 Wall thickness tolerance guide for thermoplastic extruded profiles

					PVC	
	LDPE	HIPS	PC, ABS	PP	Rigid	Flex.
Wall thickness (%, =)	10	8	8	8	8	10
Angles (Deg., =)	5	2	4	4	2	5
Profile dimensions (in., ±)						
To 0.125	0.012	0.007	00.10	0.010	0.007	0.010
0.125 to 0.500	0.025	0.012	0.020	0.015	0.010	0.015
0.500 to 1	0.030	0.017	0.025	0.020	0.015	0.020
1 to 1.5	0.035	0.025	0.027	0.027	0.020	0.030
1.5 to 2	0.040	0.030	0.035	0.035	0.025	0.035
2 to 3	0.045	0.035	0.037	0.037	0.030	0.040
3 to 4	0.065	0.050	0.050	0.050	0.045	0.065
4 to 5	0.093	0.065	0.065	0.065	0.060	0.093
5 to 7	0.125	0.093	0.093	0.093	0.075	0.125
7 to 10	0.150	0.125	0.125	0.125	0.093	0.150

Processing Effect

Processing is extremely important in regard to tolerance control; in certain cases it is the most influential factor. The dimensional accuracy of the finished part relates to the process, the accuracy of mold or die used, and the process controls, as well as the shrinkage behavior of the plastic. A change to a mold or die dimension can result in wear arising during production runs and should thus be considered.

The mold or die should also be recognized as one of the most important pieces of production equipment in the plant. This controllable, complex device must be an efficient heat exchanger and provide the part's shape. The mold or die designer thus has to have the experience or training and knowledge of how to produce the tooling needed for the part and to meet required tolerances.

A knowledge of processing methods can be useful to the designer, to help determine what tolerances can be obtained. With such high-pressure methods as injection and compression molding that use 2,000 to 30,000 psi (13.8 to 206.9 MPa) it is possible to develop tighter tolerances, but there is also a tendency to develop undesirable stresses (that is, orientations, etc.) in different directions. An example as to how tolerances can change using the same process control and injection mold can be related to the amount of plastics used to fill and pack the mold cavity.

The low-pressure processes, including contact and casting with no pressure, usually do not permit meeting tight tolerances. There are exceptions, such as certain RPs that are processed with little or no pressures. Regardless of the process used, exercising the required and proper control over it will maximize obtaining and repeating of close tolerances.

For example, certain injection-molded parts can be molded to extremely close tolerances of less than a thousandth of an inch, or down to 0.0%, particularly when filled TPs or filled TS compounds are used. To practically eliminate shrinkage and provide a smooth surface, one should consider using a small amount of a chemical blowing agent (<0.5wt%) and a regular packing procedure. Results are such that literally no change in density will occur (no visible bubbles, etc.) and the product surface will be at least as smooth if not improved. For conventional molding, tolerances can be met of ±5% for a part 0.020 in. thick, ±1% for 0.050 in., ±0.5% for 1.000 in., −0.25% for 5.000 in., and so on. Thermosets generally are more suitable than TPs for meeting the tightest tolerances.

The dimensions of the product design have to be converted into the dimensions of the mold cavity, taking the respective shrinkage into consideration. For that reason, the final decision of what kind of plastic will be used has to be made beforehand. Very often, the mold maker requires half of the tolerances permissible for the part in question for the job. This, in most cases, is not justified. Today's accuracy of metalworking permits tolerances as low as one tenth of those on the drawing, assuming the latter being reasonable for plastics.

Under difficult circumstances, it has been proved to be a good practice if critical dimensions are kept smaller in the mold first, and then being revised after a test run under production conditions thus permitting machining the cavity if required. In any case, close tolerances should be applied on such dimensions only as directly related to invariable mold dimensions. Any other dimension, which is related to a mold dimension in two different mold parts, should allow a generous tolerance.

Economical production requires that tolerances not be specified tighter than necessary. However, after a production target is met, one should mold "tighter" if possible, for greater profit by using less material. Many plastics change dimensions after molding, principally because their molecular orientations or molecules are not relaxed. To ease or eliminate the problem, one can change the processing cycle so that the plastic is "stress relieved," even though that may extend the cycle time. Also used is heat-treat, the molded part based on experience or according to the resin supplier's suggestions.

Theoretical efforts to forecast linear shrinkage have been limited because of the number of existing variables. One way to solve this problem is to simplify the mathematical relationship, leading to an estimated but still acceptable assessment. This means, however, that the number of necessary processing changes will also be reduced.

The parameters of the injection process must be provided. They can either be estimated or, to be more exact, taken from the thermal and rheological layout. The position of a length with respect to flow direction is in practice an important influence. This is so primarily for glass-filled material but also for unfilled thermoplastics. The difference between a length parallel to $(0°)$ and perpendicular to $(90°)$ the flow direction depends on the processing parameters. Measurements with unfilled PP and ABS have shown that a linear relationship exists between these points.

Regarding this relationship, when designing the mold it is necessary to know the flow direction. To obtain this information, a simple flow pattern construction can be used. However, the flow direction is not constant. In some cases the flow direction in the filling phase differs from that in the holding phase. Here the question arises of whether this must be considered using superposition.

In order to get the flow direction at the end of the filling phase and the beginning of the holding phase (representing the onset of shrinkage), an analogous model was developed that provides the flow direction at the end of the filling phase. For a flow with a Reynolds number less than 10, which is valid regarding the processing of thermoplastics, the following equation can be used: $\Delta\Phi = 0$. For a two-dimensional geometry with quasistationary conditions, this equation is valid:

$$\frac{\partial^2\Phi}{\partial x^2} + \frac{\partial^2\Phi}{\partial y^2} = 0 \tag{6-1}$$

Instead of the potential Φ, it is possible to introduce the flow-stream function ψ for a two-dimensional flow. The stream lines (ψ = constant) and the equipotential lines are perpendicular to each other. To express this, the following Cauchy-Rieman differential can be used:

$$\frac{\partial\Phi}{\partial x} = \frac{\partial\psi}{\partial y} \frac{\partial\Phi}{\partial y} = -\frac{\partial\psi}{\partial x} \tag{6-2}$$

A differential (two dimensional/quasi) equation has the same form as is used for a stationary electrical potential field,

$$\frac{\partial^2 U}{\partial x^2} + \frac{\partial^2 U}{\partial y^2} = 0 \tag{6-3}$$

as it can be realized with an unmantled molding out of resistance paper and a suitable voltage.

To control the theoretically determined flow with respect to the orientation direction, a color study was made. The comparison between flow pattern, color study, and analogous model is shown in Figs. 6.8 and 6.9. For a simple geometry the flow pattern method describes the flow direction in the filling phase as well as the holding phase (Fig. 6.8).

This description changes when a core is added and the flow is disturbed (Fig. 6.9). In this case the flow at the beginning of the holding phase differs from the flow pattern as it is shown in the color study as well as in the analogous model. Even the welding lines are broken in the holding phase so that at this place another flow direction than that in the filling phase is found. With further measurements this influence has to be tested by using more-complex moldings. Available are computer software programs that provide guide lines to melt flow behavior in the mold cavities (Chapter 5).

Recycled plastic

When plastics are granulated the probability is its processability and performance when reprocessed into any product may be slightly reduced; could be significantly reduced. Fig. 6.10 shows how properties per ASTM tests for different plastics can effect properties of

Figure 6.8 Comparison between an analogous model, flow pattern studies

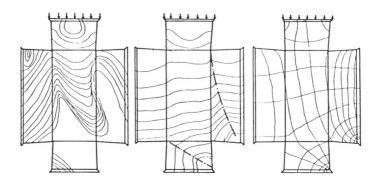

Figure 6.9 Comparison between an analogous model, a flow pattern studies with a core added

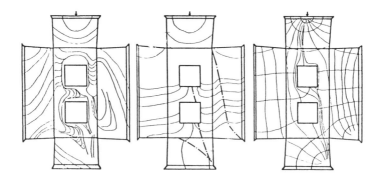

Figure 6.10 Example of the effect of recycling plastics once through a granulator

(a)   (b)   (c)

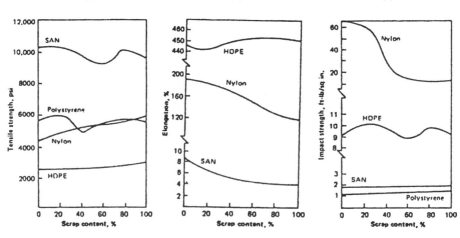

Figure 6.11 Example of the effect of recycling plastics more than once through a granulator

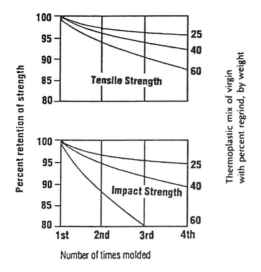

recycled (once through a granulator) plastics mixed with virgin plastics. The data presented are (a) tensile strength, (b) tensile elongation, and (c) unnotched Izod impact test. Fig. 6.11 shows effect on repeating recycling plastics where data presented are (a) tensile strength and (b). Thus it is important to evaluate what the properties of the recycled material provides. The size reduction, and particularly its uniformity, exerts a substantial influence on the quality of the recycled plastics. Recycled plastics is usually nonuniform in size so that processing with or without virgin plastics is subject to operating in a larger fabricating process window (Chapter 1).

Different approaches are used to improve performances or properties of mixed plastics such as: (1) additives, fillers, and/or reinforcements (use specific types such as processing agent, talc, short glass fibers), (2) active interlayers (crosslinking, molecular wetting), and (3) dispersing and diffusing (fine grinding, enlarging molecular penetration via melt shearing).

Most processing plants have been reclaiming/recycling reprocessable TP materials such as molding flash, rejected product, film trim, scrap, and so on during the past century. TS plastics (not remeltable) have been granulated and used as filler materials.

If possible the goal is to significantly reduce or eliminate any trim, scrap, rejected products, etc. in an industrial plant because it has already cost money and time to go through a fabricating process; granulating just adds more money and time. Also it usually requires resetting the process to handle it alone (or even when blending with virgin plastics and/or additives) because of its usual nonuniform particle sizes, shapes, and melt flow characteristics. Perhaps it was overheated during the cutting action of a granulator, etc. Keeping the scrap before/after granulating clean is an important requirement.

When fiber RPs are granulated, the lengths of the fibers are reduced. On reprocessing with virgin materials or alone, their processability and performance definitely change. So it is important to determine if the change will affect final product performances. If it will, a limit for the amount of regrind mix should be determined or no recycled RP is to be used. Consider redesigning the product to meet the recycled performance or use it in some other product.

Engineering data information source

In addition to what has been presented throughout this book such as the software on designing products in Chapter 5, this section provides source information concerning plastic material data basis. The available information worldwide has reached a volume that makes it impossible for one to review all the sources. In order to retrieve the desired or needed information, indexes and abstracts are continuously prepared by individual libraries, technical organizations, and professional societies. By the 1960s computers became available for storing and manipulating information. This lead to the creation and marketing of automated data banks.

Available for manual searching are abstracts that typically provide the name of the author, a brief abstract of the article, the title of the article, and identify where the article was published. Alphabetical author and subject indexes are usually provided, and an identifying number is assigned to refer to the abstract. Many abstracts are published monthly or more frequently. Annual cumulations are available in many cases. A comprehensive listing of abstracts and indexes can be found in Ulrich's International Periodical Directory (annual from R. R. Bowker, New York).

Most of the major indexes and abstracts are now available in machine-readable form. For a comprehensive list of databases and online vendors see Information Industry Market Place (International Directory of Information Products & Services from R. R. Bowker, New York). The names of online databases frequently differ from their paper counterparts. Engineering Index (monthly from Engineering Information Inc.) for example, offers COMPENDEX and Engineering Meetings online.

Many of the professional societies producing online databases will undertake a literature search. A society member is frequently entitled to reduced charges for this service. In addition to indexes and abstracts, periodicals, encyclopedias, and handbooks are available online. There seems to be virtually no limit to the information that can be made available online or on CD-ROM's, which can be networked in large institutions with many potential users. The high demand for quick information retrieval ensures the expansion of this service. In addition to the online indexes, several library networks and consortia, such as OCLC, the Online Computer Library Center, located in Columbus, Ohio, produce online databases. These are essentially equivalent to the catalogs of member libraries and can be used to determine which library owns a particular book or subscribes to a particular periodical.

Publication

The major emphasis on information is placed on publications and services designed to identify and obtain information. Because of space limitations references to individual works, which contain the required information, are limited to a few. The most important source of information is the primary literature. It consists mainly of the articles published in periodicals and of papers presented at conferences. New discoveries are first reported in the primary literature. It is, therefore, a major source of current information. Most engineers are familiar with a few publications, but are not aware of the extent of the total production of primary literature.

As an example there is the publication Machine Design that issues 22 per year. It covers design engineering of manufactured products across the entire industry spectrum. It offers solutions to design problems, new technology developments, CAD/CAM updates, etc. It is published by Penton Media, Inc., 1100 Superior Ave., Cleveland, OH 44114; Tel 216-696-7000; Fax 216-696-8765; website www.machinedesign.com.

Engineering Index (Engineering Information Inc. published monthly) abstracts material from thousands of periodicals and conferences. It is known as Compendex in its electronic version.

Handbooks and encyclopedias are part of the secondary literature (included in the Bibliography section). They are derived from primary sources and make frequent references to periodicals. Handbooks and encyclopedias are arranged to present related materials in an organized fashion and provide quick access to information in a condensed form.

While monographs include books written for professionals, they are either primary or secondary sources of knowledge and information. Textbooks are also part of the tertiary literature. They are derived from primary and secondary sources. Textbooks provide extensive explanations and proofs for the material covered to provide the reader with an opportunity to understand a specific subject thoroughly.

Thomas Register

Use has always been made by many of the Thomas Register of American Manufacturers. These books may occupy considerable real estate either in your office, your company's library, or somewhere in purchasing. The people at Thomas Register have developed (since 2000) two things that are of considerable help to those who have come to rely on this reference. The first are Thomas Register CD-ROMS, and the second is the Thomas Register web site. One can get at the Thomas Register web site at www.thomasregister.com.

The first time user must register to use it, but registration is free. The Thomas Register web site offers several distinct advantages over the traditional printed version. The first few advantages are obvious and not really exciting. One is simply real estate. Finding a place to store (much less use) the bound versions of the register is difficult. If you have access to the Internet, you have access to the register. The second big advantage is, in theory, how current the information is. One would assume that an electronic version would be updated more often than volumes you have sitting on your bookshelf. The improvement in storage space is met with the CD-ROM version as well.

Once registered, you can move directly to the search portion of the site, where you are allowed to search on a company name, product or service, or brand name. Selecting one of these three categories and entering the appropriate key word or words, the register quickly returns a set of broad categories. For example, searching under the word "extrusion" under products/services yielded 157 product headings. Obviously this contains a significant number of categories that are not appropriate for plastics extrusion, but serves as an example. Each product heading is reported along with a set of columns recording the number of companies found, (and now the bigger and very exciting advantages of the web site) the number of companies with on-line catalogues, the number of companies with literature requests via fax, and the number of companies with on-line ordering and links to web sites.

Selecting one of the broad product headings gets one into the listings of the individual companies themselves, where, if available, one can jump to an on-line catalogue, on-line ordering, or move to the company's web site. If none of these features are available, there is a short blurb about the company, location, phone numbers, and what type of products they offer, very similar to the "bare-bones" listing in the bound versions of the register. The designers of the Thomas web site have done an excellent job in that they split the screen when you jump to a company web site. The left-hand side of the screen gives you the Thomas Register choices of contacting the company, etc. while the right hand side is the site of the individual company.

As with any conventional desk reference, the primary means of contact with the bound versions of the Thomas Register is the telephone. With the web site, to be able to go from a search for a list of suppliers of a given item or service (either nationally or by state) to order from an on-line catalogue is a huge advantage and one would suspect, an advantage for a company that offers that option. A cursory stroll through a few randomly chosen categories shows that not everyone is offering on-line catalogues and on-line ordering. Look for these services to grow considerably as more and more people begin to rely on the Internet for goods and services.

Thomas Regional Directory Company has been in business since 1977, and publishes 19 Regional Industrial Buying Guides, in print and on CD-ROM, and now on a web site (www.thomasregional.com). The Thomas Regional Directory is listed as a "partner" to the Thomas Register.

Thomas Regional provides access to a searchable database of more than 480,000 manufacturers, distributors, and service companies organized under 4,500 product/service categories in 19 key U.S. industrial markets. As with the Thomas register, one can search by product/service or company name in the region of interest to you. As Thomas Regional points out in their own introduction.

You can also refine your search based on company type (manufacturer, distributor, manufacturer's rep, and service company), geographic location (state, city/county, area code), trade name, key words, and other specifications such as ISO 9000 certified, and minority and woman-owned businesses. View also supplier brochures, catalogs, line cards, and fax forms and contact suppliers directly via our Contact Company feature.

Thomas Regional also offers listings (by region) of trade shows and special events, including locations, dates, contacts, listings of industry and professional organizations, and government and business resources. Where available, each reference has a link to the web site of the organization in question. They will begin offering some new features that capitalize on their database of companies. As Thomas Regional points out in a press release published on their web site, industrial buyers today face the same recurring problems with the large search engines that researchers, consumers, and virtually everyone else encounters.

These searches generally produce hits in the range of thousands to millions, with far too little of it on target. For that reason, Thomas Regional will be creating a series of web sites that provide buyers with "vertical portals" to specific industries. These portals provide access to Thomas Regional's extensive databases of industrial suppliers, organized according to industry or trade. Thomas Regional is leveraging the usefulness of its content through comprehensive databases that fulfill the specific need of each industry.

Thomas Regional claims that ultimately over 90 industrial "communities" will have their own Thomas-powered web sites tailored to their interests, which will go a long way to improve the efficiency and speed of their searches. The first of these vertical web sites has been launched for the facilities management and engineering profession and may be found at www.facifitiesengineering.com.

Industry Societies

When discussing the subject of periodicals published by societies and commercial publishers, articles are identified usually by issue, and/or volume, date, and page number. Bibliographic control is excellent, and it is usually a routing matter to obtain a copy of a desired article. However, some problems exist such as periodicals that are known by more than one name, and the use of nonstandard abbreviations. Using the International Standard Serial Number (ISSN) that accurately identifies each publication solves both of these problems. With the increasing size and use of automated databases one should consider using ISSN or some other standard.

An important source of the latest information is from papers presented at conferences where the sponsoring agency is frequently a professional society (such as Society of Plastics Engineers) or a department of a university (such as the Plastics Dept., University of Massachusetts Lowell). These conferences are usually annual affairs.

Encyclopedia and Industrial Books

There are many hundreds of encyclopedias and handbooks covering science and technology. Internet sites with comprehensive catalogs of books include amazon.com and barnesandnoble.com. The date of publication should be checked before using any of these works if the required information is likely to have been affected by recent progress. The following list represents only a sampling of available works of outstanding value.

The concern with industrial health and safety has placed an additional responsibility on the designer to see that materials and products are handled in a safe manner. Sax's Dangerous Properties of Industrial Materials (Kluwer, 1996) provides an authoritative treatment of this subject. This book also covers handling and shipping regulations for a large variety of materials and products.

Designers are concerned with the interaction between humans and machines. This area has become increasingly sophisticated and specialized. Books on Human Factors have been written for the design engineer rather than the human factor specialist. The books provide the engineer with guidelines for designing products for convenient use by people.

Standards

Government agencies, professional societies, businesses, and organizations devoted almost exclusively to the production of standards produce codes, specifications, and standards. In USA the American National Standards Institute (ANSI located in New York City) acts as a clearing house for industrial standards. ANSI frequently represents the interests of USA industries at international meetings. Copies of standards from most industrial countries can be purchased from ANSI as well as from the originators.

Copies of standards issued by government agencies are available from several centers maintained by the government for the distribution of publications. Most libraries do not collect government specifications. Many of the major engineering societies issue specifications in areas related to their functions. These specifications are usually developed, and revised, by membership committees.

The American Society of Mechanical Engineers (ASME) has been a pioneer in publishing codes concerned with areas in which mechanical engineers are active. As an example in 1885 ASME formed a Standardization Committee on Pipe and Pipe Threads to provide for greater interchangeability.

A frequently used collection of specifications is the Annual Book of Standards (53-55) issued by the American Society for Testing and Materials (ASTM). Committees drawn primarily from the industry most immediately concerned with the topic prepare these standards. A member of the standards department usually prepares the standards written by individual companies. They are frequently almost identical to standards issued by societies and government agencies and make frequent references to these standards. The main reason for these in-house standards is to enable the company to revise a standard quickly in order to impose special requirements on a vendor.

The USA government is the largest publisher in the world. Most of the publications are available from the Superintendent of Documents (USA Government Printing Office, Washington, DC 20402). Publication catalogs are available on the Government Printing Office web site, GPO.gov. Increasingly, the GPO is relying on electronic dissemination rather than print. These publications are provided, free of charge, to depository libraries throughout the country. Depository libraries are obligated to keep these publications for a minimum of five years and to make them readily available to the public. The government agencies most likely to publish information of interest to engineers are probably the National Institute of Science and Technology, the Geological Survey, the National Oceanic and Atmospheric Administration, and the National Technical Information Service.

The large number of standards issued by a variety of organizations has resulted in a number of identical or equivalent standards. IHS (Information Handling Services, 15 Inverness Way East, Englewood, Co 80150) makes available virtually all standards on CD-ROM.

Engineering Information

The most comprehensive collections of engineering information can be found at large research libraries. In USA they include those in New York City, Boston, Chicago, and Los Angeles. These libraries are accessible to the public. They provide duplicating services and will answer telephoned or written reference questions. Substantial collections also exist at universities and engineering schools. These libraries are intended for use by faculty and students, but outsiders can frequently obtain permission to use these libraries by appointment, upon payment of a library fee, or through a cooperative arrangement with a public library. Special libraries in business and industry frequently have excellent collections on the subjects most directly related to their activity. They are usually only available for use by employees and the company.

Public libraries vary considerably in size, and the collection will usually reflect the special interests of the community. Central libraries, particularly in large cities, may have a considerable collection of engineering books and periodicals. Online searching is becoming an increasingly frequent service that is provided by public libraries. Regardless of the size of, a library, the reference librarian should prove helpful in obtaining materials not locally available. These services include inter-library loans from networks, issuing of courtesy cards to provide access to nonpublic libraries, and providing the location of the nearest library that owns needed materials.

Information Broker

In the past couple of decades a large number of information brokers have come into existence. For an international listing see Burwell World Directory of Information Brokers (Burwell Enterprises, Houston, TX 1996, etc.). Information brokers can be of considerable use in researching the literature and retrieving information, particularly in situations where the designer or engineer does not have the time and resources to do the searching. The larger brokers have a staff of trained information specialists skilled in online and manual searching. Retrieval of needed items is usually accomplished by sending a messenger to make copies at a library.

Most information brokers are located near research libraries or are part of an information center. The larger information brokers usually cover all subjects and offer additional services, such as translating foreign language materials. Smaller brokers, and those associated with a specialized agency, frequently offer searching in a limited number of subjects. The selection of the most appropriate information broker should receive considerable attention if a large amount of work is required or a continuing relationship is expected.

Engineering Societies and Associations

Societies and associations have exerted a strong influence on the development of the designer, engineer, and other professionals. As an example the ASME (American Society of Mechanical Engineers) publishes the periodicals in order to keep individuals informed of new developments and forward other important information. Examples of the periodicals include: (1) *Applied Mechanics Reviews* (monthly), (2) *CIME* (Computers in Mechanical Engineering, published by Springer-Veriag, New York), (3) *Mechanical Engineering* (monthly), and (4) *Transactions* (quarterly). The *Transactions* include the following areas: heat transfer, applied mechanics, bioengineering, energy resources technology, solar energy engineering, dynamic systems, measurement & control, and engineering materials and technology.

Many engineering societies have prepared a code of ethics in order to guide and protect engineers. Societies frequently represent the interests of the profession at government hearings and keep the public informed on important issues. They also provide an opportunity for continuing education, particularly for preparing for professional engineer's examinations. Examples of societies and trade associations in USA are:

Adhesives Manufacturers Assoc.
American Chemical Society
American Electronics Assoc.
American Institute of Chemical
  Engineers
American Institute of Steel
  Construction
American Mold Builders Assoc.
American
American Society of Civil
  Engineers
American Society of Heating,
  Refrigerating, & Air-
  Conditioning Engineers
American Society of Mechanical
  Engineers
Association of Rotational Molders
Composites Fabricators Assoc.
Institute of Electrical & Electronics
  Engineers
Instrument Society of America
National Association of Corrosion
  Engineers
National Electrical Manufacturers
  Association
National Fire Protection
  Association
Plastics Institute of America
Society of Automotive Engineers
Society of Plastics Engineers
Society of Plastics Industry
Underwriters Laboratories

Designs

Design books are listed in the Bibliography. They concentrate on different aspects of designing with plastics of which they have been referenced throughout the book.

Databases

Examples of hard copy databases for plastic material selections are listed in Chapter 5 and the Bibliography sections.

Websites

In addition to what has been included in this book here are additional examples of different websites that provide important services:

Reinforced plastics www.reinforcedplastics.com Elsevier/RP provides information and answers to questions concerning reinforced plastics data, latest research, buyer's guide, etc.

IBM patents website http://www.patents.IBM.com The IBM Intellectual

Property Network (IPN) has evolved into a premier Website for searching, viewing, and analyzing patent documents. The IPN provides you with free access to a wide variety of data collections and patent information.

Federal web locator http://www.infoctr.edu/fwl/ The Federal Web Locator is a service provided by the Center for Information Law and Policy and is intended to be the one stop shopping point for federal government information on the World Wide Web. This site is hosted by the Information Center at Chicago-Kent College of Law, Illinois Institute of Technology.

MAACK Business Services A Maack & Scheidl Partnership CH-8804 Au/near Zürich, Switzerland tel:+41-1-781 3040, Fax:+41-1-781 1569, http://www.MBSpolymer.com. Plastics technology and marketing business service, which organizes global conferences, and edits a range of reports and studies, which focus on important worldwide aspects of polymer research, development, production, and end uses. Provides updates on plastic costs, pricing, forecast, supply/demand, and analysis. Identified early in the cycle are trends in production, products and market segments.

Material Safety Data Sheets (MSDS) http://www.msdssearch.com/ msdssearch.com, Inc., is a National MSDS Repository, providing FREE access to over 1,000,000 Material Safety Data Sheets; the largest centralized reference source available on the Internet. msdssearch.com is dedicated to providing the most comprehensive single source of information related to the document known as a Material Safety Data Sheet (MSDS). MSDS SEARCH serves as the conduit between users of MSDSs and any reliable supplier. msdssearch.com provides access to 350K MSDSs from over 1600 manufacturers, 700K MSDSs from public access databases, links to MSDS software, services, training and product providers, links to Government MSDS information, an MSDS discussion forum where you can ask questions, and supplies MSDSs directly from manufacturers via search engine.

The Canadian Center for Occupational Health and Safety CCOHS, 250 Main Street East, Hamilton ON L8N 1H6 Canada, tel: 1-800-263-8466 (toll free in Canada only)/1-905-572-4400, Fax: 1-905-572-4500, http://www.ccohs.ca/products/databases/msds.html. Promotes a safe and healthy working environment by providing information and advice about occupational health and safety.

Training programs

An example of a training program offered to the plastic industry is presented. The Plastics Institute of America (PIA), in collaboration with the Division of Continuing Studies and Corporate Education at the University of Massachusetts Lowell, offers a series of modules providing employee training designed to enhance the knowledge, understanding and skills of mechanics and other technical staff working primarily with plastics.

Examples of their training programs follow:

Control Systems (40 hour Module)

- Safety (Ohm's Law) lockout tagout importance, machine guarding schemes
- IO devices (operations, identification, advantages and disadvantages, purposes of encoders and resolvers)
- PLCs (basic components, analog and digital domains, basic ladder logic instruction)
- HMI (password protection, HMI functions and purpose)
- Electronic Cam Switch Bank (function review)
- Control Systems Evolution (definition, examples)
- Troubleshooting and Diagnostics (mechanical and non-mechanical problems, determination of probable problem cause, mechanic's responsibilities.)
- Servo Systems (concepts and purpose, type motors and variations, explanations and applications of direct, gearbox and screw motor-to-load coupling, motor and encoder shaft alignments, servo system concepts, servo tuning, servo profiles)

Metric Measurement (12 hour Module)

- Linear Measurement (micrometers, dial calipers, surface plates, optical comparators, scientific rotation)
- Temperature (thermometers, thermocouples, RTDs)
- Hardness/Friction/Gloss/Color
- Calibration
- Measurement and inspection re: quality control
- Overview of sizes (pins, holes, identification of sizes)
- Pressure (strain gauges, piezoelectric transducers, Bourbon gauges)
- Test methods of Plastics ASTM/ISO

Engineering Drawing (20 hour Module)

- Sketching
- Print reading and interpretation
- Standard notation and symbols
- Assembly and part drawings
- Surface finishes
- Basic machining processes and expected outcomes
- Screw threads and fasteners
- Basic shop terminology

Computer Fundamentals (12 hour Module)

- Overview of computer systems
- Windows 98: help screens, mouse, pull down menus, icons, passwords, menu bars, etc.
- Introduction to database management and databases
- Navigating through databases
- Interpreting screens produced by databases
  - Interpreting database results
  - Databases
- Communicating with remote sites

Statistical Process Control (20 hour Module)

- Fundamental mathematical skills
- Target values and variances
- Process variability
- Processes that are out of control
- Use of SPC to improve the processes

Polymers (20 hour Module)

- Markets for plastics
- Commercial production of plastics
- Physical properties of plastics
- Fabrication of plastics
- Applications

Introduction to Injection Molding (16 hour Module)

- What is injection molding?
- Plasticating systems
- Clamping systems
- The electrical system

- The injection mold
- The molding process
- Process conditions
- Auxiliary equipment
- Resins-processing
- Troubleshooting
- Batch mixing
- Profile extrusions

Industrial and Molding Hydraulics (20 hour Module)

- Standards, basic fluid power law and terminology
- ANSI/ISO Circuit symbols, print interpretation
- Circuit elements and their functions
- Mechanical descriptions
- Control concepts
- Examples of molding circuits and automation circuits

Mold Design and Maintenance for Diagnostics (20 hour Module)

- Design considerations
- Mold design basics
- Cavity and core construction
- Heat transfer considerations
- Cold runner molds
- Hot runner systems
- Freeing mechanism/part ejection
- Mold maintenance

Elastomers (16 hour Module)

- General classes of elastomers
- Compounding and the rubber recipe
- Vulcanization and vulcanizing agents
- Fillers
- Processing and processability testing
- Physical testing
- Thermoplastic elastomers (TPE)

Rotational Molding (16 hour Module)

- Molds
- Equipment
- Process parameters

- Materials
- Design guidelines
- Secondary finishing
- Troubleshooting

Preventive /Predictive Maintenance (20 hour Module)

- Injection molding plasticating unit
- Heating units
- Injection molding hydraulic maintenance
- Care and maintenance of electrical components
- Safety inspection and procedures
- Storage maintenance of molds

Blow Molding (16 hour Module)

- Blow molding processes
- Materials
- Primary equipment
- Mold design
- Process controls
- Auxiliary equipment
- Troubleshooting
- Testing
- New developments

Extrusion (20 hour Module)

- Principles of extrusion
- Description of single screw extruder
- Smooth bore and grooved-feed extruders
- Blown film process
- Cast film process
- Extrusion coating
- Profile extrusion
- Materials for extrusion
- Auxiliary equipment
- Die design
- Principles process control
- Troubleshooting
- New developments

Plastics Process Control (20 hour Module)

- Principles of process control
- Instrumentation
- Data acquisition/monitoring
- Servo control for injection molding
- Control of extrusion processes
- Blow molding/parison control
- SPC/SQC
- Integrated manufacturing
- New developments

Thermoforming (20 hour Module)

- Basic process/variations
- Processing conditions
- Materials
- Mold design
- Product design
- Secondary operations
- Twin sheet forming
- Decorating
- Trimming/recycling

7 DESIGN RELIABILITY

Product design starts when one visualizes a certain material, makes approximate calculations to see if the contemplated idea is practical to meet requirements that includes cost, and, if the answer is favorable, proceeds to collect detailed data on a range of materials that may be considered for the new product. The application of appropriate data to product design can mean the difference between the success and failure of manufactured products made from any material. The available plastic test data requires an understanding and proper interpretation before an attempt can be made to apply them to the product design.

There are two important sources of information on plastics. There is the data sheet compiled by a manufacturer of the material and derived from tests conducted in accordance with standardized specifications. The other source is the description of outstanding characteristics of each plastic, along with the listing of typical applications. If suppliers' data were to be applied without a complete analysis of the test data for each property, the result could prove costly and embarrassing.

The nature of plastic materials is such that an oversight of even a small detail in its properties or the method by which they were derived could result in problems and product failure. Once it is recognized that there are certain reservations with some of the properties given on the data sheet, it becomes obvious that it is very important for the designer to have a good understanding of these properties. Thus the designer can interpret the test results in order to make the proper evaluation in selecting a material for a specific product.

Testing

When discussing testing they range from material to product testing. With no prior history or no related data available on a material or product, the usual approach is to conduct tests on the material and finished fabricated product. Choosing and testing a plastic when only a few existed that could be used for specific products would prove relatively simple, but the variety of plastics has proliferated (35,000 worldwide). Today's plastics are also more complex, complicating not only the choice but also the necessary tests. Fillers and additives can drastically change the plastic's basic characteristics, blurring the line between commodity and engineering plastics (Chapter 1). Entirely new plastics have been introduced with esoteric molecular structures. Therefore, plastic suppliers now have many more sophisticated tests to determine which plastic best suits a product design or fabricating process.

For the product designer, however, a few basic tests, such as a tensile test, will help determine which plastic is best to meet the performance requirements of a product. At times, the complex test may be required. The test or tests to be used will depend on the product's performance requirements.

To ensure quality control, material suppliers and developers routinely measure such complex properties as molecular weight and its distribution, stereochemistry, crystallinity and crystalline lattice geometry, and detailed fracture characteristics. They use complex, specialized tests such as gel permeation chromatography, wide- and narrow-angle X-ray diffraction, scanning electron microscopy, and high-temperature pressurized solvent reaction tests to develop new polymers and plastics applications.

Understanding and proper applications of the many different tests is rather an endless project. There are destructive and nondestructive tests (NDTs). Most important, they are essential for determining the performance of plastic materials to be processed and of the finished fabricated products. Testing refers to the determination by technical means properties and performances to meet product performance requirements. This action, when possible, should involve application of established scientific principles and procedures. It requires specifying what requirements are to be met. There are many different tests (thousands) that can be conducted that relate to practically any product or material requirement. Usually only a few will be applicable to meet your specific application.

A different type of evaluation is the potential of a material that comes in contact with a medical patient to cause or incite the growth of malignant cells (that is, its carcinogenicity). It is among the issues addressed in the set of biocompatibility standards and tests developed as part 3 of ISO-10993 standard that pertain to genotoxicity, carcinogenicity, and reproductive toxicity. It describes carcinogenicity testing as a means to determine the tumorigenic potential of devices, materials, and/or extracts to either a single or multiple exposures over a period of the total life span of the test animal. The circumstance under which such an investigation may be required is given in part 1 of ISO-10993.

There is usually more than one test method to determine a performance because each test has its own behavior and meaning. As an example there are different tests used to determine the abrasion resistance of materials. There is the popular ASTM Taber abrasion test. It determines the weight loss of a plastic or other material after it is subjected to abrasion for a prescribed number of the abrader disk rotations (usually 1000). The abrader consists of an idling abrasive speed controlled rotating wheel with the load applied to the wheel. The abrasive action on the circular specimen is subjected to a rotary motion.

Other abrasion tests have other types of action such as back and forth motion, one direction, etc. These different tests provide different results that can have certain relations to the performance of a product that will be subjected to abrasion in service.

With the more popular destructive testing, the original configuration of a test specimen and/or product is changed, distorted, or usually destroyed. The test provides information such as the amount of force that the material can withstand before it exceeds its elastic limit and permanently distorts (yield strength) or the amount of force needed to break it. These data are quantitative and can be used to design structural products that would withstand a certain static load, heavy traffic usage, etc.

The primary purposes of testing related to shock and vibration are to verify and characterize the dynamic response of the equipment and components thereof to a dynamic environment and to demonstrate that the final design will withstand the test environment specified for the product under evaluation (Chapter 2). Basic characterization testing is usually performed on an electrodynamic vibration machine with the unit under test hard-mounted to a vibration fixture that has no resonance in the pass band of the excitation spectrum. The test input is a low-displacement-level sinusoid that is slowly varied in frequency (swept) over the frequency range of interest. Since sweep testing produces a history of the response (displacement or acceleration) at selected points on the equipment to sinusoidal excitation over the tested excitation frequencies and displacements.

Caution is advised when using a hard-mount vibration fixture, as the fixture is very stiff and capable of injecting more energy into a test specimen at specimen resonance than would be experienced in service. For this reason, the test-input signal should be of low amplitude. In service, the reaction of a less stiff mounting structure to the specimen at specimen resonance would significantly reduce the energy injected into the specimen. If a specimen response history is known prior to testing, the test system may be set to control input levels to reproduce the response history as measured by a control accelerometer placed at the location on the test specimen where the field vibration history was measured.

Vibration-test information is used to aid in adjusting the design to avoid unfavorable responses to service excitation, such as the occurrence of coupled resonance. It is a component having a resonance frequency coincident with the resonance frequency of its supporting structure, or structure having a significant resonance which coincides with the frequency of an input shock spectrum. Individual components are often tested to determine and document the excitation levels and frequencies at which they do not perform. This type of testing is fundamental to both shock and vibration design.

For more complex vibration-service input spectra, such as multiple sinusoidal or random vibration spectra, additional testing is performed, using the more complex input waveform on product elements to gain assurance that the responses thereof are predictable. The final test exposes the equipment to specified vibration frequencies, levels, and duration, which may vary by axis of excitation and may be combined with other variables such as temperature, humidity, and altitude environments,

NDT examines material without impairing its ultimate usefulness. It does not distort the specimen and provides useful data. NDT allows suppositions about the shape, severity, extent, distribution, and location of such internal and subsurface residual stresses; defects such as voids, shrinkage, cracks, etc. Test methods include acoustic emission, radiography, IR spectroscopy, x-ray spectroscopy, magnetic resonance spectroscopy, ultrasonic, liquid penetrant, photoelastic stress analysis, vision system, holography, electrical analysis, magnetic flux field, manual tapping, microwave, and birefringence (Table 7.1).

Table 7.1 Examples of nondestructive test methods

Method	Typical Flaws Detected	Typical Application	Advantages	Disadvantages
Radiography	Voids, porposity, inclusions, and cracks	Castings, forgings, weldments, and structural assemblies	Detects internal flaws; useful on a wide variety of geometric shapes; portable; provides a permanent record	High cost; insensitive to thin laminar flaws, such as tight fatigue cracks and delaminations; potential health hazard
Liquid penetrants	Cracks, gouges, porosity, laps, and seams open to a surface	Castings, forgings, weldments, and components subject to fatigue or stress-corrosion cracking	Inexpensive; easy to apply; portable; easily interpreted	Flaw must be open to an accessible surface, level of detectability operator-dependent
Eddy current testing	Cracks, and variations in alloy composition or heat treatment, wall thickness, dimensions	Tubing, local regions of sheet metal, alloy sorting, and coating thickness measurement	Moderate cost, readily automated; portable	Detects flaws that change in conductivity of metals; shallow penetration; geometry-sensitive
Magnetic particles	Cracks, laps, voids, porosity and inclusions	Castings, forgings, and extrusions	Simple; inexpensive; detects shallow subsurface flaws as well as surface flaws	Useful for ferromagnetic materials only; surface preparation required, irrelevant indications often occur; operator-dependent
Thermal testing	Voids or disbands in both metallic and nonmetallic materials, location of hot or cold spots in thermally active assemblies	Laminated structures, honeycomb, and electronic circuit boards	Produces a thermal image that is easily interpreted	Difficult to control surface emissivity; poor discrimination
Ultrasonic testing	Cracks, voids, porosity, inclusions and delaminations and lack of bonding between dissimilar materials	Composites, forgings, castings, and weldments and pipes	Exellent depth penetration; good sensitivity and resolution; can provide permanent record	Requires acoustic coupling to component; slow; interpretation is often difficult

To determine the strength and endurance of a material under stress, it is necessary to characterize its mechanical behavior. Moduli, strain, strength, toughness, etc. can be measured microscopically in addition to conventional testing methods. These parameters are useful for design and material selection. They have to be understood as to applying their mechanisms of deformation and fracture because of the viscoelastic behavior of plastics (Chapters 1 and 2). The fracture behavior of materials, especially microscopically brittle materials, is governed by the microscopic mechanisms operating in a heterogeneous zone at the crack tip or stress raising flow.

In order to supplement micro-mechanical investigations and advance knowledge of the fracture process, micro-mechanical measurements in the deformation zone are required to determine local stresses and strains. In TPs (thermoplastics), craze zones can develop that are important microscopic features around a crack tip governing strength behavior. For certain plastics fracture is preceded by the formation of a craze zone that is a wedge shaped region spanned by oriented microfibrils. Methods of craze zone measurements include optical emission spectroscopy, diffraction techniques, scanning electron microscope, and transmission electron microscopy.

Conditioning procedures of test specimens and products are important in order to obtain reliable, comparable, and repeatable data within the same or different testing laboratories. Procedures are described in various specifications or standards such as having a standard laboratory atmosphere [50 ± 2% relative humidity, 73.4 ± 1.8F (23 ± 1C)] with adequate air circulation around all specimens. The reason for this type or other conditioning is due to the fact that the temperature and moisture content of plastics affects different properties.

Classifying Test

Properties of plastics such as physical, mechanical, and chemical are governed by their molecular weight, molecular weight distribution, molecular structure, and other molecular parameters; also the additives, fillers, and reinforcements that enhance certain processing and/or performance characteristics (Chapter 1). Properties are also effected by their previous history (includes recycled plastics), since the transormation of plastic materials into products is through the application of heat and pressure involving many different fabricating processes. Thus, variations in properties of products can occur even when the same plastic and processing equipment are used. Conducting tests such as those related to molecular characteristics and melt flow provides a means of classifying them based on test results.

Laboratory

There are different industry laboratory organizations providing testing, specifications, standards, and/or certifications. They provide updated information to meet different requirements such as aiding processors in controlling product quality, meet safety requirements, etc. Examples of important organizations include ASTM, DIN, ISO, and UL. Table 7.2 includes a list of organizations worldwide involved in preparing or coordinating testing specifications, regulations, and standards. Note that previously issued test procedures and standards are subject to change and being updated periodically. As an example ASTM issues annual publications that include all changes.

Table 7.2 Organizations involved in specifications, regulations, and standards

ASTM. American Society for Testing and Materials.	IEC. International Electrotechincal Commission.
UL. Underwriters Laboratories.	IEEE. Institute of Electrical and Electronic Engineers.
ISO. International Organization for Standardization.	ISA. Instrument Society of America.
DIN. Deutsches Instut, Normung.	JIS. Japanese Industrial Standards.
ACS. American Chemical Society.	NADC. Naval Air Development.
ANSI. American National Standards Institute.	NACE. National Association of Corrosion Engineers.
ASCE. American Society of Chemical Engineers.	NAHB. National Association of Home Builders.
ASM. American Society of Metals.	NEMA. National Electrical Manufacturers' Association.
ASME. American Society of Mechanical Engineers.	NFPA. National Fire Protection Association.
BMI. Battele Memorial Institute.	NIST. National Institute of Standards & Technology (previously the na6tional Bureau of Standards).
BSI. British Standards Institute.	
CPSC. Consumer Product Safety Commission.	
CSA. Canadian Standards Association.	NIOSH. National Institute for Occupational Safety & Health.
DOD. Department of Defense.	
DOSISS. Department of Defense Index & Specifications & Standards.	OSHA. Occupational Safety & Health Administration.
DOT. Department of Transportation.	PLASTEC. Plastics Technical Evaluation Center.
EIA. Electronic Industry Association.	PPI. Plastics Pipe Institute.
EPA. Environmental Protection Agency.	QPL. Qualified Products List.
FMRC. Factory Mutual Research Corporation.	SAE. Society of Automotive Engineers.
FDA. Food and Drug Administration.	SPE. Society of Plastics Engineers.
FTC. Federal Trade Commission.	SPI. Society of the Plastics Industry.
IAPMO. International Association of Plumbing & Mechanical Officials.	TAPPI. Technical Association of the Pulp and Paper Industry.

Available is a USA government directory list for various forms of testing worldwide that includes plastics. The National Voluntary Accreditation Program (NVLAP) endorses them. The directory is available from NIST, NVLAP Directory, A124 Building, Gaithersburg, MD 20899 USA.

Quality control

Different approaches are used in setting up QC. QC as in testing are discussed but often the least understood. Usually it involves the inspection of materials and products as they complete different phases of processing. Products that are within specifications proceed, while those that are out of specification are either repaired, recycled, or scrapped. Possibly the workers who made the out-of-spec products are notified so "they" can correct "their" mistake.

The approach just outlined is after-the-fact approach to QC; all defects caught in this manner are already present in the product being processed. This type of QC will usually catch defects and is necessary, but it does little to correct the basic problem(s) in production. One of the problems with add-on QC of this type is that it constitutes one of the least cost-effective ways of obtaining high quality products. Quality must be built into a product from the beginning of the design that follows the FALLO approach (Fig. 1.15); it cannot be inspected into the process. The target is to control quality before a product becomes defective.

Quality and Reliability

The lying down of levels of quality and reliability necessary to ensure product success and acceptability in a particular market is a cause for increasing concern. They are the most difficult aspects to quantify in absolute terms, although statistical data from company product precedents are helpful here. There are expressions used such as mean time before failure (MTBF) and mean time to repair (MTTR) that are used with mechanical, hydraulic, pneumatic, electrical, and other products. Nonetheless, some quantitative expression must be made in respect of quality and reliability at the initial design specification stage. A Company must ensure adequate feedback of any failure analysis to the design team.

Total Quality Management

In today's competitive environment, in order to survive, companies are focusing their entire organization on customer satisfaction. The approach followed for ensuring customer satisfaction is known as Total Quality Management (TQM). The challenge is to manage so that the total quality is experienced in an effective manner. This approach dates back to 1916. The beginning of TQM is during the 1940s, when such figures as W. E. Deming took an active role. In subsequent years, the

TQM approach was more widely practiced in Japan than anywhere else. In 1951, the Japanese Union of Scientists and Engineers introduced a prize, named after W. E. Deming, for the organization that implemented the most successful quality policies. On similar lines, in 1987, the USA government introduced an award related to TQM.

The consideration of quality in design begins during the specification-writing phase. Many factors contribute to the success of the quality consideration in engineering or mechanical design. Quality cannot be inspected out of a product; it must be built in. TQM is a useful tool for application during the design phase. Deming's approach to TQM involves the following fourteen-point approach:

1. Establish consistency of purpose for improving services.

2. Adopt the new philosophy for making the accepted levels of defects, delays, or mistakes unwanted.

3. Stop reliance on mass inspection as it neither improves nor guarantees quality. Remember that teamwork between the firm and its suppliers is the way for the process of improvement.

4. Stop awarding business with respect to the price.

5. Discover problems. Management must work continually to improve the system.

6. Take advantage of modem methods used for training. In developing a training program, take into consideration such items as:
   - Identification of company objectives
   - Identification of the training goals
   - Understanding of goals by everyone involved
   - Orientation of new employees
   - Training of supervisors in statistical thinking
   - Team-building
   - Analysis of the teaching need

7. Institute modem supervision approaches.

8. Eradicate fear so that everyone involved may work to their full capacity.

9. Tear down department barriers so that everyone can work as a team member.

10. Eliminate items such as goals, posters, and slogans that call for new productivity levels without the improvement of methods.

11. Make your organization free of work standards prescribing numeric quotas.

12. Eliminate factors that inhibit employee workmanship pride.

13. Establish an effective education and training program.

14. Develop a program that will push the above 13 points every day for never-ending improvement.

Quality and Design

TQM will help to improve design quality, specific quality-related steps are also necessary during the design phase. These additional steps will further enhance the product design. An informal review during specification writing may be regarded as the beginning of quality assurance in the design phase. As soon as the first draft of the specification is complete, the detailed analysis begins.

Statistics

Statistics basically is a summary value calculated from the observed values in a product or sample. It is a branch of mathematics dealing with the collection, analysis, interpretation, and presentation of masses of numerical data. The word statistic has two generally accepted meanings: (1) a collection of quantitative analysis data (data collection) pertaining to any subject or group, especially when the data are systematically gathered and collated and (2) the science that deals with the collection, tabulation, analysis, interpretation, and presentation of quantitative data.

As an example of statistical analysis of products there is statistical process control (SPC). It is an important on-line method in real time by which a production process can be monitored and control plans can be initiated to keep quality standards within acceptable limits. Statistical quality control (SQC) provides off-line analysis of the big picture such as what was the impact of previous improvements. It is important to understand how SPC operates.

There are basically two possible approaches for real-time SPC. The first, done on-line, involves the rapid dimensional measurement of a part or a non-dimensional bulk parameter such as weight and is the more practical method. In the second approach, contrast to weight, other dimensional measurements of the precision needed for SPC are generally done off-line. Obtaining the final dimensional stability needed to measure a part may take time. As an example, amorphous injection molded plastic parts usually require at least a half-hour to stabilize.

The SPC system starts with the premise that the specifications for a product can be defined in terms of the product's (customer's) requirements, or that a product is or has been produced that will satisfy those needs. Generally a computer communicates with a series of process sensors and/or controllers that operate in individual data loops.

The computer sends set points (built on which performance characteristics of the product must have) to the process controller that constantly feeds back to the computer to signal whether or not the set of points are in fact maintained. The systems are programmed to act when key variables affecting product quality deviate beyond set limits.

To target for better yields, higher quality, and increased profits, fabricators should consider the SPC and SQC techniques as standard tools for understanding, validating, and improving processes in all areas of manufacture that includes product distribution, transportation, and accounting. Using on-line software, SPC provides the close-up view; using off-line software, SQC detects differences over time. These two techniques provide two different essential functions.

Prior to the widespread implementation of supervisory control and data acquisition (SCADA) and human-machine interface (HMI) systems, most SPC and SQC was performed by quality-control departments as an off-line process. Data was collected from test stations, laboratories, etc. and statistical analysis was performed later. SCADA/HMI systems, however, have made it feasible to provide plant-floor SPC charts using data collected in real time directly from the process. Fabricators that want to standardize SPC and SQC to increase their use find they need the two following functions: (1) provide the plant floor with SPC charts and (2) make data collected by SCADA systems available for off-line analysis. Available is SPC and SQC software to support these efforts. Recognize that the bulk of SPC's value is derived from process improvements developed from offline SQC analysis.

Virtually all-classical design equations assume single-valued, real numbers. Such numbers can be multiplied, divided, or otherwise subjected to real-number operations to yield a single-valued, real number solution. However, statistical materials selection, because it deals with the statistical nature of property values, relies on the algebra of random variables. Property values described by random variables will have a mean value, representing the most typical value, and a standard deviation that represents the distribution of values around the mean value.

This requires treating the mean values and standard deviations of particular property measurements according to a special set of laws for the algebra of random variables. Extensive information can be found in statistical text. The algebra of random variables shares many elements of structure in common with the algebra of real numbers, such as the associative and cumulative laws, and the uniqueness of sum and product. In combinations of addition and multiplication, the distributive law holds true.

Testing, QC, Statistics, and People

Personnel or operators involved in testing from raw materials to fabricating products develop capability via proper training and experience. Experience and/or developing the proper knowledge are required to determine the tests to be conducted. At times, with new problems developing on-line, different tests are required that may be available or have to be developed. Unfortunately a great deal of "reinventing the wheel" can easily occur so someone should have the responsibility to be up to date on what is available.

Another unfortunate or fortunate situation exists that a very viable test was at one time developed and used within the industry. In time it was changed many times by different companies and organizations (ASTM, ISO, etc.) to meet new industries needs concerning specific requirements. One studying the potential of using that particular test may not have the access to the basic test that probably is all that is required.

Product failure

The process of analyzing designs includes the modes of failure analysis reviewed in this book. At an early stage the designer should try to anticipate how and where a design is most likely to fail. The most common conditions of possible failure are elastic deflection, inelastic deformation, and fracture. During elastic deflection a part fails because the loads applied produce too large a deflection. In deformation, if it is too great it may cause other parts of an assembly to become misaligned or overstressed. Dynamic deflection can produce unacceptable vibration and noise. When a stable structure is required, the amount of deflection can set the limit for buckling loads.

Because many plastics are relatively flexible, analysis should consider how much deflection might result from the loadings and elevated temperatures any given part might see in service. The equations for predicting such deflections should use the modulus of the material; its tensile strength is not pertinent. Usually, the most effective way to reduce deflection is to stiffen a part's wall by changing its cross-section.

Inelastic deformation causes part failure arising out of a massive realignment of the molecular structure. A part undergoing inelastic deformation does not return to its original state when its load is removed. It should be remembered that there are plastics that are sensitive to this situation and others that are not.

The existence of an elevated temperature, with or without long-term or continuous loading, would suggest the possibility that a material might exceed its elastic limits. Regarding momentary loading, the properties to consider are the proportional limit and the maximum shear stress.

The presence of fracture reflects a load that exceeds the strength of the design. The load may occur suddenly, such as upon impact, or at a low temperature, which will reduce the elongation of the material. A failure may develop slowly, from a steady, high load applied over a long time (creep rupture) or from the gradual growth of a crack from fatigue. If fracture is the expected mode of failure, analysis should examine the greatest principal stresses involved.

To ensure that a design functions as intended for the prescribed design lifetime and, at the same time, that it is competitive in the marketplace. Success in designing competitive products while averting premature mechanical failures can be reached consistently only by recognizing and evaluating all potential modes of failure that might influence the design. To recognize potential failures a designer must be acquainted with the array of failure modes observed in practice, and with the conditions leading to these failures.

A failure may be defined as the physical process or processes that take place or that combine their effects to produce a failure. In the commonly observed failures it may be noted that some failures are a single phenomena, whereas others are combined phenomena. For example, fatigue is listed as a failure, moisture is listed as a failure, and moisture fatigue is listed as another failure. These combinations are included because they are commonly observed usually resulting in a synergistic response. Failures to be reviewed only occur when they generate a set of circumstances that interferes with the proper functioning of a product. With the proper use of the plastic material required to meet the designed product performance requirements the failures reviewed will be rare, actually should not occur.

Failures in products almost always initiate at sites of local stress concentration caused by geometrical or micro-structural discontinuities. These stress concentrations, or stress raisers, often lead to local stresses many times higher than the nominal net section stress that would be calculated without considering stress concentration effects. Thinking in terms of "force flow" through a member may develop an intuitive appreciation of the stress concentration associated with a geometrical discontinuity as it is subjected to external loads.

Force and/or temperature-induced elastic deformation failure occurs whenever the elastic (recoverable) deformation in a product, brought about by the imposed operational loads or temperatures, becomes large enough to interfere with its ability to perform its intended function satisfactorily.

Yielding failure occurs when the unrecoverable deformation in a ductile product, brought about by the imposed operational loads or motions, becomes large enough to interfere with the ability of the product to not meet its performance requirements.

Ductile rupture failure occurs when the plastic deformation, in a part that exhibit ductile behavior, is carried to the extreme so that the member separates into two pieces. Initiation and coalescence of internal voids slowly propagate to failure, leaving a dull, fibrous rupture surface.

Fatigue failure refers to the sudden and catastrophic separation of a part into two or more pieces as a result of the application of fluctuating loads or deformations over a period of time. Failure takes place by the initiation and propagation of a crack until it becomes unstable and propagates suddenly to failure. The loads and deformations that typically cause failure by fatigue are far below the static failure levels. When loads or deformations are of such magnitude that more or less than about 10,000 cycles are required to produce failure, it is usually termed high-cycle fatigue or low-cycle fatigue, respectively. When a fluctuating temperature field in the part produces load or strain cycling, the process is usually termed thermal fatigue. Surface fatigue failure, usually associated with rolling surfaces in contact, manifests itself as pitting, cracking, and spalling of the contacting surfaces as a result of the cyclic contact stresses that result in maximum values of cyclic shear stresses slightly below the surface. The cyclic subsurface shear stresses generate cracks that propagate to the contacting surface.

Creep failure results whenever the deformation in a part accrues over a period of time under the influence of stress and temperature until the accumulated dimensional changes interfere with the ability of the part to perform satisfactorily its intended function. Three stages of creep are often observed: (1) transient or primary creep during which time the rate of strain decreases, (2) steady-state or secondary creep during which time the rate of strain is virtually constant, and (3) tertiary creep during which time the creep strain rate increases, often rapidly, until rupture occurs. This terminal rupture is often called creep rupture and may or may not occur depending on the stress-time-temperature conditions.

Creep buckling failure occurs when, after a period of time, the creep process results in an unstable combination of the loading and geometry of a part so that the critical buckling limit is exceeded and failure ensues.

Buckling failure occurs when, because of a critical combination of magnitude and/or point of load application, together with the geometrical configuration of a part, the deflection of the member suddenly increases greatly with only a slight change in load. This nonlinear response results in a buckling failure if the buckled member is no longer capable of performing its design function.

Stress rupture failure is intimately related to the creep process except that the combination of stress, time, and temperature is such that rupture into two parts is ensured. In stress rupture failures the combination of stress and temperature is often such that the period of steady-state creep is short or nonexistent.

Thermal relaxation failure occurs when the dimensional changes due to the creep process result in the relaxation of a pre-strained or pre-stressed member until it no longer is able to perform its intended function.

Thermal shock failure occurs when the thermal gradients generated in a part are so pronounced that differential thermal strains exceed the ability of the material to sustain them without yielding or feature.

Impact failure results when a part is subjected to non-static loads that produce in the part stresses or deformations of such magnitude that the member no longer is capable of performing its function. The failure is brought about by the interaction of stress or strain waves generated by dynamic or suddenly applied loads, which may induce local stresses and strains many times greater than would be induced by the static application of the same loads. If the magnitudes of the stresses and strains are sufficiently high to cause separation into two or more parts, the failure is called impact fracture. If the impact produces intolerable elastic or plastic deformation, the resulting failure is called impact deformation. If repeated impacts induce cyclic elastic strains that lead to initiation of a matrix of fatigue cracks, which grows to failure by the surface fatigue phenomenon, the process is called impact wear.

Brittle fracture failure occurs when the elastic deformation, in a part that exhibits brittle behavior, is carried to the extreme so that the primary plastic structure bonds are broken and the member separates into two or more pieces. Pre-existing flaws or growing cracks form initiation sites for very rapid crack propagation to catastrophic failure, leaving a multifaceted fracture surface.

Wear is the undesired cumulative change in dimensions brought about by the gradual removal of discrete particles from contacting surfaces in motion, usually sliding, predominantly as a result of mechanical action. Wear is not a single process, but a number of different processes that can take place by themselves or in combination, resulting in material removal from contacting surfaces through a complex combination of local shearing, plowing, gouging, welding, tearing, and others. Adhesive wear takes place because of high local pressure and welding at contact sites, followed by motion-induced plastic deformation and rupture of functions.

Abrasive wear takes place when the wear particles are removed from the surface by the plowing, gouging, and cutting action of the harder mating surface or by hard particles entrapped between the mating surfaces. Deformation wear arises as results of repeated plastic deformation at the wearing surfaces, producing a matrix of cracks that grow and coalesce to form wear particles. Deformation wear is often caused by severe impact loading. Impact wear is impact-induced repeated elastic deformation at the wearing surface that produces a matrix of cracks that grows.

Radiation damage failure occurs when the changes in material properties induced by exposure to a nuclear radiation field are of such a type and magnitude that the part is no longer able to perform its intended function, usually as a result of the triggering of some other failure mode, and often related to loss in ductility associated with radiation exposure.

The very broad term environment failure implies that a part is incapable of performing its intended function because of the undesired deterioration of the material as a result of chemical or electrochemical interaction with the environment. It often interacts with other failure modes such as wear or fatigue.

Spectrum Loading and Cumulative Damage

With engineering applications where fatigue is an important failure mode, the alternating stress amplitude may be expected to vary or change in some way during the service life. Such variations and changes in load amplitude, often referred to as spectrum loading, make the direct use of standard S-N curves inapplicable because these curves are developed and presented for constant stress amplitude operation

(Chapter 3). Therefore, it becomes important to a designer to have available a theory or hypothesis, verified by experimental observations, that will permit good design estimates to be made for operation under conditions of spectrum loading using the standard constant amplitude S-N curves.

It has been basically adopted by all fatigue investigators working with spectrum loading that operation at any given cyclic stress amplitude will produce fatigue damage, the seriousness of which will be related to the number of cycles of operation at that stress amplitude and also related to the total number of cycles that would be required to produce failure of an undamaged specimen at that stress amplitude. With this situation the damage incurred is permanent and operation at several different stress amplitudes in sequence will result in an accumulation of total damage equal to the sum of the damage increments accrued at each individual stress level. When the total accumulated damage reaches a critical value, fatigue failure occurs.

Although this concept is simple in principle, much difficulty is encountered in practice because the proper assessment of the amount of damage incurred by operation at any given stress level S, for a specified number of cycles N, is not straightforward. Many different cumulative damage theories have been proposed for the purposes of assessing fatigue damage caused by operation at any given stress level and the addition of damage increments to properly predict failure under conditions of spectrum loading. The fiat cumulative damage theory was proposed by Palmgren in 1924 and later developed by Miner in 1945. This linear theory, which is still widely used, is referred to as the Palmgren-Miner hypothesis or the linear damage rule. The theory may be described using the S-N plot.

Crack Growth and Fracture Mechanics

When the material behavior is brittle rather than ductile, the mechanics of the failure process are much different. Instead of the slow coalescence of voids associated with ductile rupture, brittle fracture proceeds by the high-velocity propagation of a crack across the loaded member. If the material behavior is clearly brittle, fracture may be predicted with reasonable accuracy through use of the maximum normal stress theory of failure. Thus failure is predicted to occur in the multi-axial state of stress when the maximum principal normal stress becomes equal to or exceeds the maximum normal stress at the time of failure in a simple uniaxial stress test using a specimen of the sane material.

Fatigue and Stress Concentration

Static or quasi-static loading is rarely observed in modern engineering practice, making it essential for the designer to address themselves to the implications of repeated loads, fluctuating loads, and/or rapidly applied loads. By far, the majority of engineering design projects involves products subjected to fluctuating or cyclic loads. Such loading induces fluctuating or cyclic stresses that often result in failure by fatigue. Fatigue failure investigations over the years have led to the observation that the fatigue process actually embraces two aspects of cyclic stressing or straining that are significantly different in character, and in each of which failure is probably produced by different physical mechanisms.

One aspect of cyclic loading is that for which significant plastic strain occurs during each cycle. This aspect is associated with high loads and short lives, or low numbers of cycles to produce fatigue failure, and is commonly referred to as low-cycle fatigue. The other aspect of cyclic loading is that for which the strain cycles are largely confined to the elastic range. This aspect is associated with lower loads and long lives, or high numbers of cycles to produce fatigue failure, and is commonly referred to as high-cycle fatigue.

Low-cycle fatigue is typically associated with cycle lives from 1 up to about 10^4 or 10^5 cycles. Fatigue may be characterized as a progressive failure phenomenon that proceeds by the initiation and propagation of cracks to an unstable size. Although there is not complete agreement on the microscopic details of the initiation and propagation of the cracks, processes of reversed slip and dislocation interaction appear to produce fatigue nuclei from which cracks may grow. The crack length reaches a critical dimension and one additional cycle then causes complete failure. The final failure region will typically show evidence of deformation produced just prior to final separation. For ductile materials the final fracture area often appears as a shear lip produced by crack propagation along the planes of maximum shear.

Fatigue Loading and Laboratory Testing

Important for designers involved in fatigue-sensitive product or structure, is the fatigue response of materials to various loadings that might occur throughout the design life of the part being manufactured. That is, the designer is interested in the effects of various loading and associated stress that will in general be a function of the design configuration and the operational use of the part. The simplest fatigue stress spectrum to which an element may be subjected is a zero-mean sinusoidal stress-time pattern of constant amplitude and fixed frequency, applied for a specified number of cycles. Such a stress-time pattern is often referred to as a completely reversed cyclic stress.

It is reported that the more complicated stress-time patterns are produced when the mean stress, stress amplitude, or both mean and stress amplitude change during the operational cycle. This type of quasi-random stress-time pattern might be encountered in an airframe structural member during takeoff, maneuvers, and landing. Instrumentation of existing machines, such as operational aircraft, provide some useful information to the designer if their mission is similar to the one performed by the instrumented machine. Recorded data from accelerometers, strain gages, and other transducers may in any event provide a basis from which a statistical representation can be developed and extrapolated to future needs if the fatigue processes are understood.

Basic data for evaluating the response of materials, parts, or structures are obtained from carefully controlled laboratory tests. There are fatigue-testing machines that range from being very simple to very complex. Various types of testing machines and systems are used. As an example computer-controlled fatigue testing machines are widely used in modern fatigue testing laboratories. Usually such machines take the form of precisely controlled hydraulic systems with feed-back to electronic controlling devices capable of producing and controlling virtually any strain-time, load-time, or displacement-time pattern desired.

Special testing machines for component testing and full-scale prototype testing systems are not found in the general fatigue-testing laboratory. These systems are built up especially to suit a particular need, for example, to perform a full-scale fatigue test of a bridge structure. For example, the very complex testing systems used to test a full-scale prototype, produce very specialized data applicable only to the particular prototype and test conditions used. For the particular prototype and test conditions the results are very accurate, but extrapolation to other test conditions and other pieces of hardware is usually difficult, if not impossible.

Simple smooth-specimen laboratory fatigue data are very general and can be utilized in designing virtually any part of the specimen material. To use such data in practice requires a quantitative knowledge of many pertinent differences between the laboratory and the application, including the effects of non-zero mean stress, varying stress amplitude, environment, size, temperature, surface finish, residual stress pattern, and others. Fatigue testing is performed at the extremely simple level of smooth specimen testing, the extremely complex level of full-scale prototype testing, and everywhere in the spectrum between.

Predicting Long Time Reliability

Time and effort has been used in attempting to device good/useful short-time creep tests for accurate and reliable prediction of long-term creep and stress rupture behavior. It appears, however, that really reliable creep data can be obtained only by conducting long-term creep tests that duplicate actual service loading and temperature conditions as nearly as possible. Unfortunately, designers are unable to wait for years to obtain design data needed in creep failure analysis. Therefore, certain useful techniques have been developed for approximating long-term creep behavior based on a series of short-term tests (Chapter 3). Data from creep testing may be cross-plotted in a variety of different ways.

The basic variables involved are stress, strain, time, temperature, and strain rate. Any two of these basic variables may be selected as plotting coordinates, with the remaining variables treated as parametric constants for a given curve. Three commonly used methods for extrapolating short-time creep data to long-term applications are the abridged method, the mechanical acceleration method, and the thermal acceleration method.

In the abridged method of creep testing the tests are conducted at several different stress levels and at the contemplated operating temperature. The data are plotted as creep strain versus time for a family of stress levels, all run at constant temperature. The curves are plotted out to the laboratory test duration and then extrapolated to the required design life.

In the mechanical acceleration method of creep testing, the stress levels used in the laboratory tests are significantly higher than the contemplated design stress levels, so the limiting design strains are reached in a much shorter time than in actual service. The data taken in the mechanical acceleration method are plotted as stress level versus time for a family of constant strain curves all run at a constant temperature.

The thermal acceleration method involves laboratory testing at temperatures much higher than the actual service temperature expected. The data are plotted as stress versus time for a family of constant temperatures where the creep strain produced is constant for the whole plot.

It is important to recognize that such extrapolations are not able to predict the potential of failure by creep rupture prior to reaching the creep design life. In any testing method it should be noted that creep testing guidelines usually dictate that test periods of less than 1% of the expected life are not deemed to give significant results. Tests extending to at least 10% of the expected life are preferred where feasible.

Meaning of data

The designer should not overestimate the accuracy of the data provided from different sources such as property sheets and in manuals. They are usually mean or maximum values obtained from various observations and sometimes extrapolated. Inaccuracy that can exist is partly related to variations in the processing technique and partly to the geometry of the part in question. It is evident that in order to use the data from those you perform to those from material suppliers data sheets, it is imperative to have a thorough understanding of how the data are evolved and what caution is to be exercised when applying the data to product designs or other evaluations.

They can easily be interpreted incorrectly to mean something one desires in their design approach. Interpretations are always made and provide excellent logical approaches to developing a design however they require dedicated concentrations and relationships to the basic meaning of the test. In reality tests have only certain meanings.

Safety factor

A safety factor (SF) or factor of safety (FS) (also called factor of ignorance) is used with plastics or other materials (metals, aluminum, etc.) to provide for the uncertainties associated with any design, particularly when a new product is involved with no direct historical performance record. There are no hard and fast rules to follow in setting a SF. The most basic consideration is the consequences of failure. In addition to the basic uncertainties of graphic design, a designer may also have to consider additional conditions such as: (1) variations in material property data (data in a table is the average and does not represent the minimum required in a design); (2) variation in material performance; (3) effect of size in stating material strength properties; (4) type of loading (static, dynamic, etc.); (5) effect of process (stress concentrations, residual stress, etc.); and (6) overall concern of human safety.

The SF usually used based on experience is 1.5 to 2.5, as is commonly used with metals. Improper use of a SF usually results in a needless waste of material or even product failure. Designers unfamiliar with plastic products can use the suggested preliminary safety factor guide-lines in Table 7.3 that provide for extreme safety; intended for preliminary design analysis only. Low range values represent applications where failure is not critical. The higher values apply where failure is critical. Any product designed with these guidelines in mind should conduct tests on the products themselves to relate the guidelines to actual performance. With more experience, more-appropriate values will be developed targeting to use 1.5 to 2.5. After field service of the preliminary designed products has been obtained, action should be taken to consider reducing your SF in order to reduce costs.

Table 7.3 Safety factors

Type of load	Safety factor
Static short-term loads	1 to 2.5
Static long-term loads	2 to 5
Repeated loads	5 to 15
Variable changing loads	4 to 10
Fatigue loads	5 to 15
Impact loads	10 to 15

Realistic SFs are based on personal (or others) experience. The SFs can be related to the probable consequences of failure. To ensure no failure where a product could be damaging to a person (etc.) prototype tests should be run at their most extreme service operating conditions. For instance, the maximum working load should be applied at the maximum temperature and in the presence of any chemicals that might be encountered in the end use. Impact loading should be applied at the lowest temperature expected, including what occurs during shipping and assembly. The effects of variations in plastic lots and manufacturing conditions must also be considered.

Safety Factor Example

Due to the unpredictable scheduling and high dollar costs of all weather natural testing, much of the environmental testing has been brought into laboratories or other such testing centers. Artificial conditions are provided to simulate various environmental phenomena and thereby aid in the evaluation of the test item before it goes into service under natural environments. This environmental simulation and testing does require extensive preparation and planning. It is generally desirable to obtain generalizations and comparisons from a few basic tests to avoid prolonged testing and retesting.

The type and number of tests to be conducted, natural or simulated, as usual are dependent on such factors as end item performance requirements, time and cost limitations, past history, performance safety factors, shape of specimens, available testing facilities, and the environment. Specifications, such as ASTMs' provide guidelines.

Since GRPs (glass reinforced plastics) tend not to exhibit a fatigue limit, it is necessary to design for a specific endurance, with initial safety factors in the region of 3 to 4 being commonly used. Higher fatigue performance is achieved when the data are for tensile loading with zero mean stress. In other modes of loading, such as flexural, compression, or torsion, the fatigue behavior can be worse than that in tension due to potential abrasion action between fibers if debonding of fiber and matrix occurs. This is generally thought to be caused by the setting up of shear stresses in sections of the matrix that are unprotected by some method such as having properly aligned fibers that can be applied in certain designs. Another technique, which has been used successfully in products such as high-performance RP aircraft wing structures, incorporates a very thin, high-heat-resistant film such as Mylar between layers of glass fibers. With GRPs this construction significantly reduces the self-destructive action of glass-to-glass abrasion and significantly increases the fatigue endurance limit.

With certain plastics, particularly high performance RPs, there can be two or three moduli. Their stress-strain curve starts with a straight line that results in its highest E, followed by another straight line with a lower S, and so forth. To be conservative providing a high safety factor the lowest E is used in a design however the highest E is used in certain designs where load requirements are not critical.

In many plastics, particularly the unreinforced TPs, the straight region of the stress-strain curve is not linear or the straight region of this curve is too difficult to locate. It then becomes necessary to construct a straight-line tangent to the initial part of the curve to obtain a modulus called the initial modulus. Designwise, an initial modulus can be misleading, because of the nonlinear elasticity of the material. For this reason, a secant modulus is usually used to identify the material more accurately. Thus, a modulus could represent Young's modulus of elasticity, an initial modulus, or a secant modulus, each having its own meaning and safety factors. The Young's modulus and secant modulus are extensively used in design equations.

The example of a building roof structure represents the simplest type of problem in static loading in that the loads are clearly long term and well defined. Creep effects can be easily predicted and the structure can be designed with a sufficiently large SF to avoid the probability of failure.

A seating application is a more complicated static load problem than the building example just reviewed because of the loading situation. The self-load on a chair seat is a small fraction of the normal load and can be neglected in the design. The loads are applied for relatively short periods of time of the order of 1 to 5 hours, and the economics of the application requires that the product be carefully designed with a small safety factor.

Overall, it can be stated that plastic products meet the following criteria: their functional performance meets use requirements, they lend themselves to esthetic treatment at comparatively low cost, and, finally, the finished product is cost competitive. Examples of their desirable behaviors can start with providing high volume production. Plastic conversion into finished products for large volume needs has proven to be one of the most cost-effective methods. Combining bosses, ribs, and retaining means for assembly are easily attained in plastic products, resulting in manufacturing economies that are frequently used for cost reduction. It is a case where the art and technology of plastics has outperformed any other material in growth and prosperity.

Their average weight is roughly one-eighth that of steel. In the automotive industry, where lower weight means more miles per gallon of gasoline, the utilization of plastics is increasing with every model-year. For portable appliances and portable tools lower weight helps people to reduce their fatigue factor. Lower weight is beneficial in shipping and handling costwise, and as a SF to humans (no broken glass bottles, etc.).

Throughout this book as the viscoelastic behavior of plastics has been described it has been shown that deformations are dependent on such factors as the time under load and the temperature. Therefore, when structural components are to be designed using plastics it must be remembered that the standard equations that are available for designing springs, beams, plates, and cylinders, and so on have all been derived under the assumptions that (1) the strains are small, (2) the modulus is constant, (3) the strains are independent of the loading rate or history and are immediately reversible, (4) the material is isotropic, and (5) the material behaves in the same way in tension and compression.

Since these assumptions are not always justifiable when applied to plastics, the classic equations cannot be used indiscriminately. Each case must be considered on its merits, with account being taken of such factors as the time under load, the mode of deformation, the service temperature, the fabrication method, the environment, and others. In particular, it should be noted that the traditional equations are derived using the relationship that stress equals modulus times strain, where the modulus is a constant. From the review in Chapters 2 and 3 it should be clear that the modulus of a plastic is generally not a constant. Several approaches have been used to allow for this condition. The drawback is that these methods can be quite complex, involving numerical techniques that are not attractive to designers. However, one method has been widely accepted, the so-called pseudo-elastic design method.

In this method appropriate values of such time-dependent properties as the modulus are selected and substituted into the standard equations. It has been found that this approach is sufficiently accurate if the value chosen for the modulus takes into account the projected service life of the product and/or the limiting strain of the plastic, assuming that the limiting strain for the material is known. Unfortunately, this is not just a straightforward value applicable to all plastics or even to one plastic in all its applications. This type of evaluation takes into consideration the value to use as a SF. If no history exists a high value will be required. In time with service condition inputs, the SF can be reduced if justified.

8 SUMMARY

Overview

From the initial development of plastics and particularly since the last half of the 20th century one can say it was extremely spectacular based on its growth rate but more important on how they have helped worldwide. The plastic industry is a worldwide multi-billion dollar business. Exciting discoveries and inventions have given the field of plastic products vitality. In a society that never stands still, plastics are vital components in its increased mobility.

Plastics surpassed steel on a volume basis about 1983 and by the start of this century plastics surpassed steel on a weight basis (Fig. 8.1). Plastics and a few other materials as shown in Fig. 8.1 represent about 10wt% of all materials consumed worldwide. The two major and important materials consumed are wood and construction or nonmetallic earthen (stone, clay, concrete, glass, etc.). Volumewise wood and construction materials each approach about 70 billion ft³ (2 billion m³). Each represents about 45% of the total consumption of all materials.

A continuous flow of new materials, new processing technologies (Chapter 1), and product design approaches has led the industry into profitable applications unknown or not possible in the past. What is ahead will be even more spectacular based on the continuous new development programs in materials, processes, and design approaches that are always on the horizon to meet the continuing new worldwide industry product challenges.

As an example the University of Massachusetts Lowell received patents pertaining to a method of bonding plastic components developed by Avaya, Inc., a Basking Ridge, NJ based provider of corporate net-

Figure 8.1 Estimated plastic consumption through year 2020

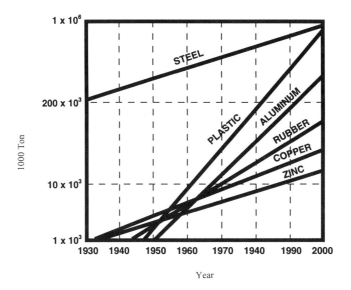

Year

working solutions and services. Reportedly valued at about $23 million, the patented technology was developed in the early 1990s for the high-speed bonding of thermoplastic parts, and has been used to assemble millions of telephones, etc. The University plans to license the technology to others for use in a wide range of commercial applications. UMass-Lowell also will commit resources to further develop the technology and incorporate it into the school's curriculum and design solutions.

Market Size

Plastic product are ranked as the 4th largest USA manufacturing industry with motor vehicles in 1st place, petroleum refining in 2nd place, and automotive parts in 3rd place. Plastic is followed by computers and their peripherals, meat products, drugs, aircraft and parts, industrial organic chemicals, blast furnace and basic steel products, beverages, communications equipment, commercial printing, fabricated structural metal products, grain mill products, and dairy products (in 16th place). At the end of the industry listings are plastic materials and synthetics in 24th place and ending in the 25th ranking is the paper mills. Fig. 8.2 provides a forecast for plastics growth to 2020 year.

Figure 8.2 Weight of plastic and steel worldwide crossed about 2000 (Courtesy of Plastics FALLO)

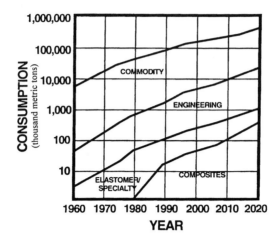

Customer

It is essential to obtain first-hand information on customer likes, dislikes, preferences, and prejudices. Eyeball-to-eyeball discussion, question and answer, and examination of competitors' trends and specifications are all useful inputs to the product designer. To a great extent, such input will depend on whether there are product line precedents already on the market or whether it is a product breaking new ground. Customer input is, nevertheless, essential to success. The degree of difficulty with which this input is obtained varies enormously from the large on-off turnkey type of project where the designer will interface directly with the customer, to the mass-produced product where one will not.

Feedback from the customer or market place should be considered. As an example it is no good incorporating a certain new design in a product that will not be accepted by customers, however when the design is valuable to the customer the skill of the salesperson is required. Examples of exploring new applications that are around us has been the fabrication of tubes, pipes, films, and others on the farm to exploring for oil in the depths of the seas.

Constraint

The constraints of current company practice should be highlighted and discussed. Is the company constrained by its previous products? If so, it is as well to know about it at the initial design stage. Possible manufacturing facility constraints (example use is to be a certain plastic and/or process), financial and investment constraints, and attitudes are very relevant. If needed are there adequate in-house facilities for research, design, development, testing, etc., including quality of personnel; perhaps outside sources will be required or are outside sources reliable.

Unfortunately constraints relate to the economic conditions with its upward and downward business trends ranging from within the USA and worldwide. Different industries including the plastics industry are effected by these recessions. Regardless of these recessions the plastics industry always continues to have good growth. As stated by Glenn L. Beall, an outspoken proponent of good plastic product design, the USA plastics industry always continues to ride out the recessions at a growth rate higher then the GDP (Gross Domestic Product).

Responsibility

The responsibilities of those involved in the World of Plastics encompass all aspects from design to fabrication as well as the functional operation in service of products. Although functional design and fabricating is of paramount importance, a product is not complete if it is functional but cannot easily be manufactured, or functional but not dependable, or if it has a good appearance but poor reliability, or the product will not fail but does not meet safety requirements. Those involved have a broad responsibility to produce products that meet all the objectives of function, durability, appearance, safety, and low cost. As an example the designer should not contend that something is now designed and it is now the manufacturing engineer's job to determine how to make it at a reasonable cost. The functional design and the production design are too closely interrelated to be handled separately.

Product designers must consider the conditions under which fabrication will take place, because these conditions affect product performance and cost. Such factors as production quantity, labor, and material cost are vital. Designers should also visualize how each product is to be fabricated. If they do not or cannot, their designs may not be satisfactory or even feasible from a production standpoint. One purpose of this book is to give designers sufficient information about manufacturing processes (with its references) so that they can design intelligently from a productivity standpoint.

Responsibility Commensurate with Ability
Recognize that people have certain capabilities; the law says that people have equal rights (so it reads that we were all equal since 1776) but some interpret it to mean equal capabilities. So it has been said via Sun Tzu, The Art of War, about 500 BC "Now the method of employing people is to use the avaricious and the stupid, the wise and the brave,

and to give responsibilities to each in situations that suit the person. Do not charge people to do what they cannot do. Select them and give them responsibilities commensurate with their abilities."

Risk

Designers and others in the plastics and other industries have the responsibility to ensure that all products produced will be safe and not contaminate the environment, etc. Recognize that when you encounter a potential problem, you are guilty until proven innocent (or is it supposed to be the reverse). So keep the records you need to survive the legal actions that can develop.

There are many risks people are subjected to in the plant, at home, and elsewhere that can cause harm, health problems, and/or death. Precautions should be taken and enforced based on what is practical, logical, and useful. However, those involved in laws and regulations, as well as the public and, particularly the news media should recognize there is acceptable risk.

Acceptable Risk
This is the concept that was developed decades ago in connection with toxic substances, food additives, air and water pollution, fire and related environmental concerns, and so on. It can be defined as a level of risk at which a seriously adverse result is highly unlikely to occur but it cannot be proven whether or not there is 100% safety. In these cases, it means living with reasonable assurance of safety and acceptable uncertainty.

Examples of this concept exists all around us such as the use of automobiles, aircraft, boats, lawnmowers, foods, medical pills and devices, water, air we breathe, news reports, and so on. Practically all elements around us encompass some level of uncertainty and risk. Otherwise as we know it would not exist.

Interesting that about 1995 a young intern at FDA made some interesting calculations. If they permitted the packaging of Coca Cola in acrylic barrier plastic bottles, and if you drank 37,000 gallons of coke per day for a lifetime, you would have a 10% risk of getting cancer. Since normal people have a 25% risk of getting cancer, reducing it to 10% was a real plus for coke (and the acrylic barrier plastic bottle). So perhaps a law should be enacted requiring that the public should drink lots of coke.

People are exposed to many risks. Some pose a greater threat than others. The following data concerns the probability over a lifetime of premature death per 100,000 people. In USA 290 hit by a car while being a pedestrian, 200 tobacco smoke, 75 diagnostic X-ray, 75 bicycling, 16 passengers in a car, 7 Miami/New Orleans drinking water, 3 lightning, 3 hurricane, and 2 fire.

DVR personal statistic (for real) based on personal knowledge of my large family, those that smoke and drank wine died close to 100 years of age. Those that did not smoke or drink died in their 60s (personal genies probably involved). Of course there were/are exceptions. So let the smokers continue to smoke and sue someone; regardless best not to smoke. Then there are other dilemmas such as exposure to asbestos, etc. that provide for interesting legal cases in USA. [After working with asbestos most of my life (now DVR at age 82) it never bothered me; however asthma has been with me since I was born except when I was in the Air Force.]

Predicting Performance

Avoiding nonstructural or structural failure can depend in part on the ability to predict performance of materials. When required, designers have developed sophisticated computer methods for calculating stresses in complex structures using different materials. These computational methods have replaced the oversimplified models of materials behavior relied upon previously. The result is early comprehensive analysis of the effects of temperature, loading rate, environment, and material defects on structural reliability. This information is supported by stress-strain behavior data collected in actual materials evaluations.

With computers the finite element analysis (Chapter 5) method has greatly enhanced the capability of the structural analyst to calculate displacement, strain, and stress values in complicated plastic structures subjected to arbitrary loading conditions.

Nondestructive testing (NDT) is used to assess a component or structure during its operational lifetime. Radiography, ultrasonics, eddy currents, acoustic emissions, and other methods are used to detect and monitor flaws that develop during operation (Chapter 7).

The selection of the evaluation method(s) depends on the specific type of plastic, the environment of the evaluation, the effectiveness of the evaluation method, the size of the structure, the fabricating process to be used, and the economic consequences of structural failure. Conventional evaluation methods are often adequate for baseline and acceptance inspections. However, there are increasing demands for more accurate characterization of the size and shape of defects that may require advanced techniques and procedures and involve the use of several methods.

Designing a good product requires a knowledge of plastics that includes their advantages and disadvantages (limitations) with some familiarity of the processing methods (Chapter 1). Until the designer becomes familiar with processing, a fabricator must be taken into the designer's confidence early in the development stage and consulted frequently during those early days. The fabricator and the mold or die designer should advise the product designer on plastic materials behavior and how to simplify the design to permit easier processability.

Design Verification

DV refers to the series of procedures used by the product development group to ensure that a product design output meets its design input. It focuses primarily on the end of the product development cycle. It is routinely understood to mean a thorough prototype testing of the final product to ensure that it is acceptable for shipment to the customers. In the context of design control, however, DV starts when a product's specification or standard has been established and is an on-going process. The net result of DV is to conform with a high degree of accuracy that the final product meets performance requirements and is safe and effective. According to standards established by ISO-9000, DV should include at least two of the following measures: (a) holding and recording design reviews, (b) undertaking qualification tests and demonstrations, (c) carrying out alternative calculations, and (d) comparing a new design with a similar, proven design.

Perfection

The target is to approach perfection in a zero-risk society. Basically, no product is without risk; failure to recognize this factor may put excessive emphasis on achieving an important goal while drawing precious resources away from product design development and approval. The target or goal should be to attain a proper balance between risk and benefit using realistic factors and not the "public-political panic" approach.

Achievable program plans begin with the recognition that smooth does not mean perfect. Perfection is an unrealistic ideal. It is a fact of life that the further someone is removed from a task, the more they are apt to expect so-called perfection from those performing it. The expectation of perfection blocks genuine communication between designers, workers, departments, management, customers, vendors, and laws (lawyers). Therefore one can define a smoothly run program as one that designs or creates a product that meets requirements, is delivered on time, falls within the price guidelines, and stays close to budget.

Perfection is never reached; there is always room for improvements as summarized in the FALLO approach (Fig. 1.15) and throughout history. As it has been stated, to live is to change and to reach perfection is to have changed often (in the right direction). Perfection is like stating that no one on "earth" is without sin.

In addition to the product the designer, equipment installer, user, and all others involved in production should all consider performing a risk assessment and target in the direction of perfection. The production is reviewed for hazards created by each part of the line when operating as well as when equipment fails to perform or complete its task. This action includes startups and shutdowns, preventative maintenance, QC/inspection, repair, etc.

Ethics

Those involved in producing products have developed guidelines for professional conduct based on the experience of many who have had to wrestle with troublesome ethical questions and situations confronted in the past. These guidelines can be found in the published codes of ethics for designers and engineers of a number of industry and technical societies such as the Industrial Designer societies.

Ergonomic

Ergonomics (also called human factors) is an applied science that makes the user central to design by improving the fit between the user and the product. There are products that have a people-machine interface during manufacture, during use in service, and if maintenance is required. Required may be height, reach, force, and operating torque that are acceptable to the user. Postures and lighting should be considered; there are products that must be a delight to use. Potential users must be consulted.

Product designs are developed to fit both the physiological and psychological needs of the user. Ergonomists examine all ranges of the human interface, from static measurements and movement ranges to users' perceptions of a product. This interface involves both software (displays, electronic controls, etc.) and hardware (knobs, grips, physical configurations, etc.) issues.

Ergonomics includes concept modeling and product design, job performance analysis, functional analysis, workspace and equipment design, computer interfaces, environment design, and so forth.

The true basis of ergonomics understands the limitations of human performance capabilities relative to product interaction. These limitations are either physical or perceptual in nature, but all address how people respond to people-made designs. Such interface analysis is crucial to establishing a safe and effective system of operation or environment for the user.

Industry studies have shown there are cost-benefit advantages in using ergonomic programs. Recognize that the cost of corrections to a poorly designed product geometrically increases throughout the development process. Therefore, human factor specialists should begin working with engineers and designers in the early stages of product development. When ergonomists are called in to fix a product that has already been sent to market and failed, costs will escalate. A manufacturer's decision to adopt an ergonomic orientation will serve to reposition its products from a commodity-based supplier to a supplier of high-value products. Integrating ergonomics into a design program ensures more comfortable, safe, and productive design solutions and a better overall product for the end-user.

Costing

A major cost advantage for fabricating plastic products has been and will continue to be their usual relatively low processing cost. The most expensive part of practically all products is the cost of plastic materials. Since the material value in a plastic product is roughly up to one-half (possibly up to 90% for certain products) of its overall cost, it becomes important to select a candidate material with extraordinary care particularly on long production runs. In production cost to fabricate usually represents about 5% (maximum 10%) of total cost.

It is a popular misconception that plastics are cheap materials; they are not. There are low cost types (commodity types) but there are also the more expensive types (engineering types) (Chapter 1). Important that one recognizes that it is economically possible to process a more expensive plastic because it provides for a lower processing cost. By far the real advantage to using plastics to produce many low-cost products is their low weight with their low processing costs.

Technical Cost Modeling

Figure 8.3 Product from designer to customer flow-chart (Courtesy of Plastics FALLO)

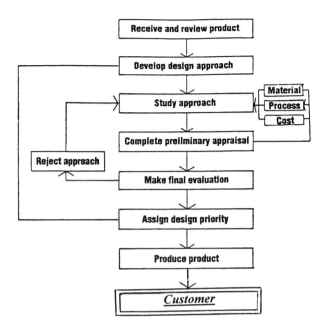

TCM has been developed as a method for analyzing the economics of alternative manufacturing processes without the prohibitive economic burden of trial-and-error innovation and process optimization. Its approach to estimating cost is not dependent on the intuition of cost-estimating individuals. It follows the conventional process modeling that ranges from design to process variables during fabrication. TCM takes all the details for each of the functions that go into designing to fabricating to delivery to the customer such as summarized in Fig. 8.3. TCM provides the means to coordinate cost estimates with processing knowledge. Included are the critical assumptions (processing rates, energy used, materials consumed, scrap, etc.) that can be made to interact in a consistent, logical, and accurate framework of economic analysis, producing cost estimates under a wide range of conditions.

TCM can establish direct comparisons between processes. In turn it determines the plastic process that is best for the production of a product without extensive expenditures of capital and time. It also determines the ultimate performance of a particular process, as well as identifying the limiting process steps and parameters.

Each of the elements that contribute to the total cost is estimated individually. These individual estimates are derived from basic principles and the manufacturing process. This reduces the complex problem of cost analysis to a series of simpler estimating problems and brings processing expertise rather than intuition to bear on solving these problems. By this approach in dividing cost into its contributing elements it takes into account that some cost elements depend upon the number of products produced annually, whereas others do not. For example, the cost contribution of the plastic is the same regardless of the number of items produced, unless the material price is discounted because of high volume. It allows for the per-piece cost of tooling that will vary with changes in production volume. These types of cost elements, which are called the variable and fixed costs, respectively, create a natural division of the elements of manufacturing product cost.

The technical cost analysis should be viewed as a philosophy, not road map. The important tenets of this philosophy are that:

1. Primary and secondary processes contribute to the cost of a finished component.

2. The total cost of a process is made up of many contributing elements.

3. These elements can be classified as either fixed or variable, depending on whether they are effected by changes in the production volume.

4. Each element can be analyzed to establish the factors and nature of the relationships that affect its value.

5. Total cost can be estimated from the sum of the elements of cost for each contributing process.

One advantage of the above philosophy over simpler cost-estimating techniques is that estimates obtained in this manner provide not only a total cost, but also quickly an understanding of the contribution of each element. This information can be used to direct efforts at cost reduction, or it can be used to perform sensitivity analyses, answering questions such as what if one of the elements should change?

Engineering and law interface

Whether engaged in R&D, manufacturing, engineering services, or technical consulting, today's engineer must be cognizant that the law imposes substantial accountability on both individual engineers and technology-related companies. The engineer can never expect to be insulated entirely from legal liability when designing a product. However, one can limit liability by maintaining a fundamental understanding of the legal concepts one is likely to encounter in the course of one's career, such as professional negligence, employment agreements, intellectual property rights, contractual obligations, and liability insurance.

Producer of a product has shown reasonable consideration for the safety, correct quantity, proper labeling, and other social aspects of the product to the consuming public. Since the 1960s these types of important concerns have expanded and been reinforced by a recognition of the consumer's right to know, as well as by concerns for conservation, ecology, antilittering, and the like.

Designer's failure to be aware of and comply with existing laws and regulations can lead to legal entanglements, fines, restrictions, and even jail sentences. In addition, there are also the penalties of costly, damaging publicity, and the loss of consumer goodwill. Unfortunately, nothing is perfect, so problems can develop, which is simply a fact of life.

Numerous safety-related and socially responsible laws have been enacted and many more are on the way. A lawsuit begins when a person (corporations, etc.) whose body or property is injured or damaged alleges that the injury was caused by the acts of another and files a complaint. The person asserting the complaint is the plaintiff; the person against whom the complaint is brought is the defendant.

Plaintiff complaint must state a cause of action (a legal theory or principle) that would, if proven to the satisfaction of the jury, permit the plaintiff to recover damages. If the cause of action asserted is negligence, then the plaintiff must prove, first, that the defendant owed the plaintiff a duty (had a responsibility toward the plaintiff, the public). Then the plaintiff must show that the defendant breached that duty and consequently, that the breach of duty by the defendant was the cause of the plaintiff's injury.

A breach of this duty of care that results in injury to persons or property may result in a tort claim, which is a civil wrong (as opposed to a criminal wrong) for which the legal system compensates the successful plaintiff by awarding money damages. To make out a cause of action in negligence, it is not necessary for the plaintiff to establish that the defendant either intended harm or acted recklessly in bringing about the harm. Rather, the plaintiff must show that the defendant's actions fell below the standard of care established by law. The standard of care or conduct that must be exercised is that the average reasonable person of ordinary prudence would follow under the same or similar circumstances. The standard of care is an external and objective one and has nothing to do with individual subjective judgment, though higher duties may be imposed by specific statutory provisions or by reason of special knowledge.

There are many examples of action to eliminate or reduce problems. As an example there is the Quality System Regulation (QSR). FDA requires details on how products such as medical devices are manufactured. The details of the process are documented so that once a product produced in USA is approved, following what was in the QSR preparation can only produce the product. No change can be made. The exact plastic composition has to be used, process control settings remain the same, etc. Literally if a waste paper basket had been identified and located in a specific location in the plant, you can not relocate, change its size, etc. It has been reported that to make a change could cost literally a million dollars. Result of the QSR regulation is too ensure the safety of a person when the medical device is used.

It has been unofficially reported that in USA there exists more liability court cases and over 85% of the lawyers worldwide are in USA. This location condition of number of cases and lawyers exists because in USA both parties (defendant and plaintiff) are innocent and if the plaintiff loses, the defendant only pays what he/she developed. Practically in the rest of the world, the law says that one side is right and the other side is wrong. But more important is the fact that if the plaintiff loses he/she pays all bills (those of the defendant, the court, and plaintiff).

Plastic material

The extent to which plastics are used in any industry in the future will depend in part upon the continued total R&D activity carried on by plastic material producers, processors, fabricators, and users in their desire to broaden the scope of plastic applications. The material producers provide the bulk of such research expenditure themselves and the rest by the additive and equipment industries that do more than the processors and fabricators. Important to plastic growth have been the continuing government projects in basic and applied research and new applications materialwise and equipmentwise, particularly the military. Their work in turn expands into the industrial industry.

Design demand

It can be said that the challenge of design is to make existing products obsolete or at least offer significant improvements. Despite this level of activity there are always new fields of products to explore. Plastics will continue to change the shape of worldwide business rapidly. Today's plastics tend to do more and cost less, which is why in many cases they came into use in the first place. Tomorrow's requirements will be still more demanding, but with sound design, plastics will satisfy those demands, resulting not only in new processes and materials but improvements in existing processing and materials.

R&D continues even more in manipulating molecules to the extent that the range of materials offered to industry will continue to present new opportunities and allow existing businesses to enjoy profitable growth. Also ahead are the different raw material sources to produce plastics that involve biotechnology. A reading of the literature and patents being issued indicates that there is a great deal of commercially oriented research being aimed at further improvement and modification into the plastic family. However recognize that the basic analysis for designing plastic products continues to be related to temperature-time-load and environment.

Unfortunately sometimes a new design concept is not accepted or may simply be ahead of its time. In 1483 Leonardo da Vinci designed what he called a spiral screw flying machine. In 1942 Igor Sikorsky developed the R4B helicopter (included plastics parts). One could say, in a joking manner, that it took 459 years to bring a designed product to market; seems a failure in materials/or perhaps the interoffice communication.

Alexander Graham Bell believed the photophone, not the telephone, was his greatest invention. His photophone carried the spoken voice by reflected sunbeams instead of wire, but did not find any practical application a century ago. Because light has 20,000 times shorter frequency than microwaves, it can carry 20,000 time more information. Only since the onset of computers has this ability been needed (includes plastic tubing, etc.). It would seem that Alex Bell was ahead of his time.

Fortunately people we know did not have to design the human body. The human body is the most complex structure ever "designed" with its so-called 2,000 parts (with certain parts being replaced with plastics). Can you imagine designing the heart (now occurring) that recirculates all the blood in the body every 20 minutes, pumping it through 60,000 miles of blood vessels, etc. Thus the designer of the human body had to be extremely creative; some of us know who designed the human body.

The past events in designing plastic products have been nothing short of major worldly achievements. Innovations and visionary provides the required high level of sophistication that is applied to problems that exist with solutions that follow. Ahead is a continuation of meeting new challenges with these innovations and idealism that continues to make plastics a dynamic and visionary industry. The statement that we are in the World of Plastics is definitely true. In fact one can say that plastic products has made life easier for all worldwide.

Plastic success

Success is related to many million of plastic products manufactured worldwide; during the start of the 21st century over 350,100 million lb (156 million tons). USA consumed over 100,000 million lb; about 90% are thermoplastics (TPs) and 10% thermoset (TS) plastics. USA and Europe consumption are each about one-third of the world total. There are well over 35,000 different types of plastic materials worldwide. However, most of them are not used in large quantities; they have specific performance and/or cost capabilities generally for specific products by specific processes that principally include many thousands of products.

Plastics are now among the nations and world's most widely used materials, having surpassed steel on a volume and weight basis. Plastic materials and products cover the entire spectrum of the world's economy, so that their fortunes are not tied to any particular business segment. Designers are in a good position to benefit in a wide variety of markets: packaging, building and construction, electronics and electrical, furniture, apparel, appliances, agriculture, housewares, luggage, transportation, medicine and health care, recreation, and so on (Chapter 4).

To meet this success what is required is a skilled designer who blends a knowledge of materials, an understanding of manufacturing processes, and imagination into successful new designs. Recognizing the limits of design with traditional materials is the first step in exploring the possibilities for innovative design with plastics. What is important when analyzing plastic designs is the ease to incorporate ergonomics and empathy that results in products that truly answers the user's needs.

With designing there has always been the need to meet engineering, styling, and performance requirements at the lowest cost. To some there may appear to be a new era where ergonomics is concerned, but this is not true. What is always new is that there are continually easier methods on the horizon to simplify and meet all the specific requirements of a design. Some designers operate by creating only the stylish outer appearance, allowing basic engineers to work within that outside envelope. Perhaps this is all that is needed to be successful, but a more in-depth approach will work better. Recognize that when you gain a property, etc. there could be a loss. Beginning with a thorough understanding of the user's needs and design toward ease of manufacture and repair. The product that emerges will then be a logical and aesthetic answer to the design challenge.

Manufacturers need to continually update their traditional design methods in order to keep pace with rapidly evolving technologies and an increasingly demanding marketplace. Consumer demand products that are increasingly faster, easier to use, and lower in cost.

Future

A continuous flow of new materials, new processing technologies, and product design approaches has led the industry into applications unknown or not possible in the past. What is ahead will be even more spectacular based on the continuous new development programs in materials, processes, and design approaches that are always on the horizon to meet the continuing new worldwide industry product challenges.

Appendix A
Abbreviations

acetal (*see* POM)
A ampere
AA acrylic acid
AAE American Assoc. of Engineers
AAES American Assoc. of Engineering Societies
AAR American Association of Railroads
ABC acrylonitrile-butadiene-styrene
ABR polyacrylate
ABS acrylontrile-butadiene-styrene
AC alternating current
AC cellulose acetate
ACES Accurate Clear Epoxy Solid
ACS American Chemical Society
ACTC Advanced Composite Technology Consortium
ad adhesive
ADC allyl diglycol carbonate (*also see* CR-39)
AEC Atomic Energy Commission
AFCMA Aluminum Foil Container Manufacturer's Assoc.
AFMA American Furniture Manufacturer's Assoc.
AFML Air Force Material Laboratory
AFPA American Forest & Paper Assoc.
AFPR Assoc. of Foam Packaging Recyclers
AGMA American Gear Mfgrs. Assoc.
AI artificial intelligence

AIA Automated Imaging Assoc.
AIAA American Institute of Aeronautics & Astronauts
AIChE American Institute of Chemical Engineers
AIMCAL Assoc. of Industrial Metallizers, Coaters & Laminators
AISI American Iron and Steel Institute
AMA American Management Association
AMBA American Mold Builders Assoc.
AMC alkyd molding compound
AMW average molecular weight
AN acrylonitrile
ANSI American National Standards Institute
ANTEC Annual Technical Conference (SPE)
APC American Plastics Council
APET amorphous polyethylene terephthalate
APF Assoc. of Plastics Fabricators
API Alliance for the Polyurethane Industry
API American Paper Institute
APM antipersonnel mine
APME Assoc. of Plastics Manufacturers in Europe
APPR Assoc. of Post-consumer Plastics Recyclers

AQL acceptable quality level
APC American Plastics Council
AR aramid fiber
AR aspect ratio
ARP advanced reinforced plastic
ASA Acrylonitrile-styrene-acrylate
ASAP as soon as possible
ASCII American Standard Code for
  Information Exchange
ASM American Society for Metals
ASME American Society of
  Mechanical Engineers
ASNDT American Society for Non-
  Destructive Testing
ASQ American Society for Quality
ASQC American Society for Quality
  Control
Assoc. association
ASTM American Society for Testing
  Materials
at wt atomic weight
atm atmosphere
ATR attenuated total reflectance
AW areal weight

bbl barrel
BDI Biodegradable Products Institute
BFRL Building & Fire Research
  Laboratory (NIST)
Bhn Brinell hardness number
BM blow molding
BMC bulk molding compound
BO biaxially-oriented
BOPP biaxially-oriented
  polypropylene
BPI Biodegradable Products Institute
bp boiling point
BR butadiene rubber
BR polybutadiene
BSI British Standards Institute
Btu British thermal unit
Buna polybutadiene
Butyl butyl rubber

C carbon
C Celsius
c centi (10^{-2})

C Centigrade (preference Celsius)
C composite
C coulomb
Ca calcium
CA cellulose acetate
CAA Clean Air Act
CAB cellulose acetate butyrate
$CaCO_3$ calcium carbonate (lime)
CAD computer-aided design
CAE computer-aided engineering
CAM computer-aided manufacturing
CAMPUS computer-aided material
  preselection by uniform standards
CAN cellulose acetate nitrate
CAP cellulose acetate propionate
CAS Chemical Abstract Service,
  division of ACS
CAS clean air solvent
CAT computer-aided testing
CBA chemical blowing agent
CBT computer-based training
CCA cellular cellulose acetate
CCV Chrysler composites vehicle
CD compact disk (disc)
CEM Consorzio Export Mouldex
  (Italian)
CFA Composites Fabricators Assoc.
CFC chlorofluorocarbon
CFECA California Film Extruders
  and Converters Association
CFE polychlorotrifluoroethylene
CIM computer integrated
  manufacturing
CLTE coefficient of linear thermal
  expansion
cm centimeter
CM compression molding
CMA Chemical Mfgrs. Assoc.
CMRA Chemical Marketing Research
  Assoc.
CN cellulose nitrate (celluloid)
CNC computer numerically
  controlled
CO carbon monoxide
CO polyepichlorohydrin
CO_2 carbon dioxide
CP Canadian Plastics

CP cellulose propionate
CPAC Center for Process Analytical
  Chemistry
CPE chlorinated polyethylene
CPET chlorinated polyethylene
  terephthalate
CPI Canadian Plastics Institute
cpm cycles/minute
CPSC Consumer Products Safety
  Commission
CPU central processing unit
CPVC chlorinated polyvinyl chloride
CR chloroprene rubber
C/R compression ratio
CR-39 allyl diglycol carbonate
CRP carbon reinforced plastics
CRT cathode ray tube
CS chlorinated solvent
CSA Canadian Standard Assoc.
CSI control system integration
CSM chlorosulfonyl polyethylene
CT carbon tetrachloride
CTFE chlorotrifluorethylene
CV coefficient of variation

d denier (preferred DEN)
d density
2-D two dimensions
3-D three dimensions
D diameter
DAIP diallyl isophthalate
DAP diallyl phthalate
DAS data acquisition system
dB decibel
DC direct current
den denier
DDR draw-down ratio
DEHP di-ethylhexyl phthalate
DGA differential gravimetric analysis
DIN Deutsches Instut, Normung
  (German Standards Commission)
DINP di-isononyl phthalate
DMA dynamic mechanical analysis
DMC dough molding compound
DN Design News publication
DNA deoxyribonucleic acid
DOD Department of Defense

DOE Department of Energy
DOE Design of Experiments
dp dew point
dP differential pressure
DP degree of polymerization
DSC differential scanning calorimeter
DSD Duales System Deutschland
  (German Recycling System)
DSQ German Society for Quality
DTA differential thermal analysis
DTGA differential thermogravimetric
  analysis
DTMA dynamic thermomechanical
  analysis
DTUL deflection temperature under
  load
DV design value
DV devolatilization
DVD Digital Versatile Disc
DVR dimensional velocity research
DVR design value resource
DVR Dominick Vincent Rosato
DVR Donald Vincent Rosato
DVR Drew Vincent Rosato
DVR Druckverformungsrest
  (compression set/German)
DVR dynamic value research
DVR dynamic velocity ratio

E elongation
E modulus of elasticity or Young's
  modulus
E_c modulus, creep (apparent)
E_r modulus, relaxation
E_s modulus, secant
EBM extrusion blow molding
EC ethyl cellulose
EC European Community
ECTFE polyethylene-
  chlorotrifluoroethylene
EDC endocrine-disrupting chemical
EDM electrical discharge machining
E/E electronic/electrical
EEC European Economic
  Community
EI modulus times moment of inertia
  (equal stiffness)

EMC electromagnetic compatibility
EMI electromagnetic interference
EMS environmental management system
EO ethylene oxide (also EtO)
EOT ethylene ether polysulfide
EP epoxy
EP ethylene-propylene
EPA Environmental Protection Agency
EPDM ethylene propylene diene monomer
EPE expandable polyethylene
EPM ethylene propylene fluorinated
EPP expandable polypropylene
EPR ethylene propylene rubber
EPS expandable polystyrene
ERP enterprise resource planning
ESC environmental stress cracking
ESCR environmental stress cracking resistance
ESD electrostatic safe discharge
ET ethylene polysulfide
ETFE ethylene terafluoroethylene
ETO ethylene oxide
EU entropy unit
EU European Union
EUPC European Assoc. of Plastics Converters
EUPE European Union of Packaging & Environment
Euro European currency
EUROMAP European Committee of Machine Manufacturers for the Rubber & Plastics Industries (Zurich, Switzerland)
EVA ethylene-vinyl acetate
E/VAC ethylene/vinyl acetate copolymer
EVAL ethylene-vinyl alcohol copolymer (tradename for EVOH)
EVE ethylene vinyl ether
EVOH ethylene-vinyl alcohol copolymer (or EVAL)
EX extrusion

F coefficient of friction

F Fahrenheit
F Farad
F force
FALLO Follow ALL Opportunities
FBF Film and Bag Federation of SPI
FC fuzzy control
FDA finite difference analysis
FDA Food & Drug Administration
FEA finite element analysis
FEP fluorinated ethylene-propylene
FFS form, fill, & seal
FLC fuzzy logic control
FMCT fusible metal core technology
FPC flexible printed circuit
fpm feet per minute
FR flame retardant
FRCA Fire Retardant Chemicals Assoc.
FRP fiber reinforced plastic
FRTP fiber reinforced thermoplastic
FRTS fiber reinforced thermoset
FS factor of safety
FS fluorosilicone
ft feet
FTIR Fourier transformation infrared
FV frictional force × velocity

g gram
G giga (10^6)
G gravity
G shear modulus (modulus of rigidity)
G torsional modulus
GAIM gas assisted injection molding
gal gallon
GB gigabyte (billion bytes)
GC gas chromatography
GD&T geometric dimensioning & tolerancing
GDP Gross Domestic Product
GF glass fiber
GFRP glass fiber reinforced plastic
gpm gallons per minute
GMP good manufacturing practice (see QSR)
GMT glass mat reinforced thermoplastic

GNP gross national product (GDP replaced GNP in US 1993 following rest of world)
GOR grill opening reinforcement
GP general purpose
GPa giga Pascal
GPC gel permeation chromatography
gpd grams per denier
GPEC Global Plastics Environmental Conference
GPPS general purpose polystyrene
gr grain
GR-S polybutadiene-styrene
GRP glass reinforced plastic
GSC gas solid chromatography
GWP global warming potential

h hour
H enthalpy
H hysteresis
H_2 hydrogen
HA hydroxyapatite
HAF high abrasion furnace
HAP hazardous air pollutant
HB Brinell hardness number
HBSE Hazard-Based Safety Engineering
HCFC hydrochlorofluorocarbon
HCl hydrogen chloride
HDBK handbook
HDPE high density polyethylene (PE-HD)
HDT heat deflection temperature
HIC Household and Industrial chemicals
HIPS high impact polystyrene
HMC high strength molding compound
HMI human machine interface
HMW-HDPE high molecular weight-high density polyethylene
H_2O water
H-P Hagen-Poiseuille
HPLC high pressure liquid chromatography
HPM hot pressure molding
HRC hardness Rockwell C (C scale)

HTS high temperature superconductor
hyg hygroscope
Hz Hertz (cycles)

I integral
I moment of inertia
IB isobutylene
IBC internal bubble cooling
IBM injection blow molding
IBM International Business Machines
IC Industrial Computing publication
ICM injection-compression molding
ID internal diameter
IEC International Electrochemical Commission
IEEE Institute of electrical & Electronics Engineers
IGA isothermal gravimetric analysis
IGC inverse gas chromatography
IIE Institute of Industrial Engineers
IIR isobutene-isoprene
IKV Institute for Plastics Processing, Aachen, Germany
in. inch
InTech Instrumentation, Systems, and Automation Society publication
ipm inch per minute
IM injection molding
IMM injection molding machine
IMPS impact polystyrene
in. inch
InTech ISA publication
I/O input/output
IOM Institute of Medicine
ips inch per second
IR infrared
IR synthetic polyisoprene (synthetic natural rubber)
ISA Instrumentation, Systems, & Automation
ISO International Standardization Organization or International Organization for Standardization
ISSN International Standard Serial Number
IT information technology

IUPAC International Union of Pure and Applied Chemistry
IV intrinsic viscosity
IVD in vitro diagnostic

J joule
J_p polar moment of inertia
JIS Japanese Industrial Standard
JIT just-in-time
JIT just-in-tolerance
JSR Japanese SBR
JSW Japan Steel Works
JUSE Japanese Union of Science & Engineering
JWTE Japan Weathering Test Center

k kilo (prefix for 10^6)
K bulk modulus of elasticity
K coefficient of thermal conductivity
K Kelvin
K Kunststoffe (plastic in German)
KB kilobyte (1000 bytes)
kc kilocycle
kg kilogram
KISS keep it short & simple
KISS keep it simple & safe
KISS keep it simple stupid
KK thousand
Km kilometer
KO knockout
kPa kiloPascal
ksi thousand pounds per square inch (psi $\times 10^3$)

l length
L litre (USA liter) for fluids only
lb pound
lbf pound-force
LC liquid chromatography
LCD liquid crystal display
LCD lowest common diameter
LCMS liquid chromatography mass spectroscopy
LCP liquid crystal polymer
L/D length-to-diameter (ratio)
LDPE low density polyethylene (PE-LD)

LED light-emitting diode
LEP Light-Emitting Polymer
LIM liquid impingement molding
LIM liquid injection molding
liq liquid
LLDPE linear low density polyethylene (also PE-LLD)
lm lumen
LMDPE linear medium density polyethylene
LNG liquefied natural gas
LOI loss on ignition
LOX liquid oxygen
LPG liquefied petroleum gas
LPM low pressure molding
LSR liquid silicone rubber

m matrix
m metallocene (catalyst)
m meter
mμ micromillimeter; millicron; 0.000001 mm
mm millimeter
μm micrometer
M mega
M million
M_b bending moment
MA maleic anhydride
MA mass spectroscopy
MAD mean absolute deviation
MAD molding area diagram
MBTS benzothiazyl disulfide
M-class Elastomer; *see* Table B-1
MD machine direction
MD mean deviation
MD&DI Medical Device & Diagnostic Industry publication
MDI methane diisocyanate
MDPE medium density polyethylene
Me metallocene catalyst
MEK methyl ethyl ketone
MF melamine formaldehyde
mfg manufacturing
MFI melt flow index
Mfr. manufacturer
MFT manufacturers for fair trade
mg milligram

μm micrometer
mg milligram
mHDPE metallocene HDPE
  (different m/plastics such as mPS,
  mPP, etc.)
MI melt index
mi mile
mike microinch (10^{-6} in.)
mil one thousand of inch (10^{-6} in.)
MIM metal powder injection molding
min minimum
min minute
MIPS medium impact polystyrene
MIT Massachusetts Institute of
  Technology
ml milliliter
mLLDPE metallocene catalyst
  LLDPE
MM billion
MMP multimaterial molding or
  multimaterial multiprocess
MO molecular orientation
MP Modern Plastics publication
MPa mega-Pascal (10^6 Pascal)
mPE metallocene polyethylene
MRPMA Malaysian Rubber Products
  Manufacturers' Assoc.
Msi million pounds per square inch
  (psi $\times 10^6$)
MSW municipal solid waste
MVD molding volume diagram
MVT moisture vapor transmission
MM billion
MW molecular weight
MWD molecular weight distribution
MWR molding with rotation

N Nano (10^{-9})
N Newton (force)
N number of cycles
N_2 nitrogen
NACE National Assoc. of Corrosion
  Engineers
NACO National Assoc. of
  CAD/CAM Operation
NAGS North America Geosynthetics
  Society

NAM National Assoc. of
  Manufacturers
NAS National Academies of Science
NASA National Aeronautics Space
  Administration
NBR butadiene acrylontrile
NBS National Bureau of Standards
  (since 1980s renamed National
  Institute Standards & Technology
  or NIST)
NC numerical control
NCP National Certification in Plastics
NDE nondestructive evaluation
NDI nondestructive inspection
NDT nondestructive testing
NEAT Nothing Else Added To it
NEMA National Electrical
  Manufacturers Assoc.
NEN Dutch standard
NeSSI new sampling/sensor initiative
NFPA National Fire Protection
  Assoc.
NISO National Information
  Standards Organization
NIST National Institute of Standards
  & Technology
nm nanometer
NMMA National Marine
  Manufacturers Assoc.
NMR nuclear magnetic resonance
NOS not otherwise specified
NPCM National Plastics Center &
  Museum
NPE National Plastics Exhibition
  (SPI)
NPFC National Publications & Forms
  Center (USA gov't)
NR natural rubber (polyisoprene)
NRTL Nationally Recognized Testing
  Laboratory
NSC National Safety Council
NST National Safety Council
NTMA National Tool & Machining
  Assoc.
NWPCA National Wooden Pallet &
  Container Assoc.
nylon (*see* PA)

Table B-1 ASTM D 1418 elastomer terminology

The ASTM standardized terminology system classifying all forms of elastomeric materials based upon the chemical composition of the polymer's backbone chain.

The "M" Class

These elastomers have saturated main polymer chains and are usually prepared from ethylene or vinyl type monomers containing one double band.

ACM   Copolymers of an acrylate and a small amount of other monomer which provides vulcanizability

ANM   Copolymers f an acrylate and acrylonitrile

CM   Chloro-polyethylene

CFM   Polychloro-trifluoro-ethylene

CSM   Chloro-sulfonly-polyethylene

EPDM   Terpolymers of ethylene, propylene and a nonconjugated diene which results in pendant unsaturation (not in the main chain)

EPM   Copolymers of ethylene and propylene

FKM   A polymer with a saturated main chain with substituents of fluorine, perfluoroalkyl, or perfluouroalkoxy

The "O" Class

These elastomers have oxygen in the main chain.

CO   Polyepichlorohydrin

ECO   Copolymer of ethylene oxide and epichlorohydrin

GPO   Copolymer of propylene oxide and allyl glycidyl ether

The "R" Class

These elastomers contain unsaturation in the main chain. The letter immediately before the "R" designates the conjugated diene which is used in its synthesis (except natural rubber.)

ABR   Copolymer of acrylate and butadiene

BIIR   Copolymer of bromoisobutene and isoprene

BR   Polybutadiene

CIIR   Copolymer of chloroisobutene and isoprene

CR   Polychloroprene

IIR   Copolymer of isobutene and isoprene

IR   Polyisoprene (synthetic only)

NBR   Copolymer of acrylonitrile and butadiene

NCR   Copolymer of acrylonitrile and chloroprene

NIR   Copolymer of acrylonitrile and isoprene

NR   Natural Rubber (poly-cis-isoprene)

PBR   Copolymer of vinyl pyridine and butadiene

PSBR   Copolymer of vinyl pyridine, styrene and butadiene

SBR   Copolymer of styrene and butadiene

SCR   Copolymer of styrene and chloroprene

SIR   Copolymer of styrene and isoprene

X   Prefix indicated carboxyl substitution

The "Q" Class

These elastomers have silicone in the main chain. Prefixes indicate the following types of substitution:

M – methyl
V – vinyl
P – phenyl
F – fluorine

The "U" Class

Elastomers with carbon, nitrogen and oxygen in the main chain – typically polyurethanes:

AU – Polyester based polurethanes
EU – Polyether based polurethanes

"Y" Designation

The "Y" prefix indicates a thermoplastic rubber which requires no vulcanization

O_2 oxygen
O_3 ozone
OCF Owens Corning Fiereglas
OD outside diameter
OEM original equipment manufacturer
OPET oriented polyethylene terephthalate
OPS oriented polystyrene
OSHA Occupational Safety & Health Administration
oz ounce

%vol percentage by volume (prefer vol%)
%wt percentage by weight (prefer wt%)
p pico (prefix for 10^{-12})
P load
P poise
P pressure
Pa Pascal
PA polyamide (nylon)
PA 6 polyamide 6 (nylon 6)
PAA polyaryl amide
PAI polyamide-imide
PAN polyacrylonitrile
PB polybutylene
PBA physical blowing agent
PBI polybenzimidazole
PBNA phenyl-*b*-naphthylamine
PBT polybutylene terephthalate
PC permeability coefficient
PC personal computer
PC plastic composite
PC plastic compounding
PC plastic-concrete
PC polycarbonate
PC printed circuit
PC process control
PC programmable circuit
PC programmable controller
PCB polychlorinated biphenyl
PCB printed circuit board
PCE perchloroethene
PCR post-consumer resin
pcf pounds per cubic foot

PCFC polychlorofluorocarbon
PCTFE polychlorotrifluoroethylene
PDFM Plastics Distributors & Fabricators Magazine
PDL Plastics Design Library publication
PE plastic engineer
PE Plastics Engineering Magazine
PE polyethylene (UK polythene)
PE professional engineer
PECTFE Polyethylene chlorotrifluoroethylene
PEEK polyetheretherketone
PEEL polyether ester
PEI polyetherimide
PEK polyether ketone
PEKK polyether ketone ketone
PEN polyethylene naphthalate
PES polyether sulfone
PET polyethylene terephthalate
PETG polyethylene terephthalate glycol
PEX polyethylene crosslinked (*see* XLPE)
PF phenol formaldehyde
PFA perfluoroalkoxy (copolymer of tetrafluoroethylene & perfluorovinylethers)
PFBA polyperfluorobutyl acrylate
PHF Plastics Hall of Fame
phr parts per hundred of rubber
pi $\pi = 3.141593$
PI polyimide
PIA Plastics Institute of America
PID proportional-integral-differential control
PIM powder injection molding
PLASTEC Plastics Technical Evaluation Center (US Army)
PLC programmable logic controller
pLED polymeric light-emitting diode
PLTA Plastics Lumber Trade Assoc.
PMC polyester molding compound (TS)
PMMA Plastics Molders & Manufacturers Assoc. (of SME)

PMMA polymethyl methacrylate (acrylic)

PMMI Packaging Machinery Manufacturers Institute

PO polyolefin

POE polyolefin elastomer

POM polyoxymethylene or polyacetal (acetal)

POSS polyhedral oligomeric silsesquioxanes

PP polypropylene

PPA polyphthalamide

ppb parts per billion

PPC polyphthalate carbonate

PPE polyphenylene ether

pph parts per hundred

ppm parts per million

ppmv parts per million by volume

ppmwt parts per million by weight

PPO polyphenylene oxide

PPPS polyphenylene sulfide sulphone

PPS polyphenylene sulfide

PPSF polyphenylsulfone

PPSU polyphenylene sulphone

PPVC plasticized polyvinyl chloride

PR public relation

PS polystyrene

PSB polystyrene butadiene rubber (GR-S, SBR)

psf pounds per square foot

PS-F polystyrene-foam

PSF polysulphone

psi pounds per square inch

PSI Polymer Search on Internet

psia pounds per square inch, absolute

psid pounds per square inch, differential

psig pounds per square inch, gauge (above atmospheric pressure)

PSU polysulfone

PT Plastics Technology publication

PT Plastics Toolbox

PTFE polytetrafluoroethylene (or TFE)

PTV pressure-temperature-volume

PUF polyurethane foam

PUR polyurethane (also PU, UP)

P-V pressure-volume (also PV)

PV pressure-velocity

PVA Polyvinyl alcohol

PVAC polyvinyl acetate

PVB polyvinyl butyral

PVC polyvinyl chloride

PVD physical vapor deposition

PVDA polyvinylidene acetate

PVDC polyvinylidene chloride copolymer

PVDF polyvinylidene fluoride

PVF polyvinyl fluoride

PVP polyvinyl pyrrolidone

PVT pressure-volume-temperature (also P-V-T or pvT)

PW Plastics World magazine (1997 became Molding Systems of SME, etc.)

QA quality assurance

QC quality control

QMC quick mold change

QPL qualified products list

QSR quality system regulation

R Rankin

R Reaumur

R Reynold's number

R Rockwell (hardness)

R_2SiO silicon

R&D research & development

R&M reliability and maintainability

R&R repeatability and reproducibility

rad That quantity of ionizing radiation that results in the absorption of 100 ergs of energy per gram of irradiated material.

radome radar dome

RAM radar absorbent material

RAM random access memory

Rapra Rubber & Plastics Research Assoc.

R&R repeatability and reproducibility

Rc Rockwell hardness C (R_c)

Re Reynold's number

Ref. reference

rf radio frequency

RFI radio frequency interference
RH relative humidity
RIM reaction injection molding
RLM reinforced liquid molding
RM rotational molding
RMA Rubber Mfgrs. Assoc.
RMPP rubber modified
   polypropylene (PP/EPDM)
RMS root mean square
ROI return on investment
RP rapid prototyping
RP reinforced plastic
RP Reinforced Plastics publication by
   Elsevier, UK
RPA Rapid Prototyping Assoc. (of
   SME)
RPM revolutions per minute
RRIM reinforced reaction injection
   molding
RT rapid tooling
RT room temperature
RTI relative thermal index
RTM resin transfer molding
RTP reinforced thermoplastic
RTS reinforced thermoset
RTV room temperature vulcanization
RV recreational vehicle
Rx radiation curing

s second
SAE Society of Automotive Engineers
SAMPE Society for the Advancement
   of Material and Process
   Engineering
SAN styrene acrylonitrile
SB styrene butadiene
SBA Small Business Administration
SBR styrene-butadiene rubber
SBS styrene butadiene styrene
SCADA Supervisor control and data
   acquisition
SCT soluble core technology
SDM standard deviation
   measurement
SES Standards Engineering Society
SF safety factor
SF short fiber

SF structural foam
s.g. specific gravity (sp. gr.)
SGMP statistical good manufacturing
   practice
SI Systeme International d'Unites
   (International System of Units)
SI silicone
SIC Standard Industrial Classification
SiO_2 silica
Sigma Greek letter σ.
SMA styrene maleic anhydride
SMC sheet molding compound
SMCAA Sheet Molding Compound
   Automotive Alliance
SME Society of Manufacturing
   Engineers
S-N stress-number of cycles
SN synthetic natural rubber
SNMP simple network management
   protocol
sp. vol. specific volume
SPC statistical process control
SPE Society of Plastics Engineers
Spec. specification
SPE Society of the Plastics Engineers
SPI Society of the Plastics Industry
sPS syndiotactic polystyrene
sq. square
SQC statistical quality control
SRI Standards Research Institute
   (ASTM)
SRIM structural reaction injection
   molding
S-S stress-strain
STP Special Technical Publication
   (ASTM)
STP standard temperature & pressure
STX stampable reinforced
   thermoplastic sheet (Azdel
   tradename)
SUV sport utility vehicle
syn synthetic

t thickness
T temperature
T time
T torque (or T_t)

T_g glass transition temperature
T_m melt temperature
T_s tensile strength
T/C thermocouple
TAC triallylcyanurate
TAPPI Technical Association of the Pulp & Paper Industry
TC thermal conductivity
TCA 1,1,1-trichloroethane
TCE trichloroethene
TCM technical cost modeling
TD thermal desorption
TD transverse direction
TDI toluene diisocyanate
TF thermoforming
TFE *see* PTFE
TFS thermoform-fill-seal
TGA thermogravimetric analysis
TGI thermogravimetric index
THF tetrahydrofuran
three D 3-dimensional (3-D)
TIME twin-screw injection molding extruder
TIR tooling indicator runout
T-LCP thermotropic liquid crystal polymer
TM transfer molding
TMA thermomechanical analysis
TMA Tooling & Manufacturing Assoc. (formerly TDI)
TMA Toy Mfgrs. of America
torr A unit of pressure equal to 1/760th of an atmosphere [mm mercury (mm Hg)]
TP thermoplastic
TPE thermoplastic elastomer
TPO thermoplastic olefin elastomer
TPP triphenylphosphine
TPR thermoplastic rubber
TPU thermoplastic polyurethane
TPV thermoplastic vulcanizate
TPX polymethylpentene
TQM Total Quality Management
TS thermoset
TS twin-screw
TSC thermal stress cracking
TSE thermoset elastomer

TUV TUV America Inc. (Testing Lab.)
TWA time weighted average
two-D 2-dimensional (2-D)
TX thixotropic
TXM thixotropic metal slurry molding
Tx toxic

UA urea, unsaturated
UD unidirectional
UDC unidirectional composite
UF urea formaldehyde
UHMWPE ultra-high molecular weight polyethylene (PE-UHMW)
UL Underwriters' Laboratory
UM University of Massachusetts
UN United Nations
UP unsaturated polyester (TS polyester)
UPVC unplasticized polyvinyl chloride
UR urethane (also PUR, PU)
URP unreinforced plastic
USA United States of America
UV ultraviolet
UVCA ultra-violet-light-curable-cyanoacrylate

V vacuum
V velocity
V volt
VA value analysis
var variable
VC vent cloth
VCM vinyl chloride monomer
VDE German Standard Institute
VE vinyl ester
VI Vinyl Institute
VLDPE very low density polyethylene
VOC volatile organic compound
VOCC volatile organic compound content
vol volume (not volatile)
vol% percentage by volume
vs. versus

w width
W watt
W wicking
W/D weight-to-displacement volume (boat hull)
WIT water-assist injection molding technology
WMMA Wood Machinery Mfgrs. of America
WP&RT World Plastics & Rubber Technology magazine
WPC wood-plastic composite
wt% percentage by weight
WVT water vapor transmission
WW 11 World War Two
WWW World Wide Web
WYSIWYG what you see is what you get

X-axis axis in plane used as 0° reference
XL cross-linked
XLPE cross-linked polyethylene (*see* PEX)
XPS expandable polystyrene
XTC glass and PET fibers

Y-axis axis in the plane perpendicular to X-axis
YPE yield point elongation

Z-axis axis normal to the plane of the X-Y axes
Z-N Ziegler-Natta
Z-twist twisting fiber direction
ZZ zigzag

Appendix B

Glossary

A-B-C-stages These letters identify the various stages of cure when processing thermoset (TS) plastic that has been treated with a catalyst; basically A-stage is uncured, B-stage is partially cured, and C-stage is fully cured. Typical B-stage are TS molding compounds and prepregs which in turn are processed to produce C-stage fully cured plastic material products; they are relatively insoluble and infusible.

Ablative A material which absorbs heat, while part of it is being consumed by heat, through a decomposition process which takes place near the surface exposed to the heat. An example is a carbon fiber-phenolic reinforced plastic that is exposed to a temperature of 1650C (3000F); it is the surface material on a reentry into the earth's atmosphere from outer space of a rocket, space vehicle, etc.

Accuracy and repeatability Most applications are concerned with repeatability that is easier to achieve than high accuracy. Repeatability deals with factors such as how closely the length of a given feed will repeat itself. Repeatability is different in that it does not include the noncumulative errors that an accuracy specification includes.

Addition polymer See **Polymer, addition**.

Adiabatic It is a change in pressure or volume without gain or loss in heat. Describes a process or transformation in which no heat is added to or allowed to escape from the system.

Air shot Also called air purge. It is the contents of a plasticator shot expelled into the air to study the characteristics of the melt; usually performed on start-up with the mold in the open position.

Air entrapment A phenomenon wherein air is occulated in a plastic giving rise to blisters; bubbles; and/or voids that are usually not desired. It can occur during fabrication. The bubbles could be due to air alone or moisture due to improper plastic material drying, compounding agent volatiles, plastic degradation, or the use of contaminated regrind. So the first step to resolving this problem is to be sure what problem exists. A logical troubleshooting approach can be used.

Algorithm An algorithm is a procedure for solving a mathematical problem.

Angel hair If plastic materials are conveyed at a faster rate than necessary, they may slide against the walls of the conveying tubes and heat up by friction, which in turn will cause them to begin to melt, producing what is called angel hair. This commonly takes place in a bend of a conveying tube due to the centrifugal force that is placed on the pellets or other forms of plastic materials, forcing them to slide along the outer periphery of the tube.

Annealing Also called hardening, tempering, physical aging, and heat treatment. The annealing of plastics can be define as a heat-treatment process directed at improving performance by removal of stresses or strains set up in the material during its fabrication. Depending on the plastic used, it is brought up to a required temperature for a definite time period, and then liquid (usually water; also use oils and waxes) and/or air cooled (quenched) to room temperature at a controlled rate. Basically the temperature is near, but below, the melting point. At the specified temperature the molecules have enough mobility to allow them to orient to a configuration removing or reducing residual stress. The objective is to permit stress relaxation without distortion of shape and obtain maximum performances and/or dimensional control.

Annealing is generally restricted to thermoplastics, either amorphous or crystalline. Result is increasing density, thereby improving the plastics heat resistance and dimensional stability when exposed to elevated temperatures. It frequently improves the impact strength and prevents crazing and cracking of excessively stressed products. The magnitude of these changes depends on the nature of the plastic, the annealing conditions, and the parts geometry.

The most desirable annealing temperatures for amorphous plastics, certain blends, and block copolymers is above their glass transition temperature (T_g) where the relaxation of stress and orientation is the most rapid. However, the required temperatures may cause excessive distortion and warping. The plastic is heated to the highest temperature at which dimensional changes owing to strain are released. This temperature can be determined by placing the plastic part in an air oven or water liquid bath and gradually raising the temperature by intervals of 3 to 5C until the maximum allowable change in shape or dimension occurs. This distortion temperature is dictated by the thermomechanical processing history, geometry, thickness, and size. Usually the annealing temperature is set about 5^0C lower using careful quality control procedures.

Rigid, amorphous plastics such as polystyrene (PS) and acrylic (PMMA) are frequently annealed for stress relief. Annealing crystalline plastics, in addition to the usual stress relief, may also bring about significant changes in the nature of their crystalline state. The nature of the crystal structure, degree of crystallinity, size and number of spherulites, and orientation control it. In cases when proper temperature and pressure are maintained during processing, the induced internal stresses may be insignificant, and annealing is not required.

Plastic blends and block copolymers typically contain other low and intermediate molecular weight additives such as plasticizers, flame-retardants, and UV or thermal stabilizers. During annealing, phase and micro-phase separation may be enhanced and bleeding of the additives may be observed. The morphologies of blends and block copolymers can be affected by processing and quenching conditions. It their melt viscosities are not matched, compositional layering perpendicular to the direction of flow may occur. As in the case of crystalline plastics, the skin may be different both in morphology and composition. Annealing may cause more significant changes in the skin than in the interior.

Applesauce surface Rough, wavy appearance on a fabricated part.

Areal weight The weight of a fiber reinforcement per unit area (width times length) of fabric or tape.

Aspect ratio It is the ratio of length (L) to diameter (D) of a material such as a fiber or rod; also the ratio of the major to minor axis lengths of a material such as a particle. These ratios can be used in determining the effect of dispersed additive fibers and/or particles on the viscosity of a fluid/melt and in turn on the performance of the compound based on L/D ratios. In reinforced plastics, fiber L/D will have a direct influence on the reinforced plastic performance.

Asymmetric It is opposite of symmetrical. Refers to an irregular shape. It is of such form or shape that no line, points, or plane exists whose opposite positions are not similar.

Asymptotic Refers to not meeting. An example is a straight line that is a limiting position of a tangent to a curve as its point of contact recedes indefinitely along an infinite branch of the curve.

Auger Refers to the action of the rotating screw in advancing the plastic going from the unmelted to melted stages.

Axis A reference line of infinite length drawn through the center of the rear of the screw shank and the center of the discharge end.

Azdel Stampable reinforced thermoplastic sheet (STX) material. STX is the original registered tradename of Azdel Inc., Shelby, NC

B-Stage See **A-B-C-stages.**

Barrel alignment It is the alignment at installation and routine maintenance checks to ensure the screw, mold, and any auxiliary equipment attached to the barrel are all aligned.

Barrel and feed unit heat control The feed-throat casting is generally water-cooled to prevent an early temperature rise of the plastics. A good starting point is to have the temperature about 110-120F (43-49C), or "warm to the touch" to help ensuring that a stable feed is developed. If the temperature rises too high, it may cause the plastic to adhere (stick) to the surface of the feed opening, causing a material-conveying problem to the screw. The over-heated plastic solidifies at the base of the hopper or above the barrel bore causing bridging whereby material no longer can enter the screw.

The problem can also develop on the screw, with plastic sticking to it, restricting forward movement of material. Over cooling the hopper can have a negative effect on performance. Reason for this action is due to its heat sink effect that pulls most of the heat from the feed zone of the barrel. The idea of hopper block cooling is primarily to prevent sticking or bridging in that area. Thus, it should not be run colder than necessary. Always control water flow in the throat cooling systems from the outlet side to prevent steam flashing and to minimize air pockets.

Barrel control transducer Thermocouple and pressure transducers are inserted in different zones of the barrel to sense melt condition; they require accuracy in proper locations and recording instrumentation.

Barrel feed housing Component of the plasticator barrel that contains the feed opening, water heating and/or cooling channels, and in certain units, contains barrel grooving to improve flow of plastics into the screw flights. If required a thermal barrier is used that is attached to the barrel.

Barrel grooved feed Grooves in the internal barrel surface in the feed section, particular for certain materials, permits considerable more friction between the solid plastic particles and the barrel surface. Result is process output increases and/or the process stability improves.

Barrel inspection To ensure proper performance, different parts of the barrel can be checked to meet tolerance requirements (usually setup by the manufacturer) and determine if any wear has occurred such as: inside diameter, straightness and concentricity, and surface condition.

Barrel wear Most barrels are made with nitrided steel or one of several types of bimetallic construction. Nitriding is the surface-hardening technique. The maximum effective depth achieved is less than 0.4 mm (0.016 in.). Once that thin surface layer is worn away, the barrel's abrasive wear resistance is essentially poor because there is only the steel substrate. Bimetallic barrels combine a structural steel exterior with an alloy inlay or a tool steel or alloy lining to improve resistance to abrasion and corrosion. In contrast to nitrided steel, bimetallic linings are uniformly hard throughout their depth. Depths are typically about 1.5 mm (0.060 in.) for centrifugal cast linings and about 6.3 mm (0.250 in.) for tool steel or alloy linings. Bimetallics are far more durable than nitrided. The main types of bimetallic barrels are tungsten carbide composites, chromium-modified iron-boron alloys, and nickel alloys.

Beam, high performance/fire resistant It has been known that steel structural beams cannot take the heat of a fire operating at and above 815C (1500F); they just loose all their strength, modulus of elasticity, etc. To protect steel from this potential environmental condition they can obtain a temporary short time protection by being covered with products such as concrete and certain plastics. To significantly extend the life of structural beams hardwood (thicker, etc.) can be used; thus people can escape even though the wood slowly burns. The more useful and reliable structural beams would be using reinforced plastics (RPs) that meet structural performance requirements with even a much more extended supporting life than wood. Because of RPs high costs to date RPs are not used in this type of fire environment.

Bell End A flange at the discharge end of the barrel that provides added strength to withstand internal pressure.

Birefringence It is the difference in the refractive indexes of two perpendicular directions in a given material such as a thermoplastic. When the refractive indexes measured along three mutually perpendicular axes are identical, they are classified as optically isotropic. When the TP is stretched, providing molecular orientation, and the refractive index parallel to the direction of stretching is altered so that it is no longer identical to that which is perpendicular to this direction, the plastic displays birefringence. Techniques of birefringence ranging from the determination of structural defects in solid plastics to more basic investigations of molecular and morphological properties are used in a wide range of applications.

Basically, birefringence is the contribution to the total birefringence of two-phase materials, due to deformation of the electric field associated with a propagating ray of light at anisotropically shaped phase boundaries. The effect may also occur with isotropic particles in an isotropic medium if they dispersed with a preferred orientation. The magnitude of the effect depends on the refractive index difference between the two phases and the shape of the dispersed particles. In thermoplastic systems the two phases may be crystalline and amorphous regions, plastic matrix and micro-voids, or plastic and filler.

Black-box A phrase used to describe a device whose method of working is ill-defined or not understood.

Blister It is a cavity or sac that deforms the surface of a material. It is usually a raised area on the parts surface caused by the pressure of gases or air inside the part that surfaces during fabrication.

Blister ring A raised portion of the root between flights of sufficient height and thickness to effect a shearing action of the melt as it flows between the blister ring and the inside wall of the barrel.

Bloom A bloom on the surface of plastics is the result of ingredients coming out of "solution" in the fabricated plastic product and migrating to its surface.

Blister It is a cavity or sac that deforms the surface of a material. It is usually a raised area on the parts surface caused by the pressure of gases or air inside the part that surfaces during fabrication.

Boltzmann principle Boltzmann, Ludwig Boltzmann superposition principle provides a basis for the description of all linear viscoelastic phenomena. Unfortunately, no such theory is available to serve as a basis for the interpretation of nonlinear phenomena, i.e., to describe flows in which neither the strain or the strain rate is small. As a result, there is no general valid formula for calculating values for one material function on the basis of experimental data from another. However, limited theories have been developed.

Born in Vienna (1844-1906) his work of importance in chemistry became of interest in plastics because of his development of the kinetic theory of gases and rules governing their viscosity and diffusion. They are known as the Boltzmann's Law and Principle, still regarded as one of the cornerstones of physical science.

Bridging When an empty hopper is not the cause of machine failure, plastic might have stopped flowing through the feed throat because of screw bridging. An overheated feed throat, or startup followed with a long delay, could build up sticky plastics and stop flow in the hopper throat. Plastics can also stick to the screw at the feed throat or just forward from it. When this happens, plastic just turns around with the screw, effectively sealing off the screw channel from moving plastic forward. As a result, the screw is said to be "bridged" and stops feeding the screw. The common solution is to use a proper rod, such as brass, to break up the sticky plastic or to push it down through the hopper without damaging the extruder.

British thermal unit Btu is the energy needed to raise the temperature of 1 lb of water 1F (0.6C) at sea level. As an example, one lb of solid waste usually contains 4500 to 5000 Btu. Plastic waste contains greater Btu than other materials of waste.

Bulk factor Ratio of the volume of a material such as a loose powder to the volume of the finished fabricated product.

Cavity Female section of the mold which generally gives the external shape to the product to be molded.

C-stage See **A-B-C-stages**.

Calculus It is the mathematical tool used to analyze changes in physical quantities, comprising differential and integral calculations. It was developed during the 17th century to study four major classes of scientific and mathematical problems of that time. (1) Find the maximum and minimum value of a quantity, such as the distance of a planet from the sun. (2) Given a formula for the distance traveled by a body in any specified amount of time, find the velocity and acceleration of the body at any instant. (3) Find the tangent to a curve at a point. (4) Find the length of a curve, the area of a region, and the volume of a solid. These problems were resolved by the greatest minds of the 17th century, culminating in the crowning achievements of Gottfried Wilhelm (Germany 1646 to 1727) and Isaac Newton (English 1642 to 1727). Their information provided useful information for today's space travel.

Capacitor There are several applications for plastics in electrical devices that use the intrinsic characteristics of the plastics for the effect on the electrical circuit. The most obvious of these is the use of plastics, particularly in the form of thin films as the dielectric in capacitors. TP polyester films such as Mylar are especially useful for this type of application because of the high dielectric strength in conjunction with a good dielectric constant. Mylar has the additional desirable feature that it is available in very thin films down to 2.5 microns.

Since the value of a capacitor is directly proportional to the area and inversely proportional to the spacing of the conductive plates, the thinner materials permit high values of capacitance in small size units. There are other materials that make good capacitors such as polyvinylidene fluoride that has a very high dielectric constant and good dielectric strength, oriented PS which makes a good capacitor for high frequencies because of its low dielectric loss constant, and others.

Carbon black A black colloidal carbon filler made by the partial combustion and/or thermal cracking of natural gas, oil, or another hydrocarbon.

Carbon fiber Polyacrylonitrile (PAN) fibers are thermally carbonized to obtain carbon fiber that is used to reinforce plastics.

Catalyst A catalyst is basically a relatively small amount of substance that augments the rate of a chemical reaction without itself being consumed; recovered unaltered in form and amount at the end of the reaction. It generally accelerates the chemical change. The materials used to aid the polymerization of most plastics are often not catalysts in the strict sense of the word (they are consumed), but common usage during the past century has applied this name to them.

There are relatively many different catalysts that are usually used for specific chemical reactions. Types include Z-N, metallocene, and others including their combinations. These different systems are available worldwide from different companies. Terms and information are used to identify the behavior of catalysts. An autocatalyst is a catalytic reaction induced by a product of the same reaction. This action occurs in some types of thermal decomposition. The catalyst benzoyl peroxide is a white, granular, crystalline solid, tasteless, faint odor of benzaldehyde, has active oxygen, and soluble in almost all-organic solvents. Use includes polymerization catalyst with different plastics such as thermoset polyester, rubber vulcanization without sulfur, embossed vinyl floor covering, etc. A catalyst carrier is a neutral material used to support a catalyst, such as activated carbon, diatomaceous earth, or activated alumina. There are fluid catalysts that are finely divided solid particles utilized as a catalyst in a fluid bed process using certain thermoset plastics.

The enzyme catalysts are organic catalysts of living cells. Because microorganisms are able to synthesize thousands of complex organic molecules, they represent an enormous catalytic potential to the industrial chemist. A remarkable aspect of enzymes is their enormous accelerated catalytic power; they can enhance reaction rates by a factor of from 10^8 to 10^{20}. Also they can function in dilute aqueous solution under moderate conditions of temperature and pH.

Catalyst, Ziegler-Natta Also called the Z-N catalyst. Karl Zeigler (1898–1973) of Germany and Giulio Natta (1903-1979) of Italy developed a catalyst for the industrial production of polyolefin plastics. Together they received the Nobel Prize for chemistry in 1963. They provided the key (Zeigler for PE; Natta for PP) at that time to a relatively simple, inexpensive, and controllable large production method. They also paved the way for the overwhelming triumph of the polyolefins in subsequent years. All this is now changing to some degree, with the metallocene catalysts. Unlike the Z-N, the new generations of catalysts can provide very simplified production capabilities that produce improvements in properties, processability, and cost.

Catalyst metallocene Catalyst Also called single site, Me and m. Metallocene catalysts achieve creativity and exceptional control in polymerization and molecular design permitting penetration of new markets and expand on of present markets. Chemists can model and predict plastic structure in a matter of days rather than years. Emphasis has been on the polyolefins (mPOs); others include PS, PE/PS, TPO, and EPDM.

Uniformity of molecular weight effectively eliminates molecular extremes resulting in a range of property improvements that are targeted to include improved mechanical, physical, and chemical properties; provide processing advantages; and lower costs. Available are uniquely synergistic combinations of complementary abilities. As an example, mPE becomes an economical material competing with the properties of nylon and thermoplastic polyester plastics. Also one can produce mLLDPE film with the same strength at a lower gauge than conventional LLDPE because of its narrow molecular weight range. These Me catalysts are more accurate in characterizing plastics than today's quality control instruments can verify.

They produce plastics that are stronger and tougher; thus less plastic is required. They process in a different manner so one has to become familiar with the processing techniques. Target is to obtain a plastic with a specific molecular weight distribution (MWD), density, melt flow rate, tensile strength, flexural modulus, or a combination of other factors. Whatever the parameter, Me catalysts allow fabricators to alter reactor temperatures, pressures, and other variables to achieve their goal. Performancewise, mPO grades, regardless of density or comonomer, can combine softness and toughness, whereas conventional POs must trade off one for the other.

These catalysts can make plastics that process well by knitting long branches into the carbon chains. They make plastics with uniform, narrow MWD, high comonomer content, very even comonomer distribution, and enormously wide choice of comonomers compared to multi-site Z-N catalysts. Comonomer choices include aromatics, styrene compounds, and cyclic olefins. Copolymers made with conventional Z-N catalysts favor ethylene and propylene. They incorporate only isolated amounts of more exotic monomers. The Me catalysts have been used to make different plastics such as PE homo-, co-, or ter-polymers from 0.865 to 0.96 density; isotactic, syndiotactic, and atactic PP; syndiotactic PS; and cyclic olefin copolymers.

As an example, blown film extruders designed to process LLDPE can process mLLDPE generally without difficulty; torque, head pressure, and motor load limitations generally do not limit film productivity. However, it is important to understand the differences arising from the different rheologies. The mLLDPE has a narrower MWD, and it thus exhibits lower shear sensitivity. The extruder would operate at higher temperatures and motor torque levels, while decreasing bubble stability and easing tensions on winding and draw down ratio. The Me with less chain branching would result in faster melt relaxation and less draw resonance. One with lower density would have greater elasticity, decreased specific rate in a grooved feed machine, and increased specific rate in a smoothbore machine, while harder to wind.

All other things being equal, they are more viscous at typical extrusion shear rates than conventional LLDPE. There is a difference between shear rheology with the same screw/barrel. The mLLDPE will extrude at a higher melt temperature profile. This action may limit output on cooling-limited lines, but it may be possible to keep line speeds constant. Result is making a thinner film having the same performance because of the better properties offered by mLLDPE. Barrel cooling can be used to reduce mLLDPE melt temperatures but it may be more desirable to optimize the extruder screw for the plastic's rheology.

Catalyst summation The new polymerization catalysts with conventional commodity feedstocks have produced a wave of new plastics that became obvious early during the 1990s. The terms used with this new technology include metallocene, single-site, constrained-geometry, and syndiotactic.

Center for Process Analytical Chemistry CPAC is an industry university consortium headquartered at the University of Washington. The team of faculty members, research staffers, visiting scientists, and graduate students conducts research at I 0 university nationwide and works with companies to develop technology. The charter focuses on chemometrics, sensors, spectroscopy, chromatography, and microflow analysis. The main objective is to develop real time measurement and relevant data handling techniques.

Channel With the screw in the barrel, it is the space bounded by the interfaces of the flights, the root of the screw, and the bore of the barrel. This is the space through which the stock (melt) is conveyed and pumped.

Chisolm's law Anytime things appear to be going better, you have overlooked something.

Chlorinated solvent Chlorinated solvents (CSs) were first produced over a century ago and came into common usage in the 1940's. Chlorinated solvents are excellent degreasing agents and they are nearly non-flammable and non-corrosive. These properties have resulted in their widespread use in many industrial processes such as cleaning and degreasing rockets, electronics and clothing (used as dry-cleaning agents), plastics, and medical devices. Chlorinated solvent compounds and their natural degradation or progeny products have become some of the most prevalent organic contaminants found in the shallow groundwater of the USA. The most commonly used chlorinated solvents are perchloroethene (PCE), trichloroethene (TCE), 1, 1, 1-trichloroethane (TCA), and carbon tetrachloride (CT).

Chromatography A technique for separating a sample material into constituent components and then measuring or identifying the compounds by other methods. As an example separation, especially of closely related compounds, is caused by allowing a solution or mixture to seep through an absorbent such as clay, gel, or paper. Result is that each compound becomes adsorbed in a separate, often colored layer.

Clean-area, fabricating Technology provides a milieu of artificial purity to protect sensitive products from air-laden particle contamination. Required measures include: (1) a workplace correctly designed for clean-air technology and suitable conduct by employees, (2) effective filtration of the air supply and carefully planned air ducting, (3) easily to clean surfaces throughout the clean-area, (4) a high degree of automation of all work operations, and (5) regular monitoring with the aid of suitable particle measuring technology.

Cleaning equipment Different equipment requires cleaning on a periodical maintenance time schedule to ensure their proper operation. Available are cleaning devices for molds, extruder dies and screen changers, molded flash, etc. that operate economically and safely removing contaminated plastics. The routine techniques used include blow torches, hot plates, hand working, scraping, burn-off ovens, vacuum pyrolysis, hot sand, molten salt, dry crystals, high pressure water, ultrasonic chemical baths, heated oil, and lasers.

Personnel have to be careful not to damage expensive tooling by spot annealing, mechanical abuse, etc. There are commercial cleaning systems used such as aluminum oxide beds (fluidized beds), salt baths, hot air ovens, and vacuum pyrolysis. As an example, the vacuum pyrolysis cleaner utilizes heat and vacuum to remove the plastic. Most of the plastic is melted and trapped. Remaining plastic is vaporized and appropriately collected in a trap.

Cleaning plastic Different techniques are used to clean fabricated products. Included is solvent, ultrasonic, blasting with dry ice (carbon dioxide) pellets, toxic chemicals, and even PCFC-based solvents, particular medical devices.

Cleanroom In the past clean rooms where left to a few, usually the larger plants or specialized operations concerned with medical or pharmaceutical products. In the mean time, processors have not been able to isolate themselves from the trend toward clean room production in order to achieve the necessary quality levels from the electronics and micro electronics industries, and lately, even from other industries such as automotive and entertainment. With careful planning, considerable savings can be made in investments and operating costs. The required degree of cleanliness, in particular, determines costs to a large extent and is directly influenced by a number factors such as the size of the room and contaminants.

The worst enemy is dust that must be eliminated with the greatest producer are human beings. The smallest dust particles are less than

0.5µm. Moreover, the number of particles depends on the type and speed of any motions. Since the continued production of dust is unavoidable, measures must be taken to reduce the total particle count. The lower the permissible amount of dust in a planned production area, the greater the resultant costs.

Cleanroom standard The US Federal Standard 209E, Airborne Particulate Cleanliness Classes in Clean-rooms and Clean Zones, is required for manufacturers who want to conform to quality system regulation. Via the industrial ISO European Community, it has been integrated with ISO. Among the more important recent changes are metrication, revision of upper confidence level (UCL) requirement, provisions for sequential sampling, and an alternative verification procedure based on determination of the concept of ultra-fine particles known as U descriptors.

Cold flow It is creep at room temperature.

Commodity & engineering plastics About 90wt% of plastics can be classified as commodity plastics, the others being engineering plastics. Commodity plastics are usually associated with the higher volume, lower priced plastics with low to medium properties. Used for the less critical parts where engineering plastics are not required. The five families of commodities LDPE, HDPE, PP, PVC, and PS account for about two thirds of all the plastics consumed. The engineering plastics such as nylon, PC, acetal, etc. are characterized by improved performance in higher mechanical properties, better heat resistance, higher impact strength, and so forth. Thus, they demand a higher price. About a half century ago the price per pound difference was at 20¢; now it is above $ 1.00. There are commodity plastics with certain reinforcements and/or alloys with other plastics that put them into the engineering category. Many TSs and RPs are engineering plastics.

Computer See Chapter 5.

Computer science & algebra The symbolic system of mathematical logic called Boolean algebra represents relationships between entities; either ideas or objects. George Boole of England formulated the basic rules of the system in 1847. The Boolean algebra has been used extensively in the fields of chemistry and engineering associated with plastics and eventually became a cornerstone of computer science.

Constant lead Also called uniform pitch screw. A screw with a flight of constant helix angle.

Contamination Any unwanted or foreign body in a material or the processing area, including air, that affects or detracts from part's quality.

Control, solid state This is the type of control system that superceded relay control. It is based on electronic components that have no moving parts and yet can, for example, provide switching action.

Controlled motion Linear guides used in different equipment provide a means of low-friction precision linear motion through an assortment of rails (round or profile), contact elements (rollers, ball bearings, or full-contact sleeves), and mounting configurations. Many types of guides exist, each engineered toward optimized performance in a specific range of applications. Therefore, various application criteria will effect linear guide incorporation. These criteria can be summarized as follows: dynamic load capacity, envelope size, mounting configurations, life, travel accuracy, rigidity, speed/acceleration, cost, and environmental considerations. The priority of these items will determine the appropriate linear guide for application.

Controller Any instrumentation such as pressures, temperatures, timers, etc. used to control and regulate the fabricating cycle.

Corona resistance Among the factors contributing to the breakdown of insulating materials is dielectric heating at high frequencies and, in many instances, the effect of corona. Corona is usually described as the partial breakdown of insulation due to the concentration of electrical stress at sharp edges, or actual breakdown of insulation when placed in series with another insulation having a different dielectric constant.

In use at DC potentials the average insulation has very little corona. However, at frequencies from 60 cycles up, the corona becomes apparent and is an important factor in the service life and breakdown of the dielectric. For use at high voltages and frequencies, for example, polyethylene is usually impregnated with insulating oil, such as silicone oil.

However, corona discharge does occur in oil, and if the oil breaks down the corona will produce higher stresses and erode away the surface of the dielectric, thereby reducing its thickness and causing a breakdown.

Most high plastic insulating materials, such as polyethylene, Mylar, polystyrene, nylon and silicone, have about the same resistance to corona. Kel-F and Teflon are particularly susceptible to the erosion effects of corona. At a frequency of 60 cycles per second, a 4.5 mil sample of polyethylene, stressed at 600 volts per mil, will average about 50 h before breakdown occurs. The time to breakdown for polystyrene is approximately 75 h, while that of Teflon is only about 6 hours.

Crack growth Crack growth behavior can be analyzed using fracture mechanics that can provide fracture toughness to prevent fracture. Fracture is a crack-dominated failure mode. For fracture to occur, a crack must somehow be created, then initiate, and finally propagates. The prevention of any of these events will prevent fracture. Cracks can be considered elastic discontinuities that can come from a variety of sources such as internal voids or dirt, and/or surface scratch, embrittlement, or weld line. Cracks can be consequences of faulty design, poor processing, and/or poor handling of raw material, assuming material arrived clean.

Crazing See **Stress whitening**.

Creep It is the time-dependent increase in strain in material, occurring under stress. Creep at room temperature is sometimes called cold flow. It is the change in dimensions of a plastic under a given stress/load and temperature over a period of time, not including the initial instantaneous elastic deformation.

Crystallinity and orientation When crystallites already exist in the amorphous matrix, orientation will make these crystallites parallel. If a plastic crystallizes too far in the melt, it may not contain enough amorphous matrix to permit orientation, and will break during stretching. (Most partially crystalline plastics can be drawn 4 to 5 times.) The degree of crystallinity is influenced by the rate at which the melt is cooled. This is utilized in the fabrication operations to help control the degree of crystallinity. The balance of properties can be slightly altered in this manner, allowing some control over such parameters as container volume, stiffness, warpage, and brittleness. Nucleating agents are available that can promote more rapid crystallization resulting in faster cycle times.

Curing It is basically to change properties of a plastic material by chemical polycondensation or addition reactions; generally refers to the process of hardening a plastic. More specifically it refers to the changing of the physical properties of a material by chemical reactions usually by the action of heat (includes dielectric heat, etc.) and/or catalyst with or without pressure. It is the process of hardening or solidification involving cross-linking, oxidizing, and/or polymerization (addition or condensation). The term curing, even though it is applied to thermoset and thermoplastic materials, is a term that refers to a chemical reaction (cross-linking) or change that occurs during its processing cycle. This reaction occurs with TS plastics or TS elastomers as well as cross-linked TPs that become TSs.

Decompression Injection and blow molding machines are fitted with decompression or suck-back. After the screw has finished rotating, it is drawn back so as to suck material away from the nozzle tip. This facility allows the use of an open nozzle. Keep the amount of suck-back as small as is practicable as the introduction of air can cause problems with some materials, for example, with PA 66 where processing temperatures are high.

Definition It is important to define words or terms, as well as abbreviations, in order to ensure that proper communication exists. Many times there can be more than one definition in order to meet different requirements as setup by different organizations, industries, legal documents, etc. In fact the definitions could have opposite or completely different meanings.

Definition, art of Providing written, graphics, etc. definitions throughout all industries (designers, fabricators, societies, trade organizations, etc.) to legal regulations (local, state, federal, worldwide, etc.) are extremely important. As an example, to enforce a law FDA, court trial, and others are obligated to interpret the language of the law. Inevitably it will sometimes stretch definitions rather "far" (as is done with the US constitution). The challenge for FDA and others is to keep them within the bounds of reason. Thus, care and a concentrated effort is required to ensure that your definition is specific, complete, concise, and not subject to change.

Deflashing Technique of removing flash from a plastic product, usually a molding. Several different methods are employed that include low temperatures of dry ice and cryogenic.

Deformation, plastic Plastics have some degree of elasticity so as long as the plastic stretches within its elastic limit, it will eventually return to its original shape. When overstressed it reaches what is known as plastic deformation where the plastic will not return to its original shape.

Denier It is the number that represents the weight in grams of 9000 meters of yarn, and is a measure of linear density. An alternative unit is the tex, representing the weight in grams of 1000 meters. Yam linear density is sometimes expressed in decitex where 1 dtex = 0. 1 tex.

Design source reduction This generally defines the design, manufacture, purchase, or use of materials or products to reduce the amount of material used before they enter the municipal solid waste stream. Because it is intended to reduce pollution and conserve resources, source reduction should not increase the net amount or toxicity of waste generated throughout the life of the product. The EPA has established a hierarchy of guidelines for dealing with the solid waste situation. Their suggestions logically include source reduction, recycling, waste-to-energy gains, incineration, and landfill. The target is to reduce the quantity of trash.

Design verification DV refers to the series of procedures used by the product development group to ensure that a product design output meets its design input. It focuses primarily on the end of the product development cycle. It is routinely understood to mean a thorough prototype testing of the final product to ensure that it is acceptable for shipment to the customers. In the context of design control, however, DV starts when a product's specification or standard has been established and is an on-going process. The net result of DV is to conform with a high degree of accuracy that the final product meets performance requirements and is safe and effective. According to standards established by ISO-9000, DV should include at least two of the following measures: (a) holding and recording design reviews, (b) undertaking qualification tests and demonstrations, (c) carrying out alternative calculations, and (d) comparing a new design with a similar, proven design.

Deviation It refers to the variation from a specified dimension or design requirement, usually defining the upper and lower limits. The mean deviation (MD) is the average deviation of a series of numbers from their mean. In averaging the deviations, no account is taken of signs, and all deviations whether plus or minus, are treated as positive. The MD is also called the mean absolute deviation (MAD) or average deviation (AD).

Deviation, root-mean-square RMS is a measure of the average size of any measurable item (length of bar, film thickness, pipe thickness, coiled molecule, etc.) that relates to the degree of accuracy per standard deviation measurement.

Devolatilization It is an important operation in the processing of plastics into products without contaminants. Since contaminants in most cases are volatile relative to their plastic, they are removed from the condensed phase by evaporation into a contiguous gas phase. Such separation processes are commonly referred to as devolatilization (DV). The plastic to be devolatilized may be in the form of a melt or particulate solid. Separation is effected by applying a vacuum or by using inert substances, such as purging with nitrogen gas or steam.

Basically one or more volatile components are extracted from the plastic. It can be either in a solid or molten state. Two types of actions occur: (1) volatile components diffuse to the plastic-vapor interface (called diffusional mass transport) and (2) volatile components evaporate at the interface and are carried away (called convective mass transport). If (1) is less then (2), the process is diffusion-controlled. This condition represents most of the plastic devolatilization processes because plastic diffusion constants are usually low.

The important relationship in diffusional mass is Fick's law. It states that in diffusion the positive mass flux of component A is related to a negative concentration of ingredients. This law is valid for constant densities and for relatively low concentrations of component A in component B. The term binary mixture is used to describe a two-component mixture. A binary diffusivity constant of one component is a binary mixture.

The diffusional mass transport is driven by a concentration gradient, as described by Fick's law. This is very familiar with Fourier's law, which relates heat transport to a temperature gradient. It is also very similar to Newton's law that relates momentum transport to a velocity gradient. Because of the similarities of these three laws, many problems in diffusion are described with similar equations. Also several of the dimensionless numbers used in heat transfer problems are also used in diffusion mass transfer problems.

Die See **Tool.**

Differential scanning calorimetry DSC is a method in which the energy absorbed or produced is measured by monitoring the difference in energy input (energy changes) into the material and a reference material as a function of temperature. Absorption of energy produces an endothermic reaction; production of energy results in an exothermic reaction. Its use includes studying processing behavior of the melting action, degree of crystallization, degree of cure, applied to processes involving a change in heat capacity such as the glass transition, loss of solvents, etc.

Dilatant Basically a material with the ability to increase its volume when its shape is changed. A rheological flow characteristic evidenced by an increase in viscosity with increasing rate of shear. The dilatant fluid, or inverted pseudoplastic, is one whose apparent viscosity increases simultaneously with increasing rate of shear; for example, the act of stirring creates instantly an increase in resistance to stirring.

Directional terminology

> **Anisotropic construction** One in which the properties are different in different directions along the laminate flat plane; a material that exhibits different properties in response to stresses applied along the axes in different directions.

Balanced construction In woven RPs, equal parts of warp and fill fibers exist. Construction in which reactions to tension and compression loads result in extension or compression deformations only, and which in flexural loads produce pure bending of equal magnitude in axial and lateral directions. It is an RP in which all laminae at angles other than O° and 90° occur in ± pairs (not necessarily adjacent) and are symmetrical around the central line.

Biaxial load A loading condition in which a specimen is stressed in two different directions in its plane, i.e., a loading condition of a pressure vessel under internal pressure and with unrestrained ends.

Bidirectional construction An R.P with the fibers oriented in various directions in the plane of the laminate usually identifies a cross laminate with the direction 90° apart.

Isotropic construction RPs having uniform properties in all directions. The measured properties of an isotropic material are independent on the axis of testing. The material will react consistently even if stress is applied in different directions; stress-strain ratio is uniform throughout the flat plane of the material.

Isotropic transverse construction Refers to a material that exhibits a special case of orthotropy in which properties are identical in two orthotropic dimensions but not the third. Having identical properties in both transverse but not in the longitudinal direction.

Nonisotropic construction A material or product that is not isotropic; it does not have uniform properties in all directions.

Orthotropic construction Having three mutually perpendicular planes of elastic symmetry.

Quasi-isotropic construction It approximates isotropy by orientation of plies in several or more directions.

Unidirectional construction Refers to fibers that are oriented in the same direction, such as unidirectional fabric, tape, or laminate, often called UD. Such parallel alignment is included in pultrusion and filament winding applications.

Z-axis construction In RP, the reference axis normal (perpendicular) to the X-Y plane (so-called flat plane) of the RP.

Disc feeder Horizontal, flat, grooved discs installed at the bottom of a hopper feeding a plasticator to control the feed rate by varying the discs speed of rotation and/or varying the clearance between discs. A scraper is used to remove plastic material from the discs.

Dispersive mixing. A mixing process in which an intrinsic change takes place in the physical character of one of the components. Agglomerates are reduced in size by fracture due to stresses generated during mixing.

Distributive mixing. Reducing the composition non-uniformity where the ingredients do not exhibit a yield stress.

Downtime See **Processing line.**

Dry cycle Number of cycles the machine can perform in 1 minute, with a mold installed, but ignoring injection, plasticizing, and dwell time. The following phases are performed by the machine during a dry cycle rate measurement: (1) mold closing and clamping, (2) nozzle-to-mold approach, (3) nozzle retraction from mold, and (4) mold opening.

Elastomer An elastomer is a rubberlike material (natural or synthetic) that is generally identified as a material that at room temperature stretches under low stress to at least twice its length and snaps back to approximately its original length on release of the stress (pull) within a specified time period. The term elastomer is often used interchangeably with the term plastic or rubber; however, certain industries use only one or the other.

Although rubber originally meant a natural thermoset elastomeric (TSE) material obtained from a rubber tree (hevea braziliensis), it identifies a TS elastomer (TSE) or thermoplastic elastomer (TPE) material. They can be differentiated by how long a material deformed requires to return to its approximately original size after the deforming force is removed and by its extent of recovery. Different properties also identify the elastomers such as strength and stiffness, abrasion resistance, solvent resistance, shock and vibration control, electrical and thermal insulation, waterproofing, tear resistance, cost-to-performance, etc. Elastomer terminology per ASTM D 1418 is shown in Table B-1 on page 487.

The natural rubber materials have been around for over a century. They will always be required to meet certain desired properties in specific products. TPEs principally continue to replace traditional TS natural and synthetic rubbers (elastomers). TPEs are also widely used to modify the properties of rigid TPs usually by improving their impact strength.

Natural rubber provides the industry worldwide with certain material properties that to date are not equaled by synthetic elastomers. They followed a process that ensures producing products. Examples include tires (with its heat build-up resistance), certain type vibrators, etc. However both synthetic TSE and TPE have made major inroads to product markets previously held by natural rubber and also expanded into new markets. The three basic processing types are conventional (vulcanizable) elastomer, reactive type, and thermoplastic elastomer.

Overall an elastomer may be defined as a natural or synthetic material that exhibits the rubberlike properties of high extensibility and flexibility. It identifies any thermoset elastomer (TSE) or thermoplastic elastomer (TPE) material. Such synthetics as neoprene, nitrile, styrene butadiene, and polybutadiene are grouped with natural rubber (NR) that are TSEs. The term's "rubber" and "elastomer" are used interchangeably.

Elastomer embrittlement The temperature at which elastomers lose their rubbery properties varies widely among elastomers. Basically, the rubbery state is maintained until the glass transition temperature (T_g) of the base polymer is reached, although in practice, elastomers become leathery as T_g is approached. Perhaps the most useful method of determining the lowest temperature at which a given vulcanizate retains elastomeric properties is a test called temperature retraction. The test is generally carried out by elongating a specimen to 75% of ultimate elongation, locking it in the elongated state, freezing it to essentially a nonelastic state, releasing the frozen specimen and allowing it to retract freely while raising the temperature at a uniform rate.

Generally the temperature at which the elastomer retracts by 10% (called TR-10) is considered to be the practical limit for low temperature performance. This type of test is especially useful for predicting the ability of an elastomer to seal at low temperatures. Of the high performance elastomers, silicone rubber has excellent low temperature properties. The brittle point for a typical vulcanizate is near –95C (-140F), and the TR-10 value is approximately –72C (-100F). As a contrast, fluoroelastomer elastomer typically have a TR-10 ranging from –18 to –23C (0 to 10F). Brittle points for general-purpose elastomers such as Buna N, neoprene, natural rubber, and SBR range from 23 to 56C (-10 to -70F). Polybutadiene elastomers, with values in the range of –107C (-160F), have perhaps the lowest brittle points of available elastomers. While brittle point (T_g) and temperature retraction evaluations serve as guides below whom elastomers are stiff and nonconformable, it does not mean that applications such as sealing cannot be maintained at lower temperatures. For example, 0 rings compressed to 90% in a tongue and groove flange have performed well at temperatures as low as –183C (-330F).

Elastomer, liquid Liquid systems for the fabrication of elastomeric items fall into three general categories: lattices, solvent cements, and liquid polymers. While the first two categories were well established when nature was the only source for elastomers, the third category of liquid polymers is essentially a product of synthetic elastomer technology.

Latex They constitute an intermediate stage of the bulk of synthetic elastomers, both from a volume and a value standpoint, which are produced today. While varying obviously in composition, the various systems are analogous to natural rubber lattices in that they are an aqueous dispersion of small particles of the particular elastomer. One of the principal advantages of latex technology resides in the fact that it is the oldest method for the fabrication of elastomer or elastomeric coated goods. Thus much of the considerable technology developed for natural rubber lattices was adaptable to synthetic elastomers as they appeared.

As lattices are subject to coagulation on shearing, compounding techniques used for solid elastomer compounding are generally inapplicable. Usually, additives required for the compounding are reduced to aqueous dispersions separately and then added to the elastomer latex. Compounded lattices have been prepared with solids content in excess of 80% and, of course, all solids levels below that figure. Items resulting from latex compounding include devices for containment of the human female form, medical and hygienic devices, and innumerable foamed objects used in transportation and as furniture.

Cement Cements also encompass the use of compounded elastomers as adhesives. They are reviewed solely as solvent dispersions of elastomer compounds. Such cements are generally prepared as needed in the fabrication of a more complicated end item. Typically, conventional elastomer compounding techniques are employed for the solid elastomer and the necessary additives, with the resulting mix blended with appropriate solvents by slow agitation.

A good example is the preparation of elastomer cements for the coating of fabrics. Desired thicknesses of cement can be continuously applied by various techniques to the fabrics followed by passage through ovens to remove the solvent and to vulcanize the elastomeric coating. For end items requiring a minimum of porosity, such as fuel diaphragm, protective clothing, and multiple coatings of the cements on fabrics are often employed.

Liquid polymer While virtually all elastomers can be prepared in sufficiently low molecular weights to exhibit fluid behavior, only those liquid polymers that can be converted to an elastomeric solid will be reviewed. It might appear that this could be accomplished by crosslinking, but this is not the case. While some crosslinking is desirable, the primary reaction required is one of chain extension that is the end-to-end attachments of the low molecular weight chains via reactive terminal groups.

In some cases this extension reaction is initiated by the use of a second reactant, and this is termed a two-component system. After mixing, the life of such systems can often be extended for considerable periods of time by low temperature refrigeration. A second type, the one-component system, utilizes an extension mechanism initiated by exposure to the atmosphere. While a number of applications for such liquid polymers exist, applications consist primarily in the areas of sealants and encapsulations.

Elastomer, thermoplastic The rapidly growing and relatively new class of thermoplastic elastomers (TPEs) (compared to natural rubbers and TSEs) differs markedly from the previous classes of elastomers, in that the processing of TPE does not involve any chemical reaction. The links between flexible molecules, which are required for rubber-like elasticity, are the result of physical interactions that operate at use temperatures, but can be suppressed for processing by raising the temperature, or with the use of a suitable solvent and restored upon cooling or drying. The various types of TPEs can generally be made in a wide range of stiffness (hardness) and thus bridge the gap between soft plastics and elastomers. Some of them are available in very soft, highly elastic grade.

Because of their thermoplastic nature, many processes can fabricate TPEs. The number of applications is rapidly increasing, as they often displace conventional vulcanizable elastomers (TSEs). TPEs offer a combination of strength and elasticity as well as exceptional processing versatility. They present creative designers with endless new and unusual product opportunities.

Quite large elastic strains are possible with minimal stress in TPEs. TPEs have two specific characteristics: their glass transition temperature (T_g) is below that at which they are commonly used, and their molecules are highly coiled as in natural TS rubber (isoprene). When a stress is applied, the molecular chain uncoils and the end-to-end length can be extended several hundred percent with minimum stress. Some TPEs have an initial modulus of elasticity of less than 10 MPa (1,500 psi); once the molecules are extended, the modulus increases.

The modulus of metals decreases with an increase in temperature. However, in stretched TPEs and particularly conventional elastomers the opposite is true, because with them at higher temperatures there is increasingly vigorous thermal agitation in their molecules. Therefore, the molecules resist more strongly the tension forces attempting to uncoil them. To resist requires greater stress per unit of strain, so that the modulus increases with temperature. When stretched into molecular alignment many elastomers can form crystals, an impossibility when they are relaxed and kinked. TPEs can be fine tuned to meet coefficient of linear thermal expansion (CLTE) required in product performance product requirements.

To date, with the exception of vehicle tires, TPEs have been replacing TS rubbers in many applications. Unlike natural TS rubbers, most TPEs can be reground and recycled, thereby reducing overall cost. The need to cure or vulcanize them is eliminated, reducing cycle times, and products can be molded to tighter tolerances. However there are TP vulcanizate (TPV) that provide property advantages. Most TPEs can be colored, whereas natural rubber is available mainly in black. TPEs also weigh 10 to 40 % less than rubber.

TPEs range in hardness from as low as 25 Shore A up to 82 Shore D (ASTM test). They span a temperature range of -34 to 177C (-29 to 350F), dampen vibration, reduce noise, and absorb shock. However, designing with TPEs requires care, because unlike TS rubber that is isotropic, TPEs tend to be anisotropic during processing as with injection molding. Tensile strengths in TPEs can vary as much as 30 to 40 % with direction.

Electret An application for plastics which uses the intrinsic properties is in electrets (a dielectric body in which a permanent state of electric polarization has been set up). Some materials such as highly polar plastics can be cooled from the melt under an intense electrical field and develop a permanent electrical field that is constantly on or constantly renewable. These electret materials find a wide range of applications that vary from uses in electrostatic printing processes, to supplying static fields for electronic devices, to some specialized medical applications where it has been found that the field inhibits clotting in vivo. An example for the material is in a microphone that has a high degree of sensitivity and the electrical waves are produced by the field variations caused by the change in spacing of an electrode to an electret.

Electrical corona discharge treatment It is a method for rendering inert plastics, such as polyolefins, more receptive to inks, adhesives, or decorative coatings by subjecting their surfaces to a corona discharge. A typical method of treating films is to pass the film over a grounded metal cylinder above that is located a sharp-edged high-voltage electrode spaced so as to leave a small gap. The corona discharge oxidizes the film by means of the formation of polar groups on reactive sites making the surface receptive to coatings, etc.

Electro-optics The liquid crystal plastics exhibit some of the properties of crystalline solids and still flow easily as liquids. One group of these materials is based on low polymers with strong field interacting side chains. Using these materials there has developed a field of electro-optic devices whose characteristics can be changed sharply by the application of an electric field.

Endotherm A process or change that takes place with absorption of heat and requires high temperature for initiation and maintenance as with using heat to melt plastics and then remove heat; as opposed to endothermic.

Endothermic Also called endoergic. Pertaining to a reaction which absorbs heat.

Energy Basically, it is the capacity for doing work or producing change. This term is both general and specific. Generally it refers to the energy absorbed by any material subjected to loading. Specifically it is a measure of toughness or impact strength of a material; as an example, the energy needed to fracture a specimen in an impact test. It is the difference in kinetic energy of the striker before and after impact, expressed as total energy per inch of notch of the test specimen for plastic and electrical insulating material [in-lb (J/m)]. Higher energy absorption indicates a greater toughness. For notched specimens, energy absorption is an indication of the effect of internal multi-axial stress distribution on fracture behavior of the material. It is merely a qualitative index and cannot be used directly in design.

Energy and bottle An interesting historical (1950s) example is the small injection blow molded whiskey bottles that were substituted for glass blown bottles in commercial aircraft; continues to be used in all worldwide flying aircraft.. At that time, just in USA, over 500×10^{12} Btu or the amount of energy equivalent to over 80×10^6 barrels of oil was reduced per year.

Engineering plastic See **Commodity & engineering plastic.**

Enthalpy It refers to the quantity of heat, equal to the sum of the internal energy of a system plus the product of the pressure-volume work performed on the system such as the action during heat processing of plastics. As a thermodynamic function, it is defined by the equation $H = U + PV$, where H = enthalpy, U = internal energy, P = pressure, and V = volume of the system.

Entropy A measure of the unavailable energy in a thermodynamic system, commonly expressed in terms of its exchanges on an arbitrary scale with the entropy of water at 0C (32F) being zero. The increase in entropy of a body is equal to the amount of heat absorbed divided by the absolute temperature of the body.

Euler equation A special case of the general equation of motion. It applies to the flow systems in which the viscous effects are negligible.

Eutectic blend It is a mixture of two or more substances that solidifies as a whole when cooled from the liquid state, without changing composition. It is the composition within any system of two or more crystalline phases that melts completely at the minimum temperature.

Exotherm It is the temperature vs. time curve of a chemical reaction or a phase change giving off heat, particularly the polymerization of thermoset plastics. The heat liberated by chemical reactions accelerated during processing. Maximum temperature occurs at peak exotherm. Some plastics such as room temperature curing TS polyesters and epoxies will exotherm severely with damaging results if processed incorrectly. As an example, if too much methyl ethyl ketone peroxide (MEK peroxide) catalyst is added to polyester plastic that contains cobolt naphthenate (promoter), the mix can get hot enough to smoke and even catch fire. Thus, an exotherm can be a help or hindrance, depending on the application such as during casting, potting, etc.

Extruder, adiabatic Also called autothermal. Describe a process or transformation in which no heat is added to or allowed to escape from the system under consideration. It is used, somewhat incorrectly, to describe a mode of a process such as an extruder in which no external heat is added to the extruder. Although heat may be removed by cooling to keep the output temperature of the melt passing through the extruder at a constant and control rate. The screw develops the heat input in such a process as its mechanical energy is converted to thermal energy.

Extruder, autogenous Some extruders operate without forced cooling or heating. This is the so-called autogenous extrusion operation; it is not to be confused with an adiabatic extruder. An autogenous process is where the heat required is supplied entirely by the conversion of mechanical energy into thermal energy. However, heat losses can occur in an autogenous process. An adiabatic process is one where there is absolutely no exchange of heat with the surroundings. An autogeneous extrusion operation can never be truly adiabatic, only by approximation.

In practice, autogeneous extrusion does not occur often because it requires a delicate balance between plastic properties, machine design, and operating conditions. A change in any of these factors will generally cause a departure from autogeneous conditions. The closer one operates to autogeneous conditions, the more likely it is that cooling will be required. Given the large differences in thermal and rheological properties of plastics, to date it is difficult to design an extruder that can operate in an autogeneous fashion with several different plastics. Therefore, most extruders are designed to have a reasonable amount of energy input from external barrel heaters.

Extruder isothermal A process where the melt stocks remains constant for a good portion in the plasticator. This type of operation is most common in small diameter screw extruders.

Fair trade vs. free trade The term *free trade* refers to foreign trade that is entirely laissez-faire – free of government regulation beyond the maintenance of the legal infrastructure necessary to facilitate proper business transactions. Manufacturers for Fair Trade [MFT, Cranesville, PA (tel. 814-756-5765) www.mftcoalition.org] believes this kind of trade often is unfair and jeopardizes freedom.

On the other hand, the term *fair trade* refers to foreign trade that is well regulated by countries for their own good first and the good of others second. It is neither unnecessarily regulated nor foolishly unregulated. It is as open or as regulated as it must be from time to time to reasonably protect the well being of the country and its citizens while facilitating foreign trade. Furthermore, because foreign trade transpires between two countries, fair trade is foreign trade that is carried on between parties from nations who show mutual good faith toward one another's rightful laws and regulations.

Fatigue It is the action that causes a failure or deterioration in mechanical properties after repeated, cyclic applications of stress. Test data provides information on the ability of a material to resist the developments of cracks, which eventually bring about failure as a result of long periods of the cyclic loading.

Feed side opening An opening that feeds the material at an angle into the side of the screw rather than the more conventional system of feeding vertically downward on he screw.

Finagle's law Once a job is fouled up, anything done to improve it makes it worst.

Fines They are very small particles, usually under 200 mesh, accompanying larger forms of molding powders, developed when granulating plastics. When plastics are extruded and pelletized, varying amounts of oversized pellets and strands are produced, along with fines. When the plastics are dewatered/dried or pneumatically conveyed, more fines, fluff, and streamers may be generated. Usually they are detrimental during processing so they are removed or action is taken to eliminate the problem during grinding scrap, etc.

Fish-eye A fault particularly in transparent or translucent plastics, such as film or sheet, appearing as a small globular mass that has not completed blended into the surrounding material. Cause includes incomplete material blending, processing variations, and/or environmental conditions that includes over stressed. Also called cat's-eye.

Flash A thin surplus web of plastic, usually occurring with thermoset plastics, attached to a molding along the parting lines, fins at holes or openings, etc.

Flight land The surface of the radial extremity of the flight constituting the periphery or outside diameter of the screw.

Flow mark Molding can cause product surface melt flow marks. Major contributor to the markings is the melt flow speed.

Foamed plastic Practically all plastics can be made into foams. When compared to solid plastics, density reduction can go from near solid to almost a weightless plastic material. There are so called plastic structural foams (SFs) that have up to 40 to 50% density reduction. The actual density reduction obtained will depend on the products' thickness, the product shape, and the melt flow distance during processing such as how much plastic occupies the mold cavity.

Ford car The gasoline powered automobile was not invented by Henry Ford. It was independently developed by Gottlieb Daimler and Karl Benz in the last decade of the 19th century. Several years latter Henry Ford invented the moving assembly line. That flash of brilliance was the means of producing cars cheaply and in great numbers.

Fossil fuels Fossil fuels (coal, crude oil or petroleum, natural gas liquids, and natural gas) are the primary sources of basic petrochemicals. About 3% are used to produce plastic materials. The most important use that consumes most of the fossil fuels is in the production of energy.

Friction By definition friction is the resistance of two surfaces sliding against one another. The coefficient of friction μ is the tangent of the angle between the gliding plane and a horizontal plane at a specified angle (α). There is a static coefficient of friction where the body on the gliding plane is at the transition of still sticking but almost gliding. The dynamic coefficient of friction is where the body is gliding. The measurement device test set up can be in a parallel-parallel mode. Signals of torque and normal force of the upper fixture are measured directly at a machine by means of an oscillograph. For calculation, the torque of the upper fixture applied by rotation of the lower turning fixture and the normal force applied by pressing the upper and lower fixture together are available. The coefficient of friction is then calculated using the relationship between torque and normal force as follows:

$$\mu = (3/2R) \times (\Gamma/F_n)$$

where   V = coefficient of friction

R = sample radius of smallest contact area

Γ = torque

F_n = normal force.

The 3/2 factor is the equivalent radius and is inserted for correction of the velocity gradient over a circular surface area

Full indicator movement FIM is a term used to identify tolerance with respect to concentricity. Terms used in the past were full indicator reading (FIR) and total indicator reading (TIR).

Fuzzy logic control Although FLC may sound exotic, it has been used to control many conveniences of modern life (from elevators to dishwashers) and more recently into industrial process control that include plastic processing such as temperature and pressure. FLC actually outperforms conventional controls because it completely avoids overshooting process limits and dramatically improves the speed of response to process upsets. These controllers accomplish both goals simultaneously, rather than trading one against another as done with proportional-integral-derivative (PID) control. However, FLC is not a cure-all because not all FLCs are equal; no more than PIDs. FL is not needed in all applications; in fact FLCs used allow them to be switched off so that traditional PID control takes over.

Geomembrane These liners chiefly provide impermeable barriers. They can be characterized as: (1) solid waste containment: hazardous landfill, landfill capping, and sanitary landfill; (2) liquid containment: canal, chemical/brine pond, earthen dam, fish farm, river/coastal bank, waste-water, and recreation; (3) mining, leach pad and tailing ponds; and (4) specialties: floating reservoir caps, secondary containment, tunnel, erosion, vapor barrier, and water purification. Plastics used include medium to very low density PE, PVC, and chlorosulfonated PE (CSPE). (The Romans used in their land and road constructions what we call geomembrane.)

Geotextile Also called geosynthetic. Geotextiles, as well as geonets, geogrids, and geomembranes, represent a major market for plastics. They appear in all manners of civil works, from roads to canals, from landfills to landscaping. They often prove more cost-effective than nature and other man-made products. The primary plastics are polyester, nylon, PP, and HDPE filaments. The fabrics are made in both woven and nonwoven varieties. The former are characterized by high-tensile, high modulus, and low-elongation traits; the latter by high-permeability and high-elongation.

Glass transition temperature Also called glass-rubber transition. Identified as T_g. Basically this important characteristic is the reversible change in phase of a plastic from a viscous or rubbery state to a brittle glassy state. T_g is the point below which plastic behaves like glass but very strong and rigid. Above this temperature it is not as strong or rigid as glass, but neither is it brittle. At T_g the plastic's volume or length increases and above it, properties decrease.. The amorphous TPs have a more definite T_g when compared to crystalline plastics. It is usually reported as a single value. However, it occurs over a temperature range and is kinetic in nature. Example of the T_g range has PE at $-125C$ and PMMA at $+105C$.

Glassy state In amorphous plastics, below the T_g, cooperative molecular chain motions are "frozen", so that only limited local motions are possible. Material behaves mainly elastically since stress causes only limited bond angle deformations and stretching. Thus, it is hard, rigid, and often brittle.

Hub It is the portion immediately behind the flight that prevents the escape of the plastic. A sealing device is used to prevent leakage of plastic back around the screw hub, usually attached to the rear of the feed section.

Hysteresis effect The hysteresis effect is a retardation of the strain when a material is subjected to a force or load (see Chapter 3).

Graft polymer It is a polymer comprising of molecules in which the main backbone chain of atoms has attached to it at various points side chains containing different atoms or groups from those in the main chain. The main chain may be a copolymer or may be derived from a single monomer.

Green strength During the processing of plastics even though the cure is not complete, the mechanical strength of certain materials allows removal from the mold and handling without tearing or permanent distortion. This characteristic is referred to as the plastic's green strength.

Hysteresis. The failure of a property that has been changed to return to its original value when the cause of the change is removed.

Inching Reduction in rate of mold closing travel just before the mating mold surfaces touch each other.

Inefficiency Does it seem that since 1776, particularly during this century, no one in any elected political job does a good job based on the opposing person seeking to be elected to that job. In fact with time passing and the expanding communication systems, there is more inefficiency occurring (that includes new developments) with elected politicians as the opponent reports. And obviously the public accepts all this inefficiency. Perhaps the public by not complaining has to be personally gaining something or is on the "take" one way or another via some government agency.

Innovator Innovators create entirely new products or business models. They offer new value to customers. rude surprises to competitors, and huge new wealth to for investors. That person has a vision, intense curiosity about the marketplace, a desperate need, brilliant intuition, and a lot of luck.

Intellectualism The Oxford English Dictionary (2001) definition is "doctrine that knowledge is wholly or mainly derived from pure reasoning." and it follows by saying that an intellectual is a "person possessing a good understanding, enlightened person".

Interpenetrating network See **Polymer, interpenetrating network**.

Investors or betters All over the world people are investing on stocks. Many of the company's employees are investing on their own company's stock as well. Are they investors or betters? Recognize betters like horseplayers try to win, not create value.

Isotactic molding Also called isotactic pressing or hot isotactic pressing (HIP). The compressing or pressing of powder material (plastic, etc.) under a gas or liquid so the pressure is transmitted equally in all directions. Examples include autoclave, sintering, injection-compression molding, elastomeric mold using hydrostatic pressure, and underwater, sintering.

Isotropic See **Directional terminology.**

Jetting Undesirable melt entering the cavity, rather than being in a parabolic melt front, the melt squirts through the gate into the cavity like a worm or a snake pattern. Causes included undersized gate and thin to thick cavity section resulting on poor control of the molded part.

Kinetic A branch of dynamics concerned with the relations between the movement of bodies and the forces acting upon them.

Kinetic theory A theory of matter based on the mathematical description of the relationship between pressures, volumes, and temperatures of gases (PVT phenomena). This relationship is summarized in the laws of Boyle's law, Charle's law, and Avogadro's law.

Latex See **Elastomer, liquid.**

Leakage resistance When dealing with low value electric currents, the leakage resistance of the insulation is also a major problem in the application of the wire. Such wire is used primarily in communications applications. The leakage of current from the wire is related to the volume resistivity of the dielectric material. In most plastics, the volume resistivity is high and in the case of the plastic most used in commercial communications wire, PE, the leakage is so low it causes no problems. When there is appreciable current leakage, the signal strength in the wire is reduced and noise from the environment is conducted into the wire to add to the loss of signal content (signal to noise ratio).

Life and work Do not run through life so fast that you forget not only where you have been, but also where you are going. Life is not a race, but a journey to be savored each step of the way. This thought can be specifically applied to many aspects of life including work and play.

Liquid crystalline polymer LCPs are best thought of as being a separate, unique class of TPs. Their molecules are stiff, rodlike structures organized in large parallel arrays or domains in both the melted and solid states. These large, ordered domains provide LCPs with characteristics that are unique compared to those of the basic crystalline or amorphous plastics. They are called self-reinforcing plastics because of their densely packed fibrous polymer chains.

Logarithm It is the exponent that indicates the power to which a number is raised to produce a given number. Thus, as an example, 1000 to the base of 10 is 3. This type of mathematics is used extensively in computer software.

Machine alignment Without proper machine installation the precision alignment built into equipment is lost when not properly supported on all its mounting points. Installation involves factors such as ground support stability, precise alignment of equipment, uniform support, and effective control of vibration. Installation and alignment has to be done with extreme accuracy. Assuming proper alignment occurs at room temperature and significant movement occurs during heat up or during operation, the causes of movement must be reconciled to prevent excessive wear or even failure of components. With plasticators the prime objective is to keep the screw and barrel centerlines coincident meeting the production line height requirement. Installation is a multi-step procedure that consists of building a foundation, setting and leveling the machine supports, and aligning the machine components to each other.

Machines not alike Just like people, not all machines may be created equal. Recognize that identical machine models, including auxiliary equipment, built and delivered with consecutive serial numbers to the same site can perform so differently as to make some completely unacceptable by the customer, assuming they were installed properly.

Maintenance cost It is estimated that the initial cost of mold construction is about 20% of the cost of the mold. Another 40% of the cost is scheduled maintenance, and the other 40% is repair due to breakdowns.

Mastication/Internal Mixer An apparatus consisting of concentric cylinders lined with teeth with the inner cylinder rotated to achieve a shredding action initially just for rubber. It produces a dough-like mass to which other materials can be readily added.

Material impurity Presence of one or more substances in another, often in such low concentrations that it cannot be measured quantitatively by ordinary analytical methods. To avoid forming microscopic cavities in a molded part, when processing TP materials it is important to maintain a minimum pressure, rather than maximum during injection of the melt. As the melt cools, the bubbles grow, which in turn can decrease mechanical and other properties of the part. The majority of the cavities formed is a result of water vapor present on the surface as well as imbedded in the plastic particles themselves. When these bubbles form on the surface, they are called splay.

Material received, checking An important factor in the production is that quality control of all types of incoming materials (plastics, steel, etc.) always conform and be checked against specifications. Unfortunately, with time after processing materials, specifications have to be changed to meet unforeseen important test.

Mathematical dimensional eccentricity The ratio of the difference between maximum and minimum dimensions on a product, such as wall thickness. It is expressed as a percentage to the maximum.

Mathematical tool Calculus is the mathematical tool used in plastic R&D programs to analyze changes in physical quantities, comprising differential and integral calculations, etc. It was developed during the 17th century to study four major classes of scientific and mathematical problems of that time. (1) Find the maximum and minimum value of a quantity, such as the distance of a planet from the sun. (2) Given a formula for the distance traveled by a body in any specified amount of time, find the velocity and acceleration of the body at any instant. (3) Find the tangent to a curve at a point. (4) Find the length of a curve, the area of a region, and the volume of a solid. These problems were resolved by the greatest minds of the 17th century, culminating in the crowning achievements of Gottfried Wilhelm (Germany 1646-1727) and Isaac Newton (English 1642-1727). Their information provided useful information for today's space travel.

Mathematician & knowledge Evariste Galois now recognized as one of the greatest 19th century mathematician, twice failed the entry exam for the

Ecole Polytechnique and a paper he submitted to the French Academy of Sciences was rejected as "incomprehensible". Embittered he turned to political activism and spent six years in prison. In 1832, at the age of 20, he was killed in a duel, reported to have arisen from a lover's quarrel, although their were those who believed that an agent provocateur of the police was involved.

Mean Arithmetical average of a set of numbers. It provides a value that lies between a range of values and is determined according to a prescribe law.

Mean absolute deviation MAD is a statistical measure of the mean (average) difference between a product's forecast and actual usage (demand). The deviations (differences) are included without regard to whether the forecast was higher than actual or lower.

Meld line It refers to a line that is similar to a weld line except the flow fronts move parallel rather than meeting head on.

Melt It is plastic in a molten or plasticated condition; it also refers to the an extruder's extrudate.

Melt deformation As a melt is subjected to a fixed stress or strain, the deformation versus time curve will show an initial rapid deformation followed by a continuous flow. When elasticity and strain are compared they provide (a) basic deformation vs. the time curve, (b) stress-strain deformation vs. time with the creep effect, (c) stress-strain deformation vs. time with the stress-relaxation effect, (d) material exhibiting elasticity, and (e) material exhibiting plasticity. The relative importance of elasticity (deformation) and viscosity (flow) depends on the time scale of the deformation. For a short time elasticity dominates, but over a long time the flow becomes purely viscous. This behavior influences processes.

Deformation contributes significantly to process-flow defects. Melts with only small deformation have proportional stress-strain behavior. As the stress on a melt is increased, the recoverable strain tends to reach a limiting value. It is in the high stress range, near the elastic limit, that processes operate.

Molecular weight (MW), temperature, and pressure have little effect on elasticity; the main controlling factor is MWD (molecular weight distribution). Practical elasticity phenomena often exhibit little concern for the actual values of the modulus and viscosity. Although MW and temperature influence the modulus only slightly, these parameters have a great effect on viscosity and thus can alter the balance of a process.

Melt fracture Also called elastic turbulence. It is the instability or an elastic strain in the melt flow usually through a die starting at the entry of the die. It leads to surface irregularities on the finished part like a regular helix or irregularly spaced ripples. Plastic's rheology influences its melt fracture behavior. Higher molecular weight plastic (with MWD) tend to have less sensitivity to its onset. This fracture can also occur in molds with complex cavities and/or improper melt flow with in the mold.

Melt index A term used that indicates how much plastic melt can be pushed through a set orifice with various conditions controlled (basically temperature, time, pressure). It represents the "flowability" of a material. The higher numbers indicate the easier flow.

Metrology The science of measurement.

Modulus of elasticity Most materials, including plastics and metals, have deformation proportional to their loads below the proportional limits. (A material's proportional limit is the greatest stress at which it is capable of sustaining an applied load without deviating from the proportionality of a stress-strain straight line.) Since stress is proportional to load and strain to deformation, this implies that stress is proportional to strain. Hooke's law, developed in 1676, follows that this straight line of proportionality is calculated as stress/strain = constant. The constant is called the modulus of elasticity (E) or Young's modulus (defined by Thomas Young in 1807 although others used the concept that included the Roman Empire and Chinese-BC.

Mold See **Tool.**

Molecular weight MW is the sum of the atomic weights of all the atoms in a molecule. It represents a measure of the chain length for the molecules that make up the polymer. Atomic weight is the relative mass of an atom of any element based on a scale in which a specific carbon atom is assigned a mass value of 12.

MW of plastics influences their properties. As an example with increasing MW properties increase for abrasion resistance, brittleness, chemical resistance, elongation, hardness, melt viscosity, tensile strength, modulus, toughness, and yield strength. Decreases occur for adhesion, melt index, and solubility.

Adequate MW is a fundamental requirement to meet desired properties of plastics. With MW differences of incoming material, the fabricated product performance can be altered; the more the difference, the more dramatic change occurs in the product. Melt flow rate (MFR) tests are used to detect degradation in products where comparisons, as an example, are made of the MFR of pellets to the MFR of products. MFR has a reciprocal relationship to melt viscosity. This relationship of MW to MFR is an inverse one; as the MW drops, the MFR increases the MW and melt viscosity is also related: as one increases the other increases.

The average molecular weight (AMW) is the sum of the atomic masses of the elements forming the molecule, indicating the relative typical chain length of the polymer molecule. Many techniques are available for its determination. The choice of method is often complicated by limitations of the technique as well as by the nature of the polymer because most techniques require a sample in solution.

Molecular weight distribution The molecular weight distribution (MWD) is basically the amounts of component polymers that go to make up a polymer. Component polymers, in contrast, are a convenient term that recognizes the fact that all polymeric materials comprise a mixture of different polymers of differing molecular weights. The ratio of the weight average molecular weight to the number average molecular weight gives an indication of the MWD. Average molecular weight information is useful; however, characterization of the breadth of the distribution is usually more valuable. For example, two plastics may have exactly the same or similar AMWs but very different MWDs. There are several ways to measure MWD such as fractionation of a polymer with broad MWD into narrower MWD fractions.

Monocoque structure Plastics provides an easy means to producing monocoque constructions such as has been done in different applications that include aircraft fuselage, automotive body, motor truck, railroad car, and houses. Its construction is one in which the outer covering "skin" carries all or a major part of the stresses. The structure can integrate its body and chassis such as in aircraft and automobiles.

Monomer Plastics are predominantly organic (carbon containing) compounds primarily made up of six elements forming a monomer such as ethylene that is a gas. Another example of a monomer is vinyl chloride that is also a gas.

Morphology It is the study of the physical form or structure of a material; the physical microstructure of a bulk polymer. Common units are lamella, spherulite, and domain. In turn there are thermoplastic (TP) and thermoset (TS) plastics. Lamella is a thin, flat scale layer of polymers. Spherulite is a rounded aggregate of radiating lamellar crystals. Spherulites exist in most crystalline plastics and usually impinge on one another to form polyhedrons. They range in size from a few tenths of a micron in diameter to several millimeters. Domain is a microphase of one polymer in a multiphase system.

Motionless mixer See **Static mixer**.

Nanocomposite Plastics derived from compounding nanofillers (clays and other particles) in polymers.

NEAT plastic Identifies a plastic with <u>N</u>othing <u>E</u>lse <u>A</u>dded <u>T</u>o. It is a true virgin polymer since it does not contain additives, fillers, etc. These are rarely used.

Newtonian flow It is a flow characteristic where a material (liquid, etc.) flow immediately on application of force and for which the rate of flow is directly proportional to the force applied. It is a flow characteristic evidenced by viscosity that is independent of shear rate. Water and thin mineral oils are examples of Newtonian flow.

Non-Newtonian flow It is a flow characteristic where materials such as plastic have basically abnormal flow response when force is applied. That is, their viscosity is dependent on the rate of shear. They do not have a straight proportional behavior with application of force and rate of flow. When proportional, the behavior has a Newtonian flow.

Nonlaminar flow Ideally, it is a melt flow in a steady, streamlined pattern in and/or out of a tool (die, mold, etc.). The melt is usually distorted, causing defects called melt fracture or elastic turbulence. To reduce or eliminate this problem, the entrance to the die or mold is tapered or streamlined.

Nucleating agent An additive, often crystalline, usually added to a crystallizing polymer to increase its rate of solidification during processing.

Nucleating agent, cell-control These agents promote symmetrical, cohesive expansion of cells within foamed polyurethane and polystyrene plastics; also to a limited extent in polyethylene and polypropylene plastics.

Nucleation, heterogeneous In the crystallization of polymers, it is the growth of crystals on vessel surfaces, dust, or added nucleating agents.

Nucleation/nucleator With polymers, any foreign additive that assists or acts as a starting site for crystallinity within the plastic. These initiators can reduce cycle time by speeding up the crystalline formations. It also identifies the addition of a gas, such as nitrogen, to the polyol in many small bubbles to assist in forming better cell structures during reaction injection molding.

Nucleation, primary The mechanism by which crystallization is initiated often by an added nucleation agent.

Nucleation, secondary The mechanism by which crystals grow.

Oil-canning Property of a panel that flexes past a theoretical equilibrium point, and then returns to the original position. This motion is analogous to the bottom of a metal oilcan when pressed and released. Part flexing can cause stress, fracturing, or undesirable melting of thin-sectioned, flat parts.

Optical sheet Black specs, bubbles/voids, die lines, surging, surface imperfections, etc. are among the major problems that processors of optical sheets (film, etc.) encounter. Majority of problems can be traced to the way the plastic was dried and handled.

Outgassing During processing certain thermoplastic and thermoset plastic compounds, particularly TSs, gas forms and has to be removed so the gas does not damage the part internal and/or external with voids, thin sections, mechanical performance, etc. Procedures exist such as providing vents, bumping, etc. When applying coatings on plastic, such as metallizing, gas release after coating can cause the coating to be stripped, blistered, etc.

Orange peel, mold So called mold orange peel occurs when a mold cavity surface cannot be cooled below the dew point of the ambient air that causes moisture condensation and a cosmetic defect on the molded product's surface when the moisture flashes to steam when it contacts the plastic melt. This is different from polishing orange peel.

Orange peel, paint A term used in the paint industry to refer to a roughened film surface due to too rapid drying.

Orange peel, polish Term for a blemish on a finished fabricated product where the result is an undesirable, uneven surface resembling orange peel. (Sometimes melt fracture is erroneously referred to as orange peel.) Usually caused by moisture in the tool that can be removed. As an example, during injection molding the problem can be eliminated by changing control settings such as using a faster fill rate, increasing first-stage pressure, increasing sprue and/or gate opening, and/or increase temperature.

Orange peel, reinforced plastic Uneven leveling of coating on reinforced plastic surfaces, usually because of high viscosity plastic melts. Simple addition of high boiling point solvent to the coating for a wetter surface is helpful, particularly when spraying.

Orientation and glassy state An important transition occurs in the structure of both crystalline and non-crystalline plastics. This is the point at which they transition out of the so-called glassy state. Rigidity and brittleness characterize the glassy state. This is because the molecules are too close together to allow extensive slipping motion between each other. When the glass transition (Tg) is above the range of the normal temperatures to which the part is expected to be subjected, it is possible to blend in materials that can produce the Tg of the desired mix. This action yields more flexible and tougher plastics.

Orientation and heat-shrinkability There are oriented heat heat-shrinkable plastic products found in flat, tubular film, and tubular sheet. The usual orientation is terminated (frozen) downstream of a stretching operation when a cold enough temperature is achieved. Reversing this operation occurs when the product is subjected to a sufficient high temperature. This reheating results in the product shrinking. Use for these products includes part assemblies, tubular or flat communication cable wraps, furniture webbing, medical devices, wire and pipe fitting connections or joints, and so on.

Orientation and mobility Orientation requires considerable mobility of large segments of the plastic molecules. It cannot occur below the glass transition temperature (Tg). The plastic temperature is taken just above Tg (Chapter 1).

Orientation, balance It is the result where stretch in the machine and transverse directions are uniform.

Orientation, biaxial Also called bi-orientation or BO. It is the stretching of material in two directions (biaxially) at right angles; along machine direction (MD) and across or transverse direction (TD). The difference in the amount of stretch in both directions varies, depending on product requirement. If they equal, then it is a balanced orientation. Small to large size lines is used. An example of one of the largest cast oriented PET film line in the world (DuPont's plant in Dumphries, Scotland built by Kampf GmbH & Co., Germany) produces 9 m wide film after MD and TD stretching. The film is wound in one piece at up to 480 m/min. A take-up roll weigh is 13 T using a high stiffness reinforced plastic carbon fiber/epoxy core. There is also a very large oriented polypropylene plastic coextruded film line (Applied Extrusion Technologies Inc., New Castle, DE). It uses massive tenter oven and turret winder built by Bruckner

Maschinenbau, Germany) that produces 10 m wide film at up to 400 m/min or 50 million lb/yr.

Orientation, cold stretching Plastics may be oriented by the so-called cold stretching; that is below its glass transition temperature (Tg). There has to be sufficient internal friction to convert mechanical into thermal energy, thus producing local heating above Tg. This occurs characteristically in the necking of fibers during cold drawing.

Orientation tenter mark A visible deformation on the side edge of a material due to the pressure from the clips and clamps; this trim is cut.

Orientation, thermal characteristic These oriented plastics are considered permanent, heat stable materials. However, the stretching decreases dimensional stability at higher temperatures. This situation is not a problem since these type materials are not exposed to the higher temperatures in service. For the heat-shrink applications, the high heat provides the shrinkage capability.

Orientation, uniaxial Also called axial orientation, mono-axial orientation, or UO.. This is the stretching only in one direction that is usually in the machine direction.

Orientation, wet stretching For plastics whose glass transition temperature (Tg) is above their decomposition temperature, orientation can be accomplished by swelling them temporarily with plasticizing liquids to lower their Tg of the total mass, particularly in solution processing. As an example, cellulose viscous films can be drawn during coagulation. Final removal of the solvent makes the orientation permanent.

Orthotropic See **Directional terminology.**

Parting line It is the line at which the two halves of the mold (plunger and cavity) meet.

Perfection Target is to approach perfection in a zero-risk society. Basically, no product is without risk; failure to recognize this factor may put excessive emphasis on achieving an important goal while drawing precious resources away from product design development and approval. The target or goal should be to attain a proper balance between risk and benefit using realistic factors and not the "public-political panic" approach.

pH An expression of the degree of acidity or alkalinity of a substance. The neutrality being at pH 7; acid solutions are less than 7, and alkaline solutions are more than 7.

Pitch It is the high molecular weight residue from the destructive distillation of petroleum and coal products. Their use includes as base materials for the manufacture of high modulus carbon fibers.

Plastic volume swept The volume of material that is displaced as the screw (or plunger) moves forward. It is the effective area of the screw multiplied by the distance of travel.

Plasticator They are a very important component in machines such as extrusion, injection molding, and blow molding by providing a melting process with its usual barrel and screw. Factors such as the proper screw design or barrel heat profile is not used, products may not meet or maximize their performance and meet their low cost requirements.

Plasticity It is the inverse condition of elasticity. The material tends to stay in its deformed shape. Occurs when stressed beyond its yield point.

Plasticizer Materials that may be added to thermoplastics to increase toughness and flexibility or to increase the ease of fabrication. These materials are usually more volatile than the plastics to which they are added.

Platen The flat surfaces of the molding press onto which the two halves of the mold are mounted. Generally positioned horizontally on compression presses, and vertically on injection machines.

Plunger It is the male portion of the mold. The plunger pushes or forces the material into any opening in the cavity. Gives the internal shape to the part being molded.

Pocket A place where a screw flight is initiated, usually starting from a cylindrical area or another flight. The feed pocket exists on most screws and is located at the intersection of the bearing and the beginning of the flight.

Poisson's ratio It is the proportion of lateral strain to longitudinal strain under conditions of uniform longitudinal stress within the proportional or elastic limit. When the material's deformation is within the elastic range it results in a lateral to longitudinal strain that will always be constant. In mathematical terms, Poisson's ratio is the diameter of the test specimen before and after elongation divided by the length of the specimen before and after elongation. Poisson's ratio will have more than one value if the material is not isotropic.

Polymer When monomers are basically subjected to a catalyst, heat, and/or pressure, the double bonds open up and the individual monomer units join 'arms' to form long chains called polymers. This process is called polymerization. Polymerization is basically the bonding of many monomer units to produce polymers.

There is a type of chemical reaction (addition or condensation) in which the molecules of a monomer are linked together to form large molecules whose molecular weight is a multiple of that of the original substance resulting in high molecular weight components.

The geometry of these chains is just as important as their chemical make-up in determining plastic processability and properties of fabricated products. Chains can be long having thousands of repeating monomer units. Short chains have fewer repeating monomer units.

Thus a polymer is a substance consisting of large molecules formed by the joining together of simple molecules that are known as monomers; where two or more monomers are involved the resultant product is known as a copolymer. As an example, polymerization of ethylene forms a polyethylene plastic or condensation of phenol and formaldehyde (with production of water) forms phenol-formaldehyde plastics.

A polymer is a pure material; they are NEAT plastics. To fabricate products practically all polymers include additives, fillers, and/or reinforcements; they are than called plastics or resins.

Polymer, addition Also called addition plastic. A plastic formed by the addition of unsaturated monomer molecules, such as olefins, with one another, without the formation of a by-product such as water. Examples are polyethylene, polypropylene, and polystyrene plastics.

Polymer, block A polymer whose molecule is made up of alternating sections of one chemical composition separated by sections of a different chemical structure or by a coupling group of low molecular weight. An example are blocks of PVC interspersed with blocks of PVAc.

Polymer, branched A polymer composed of molecules having a branch structure, chainlike between branch junctions and between each chain end and a branch junction.

Polymer chain stiffening A polymer strengthening mechanism significantly different from linear, branched or cross-linked. It has a monomer that is physically large and unsymmetrical. The ability of a chain to flex is impaired. A typical example is with polystyrene plastic.

Polymer chain transfer Refers to the termination of a growing polymer chain and the start of a new one. The process is mediated by a change transfer agent, which may be a monomer, initiator, solvent, or some species added deliberately to effect the chain transfer. Because CT occurs in all radical polymerization, it must be taken into account in any quantitative considerations of these reactions. CT always decreases the molecular weight.

Polymer characterization This is an essential step in working with polymers. As a rule, such efforts are directed toward a specific purpose. Polymers may be categorized as natural or synthetic and as homopolymers or copolymers. Copolymers and terpolymers are further classified according to the method of production and the arrangement of the monomeric units. Other methods are used. The techniques may not be uniformly applicable to all classes of polymers.

Polymer chemical reagent Polymer supported catalysts are a particular attractive type of polymer-support reactant, because a relatively small amount is used to chemically transform a relatively large amount of substrate. After the reaction, the polymer-supported species can be separated without difficulty from low molecular weight species in solution. Separation is most readily achieved if the polymer is cross-linked, since it is then insoluble in all solvents.

Polymer chemistry terminology The following is a very simplified explanation. Polymers can be made up of carbon (C), hydrogen (H), chlorine (Cl), fluorine (F), oxygen (O), and nitrogen (N). These six elements, either naturally or synthetically, first form rather simple molecules called monomers. When these monomers are subjected to a catalyst, heat, and pressure, the double bonds (or connecting arms of the monomers) open up and these individual monomer units join arms to form long chains called polymers. This process id called polymerization. Different chemical structures are formed for the different polymers. The geometry (basically they are linear, branched, and cross-linked) of their chains is just as important as their chemical make-up in determining properties.

Polymer definition This term comes from the Greek word "poly" which means many and "meras" that means parts.

Polymer, endless manufacture of A major contributor to the successful growth of plastic materials is the endless capability of the industry (chemist, engineer, etc.) in producing new polymers, additives, fillers, reinforcements, etc. as well as modifying the materials that presently exist. There is always a new developing horizon in the world of plastics.

Polymer evolution In contrast to the pioneering age of polymeric evolution, in which new plastics were determined primarily via the selection of suitable monomer components, today the number of new polymers/plastics developed primarily on the basis of known monomers is on the increase. This also applies to new plastic blends and compounds based on known plastics. Similar situation exists with catalysts.

Polymer, graft Polymer comprising of molecules in which the main backbone chain of atoms has attached to it at various points side chains containing different atoms or groups from those in the main chain. The main chain may be a copolymer or may be derived from a single monomer

Polymer, high A macromolecular substance which, as indicated by the term polymer and the name and formula by which it is identified, consists of molecules which are at least approximately multiples of the low molecular units. Molecular weight is greater than 10,000, usually composed of repeating units of low molecular weight species such as ethylene and propylene.

Polymer, homo- A plastic/polymer formed from a single monomer species.

Polymer, inter- Also called a true polymer. A particular type of copolymer in which two monomer units are as intimately distributed in the polymer molecule that the substance is essentially homogeneous in chemical composition.

Polymer, interpenetrating network IPN is a branch of blend technology; it combines two plastics into a stable interpenetrating network. There all types of blends, such as synergistic types, to meet all types of performance requirements. In true IPN's, each polymer is cross-linked to itself, but not to the other, and two polymer networks interpenetrate each other; these become thermoset (TS) plastics. In semi-IPN, only one polymer is cross-linked; the other is linear and by itself would be a thermoplastic (TP); these lend themselves to TP processing techniques. The rigidity of IPN structures increases mechanical and other properties such as chemical resistance. A polyurethane and isocyanate system is an example of a full IPN. Polymerizing an elastomeric like polysulfone within a cross-linked TS epoxy can make a semi-IPN.

Polymerization It is basically the bonding of two or more monomers to produce polymers. A chemical reaction, addition or condensation, in which the molecules of a monomer are linked together to form large molecules whose molecular weight is a multiple of that of the original substance resulting in high molecular weight components. The polymerization process has been understood since about 1930.

Polymerization, addition A chemical reaction (polymerization) in which simple molecules (monomers) are added to each other to form long chain molecules (polymers) and no byproducts are formed (water, gases, etc).

Polymerization, aqueous Vinyl polymerization with water as the medium and with the monomer present within its inherent solubility limit is a process generally called aqueous polymerization. It includes suspension polymerization in an aqueous medium. This procedure is of technical importance in preparing special plastics such as emulsifier-free latex in which the size distribution among the dispersed particles is fairly sharp.

Polymerization, block This term has been applied to both bulk polymerization (casting of polymerizing syrup) and sequence copolymerization (block copolymerization). It is only this latter that is recognized by IUPAC as true copolymerization. Confusion may be avoided to some extent by the use of the prefix "co" which implies the polymerization of more than one monomer.

Polymerization, bulk Also called mass polymerization or step-growth polymerization. It is from undiluted low molecular weight starting materials. It is the simplest and oldest method for the synthesis of macromolecules. This method has a reaction, which is relatively simple, and rapid, plastics of high purity are formed, and the plastics obtained are immediately processable. Basically, the polymerization process involves only monomer and polymerization initiator or catalyst. It is carried out in the absence of a solvent or other dispersion media. This technique is applicable to both addition and condensation polymerization. Fundamentally differences exist.

Polymerization, co- The polymerization of a mixture of two or more monomers, yields a copolymer with two different repeating units distributed along the polymer chain. The process is called copolymerization or heterpolymerization whereas when only one monomer exists, it is called polymerization. Copolymerization has practical utility for altering the properties of a homopolymer (produced from a single monomer) in a desired reaction. Copolymer properties are determined by the identities of the two monomers and their relative amounts in the copolymer.

Polymerization, condensation Also called polycondensation. A chemical reaction in which two or more molecules combine often but necessary accompanied by the separation of water or some other simple substance.

Polymerization, de- To break down the component parts of a polymerization reaction or to reverse the reaction. Such reversals occur in certain plastics when exposed to very high temperatures.

Polymerization, electron beam EB polymerization is a slow irradiating process. Monomeric materials that have been irradiated to produce plastic films include silicone, butadiene, styrene, methyl methacrylate, and epoxy. However the lack of selectivity of this energy source can result in contamination. Film properties can be tailored by varying the acceleration voltage. >coating, vacuum

Polymerization, emulsion Polymerization of monomers dispersed in an aqueous emulsion. Polymerization is in a stabilized emulsion, at least at the start of the reaction.

polymerization history The polymerization of ethylene under high pressure (about 2000 bar) was discovered in 1933 by the English chemists Fawcett and Gibson. The German scientist Ziegler discovered the alternative low pressure process (near atmospheric pressure).

Polyolefin plastic Also called olefin, olefinic plastics or olefinic resins. They represent a very large class of carbon-chain TPs and elastomers. The most important are polyethylenes and polypropylenes. They all have extensive use in many different forms and applications.

Postconsumer Identifies plastic products generated by a business or consumer that have served their intended purpose and that have been separated or diverted from solid waste for the purposes of collection, recycling, and/or disposition

Preconsumer Any material that has not made its way to the consumer. It includes scrap, waste, and rejected parts or products.

Preform Molding material that has been pressed into some desired shape for efficient loading of the mold.

Prepreg Term generally used in RPs for a reinforcement containing or combined usually with a TS liquid plastic (TPs are also used) that can be stored under controlled conditions. Reinforcement (such as fibers and/or rovings, woven and/or nonwoven fabrics, etc.) can be in different forms and patterns. The TSs is completely compounded with catalysts, etc. and partially cured to the required tack state in the B-stage. The fabricator completes the cure with heat and pressure.

Pressure, atmospheric The atmosphere (atm) is the envelope of gases (air) that surrounds the earth exerting pressure on earth with certain plastic fabricating processes taking advantage of this pressure. At various altitudes in feet, in approximate absolute pressure in psia (gauge in. of Hg), they are: sea level at 14.7 (0.0), 1000 at 14.2 (1.0), 2000 at 13.7 (2.1), 2000 at 13.2 (3.1), 4000 at 12.7 (4.1), 5000 at 12,2 (5.0), 6000 at 11.7 (6.0), etc. The pressure exerted at sea level is 14.696 psi (101.325 kPa) which will support a column of mercury (Hg) 760 mm high (about 30 in.) having a density of 13.5951 g/cm^3 at a temperature of 0C (32F) and standard gravity of 980.665 cm/s^2. This atm is a standard barometric pressure though it varies slightly with local meteorological conditions. This pressure is used in fabricating processes where only contact or very low pressure such as vacuum pressure (where atmospheric pressure is applied) is require. Those processes include certain casting, coating, and reinforced plastic systems.

Processing, Art of Processing of plastic is an "art of detail". The more you pay attention to details, the fewer hassles you will get from the process. Note that if it has been running well, it will continue running well unless a change occurs. Correct the problem, do not compensate. It may not be an easy task, but understanding what you have equipmentwise, materialwise, environmentwise, and/or peoplewise can do it.

Processing feedback It is the information returned to a control system or process to maintain the output within specific limits.

Processing fundamental Conversion or fabricating processes may be described as an art. Like all arts, they have a basis in the sciences and one of the short routes to technological improvements is a study of these relevant sciences that actually encompasses what we are exposed too in the world of plastics and described in this book as a practical and understandable approach.

Processing in-line A complete fabricating, production, or fabricating operation can go from material storage and handling, to produce the part that includes upstream and downstream auxiliary equipment, through inspection and quality control, to packaging, and delivery to destination such as warehouse bins or transportation vehicles.

Processing Intelligent What is needed is to cut inefficiency, such as the variables, and in turn cut the costs associated with them. One approach that can overcome these difficulties is called intelligent processing (IP) of materials. This technology utilizes new sensors, expert systems, and process models that control processing conditions as materials are produced and processed without the need for human control or monitoring. Sensors and expert systems are not new in themselves. What is novel is the manner in which they are tied together. In IP, new nondestructive evaluation sensors are used to monitor the development of a materials microstructure as it evolves during production in real time. These sensors can indicate whether the microstructure is developing properly. Poor microstructure will lead to defects in materials. In essence, the sensors are inspecting the material on-line before the product is produced.

Processing line downstream The plastic discharge end of the fabricating equipment such as the auxiliary equipment in an extrusion pipeline after the extruder.

Processing line downtime Refers to equipment that cannot be operate when it should be operating. Reason for downtime could be equipment being inoperative, shortage of material, electric power problem, operators not available, and so on. Regardless of reason, downtime is costly.

Processing line upstream Refers to material movement and auxiliary equipment (dryer, mixer/blender, storage bins, etc.) that exist prior to plastic entering the main fabricating machine such as the extruder.

Processing line uptime Plant operating to produce products.

Processing parameter Measurable parameters such as temperature and pressure required during preparation of plastic materials, during processing of products, inspection, etc.

Processing stabilizer Also called a flow promoter. In thermoplastics they act in the same manner as internal lubricants where they plasticize the outer surfaces of the plastic particles and ease their fusion, but can be used in greater concentrations (about 5 pph). With TS plastics they are not reactive normally and therefore reduces the rate of interactions of reactive groupings by a dilution effect. Thus easier processing may be derived mainly from the reduction in the rate at which the melt viscosity increases. At the same time the overall cross-linking density is reduced.

Processing via fluorescence spectroscopy Sensor techniques can measure the properties of plastics during processing. The intent is to improve product quality and productivity by using molecular or viscous properties of the melt as a basis for process control, replacing the indirect variables of temperatures, pressure, and time. This system analyzes the fluorescence generated in the plastic during processing and translates it into a numerical value for the property being monitored. The plastic must be doped with a small amount of fluorescence dye specific to the application. An optical fiber installed in the plasticator barrel, mold, or die scans the plastic. It is used to perform other tasks such as measuring the concentration and dispersion uniformity of filler; accuracy of 1% provides a means of optimizing residence time. It can also monitor the glass transition temperature.

Qualified products list QPL is a list of commercial products that have been pretested and found to meet the requirements of a specification.

Reactive diluent A formulation or compound containing one or more diluents that functions mainly to reduce the viscosity of the mixture.

Reactive processing Traditionally, the manufacture of products made from plastics involve two separate and distinct operations, namely reaction and processing. Polymerization reactions make monomer molecules into polymer/plastic molecules., and fabricating processing equipment transforms the plastic molecules into shapes. Reactive processing combines these two operations by conducting polymerization and polymer modification reactions in a processing machine such as reactive extruder processing (REX) and reaction injection molding (RIM).

Relief An area of the screw shank of lesser diameter than the outside diameter and located between the bearing and the spine or key-way.

Residence time It is the amount of time a plastic is subjected to heat during fabrication of virgin plastics such as during extrusion, injection and compression molding, calendering, etc. With recycled plastics, properties are affected by previous fabrication and granulating heat. This residence time can cause relatively minor to definite major undesirable or variations in properties of the plastic during the next processing step and/or the finished product. This action can occur even when the same plastic (from the same source) and same fabricating machine are used. Different thermal tests are available and used to meet specific requirements.

Residual stress It is the stress existing in a body at rest, in equilibrium, at uniform temperature, and not subjected to external forces. Often caused by the stresses remaining in a plastic part as a result of thermal and/or mechanical treatment in fabricating parts. Usually they are not a problem in the finished product. However, with excess stresses, the product could be damaged quickly or after in service from a short to long time depending on amount of stress and the environmental conditions around the product.

Responsibility The responsibilities of those involved in the World of Plastics encompass all aspects to producing products as well as people. Recognize that people have certain capabilities; the law says that people have equal rights (so it reads that we were all made equal since 1776) but some interrupt it to mean equal capabilities. So it has been said via Sun Tzu, The Art of War, about 500 BC "Now the method of employing people is to use the avaricious and the stupid, the wise and the brave, and to give responsibilities to each in situations that suit the person. Do not charge people to do what they cannot do. Select them and give them responsibilities commensurate with their abilities."

Reynold's number It is a dimensionless number that is significant in the design of any system in which the effect of viscosity is important in controlling the velocities or the flow pattern of a fluid. It is equal to the density of a fluid, times its velocity, times a characteristic length, divided by the fluid viscosity. This value or ratio is used to determine whether the flow of a fluid through a channel or passage, such as in a mold, is laminar (streamlined) or turbulent.

Rheology It concerns the response of plastics to a mechanical force; the science of a deformation and flow of a material under force.

Rheometer Also called plastometer. A rheological instrument for determining the flow properties of a plastic, usually of high viscosity or in the molten condition of thermoplastics, by forcing the melt through a die or orifice of specific size at specific temperature and pressure.

Rifled liner The barrel liner whose bore is provided with helical grooves.

Risk Designers and others in the plastics and other industry have the responsibility to ensure that all products produced will be safe and not contaminate the environment, etc. Recognize that when you encounter a potential problem, you are guilty until proven innocent (or is it the reverse). So keep the records you need to survive the legal actions that can develop.

Risk, acceptable This is the concept that has developed decades ago in connection with toxic substances, food additives, air and water pollution, fire and related environmental concerns, and so on. It can be defined as a level of risk at which a seriously adverse result is highly unlikely to occur but it cannot be proven whether or not there is 100% safety. In these cases, it means living with reasonable assurance of safety and acceptable uncertainty.

Root-mean-square See **Deviation, root-mean-square**

Rubber Natural rubber, also called elastomer, provides the industry worldwide with certain thermoset material properties that to date are not equaled by synthetic elastomers. Examples include tires (with their relative heat build-up resistance), certain type vibrators, etc. However both synthetic TSE and TPE have made major inroads to product markets previously held by natural rubber and also expanded into new markets. The three basic processing types are conventional (vulcanizable) elastomer, reactive type, and thermoplastic elastomer. More synthetic types are used than the natural worldwide.

An elastomer is a rubberlike material (natural or synthetic) that is generally identified as a material that at room temperature stretches under low stress to at least twice its length and snaps back to approximately its original length on release of the stress (pull) within a specified time period. Elastomers are characterized by a high degree of elasticity and conformability, making them ideal for use in many applications. The term elastomer is often used interchangeably with the term plastic or rubber; however, certain industries use only one or the other.

Until about 1910, the term rubber was sufficiently descriptive for most purposes. It typified natural products derived from various trees and plants that could be formed into solids of various shapes which could be bent, flexed rapidly, or stretched with the amazing ability to return to essentially the initial form. As synthetic materials emerged, particularly synthetics that were directed toward capabilities different from those of natural rubber, considerable confusion resulted as to descriptive terminology. Hence the literature was rife with such terms as rubber, rubbery, rubberlike, and similar inept descriptions. Eventually H. L. Fisher struck a major blow to this confusion and coined the term elastomer to embrace natural as well as synthetic products with those mechanical properties generally associated with natural rubber.

Subsequently, innumerable efforts, both formal and informal, have been directed to the further improvement of a more technical definition. Part of the difficulty encountered lies in the manner in which elastomers find their way to use as end items. While additives, such as stabilizers, coloring agents, and reinforcing agents, are utilized in various combinations with polymeric materials formed into fibers and various plastics, many of these polymer forms have significant value in essentially unmodified form. Polystyrene and polymethyl methacrylate is examples of thermoplastics of this type, and nylons (polyamides) and rayons (cellulosics) are examples of fibers. In contrast, rubber polymers rarely find use as end items without substantial modification.

Rubber History Columbus reportedly returned from his second voyage of 1493 to 1496 with rubber balls used in play by the natives of Haiti. Cortez is reported to have attended a game played between teams of ten or more with the objective of knocking a solid rubber ball through the stone rings at either end of the game court of the Aztec king, Montezuma II. The indefatigable historian F. J. de Torquemada recorded that the Spaniards in the New World, observing the native's use of a white liquid from a tree to coat their cloaks, to render them impervious to rain, and coated hemp or linen bags in a similar manner for the transportation of mercury. The Spaniards are also credited with fabricating personal articles from the liquid (now known as latex) such as cloaks, boots, and shoes as well as other objects for protection from the weather. However, de Torquemada said that in the sun such objects became sticky and smelled bad.

The scientific community of Europe remained essentially in ignorance of this material until in 1736 the French Academy of Science sent a survey party to Peru, under the leadership of Charles de la Condamine, on a mission concerned with his talents as a noted astronomer. La Condamine was also a naturalist and he observed the natives collecting white syrup from trees and drying it over fires. He sent samples of this rubber back to France. He must also have been something of an athlete, for he engaged in a journey of over 2000 miles before making the acquaintance of Francois Fresneau, an engineer employed by Louis XV of France.

Inevitably, Fresneau was also a naturalist who had previously seen samples of rubber but not the parent trees. While Louis XV's rule of France was something of a historical disaster, his choice of dedicated engineers was excellent. Fresneau spent 14 years in searching for the tree responsible for rubber. He subsequently found the trees and reported that turpentine could be used as a solvent for natural rubber, allowing the formation of rubber coatings and objects by solvent evaporation and eliminating the difficulties encountered in the fermentation on exposure to air of the natural latex.

Joseph Priestley, of fame as the discoverer of oxygen, noted in 1770 that gum elastic (as the natural product was then known) would rub out pencil marks from paper; hence the birth of the term rubber. The tempo of developments began to pick up quickly in 1819 as Charles Mackintosh, in a business venture involving coal tar by-products, found that coal tar naphtha was an excellent rubber solvent. For this he patented a waterproof fabric in 1823 which consisted of two layers of woolen cloth with an interlayer of rubber deposited from solution.

Minor progress in the use of rubber continued until about 1839 when Charles Goodyear found that sulfur mixed with rubber followed by heating produced remarkable increases in properties and virtually eliminated the previous deleterious effects of hot weather and sunlight. No question exists as to who made this notable discovery, but Thomas Hancock, not Goodyear, obtained the vital English patent in 1843 while Goodyear's American patent was issued in 1844. Thus Hancock died in comparative wealth in 1865, having outlived Charles Goodyear indebted at death, by five years.

hile other advances, and setbacks, were to occur, it was Goodyear's discovery that marked the beginning of the great natural rubber industry and the certain challenge of synthetics some decades later.

Screw cushion For most molding runs the amount of screw rotation must be adjusted so that there is always a pad, or cushion, of material left after the screw has finished injection. This action ensures that the screw forward time is effective and a constant hold pressure is being applied. On small injection molding machines this cushion may be 3 mm (0.118 in.); on larger machines it may be 9 mm (0.354 in.). No matter what screw cushion size is used it must be kept constant. Screw cushion size may be controlled to within 0.1 mm/0.004 in.

Shank It is the rear protruding portion of the plasticator screw to which the driving force is applied.

Simple/complex So it goes making something simple is a complex process whereas it is very simple to make something complex.

Sintering It is the forming of products from fusible plastic powders. The process involves holding the pressed powder (such as PTFE and certain nylons) products at a temperature just below its melting point for a prescribed time period based on the plastic used. Powdered particles are fused (sintered) together, but the mass as a whole does not melt. This solid-state diffusion results in the absence of a separate bonding phase. After being withdrawn, it is heated to a higher temperature to completely fuse the sintered material. This process is accompanied by increased properties such as strength, ductility, and usually density.

Skiving A specialized process for producing film is skiving. It consists of shaving off a thin film or sheet layer from a large block of solid plastic such as a round billet. Continuous film is obtained by skiving in a lathe type cutting operation which is similar to producing plywood from a tree trunk log. This process is particularly useful with plastics that cannot be processed by the usual plastic film processes, such as extrusion, calendering, or casting. PTFE is an example as it is a plastic that is not basically melt processable.

Splay Mark Also called silver streak. Fan-like or streaking on the surface of fabricated plastic parts. Causes include moisture in the plastic material, thermal degradation of plastic material, relatively cold plastic on mold or die surface due to fast an operation (called jetting), injection mold gate restriction, foam molding, etc.

Static electricity Materials are susceptible to static electricity if they are hydrophobic, resistant to water penetration. This contributes to static buildup in cold dry climates because there is too little water present to conduct the charge of static electricity away.

Static mixer Also called a motionless mixer. They are designed to achieve a homogeneous mix by flowing one or more plastic streams through geometric patterns formed by mechanical elements in a tubular tube or barrel; the mixers contain a series of passive elements placed in a flow channel. These elements cause the plastic compound to subdivide and recombine in order to increase the homogeneity and temperature uniformity of the melt. There are no moving parts and only a small increase in the energy is needed to overcome the resistance of the mechanical baffles. These mixers are located at the end of the screw plasticator such as an injection molding machine or before the screen changer and/or die of an extruder. In an extruder if a gear pump is used, the static mixer is located between the screw and gear pump. They can be used to mix different plastics and plastics with its component ingredients such as color and additives.

Statistics Deals with the collection, organization, analysis, and interpretation of numerical data.

Strain The per unit change, due to force, in the size or shape of a body referred to its original size and shape. Strain is non-dimensional but is usually expressed in unit of length per unit of length or percent. It is the natural logarithm of the ratio of gauge length at the moment of observation instead of the original cross-sectional area. Applicable to tension and compression tests.

Strength The stress required to break, rupture, or cause a failure of a substance. Basically it is the property of a material that resists deformation induced by external forces. Maximum stress occurs when a material can resist the stress without failure for a given type of loading.

Stress The intensity, at a point in a body (product, material, etc.), of the internal forces (or components of force) that act on a given plane through the point causing deformation of the body. It is the internal force per unit area that resists a change in size or shape of a body. Stress is expressed in force per unit area and reported in MPa, psi, etc. As used in tension, compression, or shear, stress is normally calculated on the basis of the original dimensions of the appropriate cross section of the test specimen. This stress is sometimes called engineering stress; it is different than true stress that takes into account the change in cross section.

Stress whitening Also called crazing. It is the appearance of white regions in a material when it is stressed. Stress whitening or crazing is damage that can occur when a TP is stretched near its yield point. The surface takes on a whitish appearance in regions that are under high stress. It is usually associated with yielding.

For practical purposes, stress whiting is the result of the formation of microcracks or crazes that is a form of damage. Crazes are not basically true fractures because they contain strings of highly oriented plastic that connect the two flat surfaces of the crack. These fibrils are surrounded by air voids. Because they are filled with highly oriented fibrils, crazes are capable of carrying stress, unlike true fractures. As a result, a heavily crazed part can carry significant stress even though the part may appear fractured.

It is important to note that crazes, microcracking, and stress whitening represent irreversible first damage to a material that could ultimately cause failure. This damage usually lowers the impact strength and other properties. In the total design evaluation, the formation of stress cracking or crazing damage should be a criterion for failure based on the stress applied.

Striation A longitudinal line in a part due to disturbance in the melt path during fabrication. It also identifies the separation of color resulting in linear effect of color variation due to incomplete mixing and/or melting of the plastic.

Surging Unstable pressure build-up in an extruder leading to variable throughput and waviness in the output product's appearance.

Synergism Arrangement or mixture of materials in which the total resulting performance is greater than the sum of the effects taken independently such as with alloying/blending plastics.

Syntactic Identifies an orderly arrangement in a compound of components, ingredients, etc. so that the product has absolutely isotropic mechanical properties.

Synthetic history Borchardat in 1879 conceived the bold idea that a comparatively simple molecule, isoprene, C_5H_8, was the foundation of natural rubber. Efforts leading to this postulation, which was correct, were based upon results of several researchers throughout the previous 50 years. Subsequent efforts were substantially devoted to the attempt to duplicate natural rubber by the polymerization of isoprene. In essence, these efforts were frustrating.

Popular belief holds the birth of synthetic elastomers to have been in about 1940 in response to the demands for strategic materials imposed by World War 11. Facts, however, contradict this. In 1912, at the Eighth International Congress of Applied Chemistry in New York, C. Duisberg, the managing director of Bayer and Company, exhibited synthetic rubber tires that had been used on the Kaiser's car. The magnitude, efficiency, and duration of the Allied blockade, posed serious problems for Germany. Germany produced 2,350 tons of synthetic rubber from 2,3-dimethyl-butadiene during the holocaust of World War I.

This monomer differed from isoprene, the basic unit of natural rubber, by the addition of one methyl (CH_3) group. The German technicians responsible for this remarkable achievement were (like Louis XV's engineer, Fresneau) dedicated and patient men. For while methyl rubber B was polymerized with metallic sodium in an atmosphere of carbon dioxide, methyl rubber H (hard) was polymerized by a process in which the monomer was stored in drums for 2 to 3 months at room temperature, and methyl rubber W was prepared from the monomer by polymerization at 70C (158F) for 3 to 6 months.

While progress was erratic until about the 1930's, synthetic elastomers were not only born, but expanding technically. It is perhaps notable that the first few products of the synthetic elastomer industry's resurgence then brought forth elastomers which offered environmental resistances vastly superior to natural products. Most notable of these were Thiokol (polysulfides) in 1929, neoprene (polychloroprene) in 1931, Buna N (copolymers of butadiene and acrylonitrile) in 1937, and butyl (copolymer of isobutylene and isoprene) in 1940. Many more were to follow.

Temperature controller, heating overshoot circuit Used in temperature controllers to inhibit temperature overshooting on warm-up.

Temperature detector, resistance RTD contains a temperature sensor made from a material such as high purity platinum wire; resistance of the wire changes rapidly with temperatures. These sensors are about 60 times more sensitive than thermocouples.

Temperature measurement Temperatures can be measured with thermocouple (T/C) or resistance temperature detector (RTD). T/Cs tend to have shorter response time, while RTDs have less drift and are easier to calibrate. Traditionally, PID controls have been used for heating and on-off control for cooling. From a temperature control point the more recent use is the fuzzy logic control (FLC). One of FLCs major advantage is the lack of overshoot on startup, resulting in achieving the setpoint more rapidly. Another advantage is in its multi-variable control where more than one measured input variable can effect the desired output result. This is an important and unique feature. With PID one measured variable affects a single output variable. Two or more PIDs may be used in a cascade fashion but with more variables they are not practical to use.

Temperature proportional-integral derivative Pinpoint temperature accuracy is essential to be successful in many fabricating processes. In order to achieve it, microprocessor-based temperature controllers can use a proportional-integrated-derivative (PID) control algorithm acknowledged to be accurate. The unit will instantly identify varying thermal behavior and adjust its PID values accordingly.

Tex A unit for expressing linear density that is equal to the mass of weight in grams of 1,000 m of fiber, filament, yarn, or other textile strand.

Thermodynamic It is the scientific principle that deals with the inter-conversion of heat and other forms of energy. Thermodynamics (thermo = heat and dynamic = changes) is the study of these energy transfers. The law of conservation of energy is called the first law of thermodynamics.

Thixotropic A characteristic of material undergoing flow deformation where viscosity increases drastically when the force inducing the flow is removed. In respect to materials, gel-like at rest but fluid or liquefied when agitated (such as during molding). Having high static shear strength and low dynamic shear strength at the same time. Losing viscosity under stress. It describes a filled plastic (like BMC) that has little or no movement when applied to a vertical plane. Powdered silica and other fillers are used as thickening agents.

Tolerance, full indicator movement FIM is a term used to identify tolerance with respect to concentricity. Terms used in the past were full indicator reading (FIR) and total indicator reading (TIR).

Tool When processing plastics some type of tooling is required. Tools include molds, dies, mandrels, jigs, fixtures, punch dies, perforated forms, etc. These tools fabricate or shape products. They fit into the overall flow chart in fabricating plastic products. The terms for tools are virtually synonymous in the sense that they have some type of female and/or negative cavity into or through which a molten plastic moves usually under heat and pressure or they are used in secondary operations such as cutting dies, stamping sheet dies, etc.

Materials of construction can be related to the two most commonly used standards that are the American Iron and Steel Institute (AISI) and the German Werkstoff material numbers and their mean (average) chemical compositions. Note that chemical compositions will always differ from data in one book to another book and from one manufacturer's tool stock list to another. As an example it will be very unlikely that the P20 steel being used will exactly match the chemical composition reported in different tables. However they will be close.

In addition to steels materials of construction include all types such as aluminum, copper, and beryllium-copper. Examples of other different types include wood and plastic for thermoforms and reinforced plastics (RPs). RP tools also include vacuum and pressure flexible bags. Operations of tools vary from fabricating solid to foamed products such as using a steam chest for producing expandable polystyrene foams.

Mold and die tools are used in processing many different plastics with many of them having common assembly and operating parts (preengineered since the 1940s) with the target to have the tool's opening or cavity designed to form desired final shapes and sizes. They can comprise of many moving parts requiring high quality metals and precision machining. As an example with certain processes to capitalize on advantages, molds may incorporate many cavities, adding further to its complexity.

Tools of all types can represent upward to one-third of the companies manufacturing investment. Metals, specifically steels, are the most common materials of construction for the rigid parts of tools. Some mold and die tools cost more than the primary processing machinery with the usual approaching half the cost of the primary machine. About 5 to 15% of tool cost are for the material used in its manufacture, design about 5 to 10%, tool building hours about 50 to 70%, and profit at about 5 to 15%.

The proper choice of materials of construction for their openings or cavities is paramount to quality, performance, and longevity (number or length of products to be processed) of tools. Desirable properties are good machinability of component metal parts, material that will accept the desired finish (polished, etc.), ability with most molds or dies to transfer heat rapidly and evenly, capability of sustained production without constant maintenance, etc. As the technology of tool enhancements continues to evolve, tool manufacturers have increasingly turned to them in the hope of gaining performance and cost advantages.

There are now a wide variety of enhancement methods and suppliers, each making its own claims on the benefits of its products. With so many suppliers offering so many products, the decision on which technology to try can be time consuming. There are toolmakers that do not have the resources to devote to a detailed study of all of these options. In many cases they treat tools with methods that have worked for them in the past, even though the current application may have different demands and newer methods have been developed. What can help is to determine what capabilities and features are needed such as hardness, corrosion resistance, lubricity, thermal conductivity, polishing, coating, and repairing.

There are many tool metals used such as D2 steel that are occasionally used in their natural state (soft) when their carbon content is 1.40 to 1.60wt%. Tool metals such as P20 that are generally used in a pre-toughened state (not fully hardened).

By increasing hardness longer tool life can often be achieved. Increased wear properties are especially critical when fabricating with abrasive glass- and mineral-reinforced plastics. This is important in high-volume applications and high-wear surfaces such as mold gates inserts and die orifices. Some plastic materials release corrosive chemicals as a natural byproduct during fabrication. For example, hydrochloric (HCl) acid is released during the tooling of PVC. These chemicals can cause pitting and erosion of untreated tools surfaces. Also, untreated surfaces may rust and oxidize from water in the plastic and humidity and other contaminants in the air.

Polishing and coating tools permit meeting product surface requirements. Improved release characteristics of fabricated products are a commonly advantage of tool coatings and surface treatments. This can be critical in applications with long cores, low draft angles, or plastics that tend to stick on hot steel in hard-to-cool areas. Coatings developed to meet this need may contain PTFE (Teflon). Also used are metals such as chrome, tungsten, or electroless nickel that provide inherent lubricity.

Torpedo An unflighted cylindrical portion of the plasticator screw usually located at the discharge end that is providing additional shear heating capabilities for certain plastics

Transistor Semiconductor device for the amplification of current required in different sensing instruments. The two principle types are field effect and junction.

Turnkey operation A complete fabrication line or system, such as an extruder with a thermoformer line with upstream and downstream equipment. Controls interface all the equipment in-line from material delivery to the end of the line handling the product for in-plant storage or shipment out of the plant.

Tyvek DuPont's trade name for a nonwoven spun bonded, tough, strong HDPE fiber sheet product. This thermoformable plastic material's use includes packaging lids, medical device packages, mailing envelopes (protects contents, etc.), wrapping around buildings to completely seal off cracks and seams to prevent drafts and cut airflow penetration between the outside and inside (allows moisture to escape from the walls, eliminating or minimizing the prospect of harmful condensation damage), etc.

Uptime See **Processing line.**

Vacuum mold cavity Press can include a vacuum chamber around or within the mold providing removal of air and other gases from the cavity(s).

Value analysis VA is an amount regarded as a fair equivalent for something, that which is desirable or worthy of esteem, or product of quality having intrinsic worth. Aside from technology developments, there is always a major emphasis on value added services. It is where the fabricator continually tries to find ways to augment or reduce steps during manufacture with the target of reducing costs.

While there are many definitions of VA, the most basic is the following formula where VA = (function of product)/(cost of the product). Immediately after the part goes into production, the next step that should be considered is to use the value engineering approach and the FALLO approach. These approaches are to produce products to meet the same performance requirements but produced at a lower cost. If you do not take this approach, then your competitor will take the cost reduction approach. VA is not exclusively a cost-cutting discipline. With VA you literally can do "it all" that includes reduce cost, enhance quality, and boost productivity.

Van der Waals force Also called intermolecular forces, secondary valence forces, dispersion force, London dispersion force, or van der Waals attraction. It is an attractive force between two atoms or non-polar molecules, which arise because a fluctuating dipole moment in one molecule induces a dipole moment in the other, and the two dipole moments then interact. They are somewhat weaker than hydrogen bonds and far weaker than inter-atomic valences. Information regarding their numerical values is mostly semi-empirical, derived with the aid of theory from an analysis of physical and chemical data.

Vent purifier The exhaust from vented plasticating barrels can show a dramatic cloud of swirling white gas; almost all of it is condensed steam proving that the vent is doing its job. However, a small portion of the vent exhaust can be other materials such as by-products released by certain plastics and/or additives and could be of concern to plant personnel safety and/or plant equipment. Purifiers can be attached (with or without vacuum hoods located over the vent opening) to remove and collect the steam and other products. The purifiers include electronic precipitators.

Virgin plastic Plastics in the form such as pellets, granules, powders, flakes, liquids, etc. that have not been subjected to any fabricating method or recycled identifies virgin plastics.

Viscoelasticity A material having this property is considered to combine the features of a perfectly elastic solid and a perfect fluid; representing the combination of elastic and viscous behavior of plastics (see Chapter 1)

Viscometer Also called a viscosimeter. This instrument is used for measuring the viscosity and flow properties of fluids such as plastic melts. Basically it is the property of the resistance of flow exhibited within a body of material. A commonly used type is the Brookfield that measures the force required to rotate a disk or hollow cup immersed in the specimen substance at a predetermined speed. Other types employ such devices as rising bubbles, falling or rolling balls, and cups with orifices through which the fluid flows by gravity. Instruments for measuring flow properties of highly viscous fluids and molten plastics are more often called rheometers.

Viscosity It is the property of resistance to flow exhibited within the body of a material. In testing, it is the ratio of the shearing stress to the rate of shear of a fluid. Viscosity is usually taken to mean Newtonian viscosity in which case the ratio of shearing stress to rate of shearing is constant. In non-Newtonian behavior, which is the usual case with plastics, the ratio varies with shearing rate. Such ratios are often called the apparent viscosities at the corresponding shear rates.

Viscosity, absolute It is the ratio of shear stress to shear rate. It is the property of internal resistance of a fluid that opposes the relative motion of adjacent layers. Basically it is the tangential force on a unit area of either of two parallel planes at a unit distance apart, when the space between the planes is filled and one of the planes moves with unit velocity in its own plane relative to the other. The Bingham body is a substance that behaves somewhat like a Newtonian fluid in that there is a linear relation between rate of shear and shearing force, but also has a yield value.

Viscosity, coefficient It is the shearing stress necessary to induce a unit velocity gradient in a material. In actual measurement, the viscosity coefficient of a material is obtained from the ratio of shearing stress to shearing rate. This assumes the ratio to be constant and independent of the shearing stress, a condition satisfied only by Newtonian fluids. Consequently, in all other cases that includes plastics non-Newtonian), values obtained are apparent and represent one point in the flow chart. Inherent viscosity refers to a dilute solution viscosity measurement where it is the ratio of the natural logarithm of the relative viscosity to the concentration of the plastic in grams per 100 ml of solvent.

Intrinsic viscosity (IV) is a measure of the capability of a plastic in solution to enhance the viscosity of the solution. IV increases with increasing plastic molecular weight. It is the limiting value at infinite dilution of the ratio of the specific viscosity of the plastic solution to the plastic's concentration in moles per liter.

IV data is used in processing plastics. As an example the higher IV of injection-grade PET plastic can be extruded blow molded; similar to PETG plastic that can be easily blow molded but is more expensive than injection molded grade PET and PVC for blow molding.

Viscoelasticity is perhaps better viewed more broadly as mechanical behavior in which the relationships between stress and strain are time dependent, as opposed to the classical elastic behavior in which deformation and recovery both occur instantaneously on application and removal of stress, respectively.

Viscosity, intrinsic IV (intrinsic viscosity) is a measure of the capability of a plastic in solution to enhance the viscosity of the solution. IV increases with increasing polymer molecular weight. It is the limiting value at infinite dilution of the ratio of the specific viscosity of the plastic solution to the plastic's concentration in moles per liter. IV data is used in processing plastics. As an example the higher IV of injection-grade PET plastic can be extruded blow molded; similar to PETG plastic that can be easily blow molded but is more expensive than injection molded grade PET and PVC for blow molding.

Volumetric efficiency The volume of plastic discharged from the machine during one revolution of the screw, expressed as a percentage of the developed volume of the last turn of the screw channel.

Vulcanization Methods for producing a material with good elastomeric properties (rubber) involves the formation of chemical crosslinks between high-molecular-weight linear molecules. The starting polymer (such as raw rubber) must be of the noncrystallizing type (NR is crystallizable), and its glass transition temperature T_g must be well below room temperature to ensure a rubbery behavior. **x-axis** The axis in the plane of a material used as 0^0 reference; thus the y-axes is the axes in the plane of the material perpendicular to the x-axis; thus the z-axes is the reference axis normal to the x-y plane. The term plane or direction is also used in place of axis.

Vulcanizing additive Thermoset elastomers must be vulcanized or crosslinked to obtain strong, dimensionally stable resilient materials. To accomplish this purpose, a formidable array of chemicals is employed. Classically, sulfur vulcanizes unsaturated elastomers such as SBR and natural rubber; however, because the rate of vulcanization is too slow for industrial applications, chemicals like benzothiazyl disulfide (MBTS) are added. The latter is typical of a class called accelerators.

Along with accelerators, yet another group of additives called activators is used. Using the above example, zinc oxide and stearic acid are activators used in conjunction with the accelerator MBTS in the sulfur vulcanization of SBR or natural rubber. The sulfur, thiazole, zinc oxide, stearic acid vulcanization system is probably the largest used combination. Generally the types and concentration of the various vulcanization systems are selected along with appropriate cure retarders so as to obtain a good cured vulcanizate in a period of 15 to 60 min at temperatures ranging from 138 to 177C (280 to 350F).

Warpage It is the dimensional distortion in a plastic part after processing. The most common cause is variation in shrinkage of the part. The major processing factors involved are flow orientation, area shrinkage, and differential cooling.

Weight, areal See **Areal weight.**

Weld line Also called weld mark, flow line, or striae. It is a mark or line when two melt flow fronts meet during the filling of an injection mold.

Well It is the space provided in the cavity block for the uncured molding powder or preform.

Whitening See **Stress whitening.**

x-**axis** The axis in the plane of a material used as 0^0 reference; thus the y-axes is the axes in the plane of the material perpendicular to the x-axis; thus the z-axes is the reference axis normal to the x-y plane. The term plane or direction is also used in place of axis.

y-**axis** A line perpendicular to two opposite parallel faces.

Yarn designation A term used to indicate the number of original singles (strands) twisted and the number of these units plied to form a yarn or cord. The first letter indicates glass composition, the second letter represents whether it is continuous or staple fiber, and the third letter indicates the diameter range of the individual fiber. As an example CD identifies type E glass with continuous fiber (C) of 0.00023 in. average fiber diameter (D).

Yarn grex A universal yarn numbering system in which the yam number is equal numerically to the weight in g/10,000 m.

Young's modulus See **Modulus of elasticity**.

z-**axis** The reference axis perpendicular to x and y axes.

Appendix C
Tradenames

Achieve Metallocene polypropylene, Exxon

Acrilan Polyacrylonitrile, AlliedSignal

Acrylate Polyethylacrylate, Cyro

Acrylite Acrylic sheet, acrylic resin, Cyro

Adiprene Polyurethane (isocyanate), Bayer

Affinity Plastomer, Dow

Aqua-Novon Waterproof/water soluble degradation system, Novon.

AquaBlok Water-blocking reinforcement, Owens Corning

Aqualift Water-based external mold releases, Franklynn

Araldite Epoxy resins and hardeners; epoxy structural adhesives, Ciba

Aramid Nylon, DuPont

Arathane Polyurethane adhesives, Ciba

Aravite Cyanoacrylate and acrylic adhesives, Ciba

Ardel Polyarylate, BP Amoco

Arimax Structural RIM resin, Ashland

Armid Aliphatic amides, Akzo

Armostat Antistatic agents, Akzo

Arnits Thermoplastic polyesters, DSM

Aroma Poly Fragrance polymer concentrates, Aroma Tech

Aseelean High-performance purging agent, Sun

Ashions Thermoplastic engineering resins, Ashley

ASP Water-washed kaolin, Engelhard

Aspun Fiber grade resins, Dow

Astrawax Amide wax, additive lubricant, AlliedSignal

Atlac Thermoset polyester, Reichhold

Attain ULDPE, Dow

Aurum Polyimide resin, Mitsui

Auto-Grader Online melt flow indexer, Brabender

Autofroth Pressurized rigid froth polyurethane foam systems, BASF

Autoguage Automatic die, Production Components Cloeren

Autopour Pressurized rigid polyurethane foam systems, BASF

Autoprofile Gauge control, Battenfeld Gloucester

Avantra Styrenic polymer resins, BASF

Azdel Stampable reinforced plastic sheet, Azdel

Azmet PBT and PET; glass fiber-reinforced composite, Azdel

Bayblend ABS/PC blend, Miles

Borax Nitrile-based barrier resin, BP Amoco

Baybiond Polycarbonate/ABS blends, Bayer

Baydur Rigid structural urethane foam, Bayer

Baydur STR Polyurethane composite solid or foam systems, Bayer

Boyertitan Titanium dioxide pigment, Bayer

Baytit Polyurethane molded foam systems, Bayer Bayflex Polyurethane elastomeric RIM systems, Bayer

Boytec Polyurethane cast elastomer systems and prepolymers, Bayer

Betabrace Reinforcing composites, Dow

Betamate Structural adhesives, Dow

Betaseal Glass bonding systems, Dow

Bi-Ply Combination mat/woven roving, Owens Corning

Bicor Biaxially oriented polypropylene, Mobil

Black Pearls Pelletized carbon black, Cabot

Blendo Resins for modifying polymers, GE

Blue Star Antimony oxide, Great Lakes

Bondmaster Structural adhesives, National Starch

Britol White mineral oil, Witco

Brushmaster Solvent-borne contact adhesives, National Starch

Budene Polybutadiene, Goodyear Tire

Buna CB Polybutadiene impact modifier, Bayer

Cab-O-Sil Amorphous fumed silica, Cabot

Cab-O-Sperse Aqueous silica dispersion, Cabot

Cadco Plastic rod, sheet, tubing, film, Hanna

Cadet Organic peroxides, Akzo

Cadon Styrene maleic anhydride resin, Bayer

Cadox Organic peroxides, Akzo

Calibre Polycarbonate resins, Dow

Capran Nylon films, AlliedSignal

Capran Unidraw Uniaxially oriented nylon film, AlliedSignal

Capron Nylon resins and compounds, AlliedSignal

Cata-Chok Urethane foam catalysts, Ferro Caytur Curing agent, catalyst, Uniroyal

Cefor Polypropylene, Shell

Celanex PBT, Hoechst

Celcon Polyacetal, Hoechst

Celogen Blowing agents, Uniroyal Chemical

Centrex Acrylonitrile-styrene-acrylic blends, Monsanto

Chernglas Glass fiber reinforcements, PTFE coated-woven, Chemfab

Chemigum Nitrile rubber powder, Goodyear

Chemlok Elastomer bonding adhesives, Lord

Chimassorb Hindered amine light stabilizers, Ciba

Chlorez Resin chloroparaffins, Dover/ICC

Cho-Sil Conductive EMI shielding material, Parker Hannifin

ChopVantage TP reinforcements, PPG

CIM System Computer integrated manufacturing, Hunkar

Class-Metor I, II, IV, Economic process management system, Hunkar

Classic Series HVLP spray guns, DeVilbiss

Cleartuf PET resin, Shell

Clysar Shrink film, DuPont

Colorcomp Custom-colored heat resins, LNP

Colartherm Zinc ferrite pigment, Bayer

Comboloop Hydrocarbon wax blends, additive lubricants, AlliedSignal

Combomet Woven roving mat, PPG

Compax Supreme Mold steel, Uddeholm

Compel Long fiber-reinforced TPs, Ticona

Contracool Air-cooled extruder, Battenfeld Gloucester

Corterra Polytrimethylene terephthalate, Shell

Corzan Chlorinated PVC, BFGoodrich

Crastin PBT PBT resin, DuPont

CRATEC Dry-use chopped strands, Owens Corning

CRATEC Plus Pelletized chopped glass fibers, Owens Corning

Cronar PET photographic film base, DuPont

Curithane Liquid catalysts, Dow

Cycolac ABS, GE

Cyrex Acrylic polycarbonate alloys, Cyro

Cyrolite Acrylic-based multipolymers, Cyro

D-Tex Intermeshing twin screw extruder, Davis- Standard

DAC Injection molding control, Hunkar

Dacron Polyethylene terephthalate, DuPont

Daltoped Elastomeric polyurethane systems, ICI

Daran Polyvinylidene chloride emulsion, Dow

Daratak Polyvinyl acetate homopolymer emulsions, Dow

Darex Styrene-butadiene latexes, Dow

Dartek Cast film & sheet, nylon, DuPont

Darvan Dispersing agents, Vanderbilt

Dechlorane Plus Fire-retardant additive, Occidental

D.E.H. Epoxy catalyst resins, Dow

Delrin Acetal resin, DuPont

Delrin ST Tough acetal homopolymer, DuPont

D.E.N. Epoxy novolac resins, Dow

Derakane Epoxy vinyl ester resins, Dow

Desmodur Polyisocyantes for coatings, Bayer

Desmophen Polyester resins for coatings, Bayer

Diens Polybutadiene, Firestone

Dion Unsaturated polyester, Reichhold

Direct Flow Manifold systems, balanced hot runner manifolds, Incoe

Dixie Clay Filler for polyesters, Vanderbilt

Dowfrost Heat transfer fluids, Dow

Dowlex LLDPE, Dow

Dowtherm Heat transfer fluids, Dow

Drapex Epoxy and polymeric plasticizers, Witco

Dri-Loc Pre-applied adhesive, Loctite

DSB Barrier feedscrew, Davis-Standard

Duradene Styrene-butadiene copolymers, Firestone

Dural Rigid PVC, AlphaGary

Duralron Polyimide, DSM

Durez Powder and flake phenolic resin, Occidental

Duro-Lam Hot-melt laminating adhesives, National Starch

Duro-O-Set Polyvinyl acetate and ethlyene, National Starch

DWF Dow window film, Dow

Dylark Styrene-maleic anhydride copolymers, Arco

Dylene Polystyrene, NOVA

Easigel Organo clay, Akzo

Eastapak, PET, Eastman

Eastar Copolyester thermoplastics; PETG, Eastman

Easy Flow Easy-processing LLDPE film resins, Union Carbide

ECRGLAS Corrosion-resistant glass fibers, Owens Corning

Ektar PCTG, PET, PETG, Eastman

Elastoflex R Semi-flexible polyurethane systems, BASF

Elvaloy Ethylene/ester/CO terpolymers, DuPont

Elvamide Nylon multipolymer resins, DuPont

Elvanol Polyvinyl alcohol, DuPont

Elvax EVA resins, DuPont

Emi-X EMI attenuated thermoplastic composites, LNP

Emiclear Conductive plastic, Toshiba

Engage Polyolefin elastomer, DuPont Dow

Epon Epoxy resins, Shell

Epotuf Epoxy resins and hardeners, Reichhold

Esperox Organic peroxide catalyst, Witco

Estane PUR resins and compounds, BFGoodrich

Estyrene Flame-retarded, Thermofil

Ethafoam Polyethylene foam, Dow

Everflex Polyvinyl acetate emulsions, Dow

Exact Metallocene plastomer, Exxon

Exceed Metallocene linear low-density PE, Exxon

Expandex Blowing agents, Uniroyal

Exxtral Thermoplastic olefin elastomer, Exxon

Fanchon Organic pigments, Bayer Corp.

Fiberfil Thermoplastic compounds, DSM

Fiberglas Glass fibers, Owens Corning

Filamid Colorants for polyamide fibers and plastics, Ciba

Gapex Nylon, Ferro

Gamaco Calcium carbonate, Georgia

Geartruder Extruder and integral gear pump, Harrel

Gemini Conical twin extruder, Davis-Standard

Geode Complex inorganic pigments, Ferro

Geon Polyvinyl chloride, Geon

Glosrod Glass-reinforced rod, Glastic

Good-Rite Antioxidants, stabilizers, BFGoodrich

GraphiGolor Cast polyvinylidene fluoride, Avery

Graphitan Graphite pigment. Ciba

GSP Calcium carbonate, ECC

GUR UHMW-PE, Ticona

H-Film Polyimide, DuPont

Hi-Fax Polyethylene (linear or high density), DuPont

Helar Ethylene-chlorotrifluoro-ethylene, Ausimont

HercuFlex Strand Polymer-impregnated glass fiber strand, PPG

Hi-Point Organic peroxide catalyst, Witco

Hipertuf PEN resin, Shell

Hipscon Conductive HIPS sheet, BFGoodrich

Hostalon, Polyethylene, UHMWPE, Ticona

HP Multipolymer Acrylic-based multipolymer sheet, Cyro

HT-1 Polyimide, DuPont

Hycer Reactive liquid polymers, BFGoodrich

Hydral Hydrated alumina, flame-retardant, smoke suppressant additive, Alcoa

Hypalon Chlorosulfonated polyethylene (HDPE), DuPont Dow

Hyperm High-performance hot-melt adhesives, National Starch

Hypol Hydrophilic PUR polymers, Dow

Hysol Adhesives and structural materials, Dexter

Hytrel Polyester, DuPont

Impax Supreme Prehardened mold steel, Uddeholm

Instant-Lok Hot-melt adhesives, National Starch

Integral Adhesive films and web, Dow

Irrathene Irradiated polyethylene, DuPont

Isobind Isocyanate binders, Dow

Isonol Polyether polyols, Dow

Isoplast Engineering thermoplastic PUR, Dow

Iupital Polyacetal (copolymer), Mitsubishi

Jeffamine Reactive polyether amines, Huntsman
JET/SPS Plastic lumber molding, Mid- Atlantic

K-Resin Styrene-butadiene, Phillips
Kadel Polyketone, BP Amoco
Kaladox PEN film, DuPont
Kalidar PEN container, fiber resins, DuPont
Kalrez Perfluoroelastomer parts, DuPont
Kapton Polyimide film, DuPont
Kel-F Monochlorotrifluoroethylene, DuPont
Kemester Fatty and non-fatty esters, Witco
Kemgard Smoke suppressants; flame retardants, Sherwin-Williams
Ketienflex Sulfonamides, Akzo
Kevlar Aramid fiber, DuPont
Kodar ASA/AES, PETG, Eastman
Korad Acrylic Weatherable surfacing for ABS, HIPS, and PVC, Polymer Extruded Products
Kraton Thermoplastic elastomer, Shell Chemicals
Kynar Polyvinylidene fluoride, DuPont

Lexan, Polycarbonate resin, film, sheet, GE
Linevol Plasticizer alcohol, Shell
Lite-Tak UV adhesives, Loctite
Lo-Vol Silica flatting agents, PPG
Lamod Copolyester elastomer, GE
Lucite Methyl methacrylate, GE
Lupersol Liquid organic peroxide, Elf Atochem
Luran Styrene-acrylonitrile copolymers, BASF
Lustran ABS Acrylonitrile-butadiene-styrene resin, Bayer
Lustran SAN Styrene-acrylonitrile resin, Bayer

Magnum ABS, Dow
Makrolon Polycarbonate resin, Bayer
Marlex HDPE< PP, Phillips
Masil Silicone, BASF
Masterblend Feeder control system, Merrick
MaxiChop Thermoplastic reinforcements, PPG
MCG Anti-Fog Prevents fog/mist on plastic surfaces, Merix
Melinar PET container resins, DuPont
Melinex Polyester film, DuPont
MF/Flow Flow analysis, Moldflow
Micro Spheres Micro pellet concentrates, Teknor
Minlon Mineral-reinforced nylon resin, DuPont
Modar Polyester resins, Ashland
Monmouth Flame-retardant, Hanna
MX Nylon Nylon-MXD6, Mitsubishi Chemical
Mylar Polyethylene glycol terephthalate film, DuPont
Mylar ECO, Recycled polyester film, DuPont

Nacor Waterborne pressure sensitive adhesives, National Starch
Neoflex Plasticizer alcohol, Shell
Neopolon Polyethylene foam, moldable, BASF
Neoprene Polychloroprene, DuPont Dow
Nicalon Silicon carbide fiber, Dow Corning
Nitropore Blowing agents, Uniroyal
Nomex Nylon, DuPont
Noryl Polyphonylene oxide resin, GE
Nypel Recycled nylon resin, AlliedSignal
Nytel Talc; PP & PVC filler, PVC, Vanderbilt

Ongard 2 Smoke & flame suppressant for PVC, Great Lakes
Optema EMA, Exxon

Orasol Dyes. Ciba
Orlon Polyacrylonitrile, DuPont
OSC Bushing Externally heated hot runner bushings, Incoe Corp.

Palatinol Plasticizers, BASF
Papi Polymeric MDI (PMOI), Dow
Paraloid Acrylic & MBS modifiers, R&H
PDI Dispersions Paste color dispersions, Ferro
Penton Chlorinated polyether, DuPont
Performance Thermoplastic reinforcements, PPG
Permyl Ultraviolet absorbers, Ferro
Perspex Acrylic polymer, ICI
Petra Post-consumer recycled PET resin, AlliedSignal
Petrothene Polyolefin plastics, Quantum
Plecolastic Polystyrene resins, Hercules
Plenco Thermoset resins and molding compounds, Plastics Engineering
Plexiglas Methyl methacrylate, AtoHaas
Plioflex Emulsion styrene-butadiene-rubbers, Goodyear Tire
Pluracol Polyether polyols, BASF
Polylite Polyester resins, Reichhold
Polypur Polyurethane compounds, Schulman
Premi-Gles Thermoset polyester-based molding compounds. Premix
Prevail, Thermoplastic resins, Dow
Prism Solid polyurethane RIM systems, Bayer
Prism Series Instant adhesives, Loctite
Procar Acrylic-coated film (two sides), Mobil
Pulse ABS/PC, Dow
Pur-Fect Lok Reactive structural adhesives, National Starch

Questra Crystalline polymers, Dow
Quindo Organic pigment, Bayer

Radel Polyethersulfone, BP Amoco
Regal Pelletized carbon black, Cabot
Regairoz Hydrogenated hydrocarbon resins, Hercules
Resi-Set Phenolic resins for industrial applications, Georgia-Pacific
Retain Polyethylene, Dow
Rexflex Flexible polyolefin, Huntsman
RIMline Polyurethane RIM systems, ICI
Rynite Thermoplastic PET resin, DuPont
Ryton Polyphenylene sulfide, Phillips

S-2 Glass High-performance glass fibers, Owens Corning
Sag Silicone antifoams, Witco
Sanres Solvent-borne urethane, BFGoodrich
Santoprene Polypropylene EPDM-based vulcanizate, Advanced Elastomer
Saran Polyvinylidene chloride resins, Dow
Saran Films Plastic films, Dow
Saranex Coextruded film, Dow
SatinGlass SMC roving, Vetrotex CertainTeed
Sclairfilm PE cast film & blown film, DuPont
Selar Nylon or PET, DuPont
Silastic Silicone, Dow Corning
Skin-Coat Bonded glass mat and various veils, Fiber Glass Industries
Spectrim Reaction molded products, Dow
SpeedBonder Anaerobic adhesive. Loctite
Spheriglass Solid glass spheres, Potters
Stycast One- and two-component encapsulants, Emerson & Cuming
Styrofoam Insulation, extruded polystyrene board, sheet, Dow
Styrolux SBS block copolymer, BASF
Styron Polystyrene resins, Dow
Styropor Polystyrene expandable beads, BASF

Surlyn Ionomer, DuPont

Tandem Two-station injection molding machines, Husky

Tedlar Polyvinyl fluoride products, DuPont

Tedlar SP Polyvinyl fluoride cast film, DuPont

Teflon FEP, TFE, and PFA fluoroplastic resins and films, DuPont

Tefzel Fluoropolymer resin and film, DuPont

Tenite Cellulose acetate butyrate, Eastman

Texin Thermoplastic urethane elastomers, Bayer

Thermogrip Hot melt adhesives, Bostik

Thermoplast Dyes, BASF

Thermx Copolyester, Eastman

Thiokol Tolysulfide, Thikol

Tinuvin Ultraviolet absorbers; hindered amine stabilizer, Ciba

TMS Antimony oxide, Great Lakes

Torlon Polyamide-imide resin, BP Amoco

TPX Polyrnethylpentene, Mitsui

Traytuf PET resin, Shell

TufRov Thermoplastic reinforcements, PPG

Type 30 Single-end glass fiber rovings, Owens Corning

Tyril SAN resins, Dow

Tyrin Chlorinated polyethylene, DuPont Dow

Tyvek Nonwoven spun bonded, tough, strong HDPE fiber sheet, DuPont

Ucar Solution vinyl resins, Union Carbide

Udel Polysulfone, BP Amoco

Ultem Polyetherimide resin, sheet, GE

Ultra Black RTV silicone, Loctite

Ultradur B PBT thermoplastic polyester, BASF

Ultramid Nylon resins, BASF

Ultrason E Polyethersulfone, BASF

Ultrason S Polysulfone, BASF

Uniprene Thermoplastic vulcanizate, Teknor

Unival HDPE blow molding resins, Union Carbide

Valox PBT film, GE

Vansil W Wollastonits, Vanderbilt

Varex Optifil P2 Blown film extrusion, Windmoeller &: Hoelscher

Vectra LCP, Hoechst

Verton Long fiber composites, LNP

Vespel Polyimide parts and shapes, DuPont

Vitafilm Extruded PVC film, Huntsman

Viton Fluoroelastomer, DuPont Dow

Vitron Fiberglass staple fiber, Johns Manville

Varanol Polyether polyols, Dow

Voratec Polyurethane rigid systems, Dow

Vydax Release agent,; DuPont

Vydyne Nylon, Monsanto

Wellamid Nylon, Wellman

Xenoy PC/PBT & PC/PET alloys, GE

XT Polymer, Acrylic-based multipolymer compounds, Cyro

Xydar, Liquid crystal polymer, BP Amoco

Z-Light Spheres Low-density hollow ceramic microspheres, 3M

Zesospheres Fine size, high-strength ceramic rnicrospheres, 3M

Zelec Antistatic agents, DuPont

Zelux Polycarbonate shapes, Westlake

Zenite LCP Liquid crystal polymers, DuPont

ZenTron High-strength single-end roving, Owens Corning

Zero Gate Valve gate systems, Incoe

Zylar Acrylics, Novacor

Zytel Nylon resins (6, 6/6, 6/12), DuPont

Zytel HTN High-temperature nylon, DuPont

Zytel ST Super-tough nylon. DuPont

Bibliography

Abbott, W. H., Statistics can be Fun, A. Abbott Publ., Chesterland, OH 44026.

Adams, H., Viktor Schreckengost and 20th Century Design, Univ. Washington Press, 2000.

Additives & Compounding Buyers' Guide, Elsevier, 2003.

Advanced Composites, Compressed Air, Sep. 1994.

Advances in Industrial Computing Technology, ISA, 1999.

Allen, R. K., Rolling Bearings, Pitman and Sons, London, 1945.

Altshuler, T. L., Fatigue: Life Predictions for Materials Selection & Guide, ASM Software, 1988.

American Gear Manufacturers Assoc., 1500 King St., Alexander, VA 22314, USA.

Ancient Engineering: Ideas and Methods that have Endured 2000 Years of Progress, Bench Mark Bulletin No. 1, Burns & McDonnell Corp., 4800 East 63rd St., Kansas City, MO 64130, tel. 816–333–4375, 1992.

Ashby, M., etal., Materials and Design, Butterworth-Heinemann, Woburn, MA, USA, 2002.

ASTM Book of Standards, Section 8: Plastics, Four Volumes, Annual Issues.

ASTM Index—Annual Book of ASTM Standards, ASTM Annual.

ASTM International Directory of Testing Laboratories, ASTM, 1999.

Auto Composites Set for Massive Growth, MP, may 8, 2000.

Automotive Parts, Polymotive Publ., Jan.-Feb. 2003

Automotive Plastics Global Markets to 2010, Margolis Polymers, Keansburg, NJ, USA, 2003

Avallone Avery, J., Gas-Assist Injection Molding, Hanser, 2001.

Bacon, D. H. et al., Mechanical Technology, 3ed Ed., IP, 2000.

Bartenev, G., et al., Friction and Wear of Polymers, Elsevier, 1981.

Bayer Design Manual, 1990.

Beall, G., Recessions Remembered, IM, Mar. 2003.

Beaumont, J. P., Revolutionizing Runner Design in Hot and Cold Runner Molds, SPE-ANTEC, May 2001.

Beaumont, J., et al., Successful Injection Molding, Hanser, 2002.

Beck, R. D., Plastics Product Design, Kluwer, 1980.

Bernhardt, A., et al., Rationalization of Molding Machine Intelligent Setting & Control, SPE-IMD Newsletter, No. 54, Summer 2000.

Best, G. C., A Formula for Certain Types of Stiffness Matrices of Structural Elements, AIAA, No. 1, 1963.

Blout, Elkan, Instant Photography, Chemical Heritage, Spring, 2003.

Bonenberger, P., The First Snap Fit Handbook, Hanser, 2000.

Branand, R., Cybersquatting and Your Rights as a "Property" Owner; Internet Corp. for Assigned Names and Numbers, MP, Nov. 2000.

Bredas, J. L. et al., Conjugated Polymeric Materials: Opportunities in Electronics, Optoelectronics, and Molecular Electronics, 1990.

Breuer, O., et al., Introducing a New Miniature Mixer for Specialty Polymer Blends and Nanoscale Composites, Univ. of Alberta, SPE-ANTEC 2003.

Brody, A. L., at al., The Wiley Encyclopedia of Packaging Technology, 2nd Edition, Wiley, 1997.

Brostow, W. and R. D. Corneliussen, Failure of Plastics, Hanser, 1986.

Brown, R. L. E., Designing and Manufacturing of Plastic Parts, Wiley, 1980.

Bucksbee, J. H., The Use of Bonded Elastomers for Energy and Motion Control in Construction, Lord Corp., 1988.

Busch, J. & F. Field, IBIS Associates, Wellesley, MA (tel. 781–239–0666), Communication 2000.

Byers, M., The Design Encyclopedia, Wiley, 1994.

Campbell, P., Plastic Component Design, IP, 1996.

Cappelletti, M., In Defense of Plastics, World Plastics Technology, 2000.

Chamis, C. C. et al., Mechanical Property Characterization of Intraply Hybrid Composites, ASTM STP 734, pp 261–280, 1981.

Chamis, C. C., Sensitivity Analysis Results of the Effects of Various Parameters on Composite Design, SPI Annual Conference, 1982.

Chemical Industries Newsletter, SRI Consulting, Menlo Park, CA 94025, Issue 1, 2000.

Colby, P. N., Plasticating Components Technology: Screw & Barrel Technology, Spirex Bulletin, Spirex Corp., 2000.

Coleman, B. D., Thermodynamics of Materials with Memory Treatise, Arch. Rat. Mech. Anal., 1964.

Colvin, R., New Software Speeds and Simplifies Plastics Mold Design Work, MP, May 2000.

Composite Systems Provide Solutions in Mass Transit, Reinforced Plastics, Mar. 2003.

Computer More than Child's Play, Inside Line, Mar, 1999.

Consumers Computers Intolerable, The Inside Line, Mar.1999.

Crawford, R. J. & J. L. Throne, Rotational Molding Technology, 2002.

Cross, N., Engineering Design Methods, Wiley, 1989.

Dealey, B., The Moldmaker's Lament, MP, Jan. 2001.

Defosse, M., Packaging Offers Big Growth Opportunities, MP, Feb. 2001.

DeGrandpre, C., et al., Twin Sheet thermoforming of a Fuel Tank with a Converted Blow Mold, National Research Council Canada, SPE-ANTEC 2003.

Deming, E W., Out of Crisis, MIT Center for Advanced Engineering Study, 1986.

DeRosa, A., Manufacturers Gain Unified Building Code, PN, Feb. 24, 2003.

Design with Wood Aircraft Structures, Bulletin ANC-18, Gov't Printing Office, 1944.

Di Giovanni, M., Flat and Corrugated Diaphragm Design Handbook, Marcel Dekker, 1982.

Domininghaus, H., Plastics for Engineers: Materials, Properties, Applications, Hanser, 1993.

Dorgham, M. & Rosato, D. V., Design Plastic Composites, Interscience Enterprises-Geneva, 1986.

Drzal, L., et al., A carbon Nanotube Alternative: Graphite Nanoplatelets as Reinforcements for

Dym, J. P., Product Design with Plastics, IP, 1982.

Eagle, S., et al., Medical Device Software Standards, MD&DI, May 2001.

Eileen Gray Architect/Design's Perspective, Univ. Washington Press, 2000

Engineering Materials Handbook, Vol. 1; Composite. ASM International, 1987.

Engineering Materials Handbook, Vol. 2: Engineering Plastics, ASM International. 1988.

EnteGreat Inc., Birmingham, AL, USA.

Entering New Era of Growth, PE, Jan. 2003.

Environmental Briefs, The Vinyl Institute, May 2000.

Ettenson, M. W. is with Ain Plastics of Michigan Inc.; graduate of Univ. Mass-Lowell; extensive experience in the plastics industry/OCF, Cadillac Plastics, etc., 2002.

Evans, R., et al., Guide to the New Microfabrication Design and Process Techniques, MDDI, Nov. 2001.

Ezrin, M., Plastics Failure Guide: Cause and Prevention, Hanser, 1996.

Facts & Figures of the U.S. Plastics Industry, SPI Annual updates.

Fatigue Design Handbook, 3rd Ed., SAE, 1997.

Faupel, J. H. et al., Engineering Design, 2nd Ed., Wiley, 1981.

Firenze, A. R., The Plastics Industry, Adaptive Instruments Corp., Hudson, MA 01749, 2000.

French, T. E., at al., Engineering Drawing and Graphic Technology, McGraw, 1986.

Friedman, E. et al., Research on High Strength, High Modulus Whisker Reinforced Plastics Composites, Summary Tech. Report Contract, AF33(615)-2997, Oct. 1966.

Garcia-Leiner, M., et al., Processing of Intractable Polymers Using High-Pressure Carbon Dioxide, SPE-ANTEC 2003.

Gastrow, H., Injection Molds, 3ed Ed., Hanser, 2003.

Glanvill, A. B. et al., Injection-Mold Design Fundamentals, IP, 1965.

Griffith, A. A., The Phenomena of Rupture and Flow in Solids, Philosophical Transactions of the Royal Society of London, Series A, 2211, 163–98, 1921

Griffith, A. A., The Theory of Rupture, Proceedings of the 1st International Congress on Applied Mechanics, p. 55, 1924.

Hallahan, F. C., Remodeling & Costlier Wood Improve Opportunities for Plastics, MP, Nov. 1994.

Halpin, J., Lecture Notes UCLA Short Course, Fiber Composites: Design, Evaluation, and Quality, 1994.

Handbook for the MetalWorking Industries, Modern Machine Shop, 2003.

Handbook of Materials Selection, Wiley, 2002.

Hannagan, T., The Use and Misuse of Statistics, Harvard Management Update, May 2000.

Harper, C. A., Modern Plastics Handbook, McGraw Hill, 2001.

Harrington, J. P., Who's Who in Plastics & Polymers, Tech. Publ. Co., 2000.

Hertz, H., Gesammelte Werke, Vol. 1, Leipzig, 1895.

Hertzberg, R. W., et al., Fatigue of Engineering Plastics, Academic Press 1980.

Hicks, T. G. et al., Standard Handbook of Engineering Calculations, McGraw Hill, 1994.

How Products are Made: Guide to Product Manufacture, Gale Research Inc., Detroit, MI 48226. 1998.

How to Use Your CAD/CAM System Effectively, MoldMaking Technology, Feb. 2001.

Ingham, P., CAD Systems in Mechanical and Production Engineering, IP, 1990.

International Directory of Designs: Design Study & Teaching Directories, 5th Ed., Penrose Press, 1993.

International Plastics Materials Selector, Condura, San Diego, CA. Annually Updated.

IQMS, Pasco Robles, CA 93446 tel. 805–227–1122 Fax 805–227–1120

Janzen, W., and Ehrenstein, G. W., Hysteresis Measurements for Characterizing the Cyclic Strain and Stress Sensitivity of Glass Fiber Reinforced PBT, SPE ANTEC, May 1989.

Jerman, R., Finding What You Need on the Internet: Two Valuable Sources, SPE Extrusion Division Newsletter, Nov. 2000.

Jones, F. D., et al., Gear Design Simplified, 3ed Ed., IP, 1961·

Juvinall, R. C., Stress, Strain and Strength, McGraw-Hill, 1967.

Keates, J., Designing it Right the First Time, IM, Feb. 2001.

Kennedy, P., Flow Analysis of Injection Molds, Hanser 1995.

Kent, R., Cost management in processing, Rapra, 2002.

Klein, M., Sports Safety and Kids, PE, Jul. 2001.

Knights, M., Hybrid Injection Clamps: The Best of Both Worlds? PT, Apr. 2003.

Kornfield, J. A., Recent Advances in Understanding Flow Effects on Polymer Crystallization, California Institute of Technology, SPE-ANTEC 2003.

Kreith, F. et al., CRC Handbook of Energy Efficiency, CRC Press, 1996.

Kutz, M., Mechanical Engineers' Handbook, Wiley, 1998.

La Mantia, F., Handbook of Plastics Recycling, Rapra, 2002.

Lawn, Garden, & Agricultural Packaging, Freedonia Group, Cleveland, OH, USA, 2003

Leaversuch, R. D., Creative Design Plays a central Role in Growth Strategy, MP, 2001.

Lee, J. K., Effect of Rheological Properties and Processing Parameters on ABS Thermoforming, Polym. Eng. Sci., 41, 2001.

Lee, S. M., International Encyclopedia of Composites, VCH, 1991.

Leventon, W., Innovations; New Materials and equipment, MDDI, Nov. 2001.

Leventon, W., Material Progress toward Better Molded Parts, MD&DI, Mar. 2002.

Lewis, P. R., Designing with Plastics, RAPRA Review Report, No 64; 1993.

Lewis, W. P., Fundamentals of Engineering Design, Prentice Hall, 1989.

Lubin, G., Handbook of Composites, VNR, 1982.

Luo, N., et al., Surface Grafting of Polyacrylamide from Polyethylene-Base Copolymer film, Clemson/Cryovac, SPE-ANTEC 2003.

Mair, M., Basic Principles of Design Vols. 1 to 4, Kluwer, 1977.

Malloy, R. A., Plastic Part Design for Injection Molding, Hanser, 1994

Mark, J. E., Polymer Data Handbook, Oxford, 1998.

Market Data Book, Annual Plastics News, Dec. 2002.

Marsh, G., Wind Energy Market Still Thriving, RP, Jul./Aug. 2001.

Marsh, G., Wright Brothers Legacy Flying High; Composites and Aircraft, Reinforced Plastics, Apr. 2003

Martyn, M. T., et al., Micromolding: Consideration of Processing Effects on Medical Materials, University of Bradford/Brunel Univ., SPE-ANTEC, 2003.

Matyjaszewski, K., et al., Handbook of Radical Polymerization, Wiley, 2002.

Metallocene Technology & Modern Catalytic Methods in Commercial Applications, PDL, 1999.

Meyers, A. R. et al., Basic machining reference Handbook, 2nd Ed., IP, 2001.

Miel, R., Auto Sales Rapidly Rising in China, PN, Mar. 3, 2003.

Modern Plastics Encyclopedia, Annual.

Mort, M., Self-Assembly Circuitry-DNA, Compressed Air, Sep. 1999.

Mortenson, M. E., Mathematics for Computer Graphics Applications, 2nd Ed., IP, 1999.

Narhi, W., Benefits of Finite Element Analysis, MP, Dec. 2000.

New Concentrates Give Special Looks to Bottles, PT, Feb. 2003.

Newberry, A. L., FEA Modeling Improves its Worth in Composites Design, RP, Oct. 2000.

Newman, T. R., Plastics as Design Form, Chilton Book Co., 1972.

Oberg, E., et al., Machinery's Handbook, 25th Ed., IP, 2000.

One Shot Car Body, Reinforced Plastics, Mar. 2003.

Palmgren, A., Ball and Roller Bearing Engineering, SKF Industries, Philadelphia, PA, 1945.

Parrington, R. J., Fractography of Metals and Plastics, PE, Dec. 2000.

Permadi, F., Et al., Development of an Environmentally Friendly Solventless Process for Electronic Prepregs, Ohio State Univ., SPE-ANTEC 2003.

Permeability and Mechanical Performance of 3–Layer EVOH/LLDPE Barrier

Films with Blended Middle Layers, National Research Council Canada, SPE-ANTEC, 2003.

Plastic Components in Vehicle Construction, Polymers in Automotive Industry, Feb. 2003.

Plastics for Aerospace Vehicles, MIL-HDBK-17A & 17B, US Supt. of Documents, GPO, 1981.

Plastics Technicians Toolbox, SPE, 2001.

Polymers, Michigan State Univ., SPE-ANTEC 2003.

Port, O., In Transportation, One Word Plastics, Business News, Mar.6, 2000.

Pottle, M., Emerging Markets and Technology in Thermoplastics Elastomers Forum, SPE-SIG, Jan. 27–29, 2002.

Progelhof, R. C., et al., Polymer Engineering Principles: Properties, Processes, Texts for Design, Hanser, 1993.

Pugh, S., Total Design, Addison-Wesley, 1991.

Puncochar, D. E., Interpretation of Geometric Dimensioning and Tolerancing, IP, 1997.

Rao, N. S. et al., Design Data for Plastics Engineers, Hanser, 1998

Rao, N. S., Design Formulas for Plastics Engineers, Hanser, 1991.

Rauwendaal, C., Polymer Extrusion, 4th Ed., Hanser, 2002.

Rauwendaal, C., SPC in Injection Molding, Hanser, 2000.

Rees, H., Mold Engineering 2nd Ed., Hanser, 2003.

Reinforced Plastics: Automotive Supplement, RP, Feb. 2003.

Reinhardt, C., Chemical Sciences in the 20th Century, Wiley-VCH, 2002.

Report Says Recycling No Panacea, MP, Apr. 2000.

Roark, R. J. et al., Formulas for Stress and Strain, 6th Ed., McGraw-Hill, 1989.

Rosato, D. V. and G. Lubin, Application of Reinforced Plastics, 4h International RP Conference, British Plastics Federation, London, UK. Nov. 25–27. 1964.

Rosato, D. V. Capt., All Plastic Military Airplane Successfully Flight Tested, Wright-Patterson AF Base, Ohio, l944.

Rosato, D. V., Advanced Engineering Design Short Course, ASME Engr. Conference, 1983.

Rosato, D. V., Blow Molding Handbook, 2nd Ed., Hanser, 2003.

Rosato, D. V., Blow Molding Handbook, Hanser, 1989.

Rosato, D. V., Capt., Theoretical Potential for Polyethylene, USAF Materials Lab., WPAFB, 1944.

Rosato, D. V., Concise Encyclopedia of Plastics (25,000 entries), Kluwer, 2000.

Rosato, D. V., Design Features That Influence Performance: Detractors, SPE-ANTEC, May 1991.

Rosato, D. V., Designing with Plastics, MD&DI, pp. 26–29, July 1983.

Rosato, D. V., Designing with Plastics, Rhode Island School of Design, Lectures 1987–1990.

Rosato, D. V., Designing with Plastics, RISD, 1984–1992.

Rosato, D. V., Designing with Plastics, SPE-IMD Newsletter, 1970–2001.

Rosato, D. V., Designing with Reinforced Composites, Hanser, 1997.

Rosato, D. V., Environmental Effects on Polymeric Materials, Vol. I.: Environment and Vol. II: Materials, Wiley, 1968.

Rosato, D. V., et al., Non-Metallic Composite Materials & Fabrication Techniques Applicable to Present & Future Solid Rocket Bodies, ARS Conference, Salt Lake City, UT, Feb. 1961.

Rosato, D. V., Extruding Plastics: A Practical Processing Handbook, Kluwer, 1998.

Rosato, D. V., Filament Winding (in Russian), Russian Publ., 1969.

Rosato, D. V., Filament Winding, Wiley, 1964.

Rosato, D. V., Fundamentals of Designing with Plastics, SPE-IMD Newsletter No. 51, 1999.

Rosato, D. V., Injection Molding Chapter: R. F. Jones book Guide to Short Fiber RP, Hanser, 1998.

Rosato, D. V., Injection Molding Higher Performance Reinforced Plastic Composites, J. of Vinyl & Additive Technology, Sep. 1996.

Rosato, D. V., Injection Molding in the 21st Century, SPE IMD Newsletter Nos. 53 & 54, 2000.

Rosato, D. V., Injection Molding, 3rd Ed., Kluwer, 2000.

Rosato, D. V., Materials Selection & Reinforced Plastics-Thermosets in Concise Encyclopedia of Polymer Science and Engineering, J. I. Kroschwitz (ed.), Wiley, 1990.

Rosato, D. V., Materials Selection Chapter in Encyclopedia of Polymer Science and Engr., Mark-Bikales-Overberger-Menges (eds.): Vol. 9, pp. 337–379, Wiley, 1987.

Rosato, D. V., Materials Selection, Polymeric Matrix Composites, Chapter in International Encyclopedia of Composites, VCH, 1991.

Rosato, D. V., Non-Woven Fibers in Reinforced Plastics, Ind. Engr. Chem., 54,8. 30–37, Sep. 1962.

Rosato, D. V., Nose Cone of First US Moon Vanguard Rocket is Made in Manheim-US, New Era-Lancaster PA Newspaper, Nov. 30, 1957.

Rosato, D. V., Outer Space Parabolic Reflector Energy Converters, SAMPE, June 1963.

Rosato, D. V., Plastic Replaces Aorta Permits Living Normal-Long Life, Newton-Wellesly Hospital-Massachusetts, USA, Mar. 1987.

Rosato, D. V., Plastics Engineering Manufacturing, SPE-IMD Newsletter, Summer, 1989.

Rosato, D. V., Plastics in Missiles, British Plastics, 348–352, Aug. 1960.

Rosato, D. V., Plastics Industry Safety Handbook, Cahners, 1973.

Rosato, D. V., Plastics Processing Data Handbook, Kluwer, 1995.

Rosato, D. V., Polymers. Processes and Properties of Medical Plastics in Synthetic Biomedical Polymers, Szycher-Robinson (eds.), CRC Press, 1980.

Rosato, D. V., Radomes, Electronic Design News, 1963

Rosato, D. V., Rosato's Plastics Encyclopedia and Dictionary, Hanser, 1993.

Rosato, D. V., Seminars presented worldwide on 21 Different Plastics Subjects from Introduction through Design Parts, Fabrication by different processes, Quality Control, Statistical Control to Marketing via University of Massachusetts-Lowell, Plastics World, ASME, General Motors Institute, SPE., SPI, China National Chemical Construction (Beijing), Hong Kong Production Centre. Singapore Institution, Open University of England, Geneva Development, and Tufts Medical University, 1974 to 1986.

Rosato, D. V., Target for Zero Defects, SPE-IMD Newsletter, Issue 26, 1991.

Rosato, D. V., Thermoset Polymers chapter in Encyclopedia of Packaging Technology, 2nd, Wiley, 1997.

Rosato, D. V., Thermosets in Encyclopedia of Polymer Science and Engr., Vol. 14, pp 350–391, Wiley, 1988.

Rosato, D. V., Weighing Out the Aircraft Market-It's New Plastics that Count, AIAA, Paper 68–320 Palm Springs, CA, Apr. 1968.

Rosato, D. V., What Molders Must Do About ANSI Safety Specifications, PW, Apr. 1978.

Rosen, S. R., Thermoforming: Improving Process Performance, SME, 2002.

Ross, E., Help! Is there an Engineer in the House? Motion Control, Jan./Feb. 2001.

Rotheiser, J. I., Joining of Plastics, Hanser, 2002.

Salvendy, G., Handbook of Human Factors, Wiley, 1997.

Schenkel, G., Plastics Extrusion Technology and Theory, Hanser, 1963

Scherr, J., Flow Analysis Gets it Right the First Time, PT, Apr. 2003.

Schindler, B. M., "Made in Japan", W. E. Deming, ASTM Standardization News. p. 88, Feb. 1983.

Schleiffer, K., Marketing vs. R&D: Spanning the Divide, MD&DI, May 2000.

Schultz, J. M., Polymer Crystallization, Oxford Univ., 2001.

Schut, J. H., Long-Fiber Thermoplastics Extend their Reach, PT, Apr. 2003.

Schwartz, R. T. and D. V. Rosato, Structural Sandwich Construction in Composite Engr. Materials, Dietz, A G. H. (ed.), pp. 165–181, MIT Press, 1969.

Schwartz, R. T., Dietz, A. G. H., and Rosato, D. V. correspondence, 1975–1985.

Shewhart, W. A., Economic Control of Quality of Manufactured Product, Van Nostrand Reinhold, 1931.

Sims, F., Engineering Formulas, Industrial Press, 1999.

Smith, P. G., The Risk of Talking about Risk, MD&DI, Mar. 2003.

SPE Guide on Extrusion Technology and Troubleshooting, SPE, 2001.

SPE Writer Advocates Burning vs. Recycling, PN, Jan. 27. 2003.

Standard for the Use of the International System of Units (SI), ASTM, 2002.

Stewart R, Flame Retardants, PE, Jan. 2003.

Stewart, R., Medical Plastics, PE, Apr. 2003.

Stewart, R., Thermoforming, PE, Feb. 2003.

Szycher, M. Biocompatible Polymers, Metals, and Composites, Technomic, 1983.

The Plastics Technician's Toolbox, SPE, 2002.

Throne, J., Technology of Thermoforming, Hanser, 1996.

Throne, J., Thermoforming 101: Basic Building Blocks of Thermoforming, SPE Thermoforming Quarterly, Vol. 22, No. 2, 2003.

Timoshenko, S. and J. M. Gere, Theory of Elastic Stability, McGraw, 1961.

Timoshenko, S., Strength of Materials, Kluwer, 1955.

Tres, P. A., Designing Plastic Parts for Assembly, 4th Ed., Hanser, 2002.

Troitzsch, J. H., International Plastics Flammability Handbook, Hanser, 1990.

Tsai, S. W. et al., Analysis of Composite Structures, NASA CR-620, 1966.

Turner, W. C., Energy Management Handbook, Fairmont Press, 1997.

U.S. Rules PVC Toys Not a Threat to Kids, PN, Mar. 3, 2003.

Vecchio, R. J., Understanding Design of Experiments, Hanser, 1997.

Votolate, G., American Design in the 20th Century, Manchester Univ. Press, 1998.

Walsh, M., Basic Polymer Principles & Applications, Thermofab@hotmail.com, 2002.

White, J. L., et al., Screw Extrusion, Hanser, 2003.

Wigotsky, V., Plastics in Electrical/Electronic, PE, May 2001.

Wilox, S. B., Applying Universal Design to Medical Devices, MD&DI, Jan. 2003.

Woodson, W. E. et al., Human Factors Design Handbook 2nd Ed., McGraw Hill, 1991.

Wulpi, D. J., Understanding How Components Fail, 2nd Edition, ASM, 1999.

Wypych, G., Plastics Failure Analysis and Prevention, PDL, 2001.

Young, W. C., Roark's Formulas for Stress and Strain, 6th Ed., McGraw-Hill, 1998.

Zhao, C. H., et al., A Fuzzy Rule Based Automatic V/P Transfer System for Thermoplastic Injection Molding, SPE-ANTEC, 1997.

Index